SOLAR PASSIVE BUILDING

SCIENCE AND DESIGN

International Series on
BUILDING ENVIRONMENTAL ENGINEERING

Volume 2
Series Editor: D. Croome, University of Bath, England

Other Pergamon Titles of Interest

BILLINGTON & ROBERTS
Building Services Engineering

BOWEN
Passive and Low Energy Alternatives
Passive and Low Energy Ecotechniques

CROOME
Noise, Buildings and People

CROOME & ROBERTS
Airconditioning and Ventilation of Buildings,
2nd Edition, Volume 1

DUNN & REAY
Heat Pipes, 2nd Edition

EGGERS-LURA
Solar Energy for Domestic Heating and Cooling

MACKENZIE-KENNEDY
District Heating: Thermal Generation and Distribution

McVEIGH
Sun Power, 2nd Edition
Energy Around the World

MEACOCK
Refrigeration Processes

OSBORNE
Fans, 2nd Edition

REAY & MACMICHAEL
Heat Pumps

ROZENBURG
English-Russian Dictionary of Refrigeration and
Low-Temperature Technology, 2nd Edition

YANNAS
Passive and Low Energy Architecture

Pergamon Related Journals *Free Specimen Copy Gladly Sent on Request*

BUILDING & ENVIRONMENT
ENERGY CONVERSION & MANAGEMENT
HABITAT
INTERNATIONAL JOURNAL OF HEAT & MASS TRANSFER
JOURNAL OF HEAT RECOVERY SYSTEMS
SOLAR ENERGY
SOLAR & WIND TECHNOLOGY

SOLAR PASSIVE BUILDING
SCIENCE AND DESIGN

M. S. SODHA, N. K. BANSAL,
A. KUMAR
Indian Institute of Technology, India
P. K. BANSAL
The Australian National University, Canberra, Australia

and

M. A. S. MALIK
World Bank, USA

PERGAMON PRESS

OXFORD · NEW YORK · TORONTO · SYDNEY
FRANKFURT · TOKYO · SAO PAULO · BEIJING

U.K.	Pergamon Press, Headington Hill Hall, Oxford OX3 0BW, England
U.S.A.	Pergamon Press Inc., Maxwell House, Fairview Park, Elmsford, New York 10523, U.S.A.
CANADA	Pergamon Press Canada, Suite 104, 150 Consumers Road, Willowdale, Ontario M2J 1P9, Canada
AUSTRALIA	Pergamon Press (Aust.) Pty. Ltd., P.O. Box 544, Potts Point, N.S.W. 2011, Australia
FEDERAL REPUBLIC OF GERMANY	Pergamon Press GmbH, Hammerweg 6, D-6242 Kronberg, Federal Republic of Germany
JAPAN	Pergamon Press, 8th Floor, Matsuoka Central Building, 1-7-1 Nishishinjuku, Shinjuku-ku, Tokyo 160, Japan
BRAZIL	Pergamon Editora Ltda., Rua Eça de Queiros, 346, CEP 04011, São Paulo, Brazil
PEOPLE'S REPUBLIC OF CHINA	Pergamon Press, Qianmen Hotel, Beijing, People's Republic of China

First edition 1986

Library of Congress Cataloging in Publication Data

Solar passive building.
(International series on building environmental engineering; v. 2)
1. Solar heating — Passive systems. I. Sodha, M. S., 1932- . II. Series.
TH7413.S6244 1986 721'.0467 85-16789

British Library Cataloguing in Publication Data

Solar passive building : science and design. —
(International series on building environmental engineering; v. 2)
1. Solar heating — Passive systems
I. Sodha, M. S. II. Series
697'.78 TH7413
ISBN 0-08-030550-4

Printed in Great Britain by A. Wheaton & Co. Ltd., Exeter

EDITOR'S PREFACE

The International Series on Building Environmental Engineering aims to examine the widening field of building services engineering and the practical implications for design covering energy, comfort, economics and services. The interaction of these factors and their role in the total design process are paramount. The first volume in the Series *Building Services Engineering: A Review of Its Development* takes a historical look at the field and describes how people have used their ingenuity to make buildings work over many centuries. It is purposely planned as an international series so that the ideas of designers, engineers and scientists working in other climates are pursued. This volume is written by a well-known group under Professor M. S. Sodha at the Centre of Energy Studies in New Delhi in conjunction with Dr Malik from the World Bank in Washington, USA.

The limitations of solely designing highly-mechanized buildings are readily apparent particularly as the lifetime of equipment is so much less than for building fabric. Space requirements, noise, maintenance and replacement costs are some of the penalties one has to pay for having sealed and flexibly controlled buildings. Recent evidence even suggests that sealed highly mechanized buildings may be less healthier to work in. There are many uncertainties here but clearly it is better to ensure that buildings are considered from the passive viewpoint before ladening them with equipment then assessing what energy balance is required to be made up by heating, ventilation or airconditioning systems for the building use under consideration. Highly-mechanized buildings are necessary sometimes but at least they should be evolved from a proper consideration of the human and climate requirements ensuring that mass, insulation, orientation, shape, built form and layout play their role in design process. The authors define the science before demonstrating how to achieve this in practice.

Derek Croome
1985

PREFACE

The fuel crisis in 1972-73 generated world-wide efforts to conserve fossil fuels and search for alternative sources of energy. Since the heating and cooling of buildings accounts for a significant part of energy consumption, the use of solar passive concepts in architecture of the building can save a considerable amount of energy, which is otherwise required to maintain thermal comfort inside the buildings by mechanical means. The conference and workshop on passive solar heating and cooling of buidlings at Albuquerque in 1976 marked the recognition of the maturity of the science of passive solar architecture. A large number of conferences and workshops have been organized since then. Scientists from various disciplines such as physics, architecture building and civil engineering and computer science have contributed to the development of solar passive science and architecture. The present book is an attempt to introduce the fundamentals of solar passive control to a variety of readers in the building design field. The exposition of the subject is simple and can be easily understood. The book has been divided into ten chapters. Each chapter can be read independent of each other, depending on the level of knowledge the reader possesses. The contents of the chapters have been made exhaustive to give an overview of the subject before the reader really starts going into the details. The book will ideally serve as a textbook to an introductory course on Solar Passive Control but it will also provide a very useful research background for those who are seeking an interpretation of the thermal performance of buildings. Chapters 4, 5 and 6 are especially recommended for them. For building designers Chapters 4, 9 and 10 are most useful. The reader is however the real judge of the utility of the book and any suggestions or critical comments on the effort are therefore most welcome and all attempts would be made to incorporate them in future.

M. S. Sodha
N. K. Bansal
P. K. Bansal
A. Kumar
M. A. S. Malik

ACKNOWLEDGMENTS

The authors gratefully thank

Scientific Community whose work has been extensively used in writing this book.

Professor O. P. Jain, former Director, IIT Delhi for encouragement and necessary facilities.

Professor A. Boettcher and Dr M. Meliss of Kerforschungsanlage Juelich, West Germany for bibliography and literature.

Mr Maheshwar Dayal, Secretary, Department of Non-conventional Energy Sources, Government of India for funding a solar passive house project and helpful discussions.

T. N. Gupta for administrative help

Lalit Agarwal for his painstaking drawings

Vijay Sodha, Shaila Bansal and Sabiha Malik for their endless patience

and to our colleagues and students with whom we had stimulating discussions.

CONTENTS

NOMENCLATURE

A	Area of wall/roof/surface, m^2
A_a	Area of sunspace/air column, m^2
A_B	Area of isothermal mass, m^2
A_e	Effective area of wall/surface, m^2
A_{eff}	Effective radiation area of the clothed body, m^2
A_g	Glazing area, m^2
a_m	Air-mass
A_N	Body (Dubois) area, m^2
A_r	Area of the receiver, m^2
A_S	Surface azimuth angle, degree
A_{Sd}	Effective area of the building for diffuse radiation, m^2
A_{SD}	Effective solar exposed area for beam radiation, m^2
A_w	Window area, m^2
A_W	Area of west wall, m^2
A_Z	Solar azimuth angle, degree
A_α	Solar altitude angle, degree
b	Breadth, m
B_h	Body height, m
C	Specific heat capacity, J/kg^oC
C_a	Specific heat capacity of air, J/kg^oC
C_c	Air conductance, $W/m^{2o}C$
C_P	Cooling power, W/m^2
C_t	Thermal conductance, $W/m^{2o}C$
C_w	Specific heat of water, J/kg^oC

DD	Number of degree-days per month, ^{o}C
d_E	Thickness of east/west wall, m
d_f	Decrement factor
d_w	Thickness of the water-layer, m
e	Eccentricity of the earth's orbit around sun
f	Dimensionless number indicating cooling efficiency
F	Geometrical shape factor
F_{12}	Configuration factor
f_s	Solar-gain factor
f_{sh}	Shading factor
f_β	Angle factor between the surface and the sky
g	Acceleration due to gravity, m/s^2
Gr	Grashof number
h	Height of the station above sea level, m
h_1	Heat-transfer coefficient between absorbing surface and air in sunspace of Trombe wall, $W/m^2 {}^oC$
h_1'	Heat-transfer coefficient between absorber plate and water in water wall, $W/m^2 {}^oC$
h_2	Convective heat-transfer coefficient between glazing and the air in sunspace, $W/m^2 {}^oC$
h_2'	Heat-transfer coefficient between water and bottom surface of water wall, $W/m^2 {}^oC$
h_b	Heat-transfer coefficient from clothed body to the air, $W/m^2 {}^oC$
h_c	Convective heat-transfer coefficient, $W/m^2 {}^oC$
H_C	Heat dissipated by convection, W
h_{co}	Convective heat-transfer coefficient for the external surface, $W/m^2 {}^oC$
h_{cw}	Convective heat-transfer coefficient between the roof surface and water, $W/m^2 {}^oC$
h_D	Mass-transfer coefficient, $kg/s\ m^2$
H_D	Moisture diffusion heat loss rate, W
h_e	Evaporative heat-transfer coefficient, $W/m^2 {}^oC$
H_e	Heat losses due to evaporation of sweat from the skin surface, W
H_F	Heating value of the fuel, W
h_{GA}	Heat-transfer coefficient between floor and sunspace, $W/m^2 {}^oC$
H_K	Heat loss by the body to the ambient (including conduction through clothing), W
H_{1d}	Sensible heat losses through respiration (due to temperature difference of the incoming and outgoing air), W
H_{1r}	Latent heat losses through respiration (due to evaporation of water), W

H_M	Metabolic heat generation rate, W
h_o	Heat loss coefficient from wall's outside surface to the ambient through glazing, $W/m^{2o}C$
h_r	Radiative heat-transfer coefficient, $W/m^{2o}C$
H_r	Heat dissipated by radiation, $W/m^{2o}C$
h_{ro}	Radiative heat-transfer coefficient for the outside surface, $W/m^{2o}C$
H_{se}	Evaporative secretion heat loss rate, W
h_{si}	Heat-transfer coefficient between the internal surface and the room air, $W/m^{2o}C$
h_{so}	Heat-transfer coefficient between the external surface and the outside air, $W/m^{2o}C$
h_{sw}	Heat-transfer coefficient between the roof surface and water above it, $W/m^{2o}C$
H_T	Net amount of the heat dissipated by the body to the Environment, W
h_{TA}	Heat-transfer coefficient between mass wall and the sunspace, $W/m^{2o}C$
h_{WA}	Heat-transfer coefficient between east/west wall and sunspace, $W/m^{2o}C$
i	Angle of incidence on any arbitrary surface, degree; iota ($= \sqrt{-1}$)
ITS	Required sweat rate in equilibrium, W
K	Thermal conductivity, W/m^oC
K_g	Thermal conductivity of glass, W/m^oC
$(k_{o_3})_\lambda$	Extinction coefficient for ozone, cm^{-1}
k_T	Hourly clearness index
L	Thickness/length of wall/roof, m
L_c	Characteristic length of the system, m
L_g	Glass thickness, m
m	Mass of evaporated water, kg
M_A	Thermal capacity of mass of air, $J/^oC$
$\dot{m}_a$	Mass flow rate of air, $kg/s\ m^2$
M_B	Thermal capacity of isothermal mass, $J/^oC$
m_d	Daily mass of the evaporated water, kg/m^2day
$\dot{m}$	Volumetric flow rate of air, m^3/s
$\dot{m}_w$	Mass flow-rate of water, kg/m^2s
M_w	Thermal capacity of water mass, $J/^oC$
n	An integer representing the number of harmonics in a Fourier series representation
N	Number of air-changes per hour, hr^{-1}
n_d	Number of the day (relative to January 1)
n_d	Number of days in a month

N_F	Number of units of the fuel
N_s	Number of sunshine hours
Nu	Nusselt number
P	Atmospheric pressure, mb
P_a	Partial vapour pressure of the ambient air, mb
P_{as}	Partial pressure of saturated water vapours at ambient temperature, mb
p_c	Coefficient which depends on clothing
Pr	Prandtl number
p_s	Permeance coefficient of skin, $kg/s.m^2.mb$
P_s	Saturated vapour pressure at skin temperature, mb
p_v	Pulmonary ventilation, kg/s
p_w	Partial pressure of saturated water vapours at water temperature, mb
q	Amount of heat-transfer, W
$\dot{Q}$	Amount of average heat loss, W
Q(t)	Amount of heat flux entering the room, W/m^2
Q_o	Daily heat-flux entered in the room; W/m^2 day
Q_c	Convective heat-transfer from water surface to the ambient, W/m^2
Q_e	Evaporative heat-transfer from water surface to the ambient, W/m^2
$\dot{Q}_G$	Amount of heat transferred to the ground, W
$\dot{Q}_I$	Amount of heat transferred to the isothermal mass, W
$\dot{Q}_m$	Monthly space heating load, W
Q_r	Evaporative heat-transfer from water surface to the ambient air, W
$\dot{Q}_r$	Total amount of heat entering the room through the opaque portion of the walls/roof, W
$\dot{Q}_v$	Amount of heat-transfer due to air ventilation, W
$\dot{Q}_w$	Amount of heat taken away by flowing water, W
$\dot{Q}_{window}$	Amount of heat transferred through window glass, W
r	Earth-sun distance on any day, km
r_o	Mean earth-sun distance, km
r_b	Ratio of instantaneous irradiance on an inclined surface to that on the horizontal surface
R_b	Ratio of monthly mean daily radiation on an inclined surface to that on the horizontal surface
r_b'	Ratio of hourly radiation on an inclined surface to that on the horizontal surface
R_{cl}	Total heat-transfer resistance from skin to outer surface of the clothed body, $W^{-1}\ m^2 {}^oC$
r_d	Radial distance, m
Re	Reynolds' number
R_h	Relative humidity

r_p	Roughness parameter
R_{si}	Thermal resistance for the internal surface which is in contact with the room air, $W^{-1}\ m^2 {}^oC$
R_{so}	Thermal resistance for the external surface which is exposed to the outside atmosphere, $W^{-1}\ m^2 {}^oC$
r_t	Radius of the tunnel, m
S	Incident solar radiation, W/m^2
$\bar{S}$	Mean solar intensity at normal incidence, W/m^2
S_c	Solar constant, W/m^2
$S_{c\lambda}$	Monochromatic extraterrestrial irradiance on a surface normal to the direction of the direct beam, $W/m^2\ \mu m$
S_d	Total diffuse radiation incident on the building, W
S_D	Incident solar flux on earth's surface, W
$S_{da\lambda}$	Diffuse spectral irradiance produced by aerosols, $W/m^2\ \mu m$
S_{dh}	Hourly diffuse radiation on a horizontal surface, W/m^2
S_{DN}	Direct radiation on a surface normal to the radiation on earth's surface, W/m^2
$S_{dr\lambda}$	Diffuse spectral irradiance due to Rayleigh scattering, W/m^2
$S_{dm\lambda}$	Diffuse spectral irradiance produced by multiple reflections, W/m^2
S_{dV}	Diffuse radiation on a vertical surface at ground, W/m^2
$S_{d\lambda}$	Monochromatic diffuse radiation arriving on the ground horizontal surface, $W/m^2\ \mu m$
$S_{d\beta}$	Diffuse radiation on a surface with tilt angle β, W/m^2
S_{eh}	Extraterrestrial irradiance on a horizontal surface, W/m^2
S_{en}	Extraterrestrial irradiance on a surface normal to the sun rays, W/m^2
S_{ehd}	Mean daily extraterrestrial radiation on a horizontal surface, kWh/m^2 day
S_{ehm}	Mean monthly extraterrestrial radiation on a horizontal surface, kWh/m^2 day
$S_{eh\lambda}$	Monochromatic extraterrestrial radiation on a horizontal surface, $W/m^2\ \mu m$
$S_{eh\omega}$	Mean hourly extraterrestrial radiation on a horizontal surface, kW/m^2
$S_{e\beta}$	Extraterrestrial irradiance on a surface inclined at angle β to the horizontal, W/m^2
$S_{e\beta d}$	Mean daily extraterrestrial radiation incident on a surface inclined at angle β, kWh/m^2 day
$S_{e\beta h}$	Hourly extraterrestrial radiation incident on a surface inclined at β, kW/m^2
$S_{e\beta t}$	Mean extraterrestrial radiation on a surface inclined at angle β to the horizontal between time t_1 and t_2, kWh/m^2
S_{gdh}	Direct solar radiation on a horizontal surface on the ground, W/m^2
$S_{gd\beta}$	Direct radiation on a surface with tilt β, W/m^2
S_{gh}	Direct radiation incident on a horizontal surface on the ground, W/m^2
S_{ght}	Hourly global radiation on a horizontal surface on earth, W/m^2

$S_{gh\lambda}$	Monochromatic direct irradiance on a horizontal surface on the earth, $W/m^2\ \mu m$
$S_{g\beta t}$	Hourly global radiation on a surface inclined at an angle β, W/m^2
$S_{g\beta\lambda}$	Monochromatic solar radiation on an inclined surface at the ground, $W/m^2\ \mu m$
S_n	Harmonic part of S, W/m^2
S_N	Solar radiation heat load, W
t	Time, s
$(T_{amr})_\lambda$	Monochromatic transmissivity due to Rayleigh scattering
$(T_{ams})_\lambda$	Monochromatic transmissivity due to scattering by suspended water droplets
t_d	Time equation in minutes
$(T_d)_\lambda$	Monochromatic transmissivity of radiation due to scattering by dust particles
T_f	Atmospheric turbidity factor
$(T_g)_\lambda$	Monochromatic transmissivity of radiation due to absorption by gases
$(T_{H_2O})_\lambda$	Monochromatic transmissivity of radiation due to absorption by water vapour
$(T_{O_3})_\lambda$	Monochromatic transmissivity of radiation due to absorption by ozone
$T_{rm\lambda}$	Monochromatic transmissivity of radiation due to Rayleigh scattering
$(T_{wp})_\lambda$	Monochromatic transmissivity of radiation due to scattering by water vapour
T_λ	Monochromatic atmospheric transmittance for direct radiation
u	Flow-velocity of the fluid, m/s
U	Overall thermal transmittance, $W/m^{2\,o}C$
U_g	Window-glass conductance, $W/m^{2\,o}C$
U_w	Wall-conductance, $W/m^{2\,o}C$
v	Velocity of air, m/s
V	Volume, kg/m^3
V_m	Mean vote
W	Rate of work done, W
W_a	Humidity ratio of inhaled air, (kg moisture/kg dry air)
W_b	Body weight, kg
W_{ex}	Humidity ratio of the expiration air, (kg moisture/kg dry air)
w_p	Water vapour contents of the atmosphere
x	Position co-ordinate, m
y	Position co-ordinate (along the length of the roof), m
Y	Admittance, $W/m^{2\,o}C$

Greek Symbols

α	Absorptivity
β	Tilt angle, degree
β_t	Coefficient of volumetric thermal expansion of the fluid, $^oC^{-1}$
γ	Longitude, degree
γ_ℓ	Local longitude, degrees
γ_s	Standard longitude, degrees
δ	Declination, degrees
$\mathcal{D}$	Thermal diffusivity, m^2/s
ε	Emissivity
ε'	Emissivity of the reflecting sheet
ε_w	Emissivity of water
ξ	albedo (or reflectivity)
	a- cloudless sky, 0.25
	cc- cloud, 0.6
	g- ground, 0.2
	s- shading devices
	$g\lambda$- monochromatic ground albedo
	so- single scattering albedo
$\bar{\xi}_g$	weighted average reflectivity of ground over entire wavelength region
η_F	Efficiency of fuel, for oil and gas furnaces between 0.5 and 0.6, and 1.0 for electricity
$\theta(x,t)$	Temperature distribution, ^oC
θ_a	Ambient air (dry-bulb) temperature, ^oC
θ_{ai}	Inside air temperature, ^oC
θ_b	Temperature base for heat load calculations, ^oC
θ_B	Isothermal mass temperature, ^oC
θ_{cl}	Temperature of the outer surface of the clothed body, ^oC
θ_{dp}	Dew-point temperature, ^oC
θ_{ex}	Temperature of the exhaled air, ^oC
θ_E	Internal surface temperature of east wall, ^oC
θ'_E	External surface temperature of east wall, ^oC
θ_f	Bulk fluid temperature, ^oC
θ_{fj}	Initial fluid temperature, ^oC
θ_{fo}	Outlet fluid temperature, ^oC
θ_g	Globe temperature, ^oC
θ_{gl}	Glass surface temperature, ^oC
θ_G	Temperature distribution in ground, ^oC
θ_j	Inlet temperature, ^oC
θ_{mrt}	Mean radiant temperature, ^oC
θ_p	Metallic surface temperature of the drum, ^oC

θ_r	Room air temperature, ^oC
θ_R	Temperature distribution in the roof, ^oC
θ_s	Skin temperature, ^oC
θ'_s	External surface temperature of the south wall, ^oC
θ_{sa}	Solair temperature, ^oC
$\bar{\theta}_{sa}$	Average solair temperature, ^oC
θ_{SA}	$\{= \theta_{sn} \exp(-i\ \psi_n)\}$ Value of nth harmonic of θ_{sa}, ^oC
θ_{si}	Inside surface temperature, ^oC
θ_{sky}	Sky temperature, ^oC
θ_{so}	Outside surface temperature, ^oC
θ_{ss}	Sunspace temperature, ^oC
θ_w	Water temperature, ^oC
θ_{wb}	Wet-bulb air temperature, ^oC
θ_w	Internal surface temperature of the west wall, ^oC
θ'_w	External surface temperature of the west wall, ^oC
θ_{wf}	Water temperature at the end of the roof, ^oC
θ_{ws}	Wall surface temperature, ^oC
$\Delta\theta$	Temperature difference, ^oC
λ	Wavelength, μm
μ	Dynamic viscosity of the fluid, kg/ms
ν	Kinematic viscosity of the fluid, m^2/s
ξ_d	Day angle, radians
ρ	Density (a-air, w-water), kg/m^3
σ	Stefan Boltzman constant ($= 5.78 \times 10^{-8}\ W/m^2\ K^4$)
τ	Transmissivity of glass,
τ_1	Fraction of energy absorbed by roof,
τ_2	Fraction of energy absorbed by water,
ϕ	Latitude, degrees
ψ	Time lag,
ω	Hour angle (for radiation calculations), Frequency (otherwise), (2π/time period), s^{-1}
ω_s	Sunrise or sunset angle for horizontal surface, degrees
ω'_s	Sunrise or sunset angle for vertical surface facing south, degrees
ω_{sr}	Sunrise angle for an arbitrary surface, degrees
ω_{ss}	Sunset angle for an arbitrary surface, degrees
F_{cl}	Ratio of the surface area of the clothed body to the surface area of the nude body
F'_{eff}	Effective radiation area factor (or configuration factor) i.e. the ratio of the effective radiation area of the clothed body to the surface area of the clothed body
L	Latent heat of vaporization of water (at 35^oC, 2.5 MJ/kg)

Chapter 1
THERMAL COMFORT

1.1 INTRODUCTION

Weather and climate influence human health and longevity. An awareness about the possible connection between them has existed since the beginning of civilization; Hippocrates (1849) probably made the first major contribution to the early thinking on this subject. Through experience and observation he recognized the seasonal nature of some illnesses and identified these as weather dependent. Huttington (1924) enumerated the effect of climate on the development of civilization, the general conclusion drawn from his studies is that human health is best and the human being is most efficient at a mean daily temperature of 18°C when the mean relative humidity is 80% (Sharma, 1977). The occurrence of early civilization in Egypt, Palestine, Sumeria, Persia and Indus Valley has been shown to be related to the climatic conditions by Markham (1947). According to him, the annual mean isotherm of 20°C passes on the world map through or close to these centres of early civilizations and people living in these regions did not have to put up a continuous struggle with the climatic elements for survival. People in Europe on the other hand were able to devote time to creative thinking only after the state of development ensured their survival in the struggle against the elements of nature. According to Brunt (1945), civilization in Western Europe made big strides after the introduction of indoor heating systems, which made the houses relatively comfortable in all weathers.

The history of shelter engineering reveals an unremitting effort by mankind to provide itself with an indoor climate to which man is best adapted. Man's preference for appropriate thermal environment is the main reason for constructing buildings. The design of buildings and the choice of building materials owe a great deal to the external climate and the thermal requirements of human beings. As a first step to systematic building design it is, therefore, necessary to appreciate the indoor conditions which are likely to be acceptable and also the conditions which have to be avoided. These conditions serve as guidelines in assessing the range of values of physical parameters in which one would feel thermally comfortable.

The most important parameters which determine the state of thermal comfort are

- activity level (heat production in the body)
- air temperature

- mean radiant temperature
- air humidity
- relative air velocity
- clothing

It is not possible to consider the effect of only one of the above parameters on the desirable thermal comfort conditions; the effect of each of them on thermal comfort depends also on other factors through the mechanism of heat exchange between the body and the surroundings.

1.2 HEAT EXCHANGE OF BODY WITH SURROUNDINGS

The purpose of the thermoregulatory system of the body is to essentially maintain a constant internal body temperature ($\sim$ 37.2°C). Excess heat generated in the metabolic processes has to be dissipated at an adequate rate. If not removed, this heat raises the body temperature above its normal value viz. 37.2°C and causes acute sensation of discomfort. Conversely, the body feels the sensation of discomfort from cold when it dissipates heat at a faster rate than being generated. In the ideal situation, the heat losses from the body exactly balance the excess metabolic heat generation so that the body temperature neither tends to increase nor decrease. Heat balance of the body under this condition can be written as

$$H_M - H_e - (H_{\ell r} + H_{\ell d}) = H_K + W, \tag{1.1}$$

The metabolic rate depends on the activity of the person and the size of the body; some typical examples expressing the metabolic rate per unit surface area of the body under different stages of activity are given in Table 1.1. The heat production is expressed per unit area of body surface, thus allowing to take into consideration people of different sizes and shape. A very good estimate of the body surface (Dubois area) is given by the Dubois equation i.e.

$$A_N = W_b^{0.425} \times B_h^{0.725} \times 0.2024, \tag{1.2}$$

Typical values for adult males and females range from 1.65 to 2.00 m^2 with 1.8 m^2 as an average figure for a single adult. The heat lost by the body to the ambient i.e. H_K, is, in turn, equal to the sum of heat dissipated by radiation, H_r and convection H_c. The net heat balance of the body can therefore be written as

$$H_M - W = H_T = H_r + H_c + H_e + (H_{\ell r} + H_{\ell d}), \tag{1.3}$$

H_T represents the net waste heat dissipated by the body to the environment. The various terms in the above equation can be evaluated in terms of various coefficients as discussed in the following sections.

1.2.1 Heat Exchange through Radiation

The radiative heat exchange between the body and the surroundings are proportional to the difference between the fourth powers of the absolute temperatures of the surrounding surfaces and the body. For a clothed body, one can therefore write the radiation heat exchange as follows

$$H_r = A_{eff}.\varepsilon.\sigma \left[(\theta_{c\ell} + 273)^4 - (\theta_{mrt} + 273)^4\right] \tag{1.4}$$

TABLE 1.1 Metabolic Rate at Different Typical Activities (Fanger, 1970)

Activity		Metabolic Rate H_M/A_N (W/m^2)	Relative Velocity in Still Air (m/s)	Mechanical Efficiency η*
Resting				
Sleeping		30	0	0
Reclining		34	0	0
Seated, quiet		42	0	0
Standing, relaxed		51	0	0
Walking				
On the level	km/hr			
	3.0	85	0.9	0
	4.0	102	1.1	0
	4.8	111	1.3	0
	5.6	137	1.6	0
	6.4	163	1.8	0
	8.0	248	2.2	0
Miscellaneous occupations				
Bakery (e.g. cleaning tins, packing boxes)		60-85	0-0.2	0-0.1
Brewery (e.g. filling bottles, loading beer boxes on to belt)		51-102	0-0.2	0-0.2
Carpentry				
Machine sawing		77	0-0.1	0
Sawing by hand		171-205	0.1-0.2	0.1-0.2
Planing by hand		240-274	0.1-0.2	0.1-0.2
Foundary Work				
Fettling (pneumatic hammer)		137	0.1-0.2	0-0.1
Tipping the moulds		171	0.1-0.2	0-0.1
Roughing (i.e. carrying 60 kg)		231	0.1-0.2	0-0.2
Tending the furnaces		291	0.1-0.2	0-0.1
Slag removal		325	0.1-0.2	0-0.1
Garage Work (e.g. replacing tyres, raising cars by jack)		94-128	0.2	0-0.1

*Defined as $\eta = W/H_M$ (see Eq. 1.22)

TABLE 1.1 (cont'd)

Activity	Metabolic Rate H_M/A_N (W/m^2)	Relative Velocity in Still Air (m/s)	Mechanical Efficiency $\eta = W/H_M$
Laboratory Work			
Examining slides	60	0	0
General laboratory work	68	0-0.2	0
Setting up apparatus	94	0-0.2	0
Locksmith	94	0.1-0.2	0-0.1
Machine Work			
Light (e.g. electrical industry)	85-103	0-0.2	0-0.1
Machine fitter	120	0-0.9	0-0.1
Heavy (e.g. Paint industry)	171	0-0.2	0-0.1
Manufacture of tins (e.g. filling, labelling and despatch)	85-171	0-0.2	0-0.1
Seated heavy limb movements (e.g. metal worker)	94	0.1-0.4	0-0.2
Shoemaker	85	0-0.1	0-0.1
Shop assistant	85	0.2-0.5	0-0.1
Teacher	68	0	0
Watch repairer	47	0	0
Vehicle driving			
Car (light traffic)	43	0	0
Car (heavy traffic)	85	0	0
Heavy vehicle (e.g. power truck)	137	0.05	0-0.1
Night flying	51	0	0
Instrument landing	77	0	0
Combat flying	103	0	0
Heavy Work			
Pushing Wheelbarrow (57 kg at 4.5 km/hr)	107	1.4	0.2
Handling 50 kg bags	171	0.5	0.2
Pick and shovel work	171-206	0.5	0.1-0.2
Digging trenches	257	0.5	0.2
Domestic Work			
House cleaning	86-146	0.1-0.3	0-0.1
Cooking	68- 85		0
Washing dishes, standing	68	0-0.2	0
Washing by hand and ironing	85-154	0-0.2	0-0.1
Shaving, washing and dressing	73	0-0.2	0

TABLE 1.1 (cont'd)

Activity		Metabolic Rate H_M/A_N (W/m^2)	Relative Velocity in Still Air (m/s)	Mechanical Efficiency $\eta = W/H_M$
Domestic Work (cont'd)				
Shopping		68	0.2-1	0
Office Work	wpm			
Typing (electrical)	30	38	0.05	0
	40	43	0.05	0
Typing (mechanical)	30	47	0.05	0
	40	51	0.05	0
Adding machine		51	0	0
Miscellaneous office work (e.g. filling, checking ledgers)		43-51	0-0.1	0
Draughtsman		51	0-0.1	0
Leisure activities				
Gymnastics		128-171	0.5-2	0-0.1
Dancing		103-188	0.2-2	0
Tennis		197	0.5-2	0-0.1
Fencing		292	0.5-2	0
Squash		308	0.5-2	0-0.1
Basketball		325	1-3	0-0.1
Wrestling		372	0.2-0.3	0-0.1

The term A_{eff} in Eq. (1.4) can be written (Fanger, 1970) as

$$A_{eff} = F_{eff} \cdot F_{c\ell} \cdot A_N \;, \tag{1.5}$$

The value of F_{eff} found by Fanger (1970) was 0.696 for sedentary body posture and 0.725 for standing posture and were seemingly independent of sex, weight, height and Dubois area. The difference between the two F_{eff} values for two postures being very small, a mean value of F_{eff} (= 0.71) can be used as a reasonable approximation in the expression for radiation heat exchange.

For the factor $F_{c\ell}$, some typical values are given in Table 1.2. The tabulated values are, however, only for certain types of clothing and more studies are required in this area. The emittance of the human skin is close to 1.0 (Hardy and Muschenheim, 1934; Mitchell, 1970) and most type of clothings have emittance value of about 0.95; a mean value of 0.97 has therefore been suggested for use (Fanger, 1970).

The mean radiant temperature is the average temperature of the surfaces of the surrounding space, weighted according to the emissivities of various surfaces and the solid angle which they subtend at the subject. Experimentally the mean radiant temperature of a given environment is estimated from measurements of the

TABLE 1.2 Data for Different Clothing Ensembles (Fanger, 1970)

Clothing Ensemble	I^*_{cl} clo	F_{cl}
Nude	0	1.0
Shorts	0.1	1.0
Typical Tropical Clothing Ensemble: Shorts, open-neck shirt with short sleeves, light socks and sandals	0.3-0.4	1.05
Apollo Constant Wear Garment (astronauts Light cotton undergarment with short sleeves and ankle length legs, cotton socks[1])	0.35	1.05
Light Summer Clothing: Long lightweight trousers, open-neck shirt with short sleeves	0.5	1.1
Light Working Ensemble: Athletic shorts, woollen socks, cotton work shirt (open-neck, and work trousers, shirt tail out (208)	0.6	1.1
US Army "Fatigues", Man's: Lightweight underwear, cotton shirt and trousers, cushion sole socks and combat boots[2]	0.7	1.1
Combat Tropical Uniform: Same general components as US Army fatigues but with shirt and trousers of cloth, wind resistant, poplin[3]	0.8	1.1
Typical Business Suit	1.0	1.15
Typical Business Suit + Cotton Coat	1.5	1.15
Light Outdoor Sportswear: Cotton shirt, trousers, T-shirt, shorts socks, shoes and single ply poplin (cotton and dacron) jacket[2]	0.9	1.15
Heavy Traditional European Business Suit: Cotton underwear with long legs and sleeves, shirt, woollen socks, shoes, suit including trousers, jacket and vest (256)	1.5	1.15-1.2
US Army Standard Cold-wet Uniform: Cotton-wool undershirt and drawers, wool and nylon flannel shirt, wind resistant, water repellent trousers and field coat, cloth mohair and wool coat liner and wool socks[3]	1.5-2.0	1.3 -1.4
Heavy Wool Pile Ensemble: (Polar weather suit)	3.0-4.0	1.3- 1.5

*I_{cl} is defined by Eq. (1.20) later.
[1]James M. Waligora, Manned Spacecraft Center, Houston, Personal communication.
[2]J. Jaax, Kansas State University, Personal communication.
[3]J. R. Breckenridge, US Army Research Institute, Natick, Personal communication.

air velocity and the globe temperature, θ_g, according to either Bedford's or Belding's formula (Givoni, 1976).

Bedford's formula

$$(\theta_{mrt} + 273)^4 = (\theta_g + 273)^4 + 8.09 \times 10^{-8} \times v^{0.5} \times (\theta_g - \theta_a), \tag{1.6}$$

Belding's formula

$$\theta_{mrt} = \theta_g + 0.24\, v^{0.25}\, (\theta_g - \theta_a), \tag{1.7}$$

1.2.2 Heat Exchange through Convection

Heat-transfer through convection from a clothed body may be expressed as

$$H_c = A_N\, F_{c\ell}\, h_c\, (\theta_{c\ell} - \theta_a), \tag{1.8}$$

where the heat-transfer coefficient, h_c, depends on the nature of the convection process. At low air velocities, free convection takes place for which Nielsen and Pedersen (1952) have suggested the following formula

$$h_c = 2.38\, (\theta_{c\ell} - \theta_a)^{0.25}, \tag{1.9}$$

this formula is in agreement with the common formulae for free convection with laminar boundary layer (Gebhart, 1961) viz. Nu = Const. $(Gr.Pr)^{0.25}$. For higher velocities, forced convection takes place and in such cases

$$h_c = 12.1\, \sqrt{v}\ , \tag{1.10}$$

for $v < 2.6\ ms^{-1}$, this equation is also in agreement with the common formulae for forced convection (Gebhart, 1961) i.e. Nu = Cons. $(Re)^{0.5}\ (Pr)^{0.33}$.

Although the velocity direction and the body position will have a certain influence on h_c, Eq. (1.10) is a reasonably good approximation for seated or standing persons, when the air flow is across the body. For moving persons (e.g. walking), the velocity used in Eq. (1.10) is the relative velocity. For most practical cases it is recommended that the free convection formula (Eq. 1.9) be used for $v < 0.1\ ms^{-1}$ and the forced convection formula (Eq. 1.10) for $v > 0.1\ ms^{-1}$.

1.2.3 Evaporative Heat Losses

Every kilogram of water that is evaporated consumes in the process about 2.5 MJ energy, the latent heat of vaporization at body temperature. Diffusion of water vapour through the skin and its subsequent evaporation from its surface is a part of the insensible perspiration, a process not subjected to thermoregulatory control. Equation for H_e, the heat loss by diffusion of water vapour and its subsequent evaporation is given by

$$H_e = H_D + H_{se},$$

where

$$H_D = L.p_s.A_N(P_s - P_a), \tag{1.11}$$

and H_{se} is given by Eq. (1.24) later. Using the analysis of Inouye *et al.* (1953), Fanger (1970) estimated a value of 1.27×10^{-7} kg/(s.m^2.mb) for the sedentary subjects under comfort conditions for the permeance coefficient P_s. From steam tables, P_s can be found as a function of various temperatures. For small temperature ranges P_s can be expressed as the linear relation

$$P_s(\theta) = R_1\theta + R_2, \tag{1.12}$$

Values of R_1 and R_2 in the temperature range (36 to 38°C) are 4.066 mb/deg and -84.2 mb respectively.

Substituting for P_s (θ) from Eq. (1.12) into Eq. (1.11) one gets

$$H_D = 0.32\ A_N \left[R_1\ \theta_s + R_2 - P_a\right], \tag{1.13}$$

1.2.4 Respiratory Heat Losses

The process of breathing is associated with simultaneous transfer of heat and water vapour to inhale air by convection and evaporation from the (mucous) lining of the respiratory tract. During the movement of the air through the respiratory tract, some heat is transferred back to the body and water is condensed but the exhaled air from the nose still contains more heat and water than the intake air in a comfortable environment, thus resulting in latent heat losses as well as dry heat losses from the body.

The latent respiration heat loss is a function of the pulmonary ventilation and the difference in water content between exhaled and inhaled air i.e.

$$H_{\ell r} = p_v\ (W_{ex} - W_a)L\ , \tag{1.14}$$

The pulmonary ventilation p_v has been seen (Asmussen and Nielsen, 1946; Liddel, 1963; Fanger, 1970) to approximately follow the relationship

$$p_v = 1.4 \times 10^{-6}\ H_M \qquad \text{(kg/s)} \tag{1.15}$$

Fanger (1970) further showed that $H_{\ell r}$ can also be very well approximated by

$$H_{\ell r} = 0.0017\ H_M\ (59 - P_a), \tag{1.16}$$

where P_a is the partial pressure of water vapour in the inhaled air (ambient air).

The dry heat loss from the body $H_{\ell d}$, resulting due to difference in the temperatures of exhaled and inhaled air can be expressed as

$$H_{\ell d} = p_v\ C_a\ (\theta_{ex} - \theta_a), \tag{1.17}$$

where C_a = 1008 J/kg °C is the specific heat of dry air at constant pressure. The temperature θ_{ex} of the exhaled air has been approximately obtained as (McCutchan and Taylor, 1951)

$$\theta_{ex} = 32.6 + 0.066\ \theta_a + 32\ Wa \tag{1.18}$$

Fanger (1970) has, however, suggested the use of a constant average value of 34°C for θ_{ex}; hence

$$H_{\ell d} = 1.4 \times 10^{-3}\ (34 - \theta_a)H_M, \tag{1.19}$$

1.2.5 Heat Conduction through Clothing

In order to define the heat transfer from the skin to the outer surface of a clothed body, Gagge *et al.* (1941) introduced the dimensionless term (for the total thermal resistance)

$$I_{c\ell} = R_{c\ell}/0.155, \tag{1.20}$$

The dry heat-transfer from skin to the outer surface of the clothed body can thus be expressed by the following formula

$$H_K = \frac{A_N}{0.155\ I_{c\ell}}(\theta_s - \theta_{c\ell}), \tag{1.21}$$

An average effective boundary conductance (inverse of the heat-transfer resistance, $R_{c\ell}$) due to heat losses from a nude body by radiation to surrounding surfaces and natural convection to the surrounding air is estimated as (O'Callaghan, 1978) $U_{c\ell}$ ($= 1/R_{c\ell}$) $= 10$ W m^{-2}K^{-1} (~ 5 W m^{-2}K^{-1} for convection plus ~ 5 W m^{-2}K^{-1} for radiation). The emissivity of clothing on a substrate has found to be less than (O'Callaghan and Probert, 1976) that usually assigned to human skin (~ 1.0), and so the presence of clothing could reduce the radiative heat-transfer coefficient and hence the overall coefficient to about ~ 7.5 W m^{-2}K^{-1}.

1.3 COMFORT EQUATION

Substituting for all the terms in Eq. (1.3) one gets

$$\begin{aligned}\frac{H_T}{A_N} = \frac{H_M(1-\eta)}{A_N} &= 0.71\sigma\ F_{c\ell}\left[(\theta_{c\ell} + 273)^4 - (\theta_{mrt} + 273)^4\right] \\ &+ F_{c\ell}\ h_c(\theta_{c\ell} - \theta_a) + 0.32\ (R_1\theta_s + R_2 - P_a) \\ &+ (H_{se}/A_N) + 1.4 \times 10^{-3}\ \frac{H_M}{A_N}(34 - \theta_a) \\ &+ 0.0017\ \frac{H_M}{A_N}(59 - P_a),\end{aligned} \tag{1.22}$$

where $\eta = W/H_M$ is the external mechanical efficiency and the magnitude (H_M/A_N) is a function of the activity of the person (see Table 1.1 for characteristic values). Comfort requires that Eq. (1.22) is satisfied in the steady state because the thermoregulatory system adjusts itself to control the physiological variables like the skin temperature, θ_s and evaporative secretion rate H_{se}. This can happen within wide limits of environmental variables and hence Eq. (1.22), though a necessary condition for thermal comfort, is far from being a sufficient condition for thermal comfort. Within these wide limits of the environmental variables for which a heat balance can be maintained, there is only a narrow interval which will create thermal comfort. For thermal comfort at a given activity level of a human being, the skin temperature and the sweat secretion rate must be within certain limits. Through a regression analysis of the experimental data for persons in thermal comfort, Fanger (1970) deduced the following relationships

$$\theta_s = 35.7 - .0275\left(\frac{H_M}{A_N}\right), \text{ for skin temperature} \tag{1.23}$$

and

$$H_{se} = 0.42\,A_N\left(\frac{H_M}{A_N} - 58\right), \text{ for heat loss due to water evaporation from the surface of the skin.} \tag{1.24}$$

It is seen that for constant comfort, the mean skin temperature decreases with increasing activity (for $(H/A_N) = 58\ W/m^2$, θ_s is 34°C and for $(H/A_N) = 174\ W/m^2$. θ_s is 31°C). For sedentary activity ($H/A_N \simeq 58\ W/m^2$) sweat secretion at thermal comfort is zero. At higher activity moderate sweat secretions are necessary.

Substituting for θ_s and H_{se} from the expressions (1.23) and (1.24) in Eq. (1.22) one can write

$$\Big[\frac{H_M}{A_N}(1-\eta) - 0.32\,[R_1'\{35.7 - 0.0275\frac{H_M}{A_N}(1-\eta)\} + R_2 - P_a]$$
$$- 0.42'\{\frac{H_M}{A_N}(1-\eta) - 58\} - 1.4\times 10^{-3}\frac{H_M}{A_N}(34-\theta_a) - 0.0017\frac{H_M}{A_N}(59-P_a)\Big]$$
$$= \left[35.7 - \frac{0.0275\,H_M}{A_N}(1-\eta) - \theta_{c\ell}\right] \Big/ 0.155\,I_{c\ell}, \tag{1.25}$$

Solving the above equation for $\theta_{c\ell}$, one gets

$$\theta_{c\ell} = 35.7 - 0.0275\frac{H_M}{A_N}(1-\eta) - 0.155\,I_{c\ell}\Big[\frac{H_M}{A_N}(1-\eta)$$
$$- 0.32'\{R_1(35.7 - 0.0275\frac{H_M}{A_N}(1-\eta)) + R_2 - P_a\}$$
$$- 0.42'\{\frac{H_M}{A_N}(1-\eta) - 58\} - 1.4\times 10^{-3}\frac{H_M}{A_N}(34-\theta_a)$$
$$- 0.0017\frac{H_M}{A_N}(59-P_a)\Big], \tag{1.26}$$

The left-hand side of Eq. (1.25) is also equal to the total radiative and convective losses from the body and can be written as

$$\Big[\frac{H_M}{A_N}(1-\eta) - 0.32'\{R_1\,(35.7 - 0.0275\frac{H_M}{A_N}(1-\eta) + R_2 - P_a\}$$
$$- 0.42'\{\frac{H_M}{A_N}(1-\eta) - 58\} - 1.4\times 10^{-3}\frac{H_M}{A_N}(34-\theta_a)$$
$$- 0.0017\frac{H_M}{A_N}(44-P_a)\Big] = \Big[0.71\sigma\,F_{c\ell}'\{(\theta_{c\ell}+273)^4 - (\theta_{mrt}+273)^4\}$$
$$+ F_{c\ell}\,h_c\,(\theta_{c\ell} - \theta_a)\Big], \tag{1.27}$$

Equation (1.27) is the desired comfort equation in which $\theta_{c\ell}$ is given by Eq. (1.26) and h_c by Eq. (1.9) or Eq. (1.10). The variable parameters occurring in Eq. (1.27) are $I_{c\ell}$, $F_{c\ell}$, H_M/A_N, η and v (through h_c), θ_a, P_a, θ_{mrt}. These depend upon clothing, activity and the environmental conditions.

1.4 THERMAL INDICES

From the comfort equation (1.27) it is apparent that the human response to the warmth or cold depends on the combined effect of mean radiant temperature θ_{mrt}, ambient temperature θ_a, humidity ($P_a = R_h P_s$) and air velocity v besides the amount and type of clothing and other physiological factors such as activity. Since it is rather difficult to rationalize clothing and activity, various attempts have been made to combine all these factors into one single factor to define the level of thermal comfort, which adequately accounting for activity and clothing, can give the designer a measure about the general quality of the environment. As a result various indices have been developed which differ in their basic approach to the problem, in the units used as the basis for expressing the combined effect of the various factors, in the range of conditions of their application, in the relative importance attributed to each of the factors, in their mutual interdependence and in the approximate expressions used for calculating various heat exchanges.

The developed thermal indices are

1. Effective temperature (E.T.)
2. Resultant temperature (R.T.)
3. Wet-bulb globe temperature (W.B.G.T.)
4. Equatorial comfort index (E.C.I.).
5. Heat stress index (H.S.I.)
6. Index of thermal stress (I.T.S.)
7. Predicted 4 hour sweat rate (P_4.S.R.)
8. Tropical summer index (T.S.I.)
9. Predicted mean value (P.M.V.)
10. Wind chill equivalent temperature (W.C.E.T.)
11. Humidex.

1.4.1 Effective Temperature

This scale was first developed by Houghton and Yaglou (1923). The factors it includes are the air temperature, the humidity and the air-velocity. Two scales were developed, one for men stripped to the waist, called the basic scale and the other for men fully clad in indoor clothing, called the normal scale of effective temperature. The unit, or basis, of E.T. index is the temperature of saturated still air with average velocity 0.12 m/s. Any combination of air temperature, humidity and air-velocity having a given value of the E.T. is supposed to produce the same thermal sensation, that is experienced in saturated still air at the same temperature as the value of the index. Bedford (1946) suggested the use of globe temperature reading instead of air temperature to account for the radiant heat. This scale is known as the corrected effective temperature (C.E.T.) scale.

Figure 1.1 represents the corrected effective temperature nomogram. C.E.T. can be obtained by connecting the appropriate points representing the dry-bulb (or globe) and wet-bulb temperatures and reading the C.E.T. as t-e intersection of this line with the relevant air velocity curve. Difference between successive points corresponds to approximately °C. The subject is thermally neutral at 20°C.

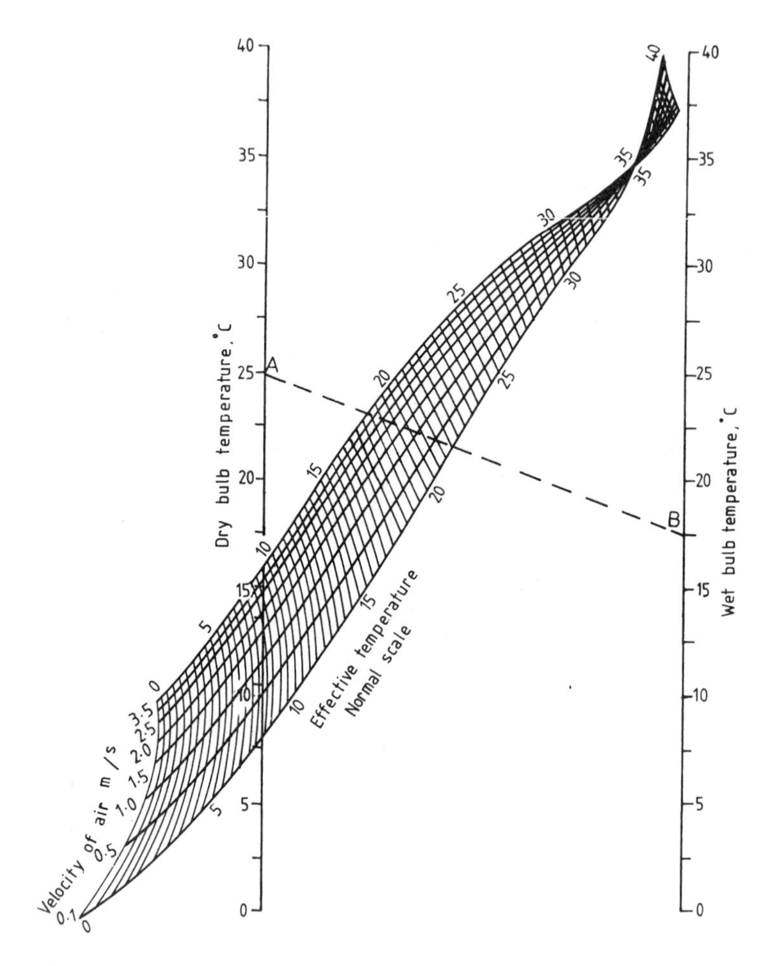

Fig 1.1. Chart of the Effective Temperature Index. (After ASHRAE *Handbook of Fundamentals*, 1981)

Example: Find the effective temperature in an enclosure with wet-bulb temperature 17.5°C and dry-bulb temperature 25°C and room air movement at 0.1 ms^{-1}.

Find point A corresponding to dry-bulb temperature of 25°C and point B corresponding to wet-bulb temperature 17.5°C. Join A with B. The point of intersection with the air movement curve corresponding to v = 0.1 ms^{-1} yields effective temperature equal to 22.0°C.

Gagge *et al.* (1974) have modified the effective temperature by considering an imaginary enclosure at 50% relative humidity (rather than 100%) in which the man will exchange the same amount of heat (by radiation, convection and evaporation) at the same skin temperature and skin wettedness, as occur in the actual environment.

Standard effective temperature previously introduced by Gagge *et al.* (1972, 1973) took into account the clothing and level of activity also.

It is expressed in terms of a uniform environment standardized at 50% relative humidity, air velocity at 0.125 ms^{-1} for sedentary level of activity (metabolic rate 58 W/m^{-2}) and intrinsic clothing at 0.6 clo value (equivalent to normal, lightweight, indoor clothing). Graphs of standardized effective temperature have been given by Markus and Morris (1980).

The effective temperature scales may be considered reasonably accurate for warm climates where heat stress is not high but it may be misleading at high levels of heat stress. This scale is based on instantaneous reaction of subjects, a phenomenon liable to change after long stay in the same environment. The results of Glickman and coworkers (1950) show an overestimation of the effect of humidity under cool and comfortable conditions.

Smith (1955) points out three main shortcomings of the effective temperature scale viz. (i) it does not make allowance for the deleterious effects of low air speeds in hot and humid conditions, (ii) it exaggerates the stress imposed by air speeds of about 0.5 to 1.5 m/s and (iii) in hot environments it underestimates the effect of humidity.

1.4.2 Resultant Temperature

Resultant temperature index developed by Missenard (1948) was based on the assumption that a more firm basis for a thermal index would be formed by experiments in which thermal equilibrium was achieved between the body and the environment, so that the effects of humidity and wind could be found. From the experimental results a nomogram was plotted as shown in Fig. 1.2, for the clothed body. For the example of the effective temperature, the value of the resultant temperature read from the nomogram comes out to be 23.5°C. The range of climatic factors covered by the resultant temperature are the air temperature between 20-45°C, the wet bulb temperature between 18 and 40°C and the wind speed between 0 and 3 m/s.

The index is not applicable to working conditions. Above 30°C, the index was found to take care adequately of the humidity and air temperature. Below 30°C, the index slightly overestimates the effect of humidity. For higher range of air velocities, the index was found to underestimate the cooling effect of air motion, while in the lower range the effect of wind was overestimated. Overall, however, the resultant temperature is in better agreement with the observed physiological response in comparison to the effective temperature scale.

1.4.3 Wet-Bulb Globe Temperature

This index was devised by Yaglou and Minard (1957) for use in the US army as a simple substitute for the effective temperature scale, making special allowance for the absorption of solar radiation by the drab olive military uniforms. W.B.G.T. is calculated according to the following weighted expression

$$\text{W.B.G.T.} = 0.2\ \theta_g + 0.1\ \theta_a + 0.7\ \theta_{wb}, \qquad (1.28)$$

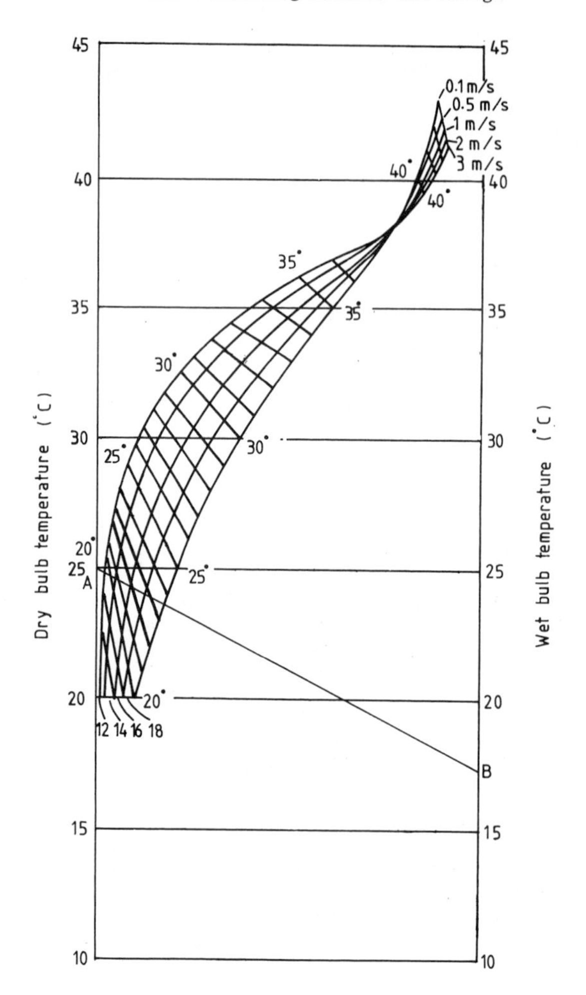

Fig. 1.2. Chart of the resultant temperature index. (after Missenard, 1948)

This index has the merit of simplicity and does not require extensive instrumentation. The index was effectively used to prescribe the limit of thermal stresses on the unacclimatized marine recruits in USA. It was recommended that the training of the raw recruits should cease when the index reached 29.4°C and all strainous activity should be discontinued regardless of the level of acclimatization when the index was 31.1°C or more. The index cannot however be used at high levels of climatic stress and has all other limitations of the effective temperature scale.

1.4.4 Equatorial Comfort Index

This index was developed by C. G. Webb (1960) from a study of the thermal comfort of persons living in Singapore. An equation was developed from the results of the analysis, expressing the observed dry-bulb temperature, vapour pressure and air speed in terms of the temperature of still, saturated air which would produce the same overall thermal sensation. The equation for E.C.I. expressed in deg C is

$$E.C.I. = 0.574\ \theta_a + 0.2033\ P_a - 1.81\ v^{0.5} + 42, \qquad (1.29)$$

The nomogram, of E.C.I., given in Fig. 1.3 is similar to that of the effective temperature and includes dry-bulb temperatures and air speed.

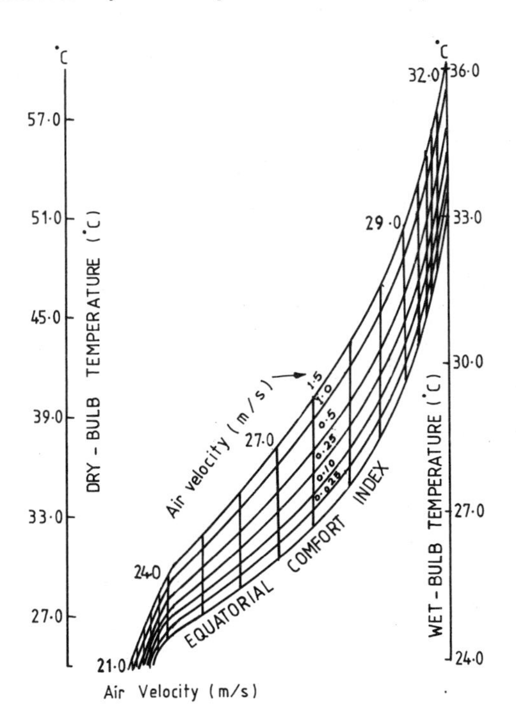

Fig. 1.3. Equatorial comfort index nomogram. (after Webb, 1960)

The E.C.I. is applicable for every limited warm, humid and oppressive conditions only where wet-bulb temperature exceeds 25°C. It is meant for indoor conditions when air temperature equals the mean radiant temperature.

1.4.5 Heat Stress Index

The heat balance equation of the human body was adopted by Belding and Hatch (1955) to include a numerical index of heat stress for the given conditions of work and climate. The evaluation of the index was based on certain assumptions namely (i) the total heat stress acting on the body (metabolism ± radiation ± convection) equals the requirement for sweat evaporation, (ii) the physiological strain imposed on the body by a given heat stress is determined by the ratio of the required evaporative cooling to the maximum evaporative capacity of the air, (iii) constant skin temperature, 35°C, is maintained when the body is subjected to heat stress and (iv) the maximum sweating capacity of an average person over an 8 hr period is approximately one litre per hour, yielding a cooling value of

694 W. Mathematically, the heat stress index is given by

$$H.S.I. = \frac{\text{Required evaporation}}{\text{Maximum evaporative capacity}} \times 100, \quad (1.30)$$

where the required evaporation

$$He_{req} = \text{metabolism} \pm \text{radiation} \pm \text{convection}$$

For the evaluation of H.S.I., following formula was used,

$$He_{req} = 100 \pm h_r (\theta_{wb} - \theta_s) \pm h_c (\theta_a - \theta_s), \quad (1.31)$$

and the maximum evaporative capacity (Sharma, 1977).

$He_{max} = 24.2\ v^{0.4}\ (60 - P_a)$, assumes that the difference between the air and skin pressures is 60 m bar (at a temperature of 35°C).

If $He_{max} > 694$, then $He_{max} = 694$. h_r and h_c in Eq. (1.31) are taken from Eqs. (1.4) and (1.10) respectively. Corresponding to skin temperature of 35°C and wall temperature of 25°C, value of h_r comes out to be 6.43 W/m^2°C.

A nomogram shown in Fig. 1.4 helps to obtain the H.S.I. for different conditions of air and radiant temperature, wind speed, humidity and metabolic level. The numerical values lie between 0 and 200. The value 0 represents the absence of the heat stress (approximately the comfort region) and 100 the upper limit for thermal equilibrium. The region of the body heating is between 100 and 200. The range of conditions covered by the index is the dry-bulb (or globe) temperature between 21-49°C, vapour pressure 4 - 60 m bar, air velocity 0.25-10.0 m/s and metabolic rate 110-550 W.

On comparing the actual measurements of total heat load with the calculated value from the theoretical formula of H.S.I. discrepancies of the order of 40% were detected by later experiments. It was suggested (Givoni, 1976) that the influence of clothing might affect all heat exchanges viz. convective, radiative and evaporative. It was also found that the H.S.I. overestimates the cooling effect of wind and the warming effect of humidity. H.S.I. has therefore been regarded inadequate for quantitative evaluation of the severity of thermal stress.

1.4.6 Index of Thermal Stress (I.T.S.)

This index was developed by Givoni and Berner-Nir (1967) to cover all the mechanisms of heat loss and heat gain by the human body, and various levels of work and clothing. The index is based on the assumption that if thermal equilibrium is possible, the rate of sweat production is sufficient to achieve the evaporative cooling required to balance the metabolic heat production and the heat exchange with the environment.

The initial version of the index of thermal stress was intended only for indoor use and one type of clothing. Later it was extended for outdoor conditions by including the effect of solar radiation. The effect of various types of clothing has also been included in it now.

The basic relationship of the I.T.S. is

$$ITS = H_{se}\ (1/f), \quad (1.32)$$

$$H_{se} = (H_M - W) \pm H_c \pm H_r, \quad (1.33)$$

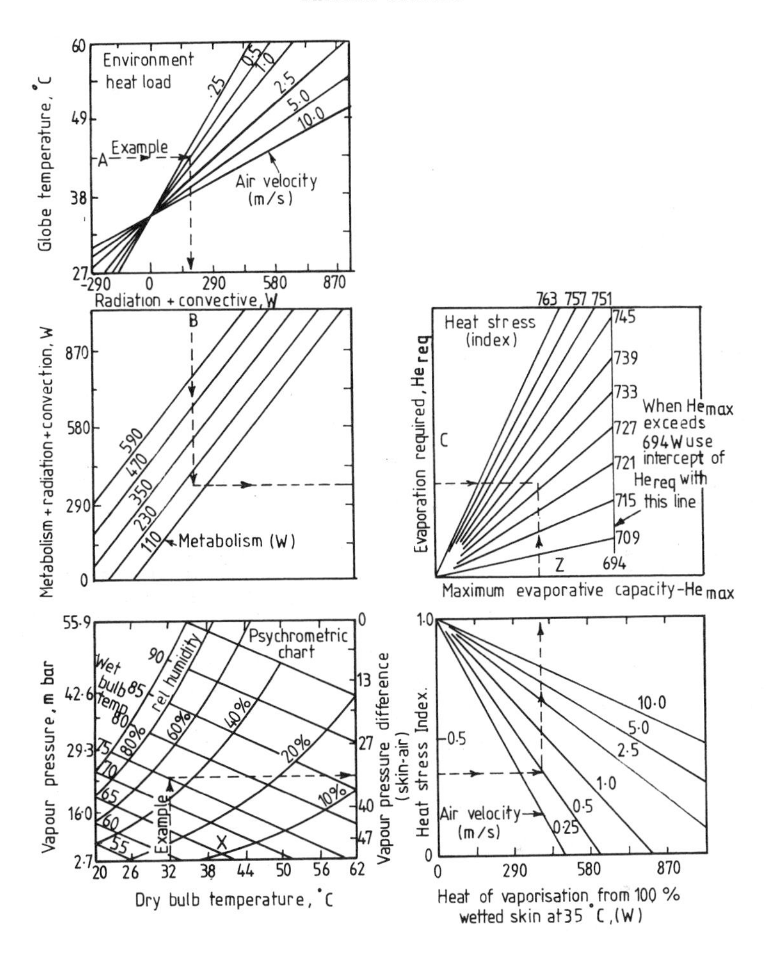

Fig. 1.4. Charts of the heat stress index. (after ASHRAE *Handbook of Fundamentals*, 1981)

where

$$\left(\frac{1}{f}\right) = \exp\left[0.6\left(\frac{H_{se}}{H_{e_{max}}} - 0.12\right)\right], \qquad (1.34)$$

and $H_{e_{max}}$, the maximum evaporative capacity of the air (W) is calculated from the expression

$$H_{e_{max}} = p_c\ v^{0.3}\ (60 - P_a), \qquad (1.35)$$

where 60 is used for water vapour pressure (mb) of skin at 35°C. The lower limit of 1/f is 1.0 and is maintained as long as the ratio ($H_{se}/H_{e_{max}}$) is below 0.12. The upper limit of 1/f is 3.5 and it is achieved when ($H_{se}/H_{e_{max}}$) reaches 2.15 and above. The values of the clothing coefficient p_c corresponding to different clothings are given in Table 1.3.

TABLE 1.3. Clothing coefficients for the Index of Thermal Stress (I.T.S.)

Clothing	α (w)	Coefficient K_{cl}	a	p_c
Semi-nude: bathing suit and hat	18.3	1.0	0.35	31.6
Light summer clothing: underwear, short-sleeved cotton shirt, long cotton trousers, hat	15.1	0.5	0.52	20.5
Military overalls over shorts	13.5	0.4	0.52	13.0

The detailed general formula of the I.T.S. was obtained by Givoni (1969) as

$$\text{I.T.S.} = \Big[H_M - 0.2\,(H_M - 11.6) \pm \alpha\, v^{0.3}\,(\theta_a - 35)$$

$$+ 1.16\, S_N\, K_{pe}\, K_{c\ell}' \{1 - a\,(v^{0.2} - 0.88)\}\Big]$$

$$\times \exp' \{0.6\,(\frac{H_{se}}{H_{e_{max}}} - 0.12)\}, \qquad (1.36)$$

where K_{pe} is a coefficient which depends on the posture and terrain and α, $K_{c\ell}$ and a depend on the clothing. Values of various coefficients are given in Tables 1.3 and 1.4.

The detailed general formula for the I.T.S. is not easily workable without the aid of several nomograms. The investigations of Givoni (1976) show that the calculated values of I.T.S. fit the experimentally observed data of the I.T.S. reasonably well; it may however be mentioned that several factors included in the definition of I.T.S. are difficult to measure and their values differ appreciably from place to place.

1.4.7 The Predicted 4 Hours Sweat Rate (P.$_4$S.R.)

This index was developed during World War II by McArdle (1947) and his colleagues. This index is based on experiments carried out over 4 hour periods and measuring the sweat rate resulting from a 4 hour exposure to the given conditions. The experiments were performed under rest conditions (metabolic level, 58 W/m^2) and working conditions (metabolic level 130 W/m^2) alternately. Nomogram representing the index is given in Fig. 1.5. The basic P.$_4$S.R. can be obtained directly from the data of the dry and wet-bulb temperatures and the air velocity. While using P.$_4$S.R. one first determines the basic 4 hour Sweat Rate (P.$_4$S.R.) from the nomogram. This corresponds to the Predicted 4 hour Sweat Rate in litres for people sitting clad in shorts in a homogeneous environment. For other conditions following corrections are applied to the basic value of the index:

(i) If the globe temperature θ_g differs from the dry-bulb temperature, the wet-bulb temperature is corrected by addition of 0.40 (θ_g - D.B.T.)^{o}C.

(ii) Another correction is applied, when the metabolic rate exceeds the resting level (62.9 W/m^2), the W.B.T. is corrected by the addition of a value, read from an insert in Fig. 1.5. For example for a metabolic rate of 116.3 W, a further 2.2^{o}C is added to W.B.T.

(iii) If the weight of the clothing is above 600 g, 0.55°C is added to the W.B.T. for every 300 g increase in clothing weight.

TABLE 1.4 Combined Solar Load Coefficient for Posture and Terrain

Posture	Terrain	K_{pe}
Sitting with back to sun	Desert	0.386
	Forest	0.379
Standing with back to sun	Desert	0.306
	Forest	0.266

The range of conditions covered by the index is between 27 - 54°C for globe temperature, 16 - 36°C for wet-bulb temperature, 0.05 - 2.5 m/s for wind speed, metabolic level between 58 - 230 W/m^2 and for two types of clothing i.e. shorts only or overall/shorts. A comparison of the experimental measurements with the values read from the nomogram of P.$_4$S.R., show that the index is able to take account of the air velocity adequately, while the effect attributed to humidity was a little smaller than that observed in the experiments. In general, however, it may be said that P.$_4$S.R. index enables reliable estimation of the overall thermal stress, manifested in the sweat rate and, within a given range of thermal sensation and under a variety of metabolic, climatic and clothing conditions.

1.4.8 Tropical Summer Index (T.S.I.)

For tropical summer conditions, Sharma (1977) has evolved this index and it is defined as the air/globe temperature of still air at 50% R.H. which produces the same overall thermal sensation as the environment under investigation. The final equation derived for T.S.I. is as follows

$$\text{T.S.I.} = 0.308\ \theta_{wb} + 0.745\ \theta_g - 2.06\ v^{\frac{1}{2}} + 0.841, \tag{1.37}$$

It was found by experiments that a value of T.S.I. equal to 27.5°C makes the subject most comfortable and successive thermal sensations will change at approximately 4.5°C intervals. The ranges and optimum values of T.S.I. for practical use are presented in Table 1.5 below.

TABLE 1.5 Ranges and Optimum Values of T.S.I. for Central Thermal Sensations

Thermal Sensation	Range	Optimum value
Slightly cool	19.0-25.0°C	22.0°C
Comfortable	25.0-30.0°C	27.5°C
Slightly warm	30.0-34.0°C	32.0°C

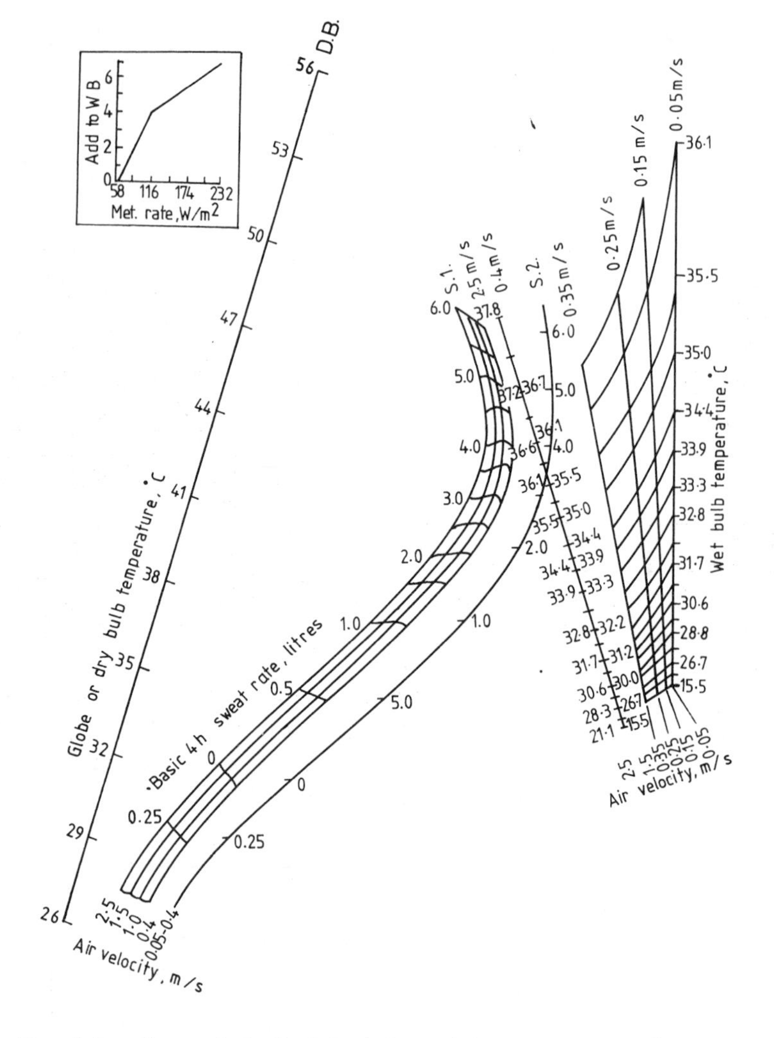

Fig. 1.5. Chart of the P.4S.R. index. (after McArdle *et al.*, 1947)

The nomogram representing T.S.I. is shown in Fig. 1.6. These represented T.S.I. values are for calm wind conditions only. The reduction in T.S.I. values for different wind speeds is also shown in Fig. 1.6.

The T.S.I. as described above is best correlated to thermal sensation of Indian subjects as compared to other thermal indices. It cannot, however, be applied to the conditions of other countries without making appropriate changes. It also does not have a mechanism to take into account the effect of clothing, activity etc.

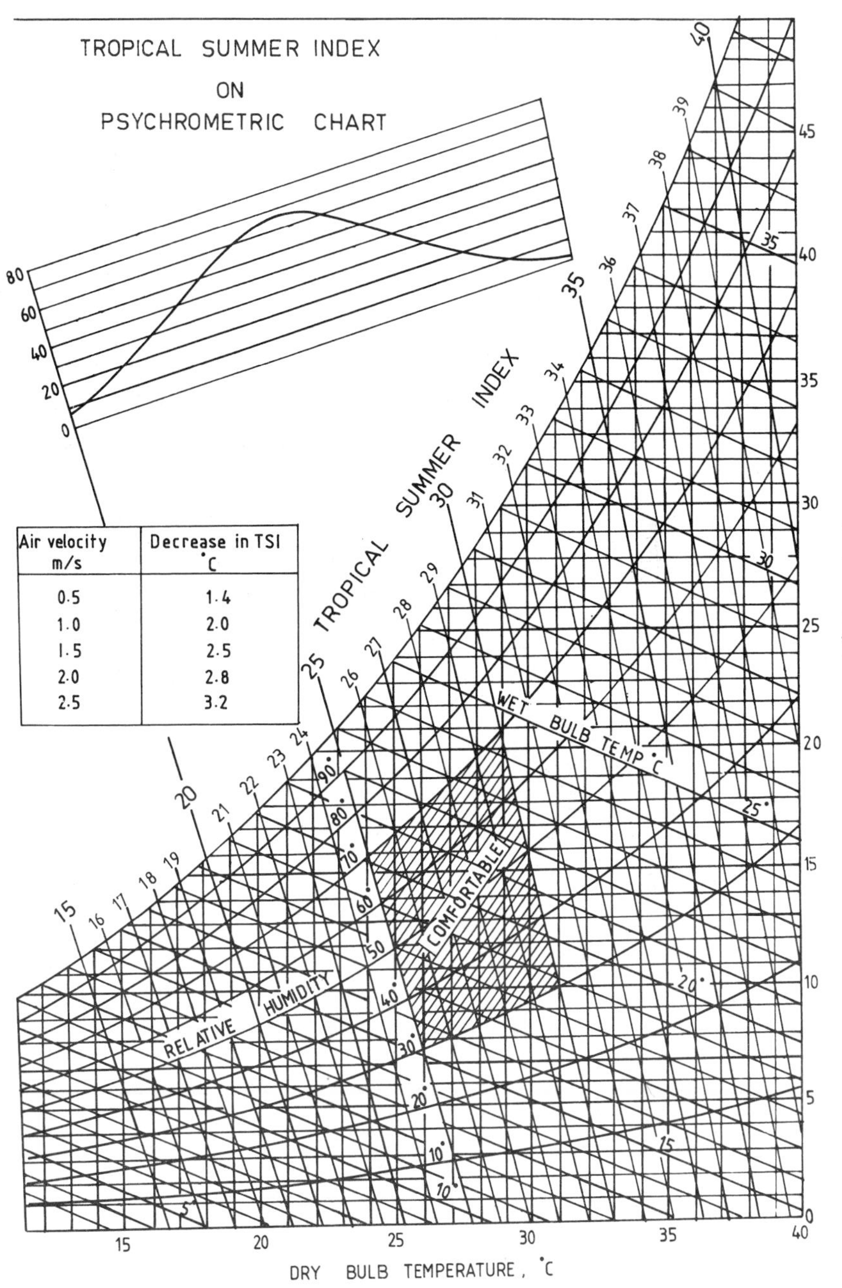

Fig. 1.6. Tropical summer index lines and comfort zone.
(after Sharma, 1977)

1.4.9 Predicted Mean Values

Fanger (1970) derived a thermal index depending upon the comfort Eq. (1.27) which makes it possible to predict the thermal sensation for any given combination of activity level, clothing and four environmental parameters. The measurement of thermal sensation was done by using the seven points of the psycho-physical scale.

- -3 Cold
- -2 Cool
- -1 Slightly cool
- 0 Neutral
- 1 Slightly warm
- 2 Warm
- 3 Hot

Using its effecting mechanism viz. vasodilatation and vasoconstriction, sweat secretion and shivering, the human body is capable of maintaining heat balance within wide limits of the environmental variables; within these wide limits, however, only a small interval can be regarded as comfortable. If one defines a quantity thermal load Q_L as the difference between the internal heat production and the heat loss to the actual environment then it can be assumed that the degree of discomfort is greater, the more Q_L deviates from the comfort equation written as

$$Q_L = \frac{H_M}{A_N}(1-\eta) - 0.32\left[R_1'\{35.7 - 0.0275\frac{H_M}{A_N}(1-\eta)\} + R_2 - P_a\right]$$

$$- 0.42'\{\frac{H_M}{A_N}(1-\eta) - 58\} - 1.4 \times 10^3 \frac{H_M}{A_N}(34 - \theta_a)$$

$$- 0.0017\frac{H_M}{A_N}(44 - P_a) - 0.71\sigma F_{c\ell}'\{(\theta_{c\ell} + 273)^4 - (\theta_{mrt} + 273)^4\}$$

$$- F_{c\ell} h_c (\theta_{c\ell} - \theta_a), \tag{1.38}$$

where $\theta_{c\ell}$ has to be found iteratively from the equation

$$\theta_{c\ell} = 35.7 - 0.0275\frac{H_M}{A_N}(1-\eta) - 0.155 I_{c\ell}\left[0.71 F_{c\ell}\right.$$

$$\times'\{(\theta_{c\ell} + 273)^4 - (\theta_{mrt} + 273)^4\}$$

$$\left. + F_{c\ell} h_c (\theta_{c\ell} - \theta_a)\right], \tag{1.39}$$

and h_c is given by Eq. (1.9) or Eq. (1.10).

For comfort conditions it is clear that Q_L is zero. In other conditions, when $Q_L \neq 0$, the effective mechanism of the body changes the mean skin temperature sweat secretion for maintaining the heat balance. The thermal sensation felt by the body in doing so can then obviously be represented by the mathematical relationship

$$v_m = F\left(Q_L, \frac{H_M}{A_N}\right), \tag{1.40}$$

where thermal sensation is expressed by the mean vote v_m on the scale as mentioned earlier.

It is to be noted that the Eq. (1.40) can be quantified only on the basis of experiments, where the subjects cast their thermal sensation votes. Based on the experimental data of Novins *et al.* (1966) for sedentary persons and of McNall *et al.* (1967) for higher activity level, Fanger (1970) found a relation between the mean vote, v_m and the ambient temperature θ_a. The data is given in Table 1.6 below.

TABLE 1.6 Connection between Mean Vote and Ambient Temperature at Four Activity Levels (After Fanger, 1970)

Activity level	H_M/A_N W/m²	I_{cl} clo	v m/s	Mean Vote v_m at R_h = 50%
Sedentary	58	0.6	0.1	$v_m = -8.471 + 0.331\ \theta_a$
Low	93	0.6	0.2	$v_m = -3.643 + 0.175\ \theta_a$
Medium	128	0.6	0.25	$v_m = -3.356 + 0.174\ \theta_a$
High	158	0.6	0.32	$v_m = -4.158 + 0.265\ \theta_a$

By noting the values of v_m for different level of activity from a large number of subjects, Fanger (1970) found the following relationship

$$v_m = \left[0.352 \exp(-0.042 \frac{H_M}{A_N}) + 0.032\right] Q_L, \tag{1.41}$$

Since the above equation is a measure of the feeling of thermal sensation, Fanger called this parameter as the "Predicted Mean Vote" (P.M.V.). Substituting the expression for Q_L from Eq. (1.38) into Eq. (1.41) one gets

$$\text{P.M.V.} = \left\{0.352 \exp\ (-0.042\ H_M/A_N) + 0.032\right\} \left[\frac{H_M}{A_N}(1-\eta) - \right.$$

$$- 0.32 \left\{ \{R_1\ (35.7 - 0.0275 \frac{H_M}{A_N}(1-\eta))\} + R_2 - P_a \right\}$$

$$- 0.42 \left\{ \frac{H_M}{A_N}(1-\eta) - 58 \right\} - 1.4 \times 10^{-3} \frac{H_M}{A_N}(34 - \theta_a)$$

$$- 0.0017 \frac{H_M}{A_N}(44 - P_a)$$

$$- 0.71\sigma F_{c\ell} \{(\theta_{c\ell} + 273)^4$$

$$- (\theta_{mrt} + 273)^4 \Big\} - F_{c\ell}\, h_c\, (\theta_{c\ell} - \theta_a)\Big], \qquad (1.41)$$

The above equation establishes the thermal index, called the Predicted Mean Vote, as a function of activity (W/m^2), clothing (clo), air temperature (^{o}C), mean radiant temperature (^{o}C), relative velocity (m/s) and humidity. The mathematical expression for P.M.V. is, however, complicated and is obviously not suitable for calculations by hand. The results of a large number of computations for various parameters have however been tabulated by Fanger (1970) and reproduced in Appendix I. The P.M.V. index can be expected to be less accurate at the combination; low clo value (nude) and high velocity, since the actual skin temperature is quite close to the air temperature in this case.

1.4.10 Wind Chill Index

A combination of temperature and wind speed governs the thermal sensation outdoors in cold climates. This combination is termed as wind chill (Landsberg, 1984). Wind chill index was first developed during the Antarctic expedition. The basis of this index is the time required to freeze a particular quantity of water under given environmental conditions. An empirical model giving the cooling power of the atmosphere has been developed i.e.

$$\text{C.P.} = 1.167\ (10\sqrt{v} + 10.45 - v)\ (33 - \theta_a), \qquad (1.42)$$

The sensation for different values of C.P. are as follows

C.P. (W/m^2)	Sensation
115 - 350	pleasant
350 - 600	cool
600 - 800	cold
800 - 1100	very cold
1100 - 1500	bitter cold
1500 and above	exposed flesh freezes

For persons at sedentary level, wind chill index is not very well understood, it has therefore been converted into a wind chill equivalent temperature (or chill factor) given in Table 1.7. In full sunshine, these equivalent temperatures get modified. For calm air this increase is about 14°C and for strong winds 7°C (Steadman, 1971).

The wind chill index is valuable for very cold climatic conditions for choosing appropriate clothing. This can prevent frostbite, disease or even death (Falconer, 1968).

1.4.11 Humidex

Developed by Canadian Weather Service, this index is most appropriate for the summers of cold climatic zones. It is defined as

$$H = \theta_a + E, \qquad (1.43a)$$

where

TABLE 1.7 Wind Chill equivalent Temperature

Wind Velocity (m/s.)	(10.0)	(5.0)	(0.0)	(-5.0)	(-10.0)	(-15.0)	(-20.0)	(-25.0)	(-30.0)	(-35.0)	(-40.0)	(-45.0)	(-50.0)
	Equivalent temperature (°C) (equivalent in cooling power on exposed flesh under calm conditions)												
calm	(10.0)	(5.0)	(0.0)	(-5.0)	(-10.0)	(-15.0)	(-20.0)	(-25.0)	(-30.0)	(-35.0)	(-40.0)	(-45.0)	(- 50.0)
2.2	(8.9)	(3.3)	(-1.7)	(-6.7)	(-12.2)	(-17.2)	(-22.8)	(-27.8)	(-33.3)	(-38.3)	(-43.9)	(-48.9)	(- 43.9)
4.5	(4.4)	(-1.7)	(-7.8)	(-13.9)	(-15.6)	(-26.1)	(-32.2)	(-38.3)	(-44.4)	(-50.6)	(-56.7)	(-62.8)	(- 68.9)
6.7	(2.2)	(-4.4)	(-10.6)	(-18.3)	(-25.0)	(-31.7)	(-38.3)	(-45.0)	(-51.7)	(-58.3)	(-65.0)	(-71.7)	(- 78.3)
8.9	(-0.0)	(-6.7)	(-13.9)	(-12.1)	(-28.3)	(-35.6)	(-42.2)	(-49.4)	(-56.7)	(-63.9)	(-71.1)	(-78.3)	(- 85.0)
11.2	(-1.1)	(-8.4)	(-16.1)	(-23.3)	(-31.1)	(-38.3)	(-45.6)	(-53.3)	(-60.6)	(-67.8)	(-75.5)	(-82.8)	(- 90.0)
13.4	(-2.2)	(-10.0)	(-17.2)	(-25.0)	(-32.8)	(-40.6)	(-47.8)	(-55.6)	(-63.3)	(-71.7)	(-78.3)	(-86.1)	(- 93.9)
15.6	(-2.8)	(-10.6)	(-18.3)	(-26.1)	(-33.9)	(-41.7)	(-49.4)	(-57.2)	(-65.0)	(-73.3)	(-80.6)	(-88.3)	(- 96.6)
17.9	(-3.3)	(-11.1)	(-19.4)	(-27.2)	(-35.0)	(-42.8)	(-50.6)	(-58.9)	(-66.1)	(-74.4)	(-82.2)	(-90.6)	(- 98.3)
20.1	(-3.9)	(-11.7)	(-19.4)	(-27.8)	(-35.6)	(-43.3)	(-51.7)	(-59.4)	(-67.2)	(-75.5)	(-83.3)	(-91.7)	(- 99.4)
22.4	(-3.9)	(-12.2)	(-20.0)	(-27.8)	(-36.1)	(-43.9)	(-52.2)	(-60.0)	(-68.3)	(-76.1)	(-84.4)	(-92.2)	(-100.0)
	Little danger				Increasing danger					Great danger			
	Danger from freezing of exposed flesh (for properly clothed persons)												

Note. For wind values of ⩽ 1 m/s, conditions are assumed to be calm.

The table indicates the limits of danger of frostbite even for appropriately dressed persons.

$$E = \frac{5}{9} (P_s - 10), \tag{1.43b}$$

with, the vapour pressure P_s (in mb) determined from the dew point temperature θ_{dp} as

$$P_s(\theta_{dp}) = 6.11 \exp\left\{ \frac{M_o^* L}{R^+} \left(\frac{1}{273.16} - \frac{1}{\theta_{dp}} \right) \right\}, \tag{1.44}$$

The degree of discomfort for various ranges of Humidex are as follows

Humidex value	Sensations
20-29	Comfortable
30-39	Warm
40-45	Hot
⩾ 46	Very Hot

The Humidex concept has been tested only for Canadian summers and it cannot be easily applied to other climatic conditions.

1.5 COMPARISON OF THERMAL INDICES

The thermal indices namely, the effective temperature, resultant temperature, wet-bulb globe temperature, equatorial comfort index have the same unit of temperature of still saturated air. The tropical summer index is also expressed in terms of the temperature of still air at 50% relative humidity. The P.4S.R. and I.T.S. indices are expressed in terms of the sweat rate under given environmental and metabolic conditions and the H.S.I. is the ratio of the evaporative cooling by the body to the maximum evaporative capacity of the body.

P.M.V. is expressed simply as a value of the predicted mean vote. The wind chill index is determined by the cooling rate and the humidex combines the effect of air temperature and evaporative cooling from the body. The properties of the indices are inherently described in their units. Saturated air temperature is inherently a factor of inconsistent physiological significance, because an increase in the temperature of the saturated air elevates its vapour pressure in a nonlinear fashion. Consequently if the saturated air temperature is used as a unit, the relationship between the index and the physiological and sensory responses is also nonlinear. Due to this nonlinear effect, it may not be possible to evaluate the difference between two climatic conditions differing in their effective temperature, resultant temperature, wet-bulb globe temperature, equatorial comfort index and the tropical summer index. As an example, increasing the effective temperature from 25 to 27°C has little effect on the subject, while an increase from 35 to 37°C makes the difference between conditions so large that it may result in a heat stroke just after some time (Givoni, 1976).

The expected sweat rate seems to be a suitable basis for the assessment of work and heat load. This index however does not give reliable prediction of discomfort due to skin wetness and care has to be taken to use it for conditions characterized by high humidity and low air velocities. In the thirties, Bedford

---

*M_o is the molecular weight of water (= 18×10^{-3} kg/mol).

+R is the gas constant (= 8.315 J/mol.K).

(1936) set up an index on the basis of a comprehensive field study amongst British industrial workers. This index, which is found from the correlation analysis, gives the observed mean vote as a function of the four environmental variables; neither clothing nor activity level were included in the index. The same applies for an index set up by van Zuilen (1953). Both Bedford's and van Zuilen's indices give, at small velocities, optimal ambient temperatures as low as 16-17°C. The index of Fanger (1970) takes into account the maximum possible variables and therefore it cannot be expressed in a very simple form. The tropical summer index is useful only for Indian conditions at sedentary activity level. Wind chill index is a good measure of weather conditions in very cold climates while humidex is good for the summer.

Range of Applicability of the Indices. In comparing range of thermal indices, a distinction should be drawn between the range of conditions covered and the zone in which physiological significance is retained.

Table 1.8 summarizes the normal range of environmental factors covered by each of these indices.

TABLE 1.8 Range of Indices

Index	Metabolic Rate (W)	Dry-Bulb Temperature (°C)	Wet-Bulb Temperature (°C)	Air Velocity (m/s)
Effective temperature	Rest only	1-43	1-43	0.10-3.5
Resultant temperature	Rest only	20-45	18-40	0.10-3.0
Wet-bulb globe temperature	Rest only	1-35	1-35	Not accounted
Equatorial comfort index	Rest only	25-55	25-55	0.25-2.5
P_4.S.R.	58-408	27-55	15-36	0.05-2.5
Heat stress index	116-583	27-60	15-35	0.25-10.0
Index of thermal stress	116-700	20-55	15-35	0.10-3.5
Tropical summer index	Rest only	10-40	5-35	0.5 -2.5
P.M.V.	58*-175W/m^2	-10-30	-10-30	0.1-1.5
Wind chill index	Rest only	-50-10	-50-10	1 ms^{-1}
Humidex	Rest only	15-40	15-40	Not accounted

*Per m^2 of Du Bois area

Based on the correlation observed between their predictions and experimental results of the physiological examinations carried out, the following conclusions are inferred for summarizing the reliability of thermal indices.

Conditions: Index	Simplicity	Rest	Activity	Conditions of heat stress
E.T.	Simple	Unsatisfactory	Unsatisfactory	Unsatisfactory
R.T.	Simple	Satisfactory	Unsatisfactory	Unsatisfactory
W.B.G.T.	Simple	Satisfactory	Satisfactory	Unsatisfactory
E.C.I.	Simple	Unsatisfactory	Unsatisfactory	Satisfactory
H.S.I.	Complicated	Unsatisfactory	Satisfactory	Unsatisfactory
I.T.S.	Complicated	Satisfactory	Satisfactory	Satisfactory
$P._4S.R.$	Simple	Satisfactory	Unsatisfactory at high activity level	Satisfactory
T.S.I.	Simple	Satisfactory for tropical climates	Unsatisfactory	Satisfactory particularly for Indian subjects
P.M.V.	Complicated	Satisfactory	Satisfactory	Satisfactory
W.C.I.	Simple	Satisfactory for very cold climates	Unsatisfactory	Not applicable
Humidex	Simple	Satisfactory for Canadian summers	Unsatisfactory	Not applicable

REFERENCES

ASHRAE (1981) American Society for Heating Refrigeration and Air Conditioning Engineers.

Asmussen, E. and Nielsen, M. (1946) Studies on the Regulation of Respiration in Heavy Work, *Acta. Physiol. Scard. 12*, 171-188.

Bedford, T. (1936) The Warmth Factor in Comfort at Work, *Re. Industr. Hlth. Res. Bd.*, No. 74, London.

Bedford, T. (1946) Environmental Warmth and Its Measurement, MRC War Memorandum No. 17, HMSO, London.

Belding, H. S. and T. F. Hatch (1955) Index for Evaluating Heat Stress in Terms of Resulting Physiological Strains, *Heat Pipe and Air Conditioning, 27*, 11: 129.

Belding, H. S. and T. F. Hatch (1956) Index for Evaluating Heat Stress in Terms of Resulting Physiological Strains, *ASHRAE Trans. 62*: 213.

Brunt, E. D. (1945) Climate and Human Comfort, *Nature 38*, 559.

Falconer, R. (1968), Wind chill, a Useful Winter Time Weather Variable, *Weather wise 27*, pp. 227-229.

Fanger, P. O. (1970) *Thermal Comfort Analysis and Applications in Environmental Engineering*, McGraw Hill, NY.

Gagge, A. P., A. C. Burton and H. C. Bazett (1941) A Practical System of Units for the Description of the Heat Exchange of Man with His Environment, *Science, 94*: 428-430.

Gagge, A. P., Y. Nishi and R. R. Gonzalez (1972) Standard effective temperature - a single index of temperature sensation and thermal discomfort, Proc. CIB Commission W45 (Human Requirements), B Res. Station 13-15 Sept.

ibid (1973) *Building Research Establishment Report, 2*, HMSO, London.

Gagge, A. P. and Y. Nishi (1974) A psychrometric chart for graphical prediction of comfort and heat tolerance, *ASHRAE Transactions, 80*, 115-130.

Gebhart, B. (1961) *Heat Transfer*, McGraw Hill, NY.

Givoni, B. (1963) Estimation of the Effect of Climate on Man: Development of a New Thermal Index, Research Report to UNESCO, Building Research Station, Technion, Haifa.

Givoni, B. (1976) *Man, Climate and Architecture*, Elsevier, Amsterdam.

Givoni, B. and Berner Nir E. (1967) Expected Sweat Rate as a Function of Metabolism in Environmental Factors and Clothing, Report to the US Department of Health, Education and Welfare, Haifa, Israel.

Glickman, N., T. Inouye, R. W. Keeton and M. K. Fahnestock (1950) Physiological Examination of the Effective Temperature Index, *ASHVE Trans., 56*, 51.

Harely, J. D. and Muschenkeim, C. (1934) The Radiation of Heat from the Human Body, IV The Emission, Reflection and Transmission of Infrared Radiation by the Human Skin, *J. Clin. Invest. 13*: 817.

Hippocrates (1849) *The General Works of Hippocrates*, Translated by Francis Adams, Sydenham Society, London.

Hardy, J. D. and C. Muschenheim (1934) The Radiation of Heat from the Human Body. IV. The Emission, Reflection and Transmission of Infra Red Radiation.

Houghten, F. C. and Yaglou, C. P. (1923) *Determining Lines of Equal Comfort, ASHVE Trans. 29*: 163.

Houghten, F. C. and C. P. Yaglou (1924) Cooling Effect on Human Being Produced by Various Air Velocities, *ASHVE Trans. 30*: 193-212.

Huttington (1924) *Civilization and Climate*, Yale University Press.

Inoye, T., F. K. Hick, R. W. Keeton, J. Losch and N. Glickmann (1953) A Comparison of Physiological Adjustment of Clothed Women and Men to Sudden Changes in Environment, *ASHVE Trans. 59*: 35-48.

Inoye, T., F. K. Hick, S. E. Telser and R. W. Keeton (1953) Effect of Relative Humidity on Heat Loss of Men Exposed to Environments of 80, 76, 72, *ASHVE Trans. 59*: 329-346.

Landsberg, H. E. (1984) *Climate and Health in Climate and Development*, ed. A. K. Biswas, Tycooly International Publishing Ltd. Dublin.

Liddell, F. D. K. (1963) Estimation of Energy Expenditure from Expired Air, *J. Appl. Physiol. 18* (I) 25-29.

Markham, F. (1947) *Climate and Energy of Nations*, Oxford University Press.

Markus, T. A. and E. N. Morris (1980) *Buildings, Climate and Energy*, Pitman Publishing Ltd., London.

MacPherson, R. K. (1960) Physiological Responses to Hot Environments, *Med. Res. Counc. Spec. Rep. Ser. No. 298*, London HMSO.

McArdle, B., W. Dunham, H. E. Holling, W. S. S. Ladell, J. W. Scott, M. L. Thomson and J. S. Weiner (1947) The Prediction of the Physiological Effects of Warm and Hot Requirements, *Med. Res. Counc. Rep. No. 47-391*, HMSO.

McCutchan, J. W. and C. L. Taylor (1951) Respiratory Heat Exchange with Varying Temperature and Humidity of Inspired Air, *J. Appl. Physiol., 4*, 121-135.

McNall Jr. P. E., J. Jaa, F. H. Rohles, R. G. Nevins and W. Springer (1967) Thermal Comfort (thermally neutral) Conditions for Three Levels of Activity, *ASHRAE Trans. 73*, I.

Missenard, F. A. (1948) Equivalences thermiques des ambiences; equivalences de passage; equivalence de sejour, *Chaleur et Industrie*, July - Aug.

Mitchell, D. (1970) Measurement of the Thermal Emissivity of Human Skin *in Vivo*, in *Physiological and Behavioral Temperature Regulation* ed. J. D. Hardy, Charles C. Thomas, Illinois, USA.

Nevins, R. G., F. H. Rohles, W. Springer and A. M. Feyerherm (1966) A Temperature Humidity Chart for Thermal Comfort of Seated Persons, *ASHRAE Trans. 72*, I: 283-291.

Nielsen, M. and L. Pedersen (1952) Studies on the Heat Loss by Radiation and Convection from the Clothed Human Body, *Acta Physiol. Scand. 27*, 272.

O'Callaghan, P. W. and S. D. Probert (1976) Thermal Properties of Clothing Fabrics, *Bldg. Serv. Engr., 44*, 71-79.

O'Callaghan, P. W. (1978) *Building for Energy Conservation*, Pergamon Press (1978).

Sharma, M. R. (1977) A Study of the Thermal Effects of Climate and Building on Human Comfort with Special Reference to India, Ph.D. Thesis, Agra University, Agra, India.

Siple, P. A. and C. F. Passel (1945) Measurement of Dry Atmospheric Cooling in Subfreezing Temperatures, *Proc. Am. Phil. Soc., 89*, 177-199.

Steadman, R. G. (1971) Indices of Wind Chill of Clothed Persons, *J. Appl. Meteorol, 10*, 674-683.

Smith, F. E. (1955) *Indices of Heat Stress*, Med. Res. Council, Memo No. 29, London.

van Zuilen, D. (1953) *Climatological Factors in Healthful Housing,* UN Publication Sales No. 1953, *8*: 22-28.

Yaglou, C. P. and W. E. Miller (1924) Effective Temperature Applied to Industrial Ventilation Problems, *ASHVE Trans., 30*, 339-364.

Yaglou, C. P. and W. E. Miller (1925) Effective Temperature with Clothing, *ASHVE Trans., 31*, 89-99.

Yaglou, C. P., W. H. Carrier, E. V. Hill, F. C. Houghten and J. H. Walter (1932) How to Use the Effective Temperature Index and Comfort Charts, *ASHVE Trans. 38*: 411-423.

Yaglou, C. P. (1947) A Method for Improving the Effective Temperature Index, *ASHVE Trans., 53*, 307.

Yaglou, C. P. and D. Minard (1957) Control of Heat Casualties at Military Training Centres, *Arch. Industr. Health, 15*, p.32.

Webb, C. G. (1960) Thermal Discomfort in an Equatorial Climate: A nomogram for the Equatorial Comfort Index, *JIHVE, 27* (10).

Chapter 2

CLIMATE, SOLAR RADIATION, BUILDING ORIENTATION AND SHADING DEVICES

2.1 CLIMATE

The purpose of natural heating and cooling methods is to provide a comfortable internal thermal environment. In this sense, the entire building and more particularly the external building envelope can be considered as a membrane which protects the indoor space from the undesirable climatic influences. At its best the external envelope will not only filter out the worst climatic elements passively but also interact with the outdoor environment to let as much radiation in as possible for the natural heating or to remove even the internally generated heat for natural cooling of buildings.

The internal thermal environment of a naturally cooled (or heated) building results from the response of the building to the changing external climatic influences which repeat themselves in daily and annual cycles. The relative importance of the daily cycle (which has a period of 24 hours) and that of the annual cycle (period 365 days) depends upon the geographical location of the site under consideration. Near the poles, the annual cycle is of maximum importance while at the equator, the daily cycle plays a major role in determining the quality of shelter needed (Fig. 2.1).

The main elements influencing the climate at any location are (i) solar radiation (its intensity, direction and duration of sunshine hours), (ii) air temperature, (iii) humidity and precipitation, (iv) wind (speed and direction) and (v) clearness of the sky. Temperature regime of any large area is determined by the amount of solar radiation which falls upon that area from one season to another. Regions which are exposed full face to the sun for a large part of the year are hot; those which receive sunshine only at low angles and for small portions of the year are cold. From a knowledge of sun earth relationship one would therefore expect the equatorial regions to be hottest and as one moves away from the equator towards poles, it gets steadily cooler. It is, however, noted that the fall of temperature from equator to pole is not uniform and numerous aberrations appear in the temperature maps.

Table 2.1 below gives the amount of extraterrestrial horizontal solar radiation received at places between the latitude 0 and 60° during various months of the year on a horizontal surface. It is observed that the maximum receipt of solar radiation on the earth's surface over the whole of the clear summer day is not at the equator but somewhere between the latitude 30° and 60°. At these latitudes, however, the winter receipt of solar radiation is low. It is noticed from the

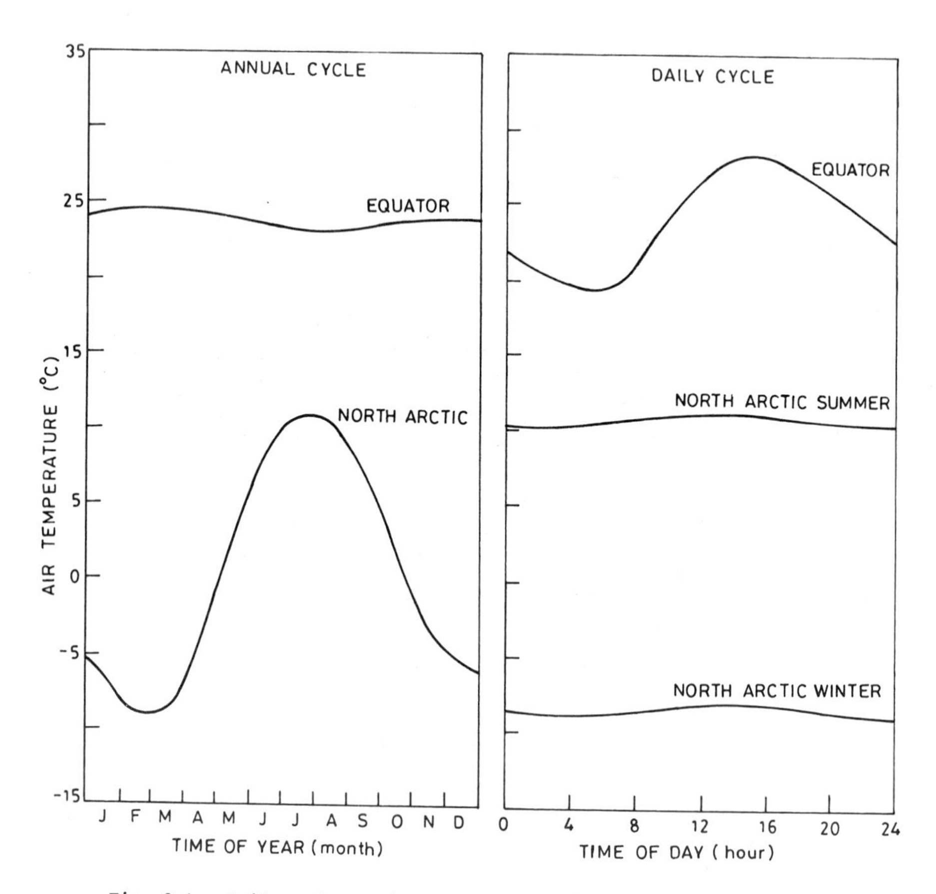

Fig. 2.1. Daily and annual temperature variations at the north pole and the equator

table that the summer radiation is intense for the middle latitudes and is very important from the point of view of building calculations.

Besides solar radiation, there are several other factors which determine the climate at a certain location. These factors have been discussed in some length below.

2.1.1 Effect of Land and Water

The presence of water in the form of ponds, lakes etc. affects the temperature at a certain location. Dry soil gets at least twice as easily heated as the same volume of water. Water also loses some of its heat by evaporation, which is not possible in the case of dry soil. A given amount of solar radiation will therefore heat dry earth to a higher temperature than it will heat water or wet earth. It is therefore apparent that the air which is in contact with dry earth

TABLE 2.1 Yearly Variation of the Extraterrestrial Horizontal Daily Insolation for North Latitudes, S_{eh} (MJm^{-2} day^{-1}), S_c = 1367 W m^{-2}

Month	North latitude (degrees)												
	0	5	10	15	20	25	30	35	40	45	50	55	60
JAN	36.32	34.31	32.09	29.66	27.05	24.28	21.39	18.40	15.34	12.26	9.21	6.27	3.54
FEB	37.53	36.22	34.65	32.83	30.77	28.51	26.04	23.41	20.62	17.71	14.72	11.68	8.64
MAR	37.90	37.58	36.98	36.10	34.94	33.53	31.85	29.94	27.80	25.46	22.92	20.22	17.37
APR	36.75	37.47	37.92	38.08	37.97	37.58	36.92	35.98	34.80	33.36	31.70	29.84	27.80
MAY	34.78	36.28	37.54	38.54	39.29	39.77	40.00	39.97	39.70	39.19	38.48	37.61	36.65
JUN	33.50	35.35	36.97	38.37	39.53	40.44	41.11	41.54	41.74	41.73	41.56	41.28	41.00
JUL	33.89	35.59	37.05	38.27	39.24	39.97	40.44	40.67	40.66	40.44	40.03	39.48	38.90
AUG	35.56	36.62	37.43	37.96	38.23	38.22	37.95	37.41	36.63	35.60	34.35	32.91	31.31
SEP	37.07	37.19	37.03	36.60	35.88	34.90	33.65	32.15	30.40	28.43	26.25	23.87	21.33
OCT	37.34	36.43	35.24	33.80	32.10	30.17	28.02	25.67	23.14	20.45	17.63	14.70	11.70
NOV	36.47	34.69	32.68	30.46	28.03	25.44	22.69	19.81	16.85	13.83	10.79	7.81	4.96
DEC	35.74	33.55	31.16	28.58	25.84	22.96	19.97	16.92	13.82	10.74	7.73	4.88	2.34
AV.	36.07	35.94	35.56	34.94	34.07	32.98	31.67	30.16	28.4	26.6	24.6	22.55	20.46

will be at a higher temperature than the air in contact with water or wet ground. The higher air temperatures are therefore generally associated with low humidities, and high humidities with only moderate air temperatures. For a particular latitude in summer, it is expected that air temperatures will be lower over large bodies of water and higher over large tracts of dry land. In winter, however, the reverse could be true since the land cools off more quickly than water, when it is not exposed to sunshine. It may therefore be concluded that annual variations in air temperature will be highest over the dry land, minimum over large bodies of water. The diurnal variations will also follow the same general pattern.

2.1.2 Effect of Wind and Air Mass Movement

The difference in the radiation received at various latitudes is one of the major causes of the general circulation of air. In the absence of the effect of rotation of earth, the circulation would take the simple form of a rising cell over the equator and a falling one over the poles.

The rotation of earth introduces two other forces superimposed on the thermal and hence the pressure induced convective movements of air. The first force is the apparent acceleration of air by virtue of earth's rotation, called the Coriolis acceleration. It results in partial deflection of a parcel of air to the right in the northern hemisphere and to the left in the southern hemisphere as it moves from poles outward toward the equator. The other force is that of the angular momentum due to which, a parcel of air which tends to conserve its velocity, will appear to move faster than the earth as it moves towards the poles creating effectively an apparent wind. This has an apparent clockwise rotation in the northern hemisphere and anticlockwise in the southern. The combination of all these gives rise to the global wind pattern displayed in Fig. 2.2.

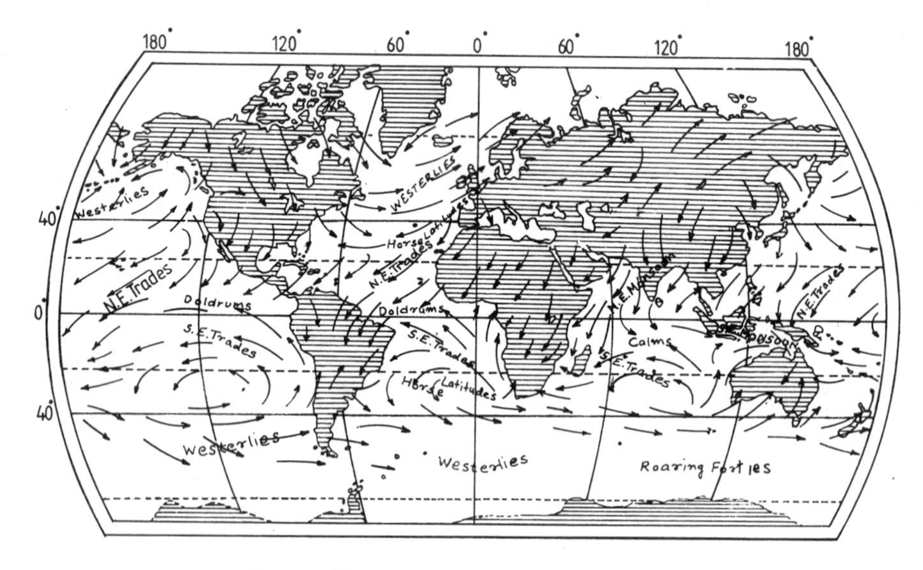

Fig. 2.2(a). Winds of the world, January

At the equator warm air moves upward and drops as cooler air in the high pressure zones of the subtropic. Simultaneously there is a division here. Part of the air flows towards the equator and a part towards the pole. At the low pressure

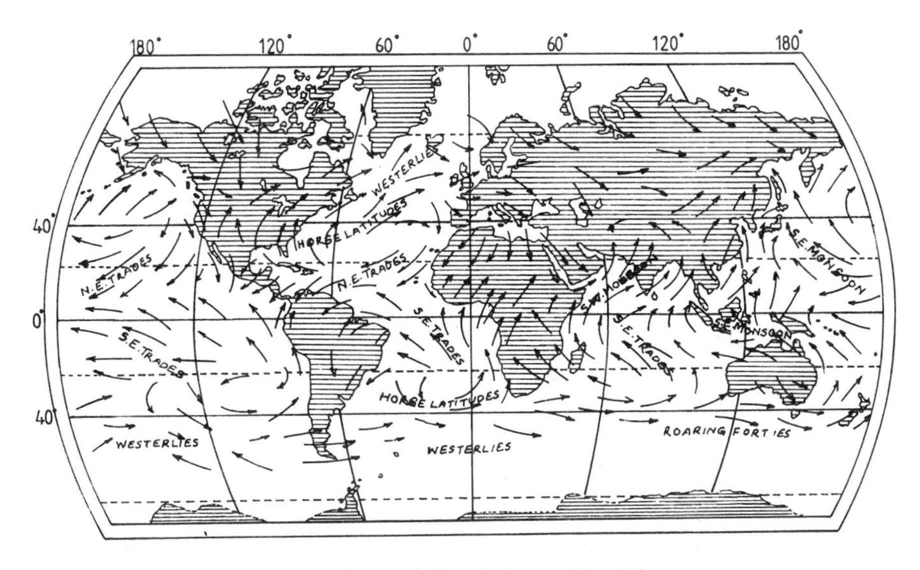

Fig. 2.2(b). Winds of the world, July

subpolar zones it rises again, thus flowing up and in both the directions (equator and poles). The rotational forces and the frictional and gravitational forces result in the atmosphere exerting a drag on the earth. This drag force alternates eastward and westward to maintain the rate of rotation constant. The tropical zones have the characteristic north-easterly and south-easterly trade winds, in the northern and southern hemsiphere respectively; at the two tropics the winds are typically light and variable in high pressure zones.

The trade winds are primarily a result of the Coriolis force, causing the wind to blow in a direction opposite to the direction of the rotation of earth. Beyond the subtropical high pressure area (up to about 60° latitude) westerlies predominate (south-westerly in northern hemisphere and north-westerly in southern hemisphere). At these regions they are influenced by the angular momentum, since the air moving north and south from about 30° latitudes with lower circumferential velocity, will appear to move faster than earth and hence in the direction of the earth's rotation. In polar regions, the movement of air is again dominated by the thermal gradients. At the poles, the air starts with nearly zero circumferential velocity and moves away from poles at low level, gradually lags behind and appears to be moving faster than the speed of earth's rotation.

Temperature difference over land and sea, day/night variations, topography etc. cause significant local variations of these patterns in space and time. Climatic conditions get severely changed by these variations. Temperatures also get affected by the altitude of the place.

In a free atmosphere, temperature diminishes with height at the rate of approximately 0.65°C per 100 m. Near the ground, one also expects the same general pattern, but a host of local factors come in to upset this particular behaviour. Cold air tends to sink into depressions in the terrain, and warm air tends to rise up the hill sides, creating actual inversion of the temperature gradient. Ground exposed to a clear sky at night loses considerable quantity of heat by radiation, thus reducing the temperature of the air in contact with it. This again produces an inversion.

With these local and often temporary exceptions, air temperature does fall with altitude at more or less the expected rate. Many of the world's highest elevations occur within or close to the geographical tropics. The equator traverses the Andes and the highlands of Africa, and passes close to the mountains' core of New Guinea with its snow line. The Himalayas and the southern part of the Rocky Mountains lie just north to the Tropic of Cancer. These elevations add to the considerable variety of climates which can be found between the latitudes of 30°N and 30°S.

Since the latitude reduces the temperature, the assumption is sometimes made that the tropical high land can be regarded as similar to regions of low elevation at higher latitudes. This is however not the case though the temperature regimes may still be similar. At the equator, there are small variations in the duration of sunlight hours over the year, whereas at higher latitudes the variations are considerable. This difference in seasonal variations of daylight hours is, of significant importance for building design calculations.

Vapour pressure of the air, measured by the amount of water vapour present in the atmosphere determines the ease with which heat can be lost by evaporation. Vapour pressure of the atmospheric air tends to be high near the equator and falls off towards the pole. There is more solar radiation at the equator, hence more evaporation and since the temperature is higher, it can hold more water. Over higher altitudes, the vapour pressure will drop if the lower temperature cools air to the point at which the vapour condenses. Winds help air movement from one place to another. Winds from the sea direction raise the vapour pressure.

The major pattern of winds (direction and force both) is of mechanical importance to the planner, architect and the builder who must erect structures capable of withstanding the stresses. From the point of view of securing natural ventilation and internal air movement also, the pattern of wind force and direction is useful.

Local topography frequently dominates the operating wind pattern. Valleys tend to channel wind to their own axis. Rising contours opposed to the wind produce up-currents on the windward side, reverse eddy currents over the crest on the leeward side, and calms at the base of the leeward side of the slope. According to Brooks (1951), wind speed begins to decrease on the windward side of a simple wind break at a distance equal to about six times the height of the barrier, and 75-80% of the full wind speed is regained at a distance 6 to 12 times the height of the barrier.

2.1.3 Classification of Climates

By the classification of the climate, one understands the general course or the conditions of weather, which prevail in a particular region. Though the classification of climates is a difficult process because of reasons like (i) the climates are made up of many weather complexes, (ii) climates of neighbouring regions influence each other considerably, (iii) irregularities of human reactions to climatic stress etc., for housing, however, it is justified to divide innumerous world climates into a limited number of classes.

Based on the temperature range and the amount of relative humidity, the climates may be classified as follows

	Mean Temperature
Hot	Over 30°C
Warm	20°C-30°C
Temperate	10°C-20°C

	Mean Temperature
Cool	Below 10°C
	Mean Vapour Pressure
Humid	Over 20 mb
Dry	Under 20 mb

The map given in Fig. 2.3 classifies various regions into different climatic zones. The subdivision is based upon the temperatures and vapour pressures prevailing in the months of January and July. Each region is denoted by four characteristics.

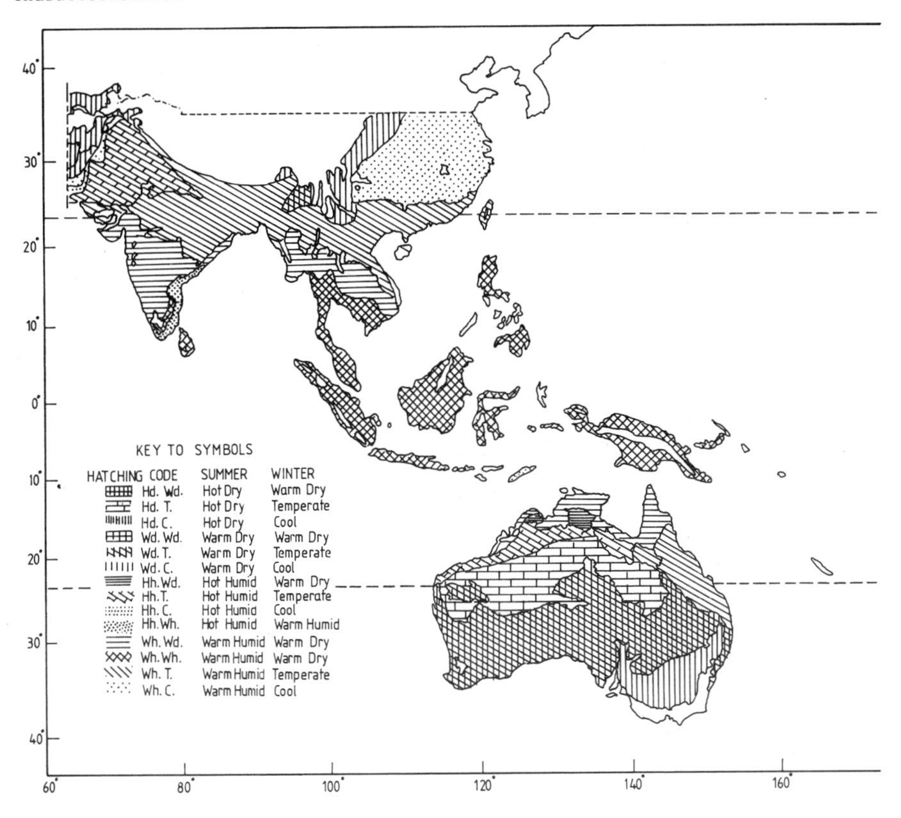

Fig. 2.3(a). Major climatic types in south-east Asia and Australia

Summer temperature	hot, warm, temperate
Summer vapour pressure	humid or dry
Winter temperature	warm, temperate or cool
Winter vapour pressure	humid or dry.

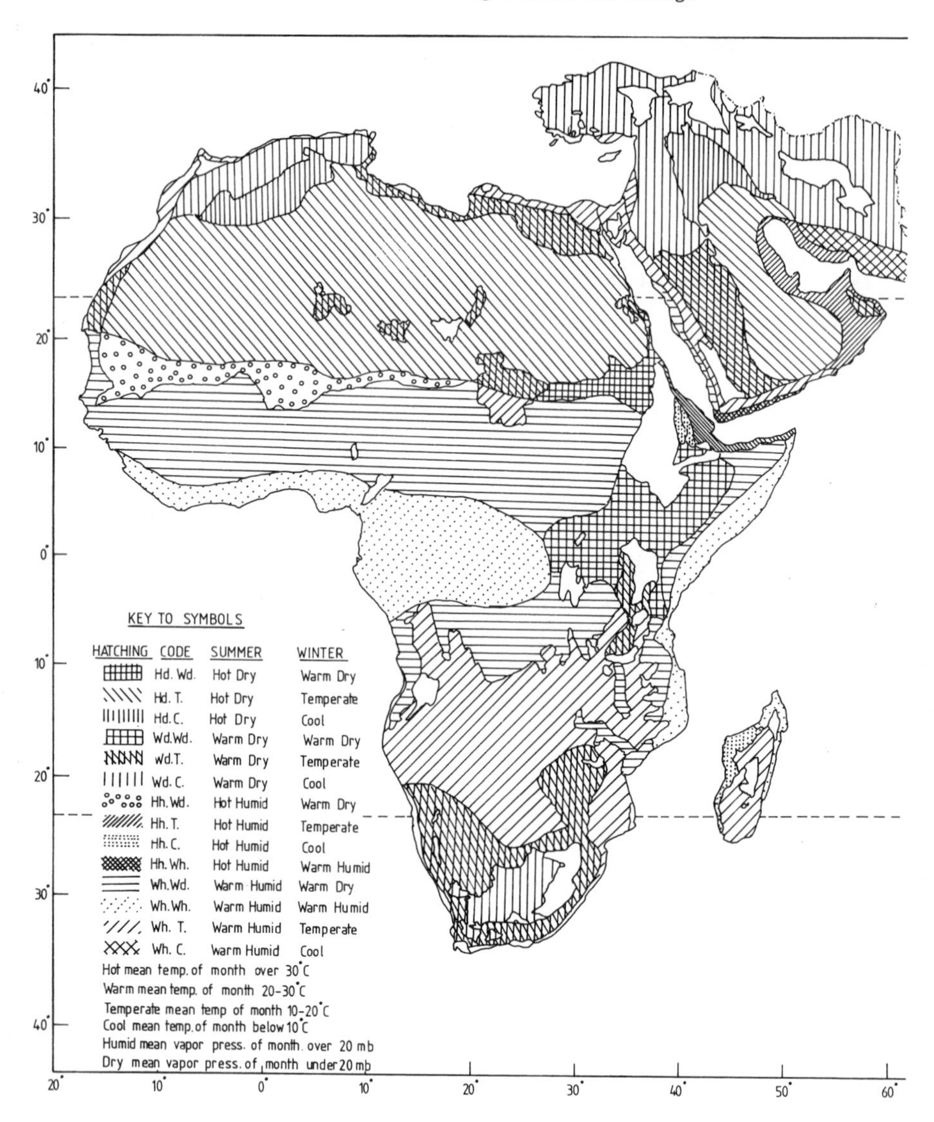

Fig. 2.3(b). Major climatic types in Africa and Arabia

Miller (1961) has classified climates into the following types

A. Hot Climates

1. Hot dry; hot deserts
2. Warm wet: equatorial and tropical marine
3. Hot dry and warm wet; tropical continental and monsoon.

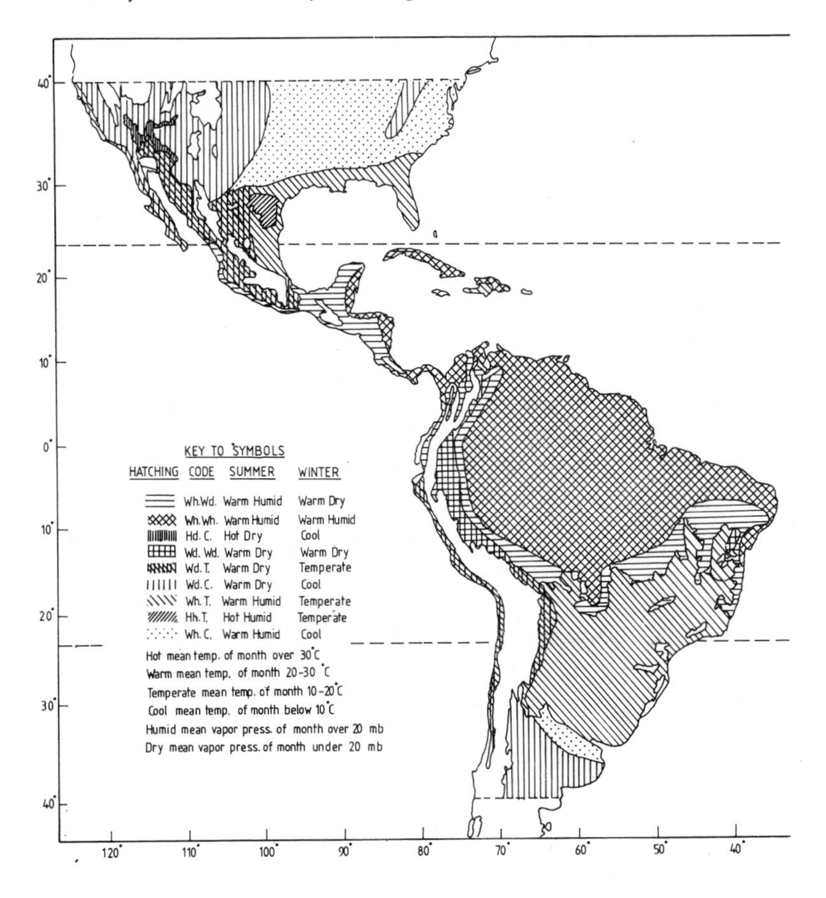

Fig. 2.3(c). Major climatic types in the Americas

B. Warm Temperature Climates

4. Western margin type
5. Eastern margin type

C. Cool Temperature Climates

6. Cool temperature
7. Cool temperature marine

D. Cold Climates

8. Cold continental: Siberian
9. Cold marine: Norwegian
10. Cold desert
11. Arctic.

For a building designer all aspects of weather are relevant viz. wind, air movement, sunshine, radiation and humidity from thermal points of view; rain, frost and humidity because of their effects on materials, problems of decay, corrosion and breakdown of finishes; wind and snow on account of the structural load they impose; rain for designing a proper drainage system. In this text we shall mainly be concerned with the thermal design of buildings for comfort and energy conservation. Solar radiation estimates are required to be made for calculating thermal loads in buildings. The next section, therefore, deals with the sun movement and radiation received on earth and on various building components.

2.2 SOLAR RADIATION

The earth revolves around the sun in an elliptical orbit, thus causing change in the distance between the sun and earth on various days of the year. Besides many other factors, the amount of solar radiation reaching the earth is inversely proportional to the square of its distance from the sun. The mean sun-earth distance (r_o) is called one astronomical unit (1 A.U.) and it is equivalent to 1.496×10^8 Km. The minimum sun-earth distance is about 0.983 A.U. and maximum approximately 1.017 A.U. At any given day of the year, the reciprocal of the square of the earth-sun distance (r) is given by the expression

$$\left(\frac{r_o}{r}\right)^2 = e = 1 + 0.033 \, \mathrm{Cos}\left(\frac{2\pi n_d}{365}\right) \tag{2.1}$$

the ratio $(r_o/r)^2$ is often called the eccentricity correction factor of the earth's orbit, e. Expression (2.1) is suitable for most engineering applications. A more accurate expression for $(r_o/r)^2$ is given by Spencer (1971) as

$$e = (r_o/r)^2 = 1.00011 + 0.34221 \, \mathrm{Cos}\, \zeta_d + 0.00128 \, \mathrm{Sin}\, \zeta_d + 0.000719 \, \mathrm{Cos}^2 \zeta_d + 0.000077 \, \mathrm{Sin}^2 \zeta_d \tag{2.2}$$

where ζ_d called the day angle is in radians, and it is calculated by the expression

$$\zeta_d = 2\pi \, (n_d - 1)/365 \tag{2.3}$$

Beyond the atmosphere of the earth and at mean sun-earth distance, the intensity of solar radiation on a surface normal to the sun rays is known as the solar constant S_c. Various measured values of solar constant vary from 1338 W/m^2 to 1368 W/m^2. The NASA value of the solar constant (NASA, 1968, ASTM 1973) is based on a weighted average of several values (Drummond and Thekaekara, 1973) and its value is 1353 W/m^2. The estimated error in this value is ± 21 Wm^{-2} or ± 1.5%. This value is called the NASA design standard and it is based on the international pyrheliometer scale (Johnson, 1954).

Since 1975 a number of measurements of solar constant have been made. The new measurements have been made using modern cavity type absolute instruments and the World Meteorological Organization has now adopted a new scale, called the world radiometric reference (W.R.R.), as a common base for all meteorological measurements. Using the new reference and the new measurements, Fröhlich and Colleagues (1981a, 1981b) recommend a value of 1367 Wm^{-2} for the solar constant S_c.

2.2.1 Solar Declination δ and Equation of Time

The earth moves around the sun in ecliptic plane and simultaneously it rotates about its polar axis, which is inclined at approximately $23\frac{1}{2}^{\circ}$ from normal to the ecliptic plane (Fig. 2.4). The earth's rotation about polar axis is responsible for diurnal variations in the incident radiation, whereas the rotation in the ecliptic plane causes seasonal changes. During rotation the angle between the polar axis and the normal to the ecliptic plane remains unchanged. The same is true for angle between the ecliptic plane and the equatorial plane. The angle between the line joining the centres of the sun and the earth to the equatorial plane changes every day. This angle is called the solar declination δ. It is

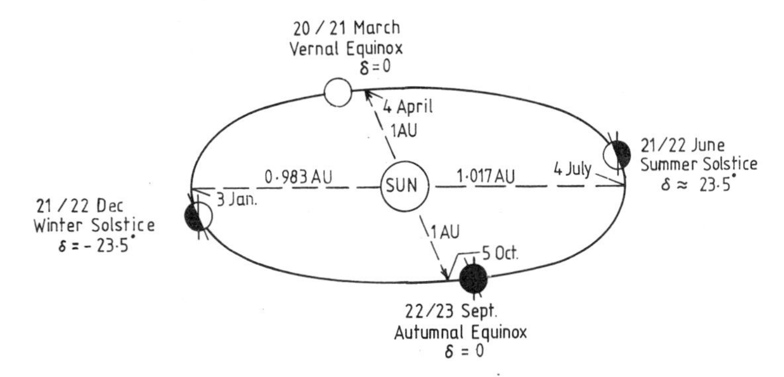

Fig. 2.4. Motion of earth around the sun

zero at the vernal and at the autumnal equinoxes and has a value approximately $+ 23.5^{\circ}$ at the summer solstice and about $- 23.5^{\circ}$ at the winter solstice. On any other day, simple formulae which are commonly used are

$$\delta = \text{Sin}^{-1} \left[0.4 \text{ Sin } \{\frac{360}{365} (n_d - 82)\}\right] \text{ in degrees} \tag{2.4}$$

obtained from Perrin de Brichambant (1975) and

$$\delta = 23.45 \text{ Sin} \left[\frac{360}{365} (n_d + 284)\right] \text{ in degrees} \tag{2.5}$$

From Cooper (1969).

Though the above expressions for solar declination are quite accurate for engineering calculations, Spencer (1971) gave the following expressions for δ in degrees

$$\begin{aligned} \delta = \frac{180}{\pi} \Big[& 0.006918 - 0.399912 \text{ Cos } \zeta_d + 0.070257 \text{ Sin } \zeta_d \\ & - 0.006758 \text{ Cos } 2\zeta_d + 0.000907 \text{ Sin } 2\zeta_d \\ & - 0.002697 \text{ Cos } 3\zeta_d + 0.00148 \text{ Sin } 3\zeta_d \Big] \end{aligned} \tag{2.6}$$

Another important quantity, which has to be taken into account for the calculation of solar radiation data is the equation of time, which represents the deviation in clock time with respect to the same position of sun and to a stationary observer on the earth. For example, if an observer facing the equator sets a clock (running at a uniform rate) at 12.00 noon, when the sun is directly over the local meridian, then after a month or so at 12.00 noon clock time, the sun may not appear exactly over the local meridian. This discrepancy in the time is measured with respect to perfect uniform terrestrial motion. Following Spencer (1971), the equation of time is given by

$$t_d = 0.000075 + 0.001868 \text{ Cos } \zeta_d - 0.032077 \text{ Sin } \zeta_d$$

$$- 0.014615 \text{ Cos } 2\,\zeta_d - 0.04089 \text{ Sin } 2\,\zeta_d \ (229.18) \qquad (2.7)$$

It is a common practice to record radiation data in terms of local apparent time (L.A.T.); which is also called the true solar time (T.S.T.). On the other hand, meteorological data such as temperature and wind speed are often recorded in terms of local clock time. While computing radiation data, it is desirable to convert local standard time (L.S.T.) (i.e. the clock time) to local apparent time given by the equation (Iqbal, 1983).

$$\text{Local apparent time} = \text{Local standard time}$$

$$+ 4\,(\gamma_s - \gamma_\ell) + t_d \qquad (2.8)$$

where γ_s is the standard longitude and γ_ℓ is the local longitude. The factor 4 arises due to the fact that a time of 4 minutes for every degree accounts for the difference between local and standard meridians. All international meridians are multiples of 15° east or west Greenwich, England. Longitude correction is positive if local meridian is east of the standard and is negative if west of the standard meridian.

2.2.2 Position of Sun Relative to Horizontal Surfaces

The position of the sun with respect to a point on the earth's surface can be described with the help of following angles (Fig. 2.5)

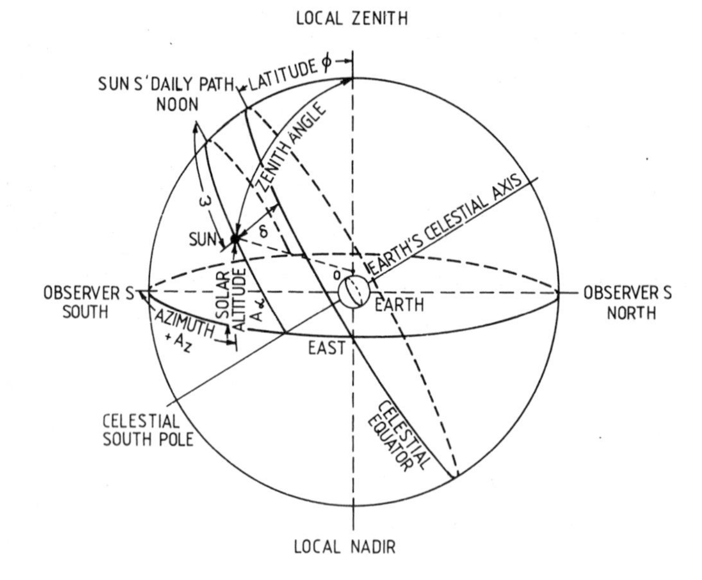

Fig. 2.5(a). Celestial sphere and sun's co-ordinates relative to observer on earth at point 0

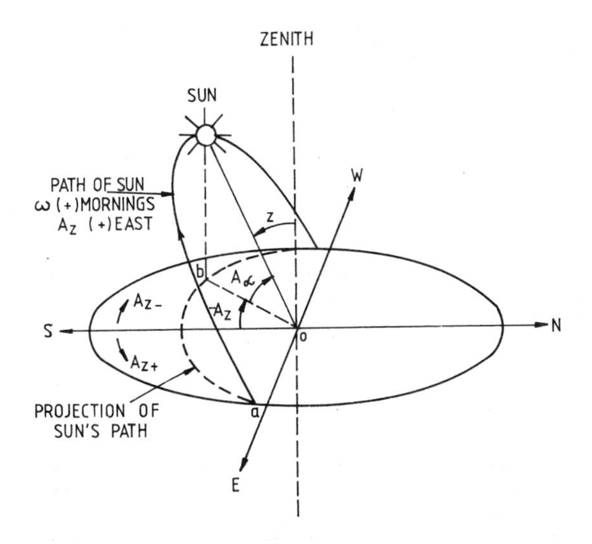

Fig. 2.5(b). Definition of sun's zenith, altitude and azimuth angles

Z	Zenith angles	in degrees
A_α	Solar altitude	in degrees; $A_\alpha = 90-Z$
ω	Hour angle	noon zero, morning +ve
ϕ	Geographical latitude,	in degrees north +ve
A_z	Solar azimuth,	in degrees, south zero, east positive.
δ	Declination,	in degrees.

The solar azimuth A_Z is the angle at the local zenith between the plane of the observer's meridian and the plane of a great circle passing through the zenith and the sun. It is measured east positive, west negative (south zero) and thus varies between 0° and $\pm 180^o$. Hour angle ω changes 15° per hour as counted from midday.

The well-known trignometric relations between the sun and a horizontal surface are

$$\text{Cos } Z = \text{Sin } \delta \text{ Sin } \phi + \text{Cos } \phi \text{ Cos } \delta \text{ Cos } \omega = \text{Sin } A_\alpha \quad (2.9)$$

and

$$\text{Cos } A_Z = (\text{Sin } A_\alpha \text{ Sin } \phi - \text{Sin } \delta)/\text{Cos } A_\alpha \text{ Cos } \phi \quad (2.10)$$

$$0^o \leq A_Z \leq 90^o \text{ , Cos } A_Z \geq 0$$

$$90^o \leq A_Z \leq 180^o \text{ , Cos } A_Z \leq 0$$

Eq. (2.9) yields the sunrise hour angle for $Z = 90^o$ i.e.

$$\text{Cos } \omega_s = - \text{Sin } \delta \text{ Sin } \phi/\text{Cos } \phi \text{ Cos } \delta \quad (2.11a)$$

or

$$\omega_s = \cos^{-1}(-\tan\delta\tan\phi) \tag{2.11b}$$

It may be noted that the sunrise and the sunset angles are exactly the same in magnitude except for the sign difference. Eq. (2.11) also helps in determining the number of sunshine hours, which is $2\omega_s$ and in terms of hours

$$N_s = \frac{2}{15}\,\mathrm{Cos}^{-1}\,(-\tan\delta\tan\phi) \tag{2.12}$$

Eq. (2.12) gives a number of interesting points.

(i) In polar region, the sun does not rise in winter and hence there is no day length when from Eq. (2.11) $\cos\omega_s > +1$. During summer there is a continuous day of about six months and hence there is no sunrise hour. At the North pole, for example, continuous duration of the polar days is reported to be 186 (Iqbal, 1983).

(ii) At the equator $\phi = 0$, and hence $\omega_s = \pi/2$ and the day's length is independent of the solar declination and it is equal to 12 hours.

(iii) At the equinoxes $\delta = 0$; therefore $\omega_s = \pi/2$ and day's length is independent of the latitude and is equal to 12 hours.

Example 2.1. Calculate (a) Zenith angle and solar azimuth at 11.00 LAT (b) the sunrise hour angle and (c) the day length, on 31 October at Delhi ($\phi = 28^\circ\ 55'$)

Solution:

On October 31, $n_d = 304$ (from January 1) and therefore using eq. (2.5),

$$\text{Declination } \delta = 23.45 \sin\left[\frac{360}{365}(304 + 284)\right] = -15.05^\circ$$

From Eq. (2.9)

(a) Zenith angle

$$Z = \cos^{-1}\left[(-0.2598)(.4835) + (.8753)(.9657)(.9659)\right] = 46.3^\circ$$

$\therefore$ Solar altitude $A_\alpha = 43.69^\circ$ and solar azimuth is obtained from eq. (2.10) i.e.

$$A_Z = \left[\frac{\mathrm{Sin}\ (43.69)\ \mathrm{Sin}\ (28.92) - \mathrm{Sin}\ (-15.05)}{\mathrm{Cos}\ (43.69)\ \mathrm{Cos}\ (28.92)}\right]$$

$$= 20.28^\circ$$

(b) Sunrise angle is given by Eq. (2.11b) i.e.

$$\omega_s = \cos^{-1}\left[-\tan(-15.05)\tan(28.92)\right.$$

$$= 81.46^\circ$$

In terms of solar time, the sun rises at $12 - \dfrac{81.46}{15}$

$= 6.57$ h or 6:34:12 (L.A.T.)

(c) The day length is

$$N_s = (2/_{15} \times \omega_s) = 10.86 \text{ h}$$

$$= 10 \text{ h } 51 \text{ m } 36 \text{ s (LAT)}$$

The values of solar altitudes A_α and solar azimuths A_Z can also be described on sunpath diagrams, given for the latitudes 29° in Fig. 2.6. For other latitudes the sunpath diagrams are given in Appendix II. A plane projection of the trajectory of the sun on a circle of unit radius describes the sunpath diagram. The outer circle in the sunpath diagrams represent solar altitudes. The radial lines indicate solar azimuths in 10° steps. The lines indicating various dates represent the altitude and azimuth at various hours and these lines are crossed by hourly time line.

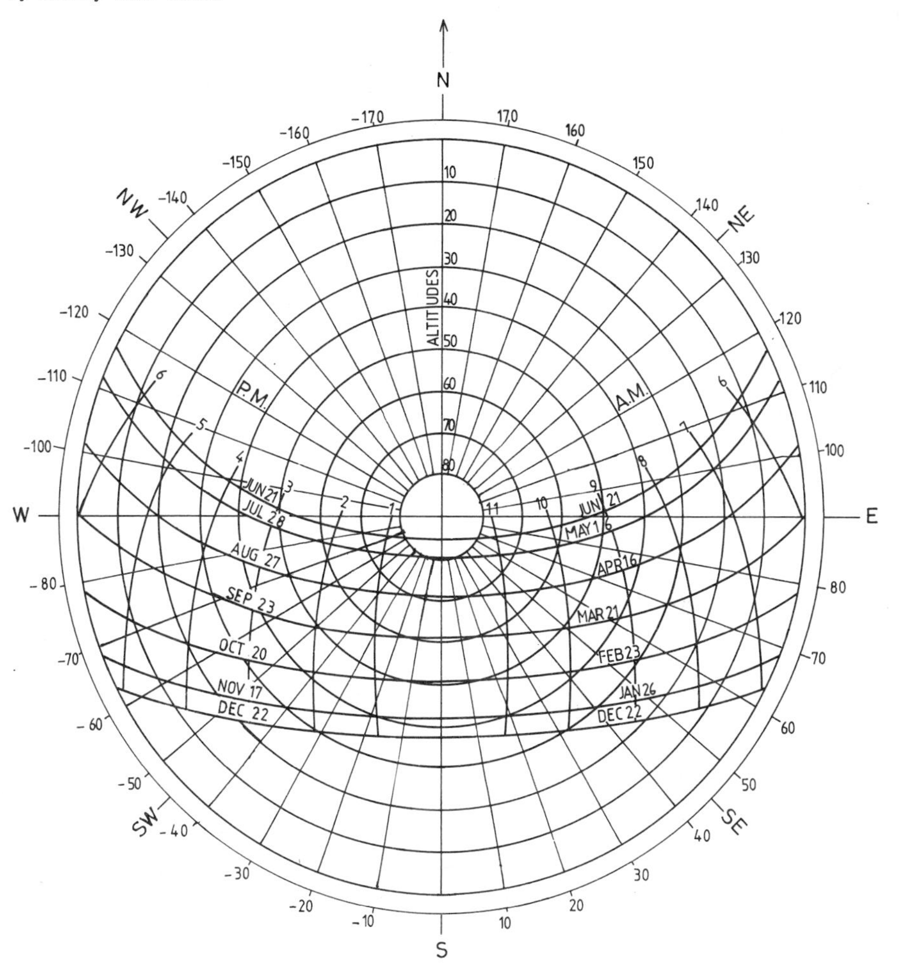

Fig. 2.6. Solar path diagrams for latitude 29°N

2.2.3 Position of Sun and Inclined Surfaces

For an inclined surface one has to prescribe the slope of the surface with respect to the horizontal (or vertical) position and its orientation in relation to the local meridian. Corresponding to Fig. 2.7 one sees that

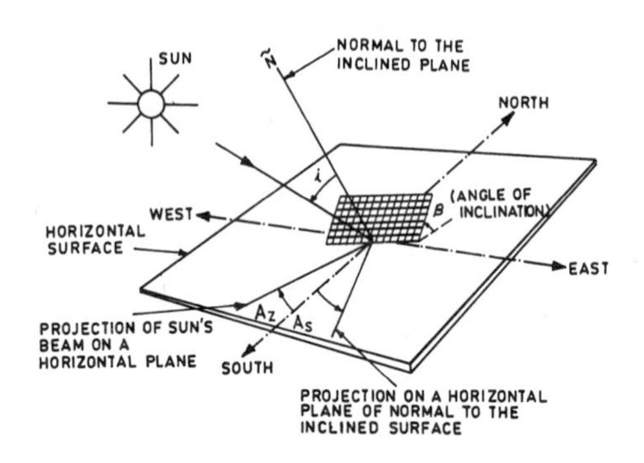

Fig. 2.7. Position of sun relative to an inclined plane

β is the slope of the surface measured from the horizontal position, in degrees.

A_s is the surface azimuth angle i.e. the deviation of the normal to the surface with respect to the local meridian, in degrees, east positive, and

i is the angle of incidence for an arbitrary oriented surface i.e. the angle between normal to the surface and earth-sun vector, in degrees.

For a surface oriented in any direction with respect to the local meridian, the following trignometric relation for the incidence angle i, has been given by Benrod and Rock (1934), Kondratyev (1969) and in detail by Coffari (1977).

$$\cos i = (\sin\phi \sin\beta - \cos\phi \sin\beta \cos A_s) \sin\delta + (\cos\phi \cos\beta + \sin\phi \sin\beta \cos A_s) \cos\delta \cos\omega + \cos\delta \sin\beta \sin A_s \sin\omega \quad (2.13)$$

or

$$\cos i = \cos\beta \cos Z + \sin\beta \sin Z \cos (A_Z - A_s) \quad (2.14)$$

Eq. (2.13) does not require a separate calculation for solar azimuth A_Z and zenith angle Z.

2.2.3.1 Surfaces facing south. For surfaces facing south $A_s = 0$ and Eq. (2.13) reduces to

$$\cos i = \sin(\phi - \beta) \sin\delta + \cos(\phi - \beta) \cos\delta \cos\omega \quad (2.15)$$

If the angle of incidence is 90° then it corresponds to sunrise or sunset. For this situation, sunrise if denoted by ω_s', one has

$$\omega_s' = \cos^{-1} [- \tan \delta \tan (\phi - \beta)] \tag{2.16}$$

The sunset angle, in magnitude, is naturally the same except for the sign difference. The following three cases are of interest now

(i) At the equinox, $\delta = 0$, Eq. (2.16) gives

$$\omega_s' = \pi/2;$$

it means the sunrise hour angle is independent of the tilt or latitude.

(ii) During the summer (in northern hemisphere), $\delta > 0$ and this gives $\omega_s > \omega_s'$, meaning that sun rises earlier on the horizontal than on an inclined surface.

(iii) During the winter (in northern hemisphere), $\delta < 0$, and this means $\omega_s' > \omega_s$. Physically this is not possible and hence a general expression for the sunrise hour angle is given by

$$\omega_s' = \min \{\cos^{-1} (- \tan \delta \tan \phi), \cos^{-1} (- \tan \delta \tan (\phi - \beta))\}; \tag{2.17}$$

where min stands for minimum.

2.2.3.2 Vertical walls/surfaces. For vertical walls, as often is the case for buildings $\beta = 90^\circ$ and hence from Eq. (2.13)

$$\cos i = - \cos \phi \cos A_s \sin \delta + \sin \phi \cos A_s \cos \delta \cos \omega + \cos \delta \sin A_s \sin \omega \tag{2.18}$$

The surface wall azimuth for various orientations is given in Table 2.2.

TABLE 2.2 Wall Orientations and Azimuths, Measured from South

Orientation	N	NE	E	SE	S	SW	W	NW
Surface azimuth deg. (A_s)	180	135	90	45	0	45	90	135

For inclined surfaces not facing true south, the sunrise angle ω_{sr} and the sunset angle ω_{ss} are not numerically (in magnitude) identical. The period of sunshine hours, i.e. the period during which the sun is seen by the surface is $|\omega_{sr} - \omega_{ss}|$ in degrees, ω_{ss} being -ve. Below an expression for ω_{sr} and ω_{ss} is developed for surfaces oriented east or west i.e. $\pm A_s$. For $i = \pi/2$, Eq. (2.13) can be written as

$$C_1 \cos \omega + C_2 \sin \omega + C_3 = 0 \tag{2.19}$$

where

$$C_1 = (\cos \phi \cos \beta + \sin \phi \sin \beta \cos A_s) \cos \delta \tag{2.20}$$

$$C_2 = \cos \delta \sin \beta \sin A_s \tag{2.21}$$

$$C_3 = (\sin \phi \cos \beta - \cos \phi \sin \beta \cos A_s) \sin \delta \tag{2.22}$$

Eq. (2.19) can be written in the form

$$C_1 \cos \omega + 2\, C_2 \sin \omega/2 \cos \omega/2 + C_3 = 0$$

or

$$C_1 \cos \omega + 2\, C_2 \left(\frac{1 - \cos \omega}{2}\right)^{0.5} \left(\frac{1 + \cos \omega}{2}\right)^{0.5} + C_3 = 0,$$

which on squaring and rearranging yields

$$(C_2{}^2 + C_1{}^2) \cos^2\omega + 2\, C_1C_3 \cos \omega + (C_3{}^2 - C_2{}^2) = 0 \tag{2.23}$$

Eq. (2.23) yields the solution

$$\cos \omega = \frac{- mn \pm \sqrt{m^2 - n^2 + 1}}{m^2 + 1}, \tag{2.24}$$

where

$$m = \frac{\cos \phi}{\sin A_s \tan \beta} + \frac{\sin \phi}{\tan A_s}, \tag{2.25}$$

and

$$n = \tan \delta \left(\frac{\sin \phi}{\sin A_s \tan \beta} - \frac{\cos \phi}{\cos A_s}\right) \tag{2.26}$$

Since the sunrise or sunset hour angle for a tilted surface can never be greater than for a horizontal surface, the two angles for surfaces oriented east or west can be expressed as; for $A_s > 0$, surface oriented toward east:

$$\omega_{sr} = \min \left[\omega_s, \cos^{-1}\left(\frac{- mn - \sqrt{m^2 - n^2 + 1}}{m^2 + 1}\right)\right], \tag{2.27}$$

and

$$\omega_{ss} = - \min \left[\omega_s, \cos^{-1}\left(\frac{- mn + \sqrt{m^2 - n^2 + 1}}{m^2 + 1}\right)\right] \tag{2.28}$$

for $A_s < 0$, surface oriented towards west:

$$\omega_{sr} = \min \left[\omega_s, \cos^{-1}\left(\frac{- mn + \sqrt{m^2 - n^2 + 1}}{m^2 + 1}\right)\right] \tag{2.29}$$

and

$$\omega_{ss} = - \min \left[\omega_s, \cos^{-1}\left(\frac{- mn - \sqrt{m^2 - n^2 + 1}}{m^2 + 1}\right)\right] \tag{2.30}$$

Example 2.2. Calculate the sunrise and sun hour angles on 11 December for $\beta = 30^o$, for a flat plate collector at New Delhi (28° 55'N) oriented 30° east from its south position.

Solution:

From Eq. 2.5

$$\delta = 23.45 \sin \frac{360}{365} (334 + 284)$$

$$= -21.96^{\circ}$$

We first calculate, ω_s the sunrise hour angle for a horizontal surface. From Eq. (2.11) i.e.

$$\omega_s = \cos^{-1} [- \tan (-21.96) \tan (28.92)]$$

$$= 77.13^{\circ}$$

The collector is oriented 30° east of south and hence $A_s = +30^{\circ}$.

(a) The collector is inclined at an angle $\beta = 30^{\circ}$ from the horizontal. From Eqs. (2.25) and (2.26)

$$m = \frac{\cos 28.92}{\sin 30 \tan 30} + \frac{\sin 28.92}{\tan 30}$$

$$= 3.8696$$

$$n = \tan (-21.96) \left[\frac{\sin 28.92}{\sin 30 \tan 30} - \frac{\cos 28.92}{\cos 30}\right]$$

$$= -0.0581$$

The second term on R.H.S. of Eq. (2.27) is therefore

$$\cos^{-1} \left(\frac{3.8696 \times 0.0581 - \sqrt{3.8696^2 - 0.0581^2 + 1}}{3.8696^2 + 1} \right)$$

Therefore $\omega_{sr} = \min (77.13, 76.34)$

$$= 76.34^{\circ}$$

Similarly $\omega_{ss} = -\min (77.13, 74.67)$

$$= -74.67^{\circ}$$

2.2.4 Extraterrestrial Irradiation

2.2.4.1 Horizontal surface. If S_{en} is the extraterrestrial irradiance on a surface normal to the rays from the sun, where

$$S_{en} = S_c (r_o/r)^2 = S_c e \qquad (2.31)$$

then the irradiance on a horizontal surface can be written as

$$S_{eh} = S_{en} \cos Z, \qquad (2.32)$$

Using Eq. (2.9) one can write

$$S_{eh} = S_c e (\sin \delta \sin \phi + \cos \delta \cos \phi \cos \omega) \qquad (2.33)$$

In order to calculate the hourly irradiation, the amount of radiation incident in time interval dt is

$$d\,S_{eh} = S_c e \cos Z\, dt \tag{2.34}$$

where time dt is expressed in terms of hour angle by the relation

$$\frac{d\omega}{dt} = \frac{2\pi}{24\text{ h}} = \frac{2\pi}{24 \times 3600}\text{, (rad/s),}$$

hence

$$d\,S_{eh} = \left(\frac{12 \times 10^{-3}}{\pi}\right) S_c e\,(\sin\delta \sin\phi + \cos\delta \cos\phi \cos\omega)\,d\omega\text{, (kwh/m}^2\text{)} \tag{2.35}$$

To obtain the hourly radiation, the above expression is integrated over a period of 1 hour centred around the hour angle ω_i

$$S_{eh\omega} = \left(\frac{12 \times 10^{-3}}{\pi}\right) S_c e \int_{\omega_i - \frac{\pi}{24}}^{\omega_i + \frac{\pi}{24}} (\text{Sin}\,\delta\,\text{Sin}\,\phi + \text{Cos}\,\delta\,\text{Cos}\,\phi\,\text{Cos}\,\omega)d\omega$$

$$= 10^{-3} S_c e \left[\text{Sin}\,\delta\,\text{Sin}\,\phi + \frac{24}{\pi}\text{Sin}\,(\pi/24)\,\text{Cos}\,\delta\,\text{Cos}\,\phi\,\text{Cos}\,\omega_i\right] \tag{2.36}$$

since $(24/\pi)\,\text{Sin}\,(\pi/24) = .9972 \approx 1$, one can write

$$S_{eh\omega} = 10^{-3} S_c e \left[\sin\delta \sin\phi + \cos\delta \cos\phi \cos\omega_i\right] \tag{2.37}$$

Using the definition of hour angle ω_s corresponding to the sunset (Eq. 2.12), one can write

$$\cos Z = \cos\delta \cos\phi\,(\cos\omega - \cos\omega_s) \tag{2.38}$$

and consequently Eq. (2.37) reduces to

$$S_{eh\omega} = S_c e \cos\delta \cos\phi\,(\cos\omega_i - \cos\omega_s) \tag{2.39}$$

Eq. (2.35) can be used to calculate the average extraterrestrial radiation over any time interval. Let S_{ehm} is the extraterrestrial monthly average hourly radiation on a horizontal surface and will be calculated from the expression

$$S_{ehm} = \frac{1}{n_2 - n_1} \sum_{n_1}^{n_2} S_{eh\omega}\text{,} \tag{2.40}$$

where n_1 and n_2 are the day numbers at the beginning and end of the month respectively. It is interesting to note that in each month there is a particular declination on which the extraterrestrial radiation is identical to its monthly average values. Table 2.3 gives these declinations for a latitude of 35°N.

Diurnal variation of the extraterrestrial hourly radiation on a horizontal surface at two latitudes is given in Fig. 2.8. The irradiation during a day, from sunrise (sr) to sunset (ss) is given by the area under the curves i.e.

TABLE 2.3 Characteristic Declination δ, on which the Extraterrestrial Irradiation is Identical to its Monthly Average Value (ϕ = 35°) (After Iqbal, 1983)

Month	Date	δ (degrees)	Day Number n_d
Jan.	17	- 20.84	17
Feb.	14	- 13.32	45
March	15	- 2.40	74
April	15	+ 9.46	105
May	15	+ 18.78	135
June	10	+ 23.04	161
July	18	+ 21.11	199
Aug.	18	+ 13.28	230
Sept.	18	+ 1.97	261
Oct.	19	- 9.84	292
Nov.	18	- 19.02	322
Dec.	13	- 23.12	347

$$S_{ehd} = \int_{sr}^{ss} S_{eh}\, dt \qquad (2.41)$$

$$= 2 \int_{o}^{ss} S_{eh}\, dt \qquad (2.42)$$

Assuming that e remains constant during a day and converting time in terms of hour angle ω, one obtains

$$S_{ehd} = \frac{10^{-3} \times 24}{\pi} S_c e \int_{o}^{\omega_s} (\sin\delta \sin\phi + \cos\delta \cos\phi \cos\omega)\, d\omega \qquad (2.43)$$

or

$$S_{ehd} = \frac{10^{-3} \times 24}{\pi} S_c e \left[\frac{\pi}{180} \omega_s \text{ in } \delta \sin\phi + \cos\delta \cos\phi \sin\omega_s\right] \qquad (2.44)$$

being measured in degrees.

Combining Eqs. (2.11) and (2.44), S_{ehd} can also be written as

$$S_{ehd} = \frac{24 \times 10^{-3}}{\pi} S_c e \sin\phi \sin\delta \left(\frac{\pi}{180} \omega_s - \tan\omega_s\right) \qquad (2.45)$$

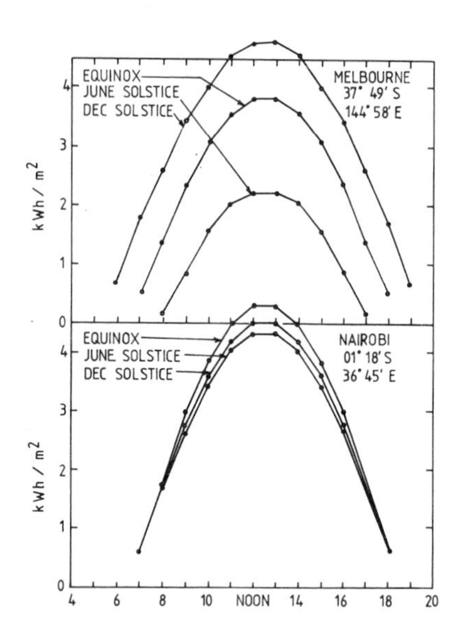

Fig. 2.8. Diurnal variation of the extraterrestrial hourly irradiation on a horizontal surface

or

$$S_{ehd} = \frac{24 \times 10^{-3}}{\pi} S_c e \cos\phi \cos\delta \left[\sin\omega_s - \frac{\pi}{180}\omega_s \cos\omega_s\right] \tag{2.46}$$

Eqs. (2.45) and (2.46) are not valid at the equator (ϕ = 0) and at the poles (ϕ = 90°) respectively.

Example 2.3. Calculate the extraterrestrial daily irradiation on 15 February of a horizontal surface at New Delhi (28° 55'N).

Solution

Latitude $\phi = 28.92°$

On 15 February e = 1.0256 from Eq. (2.1)

$$\text{and } \delta = 23.45 \sin \frac{360}{365}(46 + 284)$$

$$= -\ 13.29°$$

From Eq. (2.12)

$$\omega_s = \cos^{-1}(-\tan(-13.29)\tan(28.92))$$

$$= 82.5°$$

Using Eq. (2.45)

$$S_{ehd} = \frac{24 \times 1367 \times 1.0256}{\pi} \left[\sin (28.92) \sin (-13.29) \left\{ \frac{\pi}{180} \times 82.5 - \tan 82.5 \right\} \right] \times 10^{-3}$$

$$= \frac{1}{\pi} [0.4836 \times 0.2299 \times 6.1558 \times 24 \times 1367 \times 1.0256]$$

$$= 7.33 \text{ kWh/m}^2 \text{ day}$$

Two special cases of extraterrestrial daily radiation on a horizontal surface are of interest

(i) At the poles, during the summer there is no sunset and sunrise. Hence $\omega_s = \pi$ yielding

$$S_{ehd} = \frac{24 \times 10^{-3}}{\pi} S_c e \sin \phi \sin \delta \, (\pi^2/180) \tag{2.47}$$

(ii) At the equator $\phi = 0$ and hence $\omega = \pi/2$ yielding

$$S_{ehd} = \frac{24 \times 10^{-3}}{\pi} S_c e \cos \delta \tag{2.48}$$

2.2.4.2 Inclined surface. In general the extraterrestrial radiation on a surface inclined at an angle β from the horizontal is given by

$$S_{e\beta} = S_c e \cos i \tag{2.49}$$

where i is the angle of incidence given by Eq. (2.13). The total irradiation between hour angles ω_1 and ω_2 is given by

$$S_{e\beta t} = \frac{0.012}{\pi} S_c e \int_{\omega_1}^{\omega_2} [(\sin \phi \cos \beta - \cos \phi \sin \beta \cos A_s) \sin \delta + (\cos \phi \cos \beta + \sin \phi \sin \beta \cos A_s) \cos \delta \cos \omega + \cos \delta \sin \beta \sin A_s \sin \omega] \, d\omega \tag{2.50}$$

When integrating the above equation, it should be borne in mind that neither of these values of ω_1 and ω_2 exceeds the sunrise (or sunset) hour angle for tilted surfaces. Assuming again ω_i corresponding to middle of the house, the radiation over a period of π hour is obtained with $\omega_1 = \omega_i - \pi/24$ and $\omega_2 = \omega_i + \pi/24$, yielding

$$S_{e\beta h} = 10^{-3} S_c e \left[(\sin \phi \cos \beta - \cos \phi \sin \beta \cos A_s) \sin \delta + (\cos \phi \cos \beta + \sin \phi \sin \beta \cos A_s) \cos \delta \cos \omega_i + \cos \delta \sin \beta \sin A_s \sin \omega_i \right]; \tag{2.51}$$

since $(24/\pi) \sin (\pi/24) = 1.0$

Example 2.4. Find out the extraterrestrial hourly radiation on an arbitrarily oriented surface inclined at (30°N) at *hour ending at 10.00* (L.A.T.) on July 4.

Solution

In this example $A_s = 20^o$, $\beta = 30^o$, $\delta = 22.98^o$

$\phi = 34^o$, $\omega_i = 37.5^o$ and $e = 0.9666$.

Substituting in Eq. (2.51) one obtains

$$S_{e\beta h} = 10^{-3} \times 1367 \times 0.9666 \left[\{\sin(34)\cos(30) - \cos(34)\sin(30)\cos(20)\}\right.$$
$$\sin(22.98) + \{\cos(34)\cos(30) + \sin(34)\sin(30)\cos(20)\}.$$
$$\cos(22.98)\sin(37.5) + \cos(22.98)\sin(30)\sin(20).$$
$$\left.\sin(37.5)\right] \text{ kWh/m}^2\text{h}$$
$$= 1.321\left[(0.4842 - 0.3895)\, 0.3904 + (0.7179 + 0.2627)\right.$$
$$\left. 0.5605 + 0.0958\right]$$
$$= 0.901 \text{ kWh/m}^2\text{h} = 901 \text{ W/m}^2.$$

As a special case for a surface facing equator (true south) i.e. $A_s = 0$, the hourly radiation comes out to be

$$S_{e\beta h} = 10^{-3} S_c e \left[\sin\delta \sin(\phi - \beta) + \cos\delta \sin\omega_i \cos(\phi - \beta)\right] \quad (2.52)$$

It may be noted that the monthly average extraterrestrial hourly radiation can be calculated by computing $S_{e\beta h}$ for $\delta = \delta_c$, where δ_c is the characteristic declination (Table 2.3).

For duration, shorter than an hour i.e. between t_1 and t_2, the radiation has to be calculated by the integration of $S_{e\beta}$ yielding

$$S_{e\beta t} = 10^{-3} S_c e \left[\sin\delta \sin(\phi - \beta)(t_2 - t_1) + \frac{12}{\pi}\cos\delta \cos(\phi - \beta)\{\sin(15t_1) - \sin(15t_2)\}\right] \quad (2.53)$$

where the times t_1 and t_2 in hours are counted from midnight. These times should naturally lie between sunset and sunrise hours.

The daily irradiation on an inclined surface is given by the integration of Eq. (2.49) between sunset and sunrise hours i.e.

$$S_{e\beta d} = \frac{12 \times 10^{-3}}{\pi} S_c e \int_{\omega_{sr}}^{\omega_{ss}} \cos i \, d\omega \quad (2.54)$$

An integration yields the result

$$S_{e\beta d} = \frac{12 \times 10^{-3}}{\pi} S_c e \left[\{\sin\phi \sin\delta \cos\beta - \sin\delta \cos\phi \sin\beta \cos A_s\}\right.$$
$$|\omega_{ss} - \omega_{sr}|(\pi/180) + (\cos\phi \cos\beta + \sin\phi \sin\beta \cos A_s).$$
$$\cos\delta\,(|\sin\omega_{ss} - \sin\omega_{sr}|) + \cos\delta \sin\beta \sin A_s (\cos\omega_{ss} - \cos\omega_{sr}$$

$$(\cos \omega_{ss} - \cos \omega_{sr})\Big] \tag{2.55}$$

Example 2.5. For the example 2.4, calculate the mean daily extraterrestrial radiation.

Solution

From Eq. (2.12)

$$\omega_s = \cos^{-1} (- \tan \delta \tan \phi) = \cos^{-1} (- \tan 22.98 \tan 34)$$

$$= 106.6^o$$

In order to find out ω_{sr} and ω_{ss}, one uses Eqs. (2.25) and (2.26) and subsequently (2.27) and (2.28) yielding $\omega_{sr} = \min (106.6, 108^o) = 106.6^o$

and $\omega_{ss} = - \min (106.6, 88.2) = - 88.2^o$

Substituting in Eq. 2.55 one gets

$$S_{e\beta d} = 7.893 \text{ kWh/m}^2 \text{ day.}$$

2.2.4.3 Calculation of R_b. R_b is defined as the ratio of the mean daily radiation on an inclined plane to that on the horizontal plane in the absence of the earth's atmosphere and hence given by the expression

$$R_b = S_{e\beta d}/S_{ehd}$$

Using Eqs. (2.55) and Eq. (2.44), one gets

$$R_b = 2 \left[\frac{\pi}{180} \omega_s \sin \delta \sin \phi + \cos \delta \cos \phi \sin \omega_s\right]^{-1} \Big[\{\sin \phi \sin \delta \cos \beta$$

$$- \sin \delta \cos \phi \sin \beta \cos A_s\} (|\omega_{ss} - \omega_{sr}|) (\pi/180)$$

$$+ (\cos \phi \cos \beta + \sin \phi \sin \beta \cos A_s) \cos \delta (|\sin \omega_{ss} - \sin \omega_{sr}|)$$

$$+ \cos \delta \sin \beta \sin A_s (|\cos \omega_{ss} - \cos \omega_{sr}|)\Big] \tag{2.56}$$

For a surface facing true south ($A_s = 0$), $S_{e\beta d}$ is given by the expression

$$S_{e\beta d} = \frac{24 \times 10^{-3}}{\pi} S_c e \int_0^{\omega_s'} \cos i \, d\omega$$

$$= \frac{24 \times 10^{-3}}{\pi} S_c e \Big[\frac{\pi}{180} \omega_s' \sin \delta \sin (\phi - \beta)$$

$$+ \cos \delta \cos (\phi - \beta) \sin \omega_s'\Big] \tag{2.57}$$

In this case one gets

$$R_b = \left[\frac{(\pi/180)\omega_s' \sin \delta \sin (\phi - \beta) + \cos \delta \cos (\phi - \beta) \sin \omega_s'}{(\pi/ 180)\omega_s \sin \delta \sin \phi + \cos \delta \cos \phi \sin \omega_s}\right] \tag{2.58}$$

The ratio of the instantaneous irradiance on an inclined surface to that on a horizontal surface in the absence of earth's atmosphere is denoted by r_b and hence

$$r_b = \frac{S_{e\beta}}{S_{eh}} = \frac{\cos i}{\cos Z} \tag{2.59}$$

To get the value of r_b for an interval of 1 hour, one has to integrate the above formula for 1 hour intervals i.e.

$$r'_b = \left[\int_{\omega_i-\pi/24}^{\omega_i+\pi/24} \cos i \, d\omega\right] \Big/ \left[\int_{\omega_i-\pi/24}^{\omega_i+\pi/24} \cos Z \, d\omega\right] \approx \frac{\cos i}{\cos Z} \tag{2.60}$$

2.2.5 Spectral Distribution of Solar Radiation

In outer space the spectral distribution of solar radiation (Fig. 2.9) corresponds to the energy output of a blackbody at a temperature of about 6000°K. The invisible ultraviolet region with wavelengths between 0.29 and 0.40 μm contains about 9% of the total energy while the visible region between 0.4 and 0.7 μm contains 40% and the infrared region from 0.7 to 3.5 μm carries the remaining 51%. The peak intensity, about 2074 $Wm^{-2}\mu m^{-1}$ is reached at 0.49 μm in the green portion of the visible spectrum.

Attenuation of solar radiation takes place in traversing through the sky mainly by two physical processes namely the scattering and the absorption. The extent of attenuation is determined by the actual composition of the atmosphere and by the length of the atmospheric path traversed by the sun rays. This length is expressed in terms of the air mass a_m.

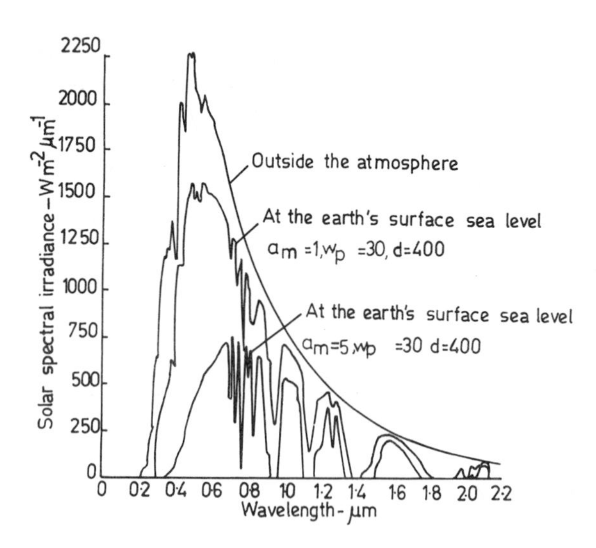

Fig. 2.9. Solar spectral energy distribution curves above the earth's atmosphere and at earth's surface (sea level) for air masses 1 and 5

Ignoring earth's curvature and assuming the atmosphere to be nonrefractive and completely homogeneous, it is easily seen that a_m = sec Z (Fig. 2.10). Because of the curvature of the earth and the refraction of the real atmosphere, the above expression of air mass has error, which is 0.25% at Z = 60° and increases to 10% at Z = 85°. More accurate expression for the air mass was evolved by Kasten (1966), i.e.

$$a_m = \left[\cos Z + 0.15\ (93.885 - Z)^{-1.253}\right]^{-1} \tag{2.61}$$

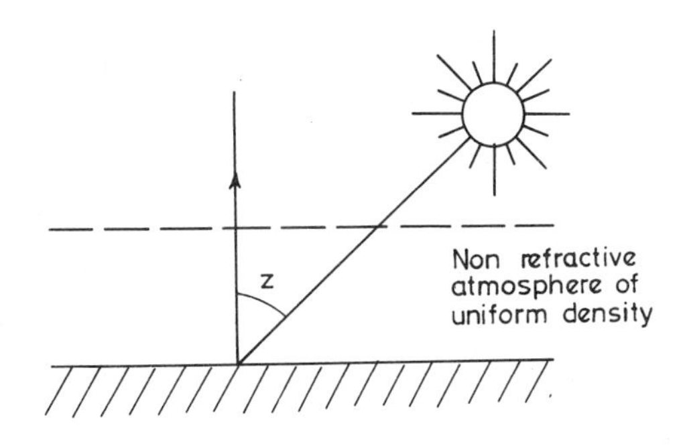

Fig. 2.10. Air mass and zenith angle

For ground level measurements, this formula is accurate and better than 0.1% for zenith angles up to 86°. At Z = 89.5°, the deviation of 1.25% occurs, which is the largest one. Another restriction on Eq. (2.61) is that of the pressure. This equation is applicable to a standard pressure of 1013.25 mb at sea level. At any other place for which the atmospheric pressure P is different from the normal sea level pressure P_o, like hill stations where the atmosphere is reduced, the corresponding air mass (a_m) is obtained from the equation,

$$a_m P = a_{m_o} P_o \tag{2.62}$$

a_{m_o} being the air mass at normal sea level pressure. The pressure above sea level can, with good accuracy, be obtained from the equation

$$\left(\frac{P}{P_o}\right) = \exp\ (-\ 0.0001184\ h) \tag{2.63}$$

h being the height of the station above sea level.

2.2.6 Attenuation of Direct Solar Radiation

In traversing through the earth's atmosphere, a part of the incident energy is removed by scattering and a part by absorption. A part of the scattered radiation goes back to space and a portion reaches the earth as diffused radiation (Fig. 2.11). The radiation arriving on the ground directly in line from the solar disc is called direct or beam radiation.

Attenuation of direct radiation takes place due to the presence of water vapour, ozone aerosoles, and the pressure exerted by air molecules. The monochromatic transmittance due to direct radiation can be written as

$$T_\lambda = \left[(T_{am})_\lambda\ (T_{wp})_\lambda\ (T_d)_\lambda\ (T_{\sigma_3})_\lambda\ (T_{H_2O})_\lambda\ (T_g)_\lambda\right] \tag{2.64}$$

The above expression is based on the experimental data of Fowle, Abbot and others and it was obtained by Moon (1940). The first three terms refer to the effect of scattering by air molecules (am), water vapour (w_p), dust particles (d) and the last three refer to the effect of absorption by ozone, water vapour and other gases respectively.

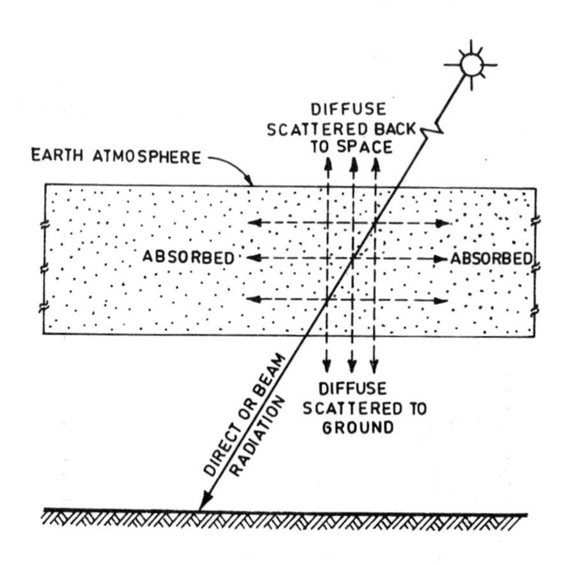

Fig. 2.11. Distribution of direct, diffuse, and absorbed solar radiation

Monochromatic transmissivity, corresponding to a particular wavelength, can be calculated for various atmospheric conditions and air masses from the above equation (2.64). By multiplying the spectral radiation intensity outside the earth's atmosphere with the monochromatic transmissivity of the atmosphere, one obtains the monochromatic direct radiation intensity ($S_{gh\lambda}$) at the earth's surface i.e.

$$S_{gh\lambda} = T_\lambda \; S_{eh\lambda}$$

Figure 2.9 showed some typical results of the solar spectral distribution energy curves on clear days at sea level for air masses 1 and 5 for an assumed atmospheric condition of w_p = 30 mm; d = 400 particles/cm^3 and O_3 = 2.5 mm. The importance of air mass in reducing the solar radiation intensity is clearly seen in the figure. The next section deals with the transmittance of the atmosphere due to various processes of scattering and absorption.

2.2.6.1 Scattering. The scattering effects depend on the particle size and hence these are different for the air molecules, water vapour molecules and the dust particles. For particles with radius less than 0.1 λ, Rayleigh scattering theory is applicable and for such cases extinction coefficient is proportional to λ^{-4}. Scattering by dust free air is the basic atmospheric effect. Dust particles scatter radiation even at wavelengths greater than 1 μm and this depletes the solar energy appreciably. For particle size between 0.1 λ and 25 λ, more complicated Mei scattering theory is applicable and the extinction coefficient in such cases is proportional to $\lambda^{-\Gamma}$, where Γ lies between 1 and 2; Γ is usually taken as 1.3. For still larger particles, such as suspended water droplets, laws of geometrical optics can be used to describe the scattering effects.

Because of the variation of the Rayleigh scattering coefficient with λ^{-4}, the spectral transmittance of air molecules rapidly increases with wavelength (Fig. 2.12) and decreases with increasing optical air mass. The transmittance due to Rayleigh scattering is expressed as (Iqbal, 1983)

$$(T_{amr})_\lambda = \exp\,(-0.008735\ \lambda^{-4.08}\ a_m) \tag{2.65}$$

It is evident from the Fig. 2.12 and from Eq. (2.65), that the scattering is predominant for short wave radiation only.

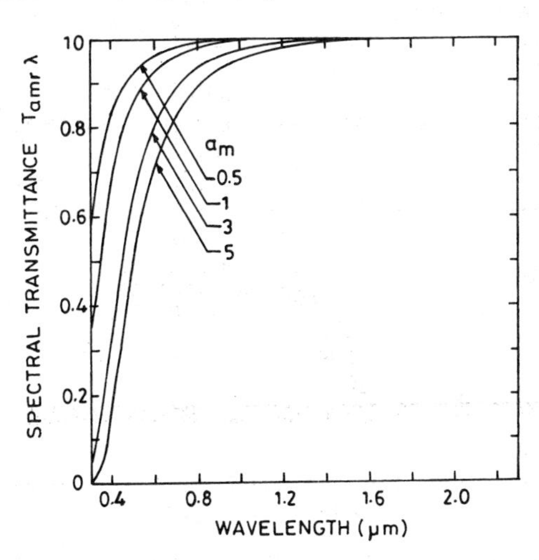

Fig. 2.12. Rayleigh spectral transmittance as a function of air mass (after Iqbal, 1983)

The spectral transmittance for water and dust particles is less sensitive to wavelength than to Rayleigh scattering. According to Iqbal (1983), the transmittances due to water vapour content and the dust content are estimated by using the expressions

$$(Tw_p)_\lambda = \exp\left[-0.008635\ \lambda^{-2}\ w_p\ a_m\right] \tag{2.66}$$

$$(T_d)_\lambda = \exp\left[-0.08128\ \lambda^{-0.75}\ \frac{d}{800}\ a_m\right] \tag{2.67}$$

Due to suspended water droplets, the attenuation of the radiation is proportional to $\lambda^{-\Gamma}$ and the transmissivity can be written as

$$(T_{ams})_\lambda = \exp\,(-\ T_f\ \lambda^{-\Gamma}\ a_m) \tag{2.68}$$

where T_f, the turbidity factor is given in Table 2.4 depending on atmospheric conditions. The total transmissivity due to scattering $(T_{am})_\lambda$ is then given by the product of expressions (2.65) and (2.68) i.e.

$$(T_{am})_\lambda = (T_{amr})_\lambda\ (T_{ams})_\lambda \tag{2.69}$$

TABLE 2.4 Parameters for Various Degrees of Atmospheric Cleanliness

Atmosphere	T_f	Γ	Visibility (Km)
Clean	0.00	1.30	340
Clear	0.10	1.30	28
Turbid	0.20	1.30	11
Very turbid	0.40	1.30	5

2.2.6.2 Absorption. Oxygen, ozone, water vapour and carbon dioxide are the main absorbing components of the atmosphere; the absorption takes place in the selected absorption bands. The extreme ultraviolet spectrum in the range 0.12-0.18 μm is absorbed by the oxygen. In the range 0.20 and 0.33 μm it is mainly absorbed by ozone which also absorbs the radiation in the visible region between 0.44 and 0.76 μm. The ozone layer is located at altitudes between 10 and 50 Km, with a maximum concentration at 25 Km. Due to the absorption of radiation by ozone and oxygen, the earth receives almost negligible radiation up to 0.3 μm.

In the infrared region, water vapour has a number of absorption bands, namely 0.93, 1.13, 1.42, 1.47, 1.9, 2.7, 3.2 and 6.3 μm. For carbon dioxide the absorption bands are 1.6, 2.0, 2.7 and 4.3 μm. A significant percentage of sun's radiation in the infrared region is cut off by these absorptions by the water vapour and the carbon dioxide. Water vapour content decreases rapidly with height up to 4 Km and becomes almost negligible above the tropopause, the range of troposphere. The total precipitable water content of the atmosphere varies from place to place, with seasons and from day to day at any given place.

Absorption of electromagnetic radiation by ozone has been investigated by a number of researchers. If $k_{o_3\lambda}$ is the extinction coefficient* for ozone than one can write (Iqbal, 1983)

$$(T_{o_3})_\lambda = \exp\,(-\,k_{o_3\lambda}\,\ell\,a_m) \qquad (2.70)$$

where ℓ is the amount of ozone in cm (NTP). Values of extinction coefficients for various values of λ are given in Table 2.5. The spectral transmittance $(T_{H_2O})_\lambda$ for absorption due to water vapour has been shown to be given by (Iqbal, 1983).

$$(T_{H_2O})_\lambda = \exp\left[-\,0.2385\,k_{w\lambda}\,w_p\,a_m/(1 + 20.07\,k_{w\lambda}\,w_p\,a_m)^{0.45}\right] \qquad (2.71)$$

the extinction coefficients $k_{w\lambda}$'s being given in Table 2.6. The transmittance due to absorption of radiation by CO_2, O_2 etc., is written as

$$(T_g)_\lambda = \exp\left[-\,1.41\,k_{g\lambda}\,a_m/(1 + 118.93\,k_{g\lambda}\,a_m)^{0.45}\right] \qquad (2.72)$$

The attenuation coefficients $k_{g\lambda}$'s are presented in Table 2.7.

---

*If a_m is the optical path length, k_λ the monochromatic extinction or attenuation coefficient, $S_{eh\lambda}$ extraterrestrial monochromatic irradiance then after traversing through the atmosphere, the irradiance becomes

$$S_{gh\lambda} = S_{eh\lambda}\,\exp\,(-k_\lambda\,a_m)$$

TABLE 2.5 Spectral Absorption Coefficient $K_{o3\lambda}$ for Ozone (After Leckner, 1978)

λ (μm)	$k_{o3\lambda}$ (cm^{-1})	λ (μm)	$k_{o3\lambda}$ (cm^{-1})	λ (μm)	$k_{o3\lambda}$ (cm^{-1})
0.290	38.000	0.485	0.017	0.595	0.120
0.295	20.000	0.490	0.021	0.608	0.125
0.300	10.000	0.495	0.025	0.605	0.130
0.305	4.8000	0.500	0.030	0.610	0.120
0.310	2.700	0.505	0.035	0.620	0.105
0.315	1.350	0.510	0.040	0.630	0.090
0.320	0.800	0.515	0.045	0.640	0.079
0.325	0.380	0.520	0.048	0.650	0.067
0.330	0.160	0.525	0.057	0.660	0.057
0.335	0.075	0.530	0.063	0.670	0.048
0.340	0.040	0.535	0.070	0.680	0.036
0.345	0.019	0.540	0.075	0.690	0.028
0.350	0.007	0.545	0.080	0.700	0.023
0.355	0.000	0.550	0.085	0.710	0.018
0.445	0.003	0.555	0.095	0.720	0.014
0.450	0.003	0.560	0.103	0.730	0.011
0.455	0.004	0.565	0.110	0.740	0.010
0.460	0.006	0.570	0.120	0.750	0.009
0.465	0.008	0.575	0.122	0.760	0.007
0.470	0.009	0.580	0.120	0.770	0.004
0.475	0.012	0.585	0.118	0.780	0.000
0.480	0.014	0.590	0.115	0.790	0.000

The monochromatic direct spectral irradiance $S_{gh\lambda}$ on a horizontal surface on earth is then determined by the extraterrestrial monochromatic irradiation $S_{c\lambda}$, the values of which are given in Table 2.8 i.e.

$$S_{gh\lambda} = e\, S_{c\lambda}\, T_{\lambda} \cos Z \tag{2.73}$$

Example 2.6. Calculate the direct spectral irradiance on a horizontal surface for $\lambda = 0.7$ μm under cloudless sky conditions $\beta = 0.1$, $\Gamma = 1.3$ on January 15 at New Delhi (28° 55') at 11 am, with precipitable water content at 1 cm and ozone layer thickness 0.1 cm.

Solution

$$\delta = -21.2, \qquad \phi = 28.92^{\circ}$$

$$Z = \cos^{-1}\left[\sin(-21.2)\sin 28.92 + \cos(-21.2)\cos(28.92)\cos 15\right] = 52.16$$

TABLE 2.6 Spectral Absorption Coefficients of Water Vapour (After Leckner, 1978)

λ(um)	k_{w_λ} (cm^{-1})	λ(um)	k_{w_λ} (cm^{-1})	λ(um)	k_{w_λ} (cm^{-1})
0.69	0.160E - 01	0.93	0.270E + 02	1.85	0.220E + 04
0.70	0.240E - 01	0.94	0.380E + 02	1.90	0.140E + 04
0.71	0.125E - 01	0.95	0.410E + 02	1.95	0.160E + 03
0.72	0.100E + 01	0.96	0.260E + 02	2.00	0.290E + 01
0.73	0.870E + 00	0.97	0.310E + 01	2.10	0.220E + 00
0.74	0.610E - 01	0.98	0.148E + 01	2.20	0.330E + 00
0.75	0.100E - 02	0.99	0.125E + 00	2.30	0.590E + 00
0.76	0.100E - 04	1.00	0.250E - 02	2.40	0.203E + 02
0.77	0.100E - 04	1.05	0.100E - 04	2.50	0.310E + 03
0.78	0.600E - 03	1.10	0.320E + 01	2.60	0.150E + 05
0.79	0.175E - 01	1.15	0.230E + 02	2.70	0.220E + 05
0.80	0.360E - 01	1.20	0.160E - 01	2.80	0.800E + 04
0.81	0.330E + 00	1.25	0.180E - 03	2.90	0.650E + 03
0.82	0.153E + 01	1.30	0.290E + 01	3.00	0.240E + 03
0.83	0.660E + 00	1.35	0.200E + 03	3.10	0.230E + 03
0.84	0.155E + 00	1.40	0.110E + 04	3.20	0.100E + 03
0.85	0.300E - 02	1.45	0.150E + 03	3.30	0.120E + 03
0.86	0.100E - 04	1.50	0.150E + 02	3.40	0.195E + 02
0.87	0.100E - 04	1.55	0.170E - 02	3.50	0.360E + 01
0.88	0.260E - 02	1.60	0.100E - 04	3.60	0.310E + 01
0.89	0.630E - 01	1.65	0.100E - 01	3.70	0.250E + 01
0.90	0.210E + 01	1.70	0.510E + 00	3.80	0.140E + 01
0.91	0.160E + 01	1.75	0.400E + 01	3.90	0.170E + 00
0.92	0.125E + 01	1.80	0.130E + 03	4.00	0.450E - 02

$a_m = 1.6$

From Table 2.8 $S_{c_\lambda} = 1427.5\ Wm^{-2}\ \mu m^{-1}$

$e = 0.9684$ for 15 January

Rayleigh scattering transmittance

$$(T_{am})_\lambda = \exp(-0.008735 \times 0.7^{-4.08} \times 1.6)$$

$$= 0.9418$$

$$(T_{ams})_\lambda = \exp(-\ 0.1 \times 0.7^{-1.3} \times 1.6)$$

$$= .7753$$

TABLE 2.7 Spectral Absorption Coefficients of Uniformly Mixed Gases (After Leckner, 1978)

λ(μm)	$k_{g\lambda}$	λ(μm)	$k_{g\lambda}$	λ(μm)	$k_{g\lambda}$
0.76	0.200E + 01	1.75	0.100E - 04	2.80	0.150E + 03
0.77	0.210E + 00	1.80	0.100E - 04	2.90	0.130E + 00
		1.85	0.145E - 03	3.00	0.950E - 02
1.25	0.730E - 02	1.90	0.710E - 02	3.10	0.100E - 02
1.30	0.400E - 03	1.95	0.200E + 01	3.20	0.800E + 00
1.35	0.110E - 03	2.00	0.300E + 01	3.30	0.190E + 01
1.40	0.100E - 04	2.10	0.240E + 00	3.40	0.130E + 01
1.45	0.640E - 01	2.20	0.380E - 03	3.50	0.750E - 01
1.50	0.630E - 03	2.30	0.110E - 02	3.60	0.100E - 01
1.55	0.100E - 01	2.40	0.170E - 03	3.70	0.195E - 02
1.60	0.640E - 01	2.50	0.140E - 03	3.80	0.400E - 02
1.65	0.145E - 02	2.60	0.660E - 03	3.90	0.290E + 00
1.70	0.100E - 04	2.70	0.100E + 03	4.00	0.250E - 01

Using Eq. (2.66), (2.70), (2.71)

$$(Tw_p)_\lambda = \exp(-\ 0.008635 \times 0.7^{-2} \times 1 \times 1.6) = 0.9722$$

$$(T_{O_3})_\lambda = \exp(-\ 0.023 \times 0.1 \times 1.6) = 0.9963$$

$$(T_{H_2O})_\lambda = \exp\left[\frac{-\ 0.2385 \times 0.024 \times 1 \times 1.6}{(1 + 20.07 \times 0.024 \times 1 \times 1.6)^{0.45}}\right] = 0.9929$$

$$k_{g\lambda} = 0, \ Tg_\lambda = 1$$

$$T_\lambda = .9418 \times .7753 \times .9722 \times .9963 \times .9929$$

$$= 0.7022$$

Hence $S_{gh\lambda} = 1427.5 \times 0.9684 \times .7022 \times .613$

$= 595\ \text{W/m}^2\ \mu\text{m}.$

Exercise: Calculate the irradiance for the same data on 15 June.

2.2.7 Diffuse Radiation

Diffuse radiation is generated by scattering effects of air molecules and aerosols. A portion of Rayleigh and aerosol scattered diffuse radiation reaches the ground after the first pass through the atmosphere. This radiation gets partly reflected by the ground to the atmosphere, which reflects it back to the ground. Hence the diffuse radiation, $S_{d\lambda}$, arriving on the ground consists of

TABLE 2.8 Extraterrestrial Solar Spectral Irradiance at Mean Sun-Earth Distance (WRC spectrum) (After Frölich and Wehrli, 1981)

λ	$S_{c\lambda}$	$\sum_{o}^{\lambda} S_{c\lambda}$	$P_{o,\lambda}^{*}$	λ	$S_{c\lambda}$	$\sum_{o}^{\lambda} S_{c\lambda}$	$P_{o,\lambda}$
0.250	54.56	2.51	0.18	0.475	2016.25	247.45	18.10
0.255	91.25	2.84	0.21	0.480	2055.00	257.62	18.85
0.260	122.50	3.47	0.25	0.485	1901.26	267.64	19.58
0.265	253.75	4.29	0.31	0.490	1920.00	276.00	20.26
0.270	275.00	5.70	0.42	0.495	1965.00	286.59	20.97
0.275	212.50	6.87	0.50	0.500	1862.52	296.37	21.68
0.280	162.50	7.86	0.57	0.505	1943.75	305.80	22.37
0.285	286.25	9.12	0.67	0.510	1952.50	315.50	23.08
0.290	535.00	10.97	0.80	0.515	1835.01	325.05	23.78
0.295	560.00	13.91	1.02	0.520	1802.49	333.79	24.42
0.300	527.50	16.54	1.21	0.525	1849.99	343.33	25.12
0.305	557.50	19.26	1.41	0.530	1947.49	352.67	25.80
0.310	602.51	22.13	1.62	0.535	1926.24	362.34	26.51
0.315	705.00	25.51	1.87	0.540	1857.50	371.87	27.20
0.320	747.50	29.09	2.13	0.545	1895.01	381.22	27.89
0.325	782.50	32.70	2.39	0.550	1902.50	390.72	28.58
0.330	997.50	37.51	2.74	0.555	1885.00	400.17	29.27
0.335	906.25	42.34	3.10	0.560	1840.02	409.42	29.95
0.340	960.00	46.79	3.42	0.565	1850.00	418.71	30.63
0.345	877.50	51.45	3.76	0.570	1817.50	427.94	31.31
0.350	955.00	55.89	4.09	0.575	1848.76	437.11	31.98
0.355	1044.99	61.08	4.47	0.580	1840.00	446.22	32.64
0.360	940.00	65.72	4.81	0.585	1817.50	455.44	33.32
0.365	1125.01	71.01	5.20	0.590	1742.49	464.21	33.96
0.370	1165.00	76.92	5.63	0.595	1785.00	473.16	34.61
0.375	1081.25	82.15	6.01	0.600	1720.00	481.98	35.26
0.380	1210.00	88.32	6.46	0.605	1751.25	490.71	35.90
0.385	931.25	93.11	6.81	0.610	1715.00	499.35	36.53
0.390	1200.00	98.30	7.19	0.620	1715.00	516.51	37.79
0.395	1033.75	103.61	7.58	0.630	1637.50	533.22	39.01
0.400	1702.49	109.81	8.03	0.640	1622.50	549.73	40.22
0.405	1643.75	118.40	8.66	0.650	1597.50	565.79	41.39

*$P_{o,\lambda}$ is the percentage of the solar constant associated with wavelength λ, i.e. $\sum_{o}^{\lambda} S_{c\lambda} = (P_{o,\lambda} * S_c)/100$.

TABLE 2.8 (Cont'd)

λ	$S_{c\lambda}$	$\sum_{o}^{\lambda} S_{c\lambda}$	$P_{o,\lambda}$	λ	$S_{c\lambda}$	$\sum_{o}^{\lambda} S_{c\lambda}$	$P_{o,\lambda}$
0.410	1710.00	126.68	9.27	0.660	1555.00	581.10	42.51
0.415	1747.50	135.37	9.90	0.670	1505.00	596.65	43.65
0.420	1747.50	143.98	10.53	0.680	1472.50	611.50	44.73
0.425	1692.51	152.69	11.17	0.690	1415.02	625.86	45.78
0.430	1492.50	160.74	11.76	0.700	1427.50	640.28	46.84
0.435	1761.25	168.74	12.34	0.710	1402.50	654.28	47.86
0.440	1755.02	177.59	12.99	0.720	1355.00	668.10	48.87
0.445	1922.49	187.02	13.68	0.730	1355.00	681.84	49.88
0.450	2099.99	196.86	14.40	0.740	1300.00	695.28	50.86
0.455	2017.51	207.15	15.15	0.750	1272.52	708.17	51.81
0.460	2032.49	217.29	15.90	0.760	1222.50	720.62	52.72
0.465	2000.00	227.59	16.65	0.770	1187.50	732.70	53.60
0.470	1979.99	237.50	17.37	0.780	1195.00	744.52	54.47
0.790	1142.50	756.23	55.32	1.900	136.01	1273.42	93.16
0.800	1144.70	767.69	56.16	1.950	126.00	1280.08	93.64
0.810	1113.00	779.02	56.99	2.000	118.50	1286.30	94.10
0.820	1070.00	789.93	57.79	2.100	93.00	1296.72	94.86
0.830	1041.00	800.50	58.56	2.200	74.75	1305.00	95.47
0.840	1019.99	310.77	59.31	2.300	63.25	1312.05	95.98
0.850	994.00	820.96	60.06	2.400	56.50	1317.96	96.41
0.860	1002.00	830.85	60.78	2.500	48.25	1323.16	96.80
0.870	972.00	840.61	61.49	2.600	42.00	1327.66	87.12
0.880	966.00	850.39	62.21	2.700	36.50	1331.57	97.41
0.890	945.00	859.94	62.91	2.800	32.00	1334.98	97.66
0.900	913.00	869.26	63.59	2.900	28.00	1337.97	97.88
0.910	876.00	878.16	64.24	3.000	24.75	1340.60	98.07
0.920	841.00	886.81	64.87	3.100	21.75	1342.93	98.24
0.930	830.00	895.10	65.48	3.200	19.75	1344.99	98.39
0.940	801.00	903.27	66.08	3.300	17.25	1346.84	98.53
0.950	778.00	911.18	66.66	3.400	15.75	1348.48	98.65
0.960	771.00	918.90	67.22	3.500	14.00	1349.96	98.76
0.970	765.00	926.58	67.78	3.600	12.75	1351.30	98.85
0.980	769.00	934.21	68.34	3.700	11.50	1352.51	98.94
0.990	762.00	941.88	68.90	3.800	10.50	1353.60	99.02
1.000	743.99	949.41	69.45	3.900	9.50	1354.59	99.09
1.050	665.98	984.76	72.04	4.000	8.50	1355.49	99.16
1.100	606.04	1016.27	74.35	4.100	7.75	1356.31	99.22

TABLE 2.8 (Cont'd)

λ	$S_{c\lambda}$	$\sum_o^\lambda S_{c\lambda}$	$P_{o,\lambda}$	λ	$S_{c\lambda}$	$\sum_o^\lambda S_{c\lambda}$	$P_{o,\lambda}$
1.150	551.04	1045.16	76.46	4.200	7.00	1357.05	99.27
1.200	497.99	1071.43	78.38	4.300	6.50	1357.72	99.32
1.250	469.99	1095.66	80.15	4.400	6.00	1358.33	99.37
1.300	436.99	1117.96	81.78	4.500	5.50	1358.89	99.41
1.350	389.03	1138.51	83.29	4.600	5.00	1359.40	99.45
1.400	354.03	1156.97	84.64	4.700	4.50	1359.86	99.48
1.450	318.99	1173.91	85.88	4.800	4.00	1360.29	99.51
1.500	296.99	1189.28	87.00	4.900	3.75	1360.69	99.54
1.550	273.99	1203.52	88.04	5.000	3.47	1361.04	99.57
1.600	247.02	1216.48	88.99	6.000	1.75	1363.50	99.75
1.650	234.02	1228.52	89.87	7.000	0.95	1364.79	99.84
1.700	215.00	1239.74	90.69	8.000	0.55	1365.52	99.89
1.750	187.00	1249.69	91.42	9.000	0.35	1365.96	99.93
1.800	170.00	1258.55	92.07	10.000	0.20	1366.24	99.95
1.850	149.01	1266.42	92.64	25.000	0.12	1366.97	100.00

(After Frölich and Wehrli, 1981)

essentially three parts

$S_{dr\lambda}$ = diffuse spectral irradiance produced by Rayleigh scattering that arrives on the ground after the first pass through the atmosphere.

$S_{da\lambda}$ = diffuse spectral irradiance produced by aerosols (bigger water droplets) that arrives on the ground after the first pass through the atmosphere.

and $S_{dm\lambda}$ = diffuse spectral irradiance produced by multiple reflections.

The diffuse spectral irradiance due to Rayleigh scattering can be estimated by the following expression

$$S_{dr\lambda} = S_{c\lambda} \cos Z \, (T_s)_\lambda \, [0.5 \, (1 - (T_{amr})_\lambda) \, (T_{ams})_\lambda], \tag{2.74}$$

where

$$(T_s)_\lambda = (T_{H_2O})_\lambda \, (T_g)_\lambda \, (T_{O_3})_\lambda \tag{2.75}$$

The diffuse spectral irradiance produced by aerosols is estimated by the expression

$$S_{da\lambda} = S_{c\lambda} \cos Z \, (T_s)_\lambda \, [f_c \, \zeta_{so} \, (1 - (T_{ams})_\lambda) \, (T_{amr})_\lambda] \tag{2.76}$$

where ζ_{so}, the single scattering albedo, is the ratio of the energy scattered by aerosols to total attenuation under the first impingement by direct radiation. The magnitude of ζ_{so} depends on the material, shape, size and optical properties of aerosol particles and hence the determination of ζ_{so} is extremely complex. It is common to assign a value between 0.7 and 1.0 for ζ_{so}; usually for urban-industrial regions $\zeta_{so} \approx 0.6$ and for rural agricultural regions $\zeta_{so} \approx 0.9$.

The constant f_c in Eq. (2.76) comes because of the fact that only a portion of the total scattered radiation reaches the ground. f_c is a function of the zenith and its values are given in Table 2.9.

TABLE 2.9 Values of f_c for different Z's

Z	0	10	20	30	40	50	60	70	80	85
f_c	0.92	0.92	0.90	0.90	0.90	0.85	0.78	0.68	0.60	0.50

The contribution of multiple reflections on the spectral diffused irradiance is small. In the event of snow cover however, the multiple reflections assume considerable importance. $S_{dm\lambda}$ has been approximated to be given by

$$S_{dm\lambda} = S_{c\lambda} \cos Z \left(\frac{\zeta_{g\lambda}\, \zeta^*_{a\lambda}}{1 - \zeta_{g\lambda}\, \zeta^*_{a\lambda}} \right), \tag{2.77}$$

where

$$\zeta^*_{a\lambda} = (T_s)_\lambda \, [0.5\,(1 - (T_{amr})_\lambda)\,(T_{ams})_\lambda + (1 - f_c)\,\zeta_{so}\,(1 - (T_{ams})_\lambda)\,(T_{amr})_\lambda] \tag{2.78}$$

and $\zeta_{g\lambda}$ is the ground albedo often taken to be 0.2.

2.2.8 Global Irradiance

2.2.8.1 Horizontal surface. The spectral global irradiance is sum of the direct and diffuse radiation and hence given by

$$(S_{ght})_\lambda = S_{gh\lambda} + S_{d\lambda} \tag{2.79}$$

For most applications of solar energy, amount of solar radiation over a certain band width or over the complete solar spectrum is required and therefore one has to integrate over the desired wavelength interval or over the complete solar spectral range respectively, i.e.

$$S_{ght} = \sum_{\lambda=o}^{\infty} (S_{gh\lambda} + S_{d\lambda})\, d\lambda \tag{2.80a}$$

2.2.8.2 Arbitrarily oriented surface. The spectral irradiance on any arbitrary surface oriented at angle A_s and tilted at angle β with the horizontal is given by

$$S_{g\beta\lambda} = e\, S_{c\lambda}\, T_\lambda \cos i + S_{d\beta\lambda} \tag{2.80b}$$

where cos i is given by Eq. (2.13).

To calculate the total energy incident on a surface tilted at an angle β one has to integrate $S_{g\beta\lambda}$ over the entire wavelength region. The integration $\int S_{g\beta\lambda}\, d\lambda$ over specific bands yields the energy contained in various wavelength regions. This fraction of energy in various colour bands is given in Table 2.10. It is seen that as the zenith angle Z increases, the fraction in the infrared region increases. The total diffuse radiation in various wavelength regions i.e. $\int S_{d\lambda}\, d\lambda$ is given in Table 2.11. When the atmosphere is clean, the maximum energy is in the ultraviolet and violet portion of the spectrum. With increase in turbidity, the fraction in the infrared increases.

2.2.9 Approximate Methods: Method of ASHRAE

2.2.9.1 Direct Radiation. Besides calculating radiation by Eq. (2.80), which leads to precise results, approximate methods are also available. The simple procedure usually adopted by the engineering and architectural community is the one that is known as ASHRAE algorithm (Farber and Morrison 1977, ASHRAE 1982).

The method of ASHRAE is based essentially on the work of Moon (1940), Threlkeld and Jordan (1958), Threlkeld (1963) and Stephenson (1967). Based on various measurements and the work of several other researchers on attenuation coefficients for various atmospheric constituents, Moon found out the following transmittances

Rayleigh scattering $$(T_{rm})_\lambda = \exp\,(-0.00885\,\lambda^{-4}\, a_m) \qquad (2.81)$$

Water vapour scattering $$(T_{wp})_\lambda = \exp\,(-0.008635\,\lambda^{-2}\, w_p\, a_m) \qquad (2.82)$$

Dust scattering $$(T_d)_\lambda = \exp\left[-0.08128\,\lambda^{-0.75}\,\frac{d}{800}\, a_m\right] \qquad (2.83)$$

Ozone absorption $$(T_{o3})_\lambda = \exp\,(-\,k_{g\lambda}\,\ell\, a_m) \qquad (2.84)$$

Writing the spectral transmittance of the atmosphere as

$$T_\lambda = (T_{rm})_\lambda\,(T_{wp})_\lambda\,(T_d)_\lambda\,(T_{o3})_\lambda \qquad (2.85)$$

and taking $a_m = \sec Z$, Moon calculated the spectral as well as wavelength integrated values of the cloudless sky direct normal irradiance by utilizing a solar constant value of 1322 W/m^2.

Threlkeld and Jordan utilized the technique of Moon to calculate the broad band normal incidence direct radiation for basic atmospheric conditions with 0.25 (NTP) cm ozone, 200 dust particles per cubic centimetre and a variable water content. This data corresponds to a typical nonindustrial mid latitude conditions except for ozone thickness. Hourly values of the normal incident radiation on first day of each month for four latitudes viz. 30°, 36°, 42° and 48° were put in the form of clear day insolation diagrams. It was proposed to multiply the values on these curves by a clearness index (k_T) defined as

$$k_T = \frac{S_{gdh}\ \text{(Calculated with local mean clear day water vapour)}}{S_{gdh}\ \text{(Calculated with water vapour according to basic atmosphere)}} \qquad (2.86)$$

The local clear day water vapour was defined as 85% of its monthly average value.

It is, however, apparent that more useful data would have been generated, if k_T had been obtained as a ratio of the measured to calculated direct normal irradiance. Threlkeld and Jordan (1957) attempted to incorporate this, but they were not able to find the required data.

TABLE 2.10 Direct Normal Irradiance within Various Bands*

Z	T_λ	S	Fraction of direct energy colours, (μm)							
			uv	Violet	Blue	Green	Yellow	Orange	Red	ir
(Degrees)		(W m^{-2})	0.39	0.39-0.455	0.455-0.492	0.492-0.577	0.577-0.597	0.597-0.622	0.622-0.77	0.77
0.0	0.0	1053.30	0.04	0.08	0.06	0.13	0.03	0.04	0.18	0.45
60.0	0.0	934.01	0.02	0.06	0.06	0.13	0.03	0.04	0.19	0.47
70.0	0.0	853.01	0.01	0.05	0.05	0.13	0.03	0.04	0.19	0.49
80.0	0.0	689.18	0.00	0.03	0.04	0.11	0.03	0.04	0.20	0.55
85.0	0.0	525.18	0.00	0.01	0.02	0.07	0.02	0.03	0.20	0.64
0.0	0.1	895.12	0.03	0.07	0.06	0.12	0.03	0.04	0.18	0.48
60.0	0.1	688.18	0.01	0.05	0.05	0.11	0.03	0.04	0.18	0.53
70.0	0.1	558.62	0.01	0.03	0.04	0.10	0.03	0.04	0.18	0.58
80.0	0.1	334.38	0.00	0.01	0.02	0.06	0.02	0.03	0.17	0.69
85.0	0.1	166.73	0.00	0.00	0.00	0.02	0.01	0.01	0.12	0.83
0.0	0.2	766.07	0.02	0.06	0.05	0.12	0.03	0.04	0.17	0.51
60.0	0.2	518.89	0.01	0.03	0.04	0.10	0.02	0.03	0.17	0.59
70.0	0.2	381.62	0.00	0.02	0.03	0.08	0.02	0.03	0.17	0.66
80.0	0.2	181.35	0.00	0.00	0.01	0.03	0.01	0.02	0.13	0.80
85.0	0.2	67.60	0.00	0.00	0.00	0.01	0.00	0.00	0.06	0.93
0.0	0.3	659.98	0.02	0.05	0.04	0.11	0.03	0.04	0.17	0.56
60.0	0.3	399.40	0.01	0.02	0.03	0.08	0.02	0.03	0.16	0.65
70.0	0.3	270.33	0.00	0.01	0.02	0.06	0.02	0.02	0.15	0.72
80.0	0.3	107.15	0.00	0.00	0.02	0.01	0.01	0.01	0.09	0.88
85.0	0.3	32.36	0.00	0.00	0.00	0.00	0.00	0.00	0.02	0.97

*w_p = 2 cm, O_3 = 0.35 cm (NTP), Γ_{so} = 1.3, S_c = 1367 W/m^{-2}, extraterrestrial spectrum see Table 2.8

TABLE 2.11 Diffuse Horizontal Irradiance within Various Bands

Z	T_λ	S_d	Fraction of diffuse energy in different colours, (μm)							
			uv	Violet	Blue	Green	Yellow	Orange	Red	ir
(Degrees)		(W m^{-2})	0.39	0.39-0.455	0.455-0.492	0.492-0.577	0.577-0.597	0.597-0.622	0.622-0.77	0.77
0.0	0.0	70.06	0.29	0.26	0.13	0.16	0.02	0.03	0.07	0.04
60.0	0.0	52.96	0.25	0.26	0.14	0.17		0.03	0.08	0.04
0.0	0.1	205.03	0.12	0.16	0.10	0.18	0.03	0.04	0.16	0.21
60.0	0.1	129.69	0.08	0.14	0.10	0.18	0.04	0.05	0.17	0.24
0.0	0.2	316.94	0.09	0.14	0.10	0.18	0.04	0.05	0.17	0.25
60.0	0.2	185.70	0.05	0.11	0.09	0.18	0.04	0.05	0.19	0.30
0.0	0.3	409.95	0.07	0.13	0.09	0.18	0.04	0.05	0.18	0.27
60.0	0.3	226.90	0.04	0.10	0.08	0.17	0.04	0.05	0.20	0.33

*w_p = 2 cm, O_3 = 0.35 cm (NTP), Γ_{so} = 1.3, a_m = 1, $\bar{\zeta}_g$ = 0.2, S_c = 1367 W m^{-2}, extraterrestrial spectrum from Table 2.8.

From the clear day insolation curves of Threlkeld and Jordan (1957), Stephenson (1967), developed the following simple expression

$$S_{gdh} = (k_T)\ A \exp\ (-\ B \sec Z) \tag{2.87}$$

Where the coefficients A and B were obtained by fitting this equation to the proposed standard clear day insolation curves of Threlkeld and Jordan. Monthly variations of the constants A and B are given in Table 2.12. A is, in fact, the apparent extraterrestrial irradiance given by

$$A = S_c\ e \tag{2.88}$$

The variable B represents an overall broad band value of the atmospheric attenuation coefficient.

Eq. (2.87) is actually for the standard pressure and for other pressures one should use the expression

$$S_{gdh} = (k_T)\ A \exp\ [-\ B \sec Z\ (P/P_o)] \tag{2.89}$$

P being the actual pressure at the station and P_o the standard pressure.

2.2.9.2 Diffuse Radiation

Diffuse Sky Radiation. The intensity of the diffuse radiation on a horizontal surface may be between 50 - 350 Wm^{-2} depending upon the atmospheric and cloud-cover conditions. On a clear day the diffuse radiation is only 10-15% of the total radiation, while on partly cloudy days it can be 30-45% of the total radiation received on a horizontal surface. The diffuse radiation is essentially in the ultraviolet and the visible region (between 0.29 and 1 μm) and its percentage in different wavelength regions has already been given in Table 2.11. The diffuse radiation from clear sky that falls on any surface can be approximated by

$$S_{d\beta} = k\ S_{gd\beta}\ f_\beta \tag{2.90}$$

where the factor k (ratio of diffused to max. direct radiation) has values given in Table 2.12 for various months. If the diffuse radiation is assumed to be uniformly distributed over the hemispherical sky dome then the angle factor f_β in Eq. (2.90) is given by

$$f_\beta = \left(\frac{1 + \cos\ \beta}{2}\right) \tag{2.91}$$

Constant k, apart from depending on month, also depends on the condition of the sky; the values in Table 2.12 correspond to the conditions of clear sky whereas for other sky conditions k can be approximately modified from the values given in Table 2.13.

Erbs *et al.* (1982) recently used hourly pyrheliometer and pyranometer data from 4 US locations to establish a relationship between the hourly diffuse radiation and the hourly clearness index k_T, defined as the ratio of the hourly global radiation to the hourly extraterrestrial radiation on a horizontal surface. These authors developed a relation between (S_{dh}/S_{ght}) and k_T as

$$\frac{S_{dh}}{S_{ght}} = 1.0 - 0.09\ k_T\ ;\ k_T \leqslant 0.22$$

TABLE 2.12 Constants A, B, and k* for Calculation of Solar Irradiance According to ASHRAE and the Revised Recommended Values

Date	ASHRAE			Revised recommended value		
	A ($W\ m^{-2}$)	B $(air\ mass)^{-1}$	k (dimensionless)	A ($W\ m^{-2}$)	B $(air\ mass)^{-1}$	k (dimensionless)
21 Jan.	1230	0.142	0.058	1202	0.141	0.103
21 Feb.	1215	0.144	0.060	1187	0.142	0.104
21 March	1186	0.156	0.071	1164	0.149	0.109
21 April	1136	0.180	0.097	1130	0.164	0.120
21 May	1104	0.196	0.121	1106	0.177	0.130
21 June	1088	0.205	0.134	1092	0.184	0.137
21 July	1085	0.207	0.136	1093	0.186	0.138
21 Aug.	1107	0.201	0.122	1107	0.182	0.134
21 Sept.	1151	0.177	0.092	1136	0.165	0.121
21 Oct.	1192	0.160	0.073	1136	0.152	0.111
21 Nov.	1221	0.149	0.063	1190	0.144	0.106
21 Dec.	1233	0.142	0.057	1204	0.141	0.103

*used for calculating diffuse radiation later.

TABLE 2.13 Diffuse Radiation on a Horizontal Surface

Sky condition	Ratio of actual direct to max. direct radiation	Ratio of diffuse to max. direct radiation k
Clear	1.0	0.12
Clear, slightly hazy	0.8	0.25
Hazy	0.60	0.35
Overcast	0.40	0.55

$$\frac{S_{dh}}{S_{ght}} = 0.9511 - 0.1604\ k_T + 4.388\ k_T^2 + 12.336\ k_T^4\ ;\ 0.22 < k_T \leqslant 0.8$$

$$\frac{S_{dh}}{S_{ght}} = 0.165\ ;\ k_T > 0.8 \qquad (2.92)$$

Ballantyne (1967) expressed the diffuse radiation on a horizontal surface as a power series of solar altitude i.e.

$$S_{dh} = C_1 \sin A_\alpha + C_2 \sin^2 A_\alpha + C_3 \sin^3 A_\alpha$$

$$+ C_4 \sin^4 A_\alpha + C_5 \sin^5 A_\alpha \qquad (2.93)$$

Based on the measured data, the constants C_1----C_5 have to be determined for each place. For Melbourne the values obtained by Ballantyne are $C_1 = 438$, $C_2 = -1306$, $C_3 = 2259$, $C_4 = -1893$ and $C_5 = 618$.

Diffuse radiation on vertical and inclined surface. Based on the measurements of Parmelea (1954), Lim (1980) obtained the following equation for the diffuse radiation on vertical surfaces

$$S_{dv} = k_1 \sin A_\alpha + k_2 \sin^2 A_\alpha \cos A_\alpha \cos i\ \zeta'$$

with $\zeta' = 1$ for $0 \leqslant i \leqslant 90^\circ$

$= 0$ for $i > 90^\circ$ (2.94)

Values of the constants depend on the atmospheric conditions. Parmelea (1954) obtained the values as $k_1 = 165$ Wm^{-2}, $k_2 = 85$ Wm^{-2}, while Spencer (1965) at Melbourne obtained the values as 208 Wm^{-2} and 69 Wm^{-2} for k_1 and k_2 respectively.

Threlkeld's equation adopted by ASHRAE gives the diffuse radiation on vertical surface in terms of the diffuse radiation on horizontal surface and the angle of incidence at the vertical surface. The expression is

$$S_{dv} = S_{dh} \left[0.55 + 0.437 \cos i + 0.313 \cos^2 i\right]$$

for $\cos i > -0.2$;

$$S_{dv} = 0.45\ S_{dh};\ \cos i \leqslant -0.2 \tag{2.95}$$

To estimate the diffuse radiation on inclined surfaces, one can approximately use the formula

$$S_{d\beta} = S_{dh}\ \frac{\cos^2\beta}{2} \tag{2.96}$$

2.2.9.3 Reflected Radiation. The amount of reflected radiation incident on a surface is a function of the average reflectivity of the nearby surfaces. The average reflectivity is, in turn, a function of orientation, colour, material of the surface, texture of the surface etc. A precise calculation of the amount of radiation reflected from the ground on to a surface (say wall) is difficult to determine, but if an estimate of the weight average reflectivity $\bar{\zeta}_g$ is made then one can write

$$S_{RW} = \frac{1}{2}\ \bar{\zeta}_g\ S_{ght} \tag{2.97}$$

where S_{RW} is the radiation reflected on the wall as a result of total incident radiation S_{ght} on the horizontal surface. Equation (2.97) assumes that all the reflected radiation is perfectly diffused or reflected equally in all the directions. Therefore, a vertical surface, regardless of the direction it faces, sees only one half of the radiation reflected from a horizontal surface. On a surface with slope β, diffuse radiation as a result of ground reflection with reflectivity $\bar{\zeta}_g$ can be written as

$$S_R = \frac{1}{2}\ \bar{\zeta}_g\ S_{ght}\ \sin^2\beta \tag{2.98}$$

Total diffuse radiation on the surface is naturally the sum of the expression (2.96) and (2.98) i.e.

$$S_{d\beta} = S_{dh}\ \frac{\cos^2\beta}{2} + \bar{\zeta}_g\ S_{ght}\ \frac{\sin^2\beta}{2} \tag{2.99}$$

Estimated average reflectivity of some materials has been tabulated in Chapter 4.

2.2.9.4 Global radiation. Once the direct and diffuse radiation are known the global radiation is calculated from the expression

$$S_{g\beta t} = S_{gd\beta} + S_{d\beta} \tag{2.100}$$

The calculation of solar irradiance by ASHRAE method gives deviations ⩾ 6%. The main reasons for these deviations are

(i) the solar constant used (1322 W/m^2) and the extraterrestrial irradiance values employed by Moon are outdated.

(ii) the attenuation coefficients used by Moon were empirically chosen functions and were very site specific.

(iii) the experimental data were based on the use of old instruments.

(iv) for broadband irradiance ℓ_n (S_{gdh}) versus a_m (= sec Z) graph is not found to be linear as predicted by Eq. (2.87).

(v) in diffuse radiation, there is no provision to incorporate ground albedo and hence contribution by multiple reflections can not be varied.

(vi) there is no separate provision to incorporate aerosol generated diffuse irradiance.

Iqbal (1983) has compared the results of ASHRAE algorithm with the calculations of a better model presented in the next section and called parametric model and found that the difference in the two results are less than 6%. If one slightly modifies the constants A and B, as given in Table 2.12 the comparison becomes better.

Though there is a large difference in the values calculated for the diffuse radiation by ASHRAE procedure and the parametric model (sometimes as large as 80%), corrected k (Table 2.12) give better results at least during summer months.

2.2.9.5 Radiation on inclined surfaces. Since the radiation on inclined surfaces is of extreme importance in many engineering calculations, this section tries to summarize the calculation procedure for total incident radiation on an arbitrarily inclined surface.

Consider a plane surface inclined at angle β (Fig. 2.13). The total radiation incident on this surface consists of the following parts

(i) direct radiation

(ii) sky diffuse radiation

(iii) radiation reflected from the ground.

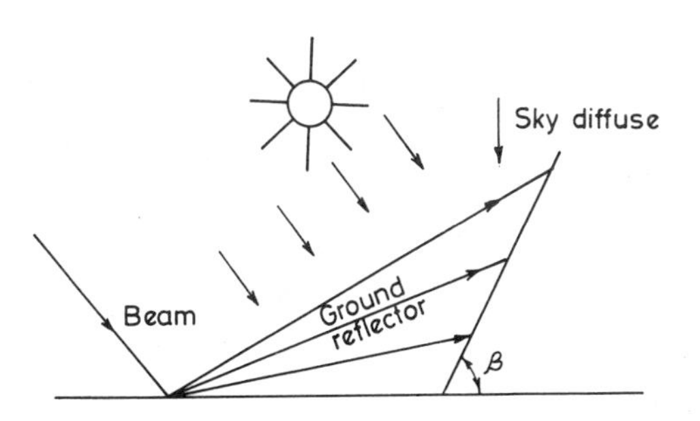

Fig. 2.13. Inclined surface and various components of radiation incident on it

The instantaneous or hourly direct radiation on a surface inclined at angle β, can be calculated from its value at the horizontal surface by using the formula.

$$S_{gd\beta} = S_{gdh}\left(\frac{S_{e\beta d}}{S_{ehd}}\right) = S_{ghd} \times r_b \tag{2.101}$$

where r_b is evaluated from expressions (2.59) and (2.60).

The daily mean radiation on an inclined surface can be calculated from the mean beam radiation on a horizontal surface i.e.

$$\bar{S}_{g\beta d} = \bar{S}_{ghd} \times R_b \tag{2.102}$$

Where R_b is defined in Eq. (2.58). The total radiation on an inclined surface can, therefore, be written as

$$S_{g\beta t} = S_{gd\beta} + S_{d\beta} + S_{\beta\zeta g} \qquad (2.103)$$

The ground reflected radiation $S_{\beta\zeta g}$ is estimated by using the relation

$$S_{\beta\zeta g} = \frac{1}{2}\,(S_{g\beta d} + S_{\beta d})\,\zeta_g\,(1 - \cos\beta) \qquad (2.104)$$

2.3 Building Orientation

It has long been recognized that the buildings are oriented with respect to the direction of prevailing winds and earth-sun angular relationship. Orientation of the building for sun essentially means that the building should receive as much solar radiation in winter as possible and as little as possible in summer. Orientation is also crucial to the amount of daylight received in building and these values are mainly determined by latitude, altitude (particularly in urban areas), local atmospheric conditions at a point in time and topography. Another aspect for consideration is the wind direction, when one considers the orientation. For evaluating the effects of wind on human comfort, one should consider the annual and monthly variations of wind prevalence, its velocity and temperature along with its direction. The effect of prevailing winds both outside and inside the buildings has to be considered.

Arumi (1977) studied theoretically the effect of proper window sizing as well as proper selection of external surface to volume ratio on energy savings for heating cooling and lighting in the periphery zones of buildings.

Figure 2.14 gives (Arumi, 1977), the sky illumination values for various times of the day and for various sections of the sky. Various components of the illumination, namely (i) direct, (ii) diffuse and (iii) reflected are taken into account and shown separately in the figure. It is seen that east and west elevations will generally enjoy the same daylight levels in the forenoon and afternoon respectively. Also east-west orientated surfaces normally receive more and higher levels of daylight than north-south directions.

Detailed calculations by Arumi (1977) showed that as the window area increases relative to the wall area, several trade offs begin to occur. First and most obvious is the decreased reliance on artificial lighting and hence the component of the total energy consumption. However, less lighting means decreasing internal heat generation due to lighting. The internal heat generation due to electric lighting can be desirable whenever the reduction in heating cost during the winter is greater than the increase in cooling cost. Figure 2.15 shows the annual energy consumption for heating, cooling and lighting per unit area of floor as a function of the size of window on the wall for a south-facing room with surface to volume ratio of 0.33 m^{-1}. The total energy consumption shows a well defined minimum when the window area is 25% of the wall area. The windows were assumed to be shaded from direct sun from March up to September (at Austin, Texas). The energy consumption for heating and cooling goes down at first as the window area increases. The reduction comes from a lowering of cooling cost due to reduction of internal heat generation. Further increase in the window area increases the air-conditioning cost (heating and cooling both).

Another factor affecting energy consumption is the surface to volume ratio. Figure 2.16 shows the total energy consumption (heating, cooling and lighting) as a function of the percentage of glass on the wall with changing values of surface to volume ratio. For each value of the surface to volume ratio, there is an optimum window area. Conversly, there is an optimum value of surface to volume ratio for a given window area to wall area ratio as shown in Fig. 2.17. For walls with 40% window area the energy consumption shows the expected increase as the surface to volume ratio increases; otherwise it shows the existence of an optimum surface to volume ratio, including the case with no windows.

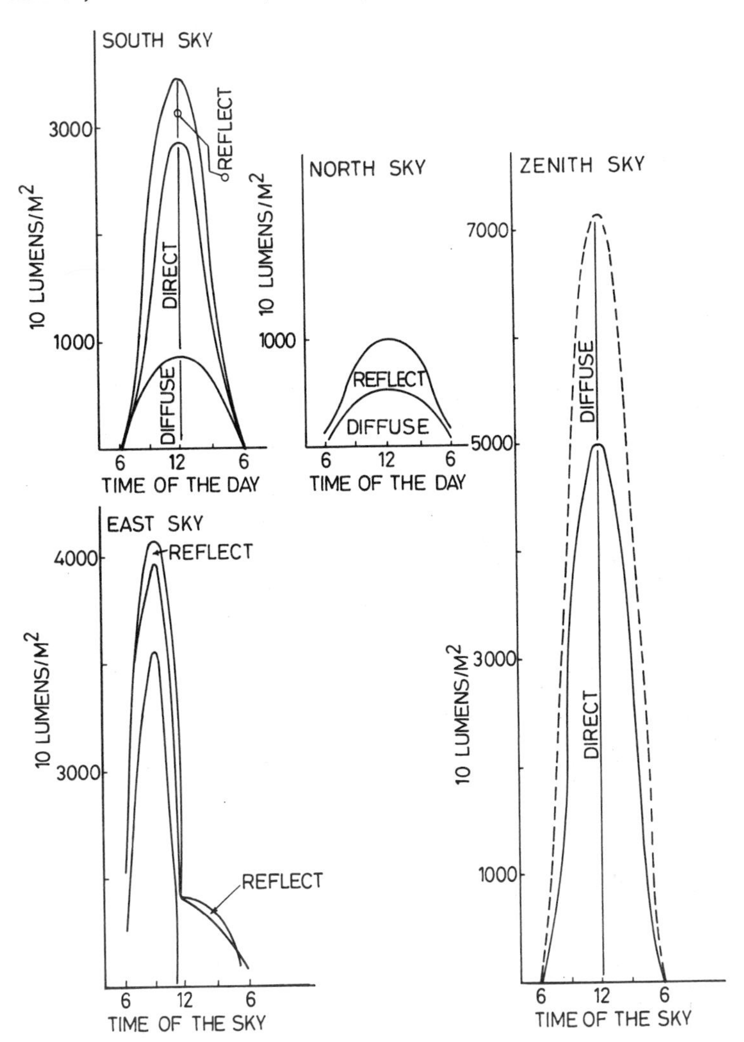

Fig. 2.14. Illumination level towards various orientations (Arumi, 1977)

The importance of the size, shape and orientation of the buildings to reduce solar heat gain in the summer has been investigated intensively by Neubauer (1968).

The results have shown that when a rectangular building is oriented east-west (large exposures on the north-south walls), the interior air temperature of the building is considerably less than if the same shape rectangular building were oriented north-south (large exposures on east and west walls). When a rectangular building oriented north-south has a light reflective roof, the benefit of the reflective roof is not appreciable as compared to the tremendous increase of solar radiation incident on the walls.

There are huge variations in the amount of solar radiation incident on the

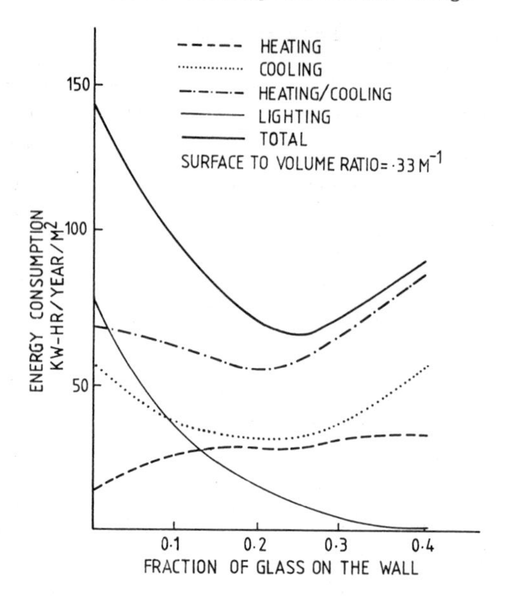

Fig. 2.15. Annual energy consumption for heating, cooling and lighting per unit area of floor as a function of amount of window on the wall for a south-facing room with surface to volume ratio of (= 0.33 m^{-1}). The total energy consumption shows a well defined minimum when the window area is 25% of the wall area. The windows are shaded from direct sun from March through September. (Adapted from Arumi, 1977).

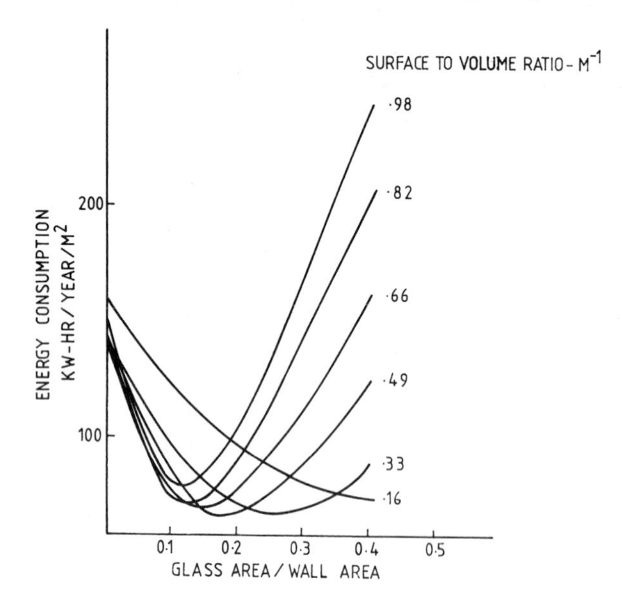

Fig. 2.16. Annual energy consumption for heating, cooling and lighting as a function of glass to total wall area with changing values of surface to volume ratio (Arumi, 1977).

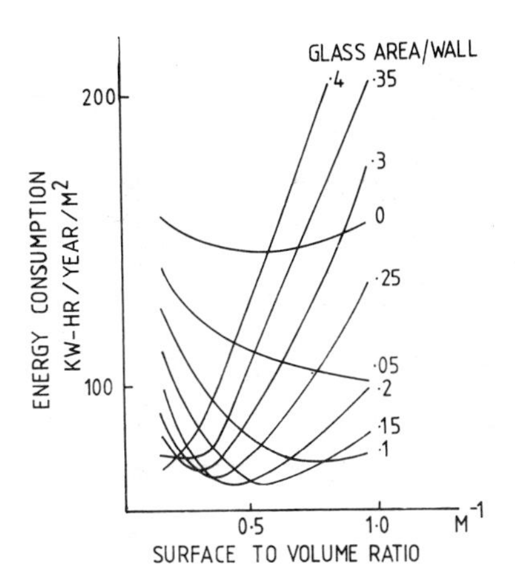

Fig. 2.17. Annual energy consumption for cooling, heating and lighting as a function of surface to volume ratio parameters of the relative window. (Arumi, 1977)

different surfaces of a building. Figure 2.18 shows graphically these variations at a latitude of 29°N (extraterrestrial). For other latitudes there will be small differences.

It is seen that a southern wall* receives much more radiation in December than in June. South-eastern and south-western walls also receive higher level of radiation in winter, but annual variations are of smaller order of magnitude than the variations on the southern wall. The radiation of east and west oriented wall remains nearly constant, irrespective of the season.

Breezes affect heat gains and losses in a building in two ways: (a) increased wind speed increases the infiltration rate by forcing air through cracks and crevices in the building and (b) increased wind speed reduces the thermal resistance of a building section by increasing the heat-transfer between external component's surface and the air film. The infiltration effects of wind and breezes are much more significant than the thermal resistance effects on the heating and cooling loads of a building. The influence of air infiltration on calculated heating and cooling loads of a building is a direct function of the accuracy with which air infiltration rates are determined. The rate of air flow is approximately given as (Dryden, 1975)

$$\dot{m}_v \alpha \; A_p \; (\Delta p)^n;$$

Δp being the velocity pressure and $\dot{m}_v$ is the rate of air flow (m^3s^{-1}) and A_p is the area of the aperture (m^2). The value of the constant of proportionality ($\simeq$ 0.5-0.7) and the values of the index n ($\simeq$ 0.5) depends on the characteristic of the component involved (O'Callaghan, 1978). The static pressure on the windward side of a building ranges from about 0.5 to 0.8 times the velocity pressure ($= \rho \cdot v^2/2$) of the undisturbed free flow of wind. On the leeward side, an external negative pressure of approximately 0.3-0.4 times the free flow velocity pressure is induced. The local pressures on the sides of the building and parallel to the wind direction, depend partly on the velocity gradients and the aerodynamic, characteristics of surfaces and vary in a complicated way. These

*North wall in the southern hemisphere.

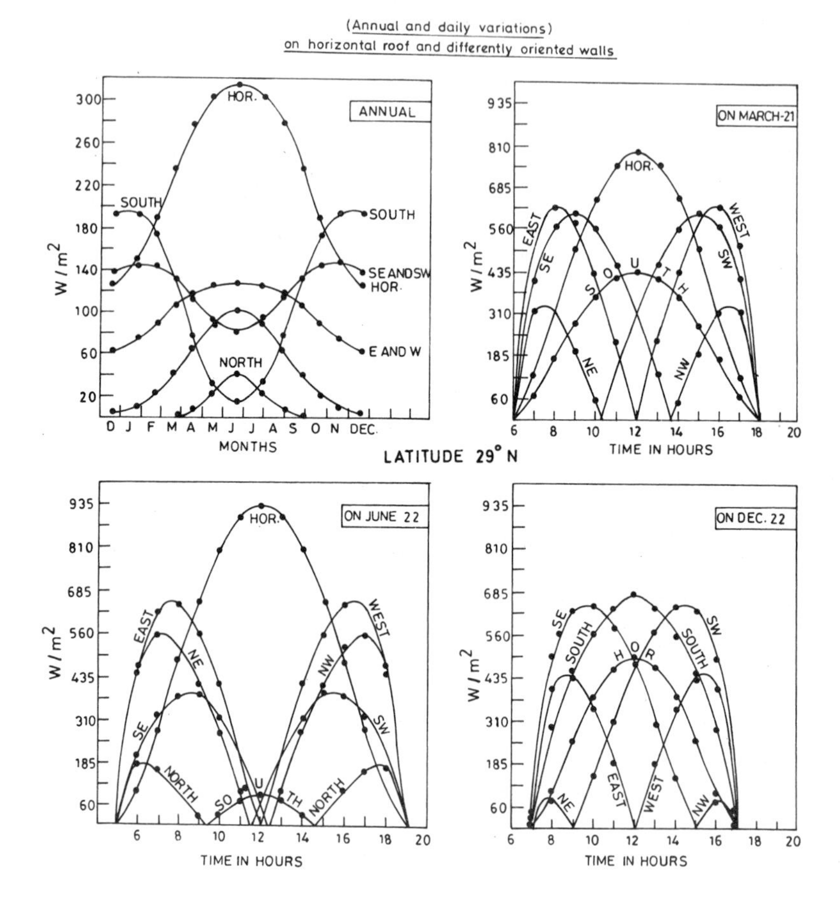

Fig. 2.18. Direct solar radiation incident on clear days

are affected by the height, size and shape of the building and the proximities of the neighbouring structures. Mean pressure differences across buildings of different heights for a free wind speed of 9 ms^{-1} are given in Table 2.14.

Natural infiltration in most existing buildings is often far greater than the minimum ventilation requirements. To control the arising heating/cooling loads, as a result of ventilation, infiltration must be eliminated by draught proofing and associated isolating techniques. Outside air must be regulated and restricted to fulfil the ventilation requirements. Generally 21 m^3 h^{-1} of fresh air per person is advised (O'Callaghan, 1978) to prevent vitiation and palpable body odour. Though exact required ventilation rates will depend upon the rates of metabolism, combustion in open fires, cigarette smoking and cooling, and odour and contaminant production, minimum recommended total ventilation rates are tabulated in the Table 2.15.

Parameters that are suspected to influence air infiltration rates are wind speed, wind direction, orientation, cracks sizes, window and door openings, porosity of

TABLE 2.14 Pressure Differences Due to Wind Effects (Dryden, 1975)

Building height (m)	Mean pressure difference across a building ($N\ m^{-2}$)		
	Open country ($9\ ms^{-1}$)	Suburban ($3.5\ ms^{-1}$)	City Centre ($3\ ms^{-1}$)
10	58	21	6
20	70	31	11
30	78	38	15
40	85	44	21
50	90	49	23
60	95	55	26
70	100	59	31
80	104	63	34

TABLE 2.15 Recommended Total Ventilation Rate (Dryden, 1975)

Occupancy known		Occupancy unknown	
Type	Air changes (m^3/s per person)	Type	Air changes/ hour
Homes	0.012	Offices	3-8
Schools, theatres	0.014	Engine rooms	4
Factories, shops	0.016-0.028	Garages	5
Hospitals	0.019-0.047	Baths	5-8
		Lavatories	5-10
		Restaurants	5-10
		Cinemas, theatres	5-10
		Kitchens	10-40

walls etc. A detailed research investigation was conducted by Sepsy *et al.* (1978) for the Electric Power Research Institute to develop a computerized simulation of infiltration as influenced by all governing parameters. After carefully studying nine residences and also analysing experimental data of previous studies, the authors concluded that air infiltration could not be analysed by considering wind direction components, as originally thought, but air infiltration occurs in a more general pattern which can be characterized in simpler models. The conclusion, that the wind direction was not a governing factor in air infiltration, is applicable only to residential structures. Wind direction could be a governing factor, for example, in agricultural structures that are usually long and narrow.

2.4 SHADING DEVICES

Modern buildings are characterized by the widespread use of glass on building facades. This along with lightweight structures has caused the problems of overheating even in cold or temperate countries. The effect of glazing in admitting the incident radiation depends on the spectral properties of glass and on the shading devices which may be provided internally as well as externally. The objectives which a building designer usually strives to meet are

1. to minimize heat gain during summer while allowing winter sun to come in,

2. to prevent the direct rays of the sun from falling on any light-coloured surface that is visible to normal occupants of the room,

3. to allow natural light to enter in such a way that it can be diffused as evenly as possible over the whole room,

4. to interfere as little as possible with the view from the window.

Shading devices to meet the above objectives can be fixed, adjustable or retractable and can be of a variety of architectural shapes and geometrical configurations. Internal shading devices include venetian blinds, roller blinds and curtains. Usually these can be lifted, rolled or drawn back from the window. External shading devices include shutters, awnings, overhangs and a variety of louvres; vertical, horizontal or a combination of both (egg-crate).

The position of shading devices which may be adjusted, their reflectivity ζ_s, angles between slats determine the amount of solar flux transmitted through a glazed window. For different values of these parameters, the fraction of the total incident radiation transmitted inside is given in Table 2.16.

TABLE 2.16 Fraction of Solar Heat Gains through a Window with Different Shading Factors. (After Givoni, 1976)

	s^+	ζ_s	$Qin^*/S_{g\beta}$	s	ζ_s	$Qin^*/S_{g\beta}$
		0.2	0.43			
Internal	30	0.4	0.57	45	0.4	0.39
		0.6	0.66			
		0.2	0.18		0.2	0.13
External	30	0.4	0.15	45	0.4	0.10
		0.6	0.089		0.6	0.081

*The net heat flux through a shaded window comprises of two parts viz. the radiation that is transmitted directly and the radiation that is first absorbed and subsequently transmitted (as longwave radiation) to the living space. Usually one-third of the total absorbed radiation gets into the room.

+Slope in degrees.

Various types of adjustable shading devices have different efficiency, which is measured in terms of the percentage of heat gain through unshaded ordinary glass. The results of a study by Givoni (1976) are given in Table 2.17 and one can draw the following conclusions from this table:

(i) External arrangement of shading is much more effective than the internal arrangement.

(ii) If the shading is of darker colour, then the difference between external and internal shading increases.

(iii) With efficient shading, such as external shutters, it is possible to eliminate more than 90% of the heating effect of solar radiation.

Fixed shading devices, on various windows have to be properly sized before the construction of the building. Pioneer work in the design of sun shading devices was done by Olgyay and Olgyay (1957). According to them, there are in general two basic types of fixed shading devices viz. (i) vertical shading element called fin and (ii) horizontal shading element called overhangs.

A shadow angle protractor (Fig. 2.19) is the tool that can be used to operate shade or exposure. From the point of view of shading, one has to define horizontal and vertical shadow throws. A pin of unit length fixed normally on a wall, in general, casts an inclined shadow. The horizontal component of this shadow through the foot of the pin is called the horizontal shadow throw. The vertical component is, similarly, called the vertical shadow throw. Whereas the vertical shadow throw can only be downward, the horizontal shadow throw can either be to the right or to the left defined with respect to an observer facing the wall, the horizontal shadows are therefore identified by the suffix R or L.

SHADOW ANGLE PROTRACTOR

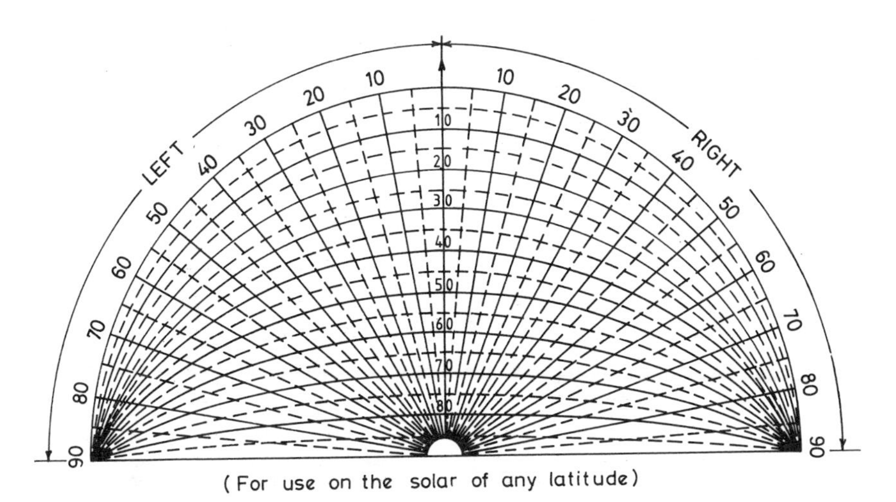

Fig. 2.19. Place centre to centre with protractor on the side to which the wall faces and its base line along the wall direction. Read the horizontal shadow angle along the radial line and the vertical shadow angle along the curve passing through the sun's position at the line.

TABLE 2.17 Shading Factors of Various Glass-shading Combinations (per cent of Heat gain through Unshaded Ordinary Glass) (After Givoni, 1976)

	Data						
	Computed*	Measured**	Measured**	Computed*	Measured**	Measured**	Measured**
	Combination of glass and						
Shading absorptivity	Internal shading slats at 45°	Internal shading slats at 45°	Internal shading slats at 45°	External shading slats at 45°	External shading slats at 45°	Roller shade	Cloth curtain
0.2	40.3	40	-	12.8	-	-	White
0.4	51	51.0	White cream 56	10.2	10	White cream 41.0	-
0.6	62.0	61	Average colour 65	8.05	-	Average colour 62.0	-
0.8	-	71	Dark colour 75	-	-	Dark colour 81.0	Dark colour 64.0
1.0	83	Black 80	-	5.0	-	-	-

* = Computations based on conditions in Israel, on July 21 at 2 p.m.

** = Measured at the ASHRAE Research Laboratory, Cleveland, USA.

The angle subtended by the horizontal shadow throw at the top of pin is the horizontal shadow angle as read by the protractor. Similarly, the angle subtended by the vertical shadow throw at the top of the pin is the vertical shadow angle. For a pin of unit length, the relation between the shadow throws and the corresponding shadow angles can therefore be expressed by the equations (Fig. 2.20).

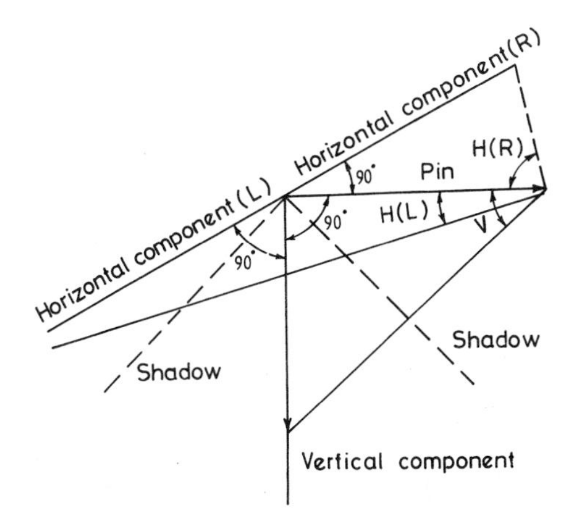

Fig. 2.20. Position of pin, wall, vertical and horizontal shadow angles

$$\tan H_a = h_t$$

$$\tan V_a = v_t,$$

where H_a and V_a are the horizontal and vertical shadow angles, and h_t and v_t are the corresponding shadow throws. The use and scope of solar charts and shadow throws is in the design of external shading devices to control the solar radiation coming through various windows. Though a complex computer model has been developed by Shaviv (1975) for the design of fixed external shades but below we illustrate a simpler method to size the fixed overhangs.

2.4.1 Fixed External Overhangs

Roof overhangs and horizontal or vertical projections over the windows can considerably reduce solar heat gain through windows. The method of calculating the desired projection is explained in the example below.

Example 2.7 It is required to cover a window of height 2.5 m against direct sunlight by a horizontal overhang fixed 150 mm above the upper edge of the window. The window panes are 300 mm behind the outer surface of the wall. The wall faces north-west at a place on latitude 29°N. Full coverage is to be obtained on May 16, at 2.30 p.m.

The shadow protractor can be used to find the vertical and horizontal shadow throws. The protractor* is placed centre to centre on the solar chart of the latitude 29°N, with its base line in the direction of the wall, south-west - north-east, and its central line in the direction to which the wall faces i.e.

---

*Transport protractor is provided at the back of the book.

north-west. Shadow throws on walls of various orientations, on this latitude, are read as

Height of the window	= 2.5 m
Level of the overhang above the top edge of the window	= 0.15 m
Total vertical height to be covered	= 2.65 m
Vertical shadow angle at 2.30 p.m. on May 16 for wall facing north-west	= 65°
Vertical shadow throw = tan 65	= 2.14
Required projection of the overhang	$= \frac{2.65}{2.14} = 1.24$ m

The outer surface of the wall is 0.30 m in front of the window panes. Hence the effective required projection of the louver in front of the wall is 1.24-0.30 = 0.94 $\approx$ 1.0 m approximately. If the width of the window is 1.00 m, what should be the extension of the horizontal louver to cover the window completely?

The horizontal shadow angle as read by the protractor is $H_a = 50^{o}L$ yielding h_t = tan 50 = 1.19L as the value for the horizontal shadow throw. For an overhang projecting 1.24 m in front of the window panes, the horizontal shadow throw is 1.24 x 1.19 = 1.476 m $\approx$ 1.5 m to the left.

The shadow cast by the overhang of the same length as the width of the window is a parallelogram as shown in the Fig. 2.21 representing a vertical section containing the window panes. In this case even though the shadow extends to the lower edge of the window, a triangular strip on the right side is left uncovered. To cover this also, the overhang should extend at least 1.5 m to the right beyond the width of the window. Alternatively a vertical fin of the same width at the right can be used in conjunction with the horizontal louvre.

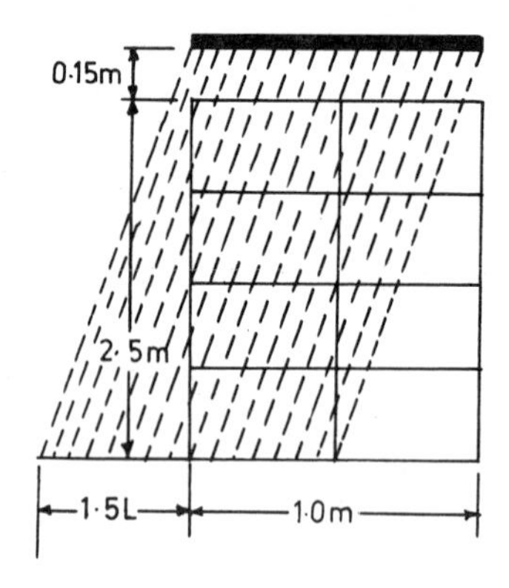

Fig. 2.21.

2.4.2 Inclined Overhangs

It is possible to obtain additional shading coverage from a overhang by inclining

TABLE 2.18 Shadow-throws for Inclined Louvres

Angle of inclination	Shadow angle*															
	5°	10°	15°	20°	25°	30°	35°	40°	45°	50°	55°	60°	65°	70°	75°	80°
0°	.09	.18	.27	.36	.47	.58	.70	.84	1.0	1.19	1.43	1.73	2.14	2.57	3.73	5.67
5°	.18	.26	.36	.45	.56	.67	.78	.92	1.08	1.27	1.51	1.81	2.22	2.83	3.80	5.73
10°	.26	.35	.44	.53	.64	.75	.86	1.00	1.16	1.35	1.58	1.88	2.28	2.88	2.85	5.76
15°	.35	.43	.52	.61	.71	.82	.94	1.07	1.23	1.41	1.64	1.93	2.33	2.92	3.86	5.73
20°	.43	.51	.60	.68	.78	.89	1.00	1.13	1.28	1.46	1.69	1.97	2.35	2.93	3.85	5.67
25°	.51	.59	.67	.75	.85	.95	1.06	1.18	1.33	1.50	1.72	1.99	2.36	2.92	3.80	5.56
30°	.58	.66	.73	.81	.91	1.00	1.11	1.22	1.37	1.53	1.74	2.00	2.35	2.88	3.73	5.41
35°	.65	.72	.80	.87	.96	1.05	1.15	1.26	1.39	1.55	1.75	1.99	2.33	2.83	3.63	5.22
40°	.71	.78	.85	.92	1.00	1.09	1.18	1.29	1.41	1.56	1.74	1.97	2.28	2.75	3.50	4.99
45°	.77	.83	.90	.96	1.04	1.12	1.20	1.30	1.41	1.55	1.72	1.93	2.22	2.65	3.34	4.72
50°	.82	.88	.94	1.00	1.07	1.14	1.22	1.31	1.41	1.53	1.69	1.88	2.14	2.53	3.16	4.41
55°	.87	.92	.97	1.03	1.09	1.15	1.22	1.30	1.39	1.50	1.64	1.81	2.05	2.40	2.96	4.07
60°	.91	.96	1.00	1.05	1.10	1.16	1.22	1.29	1.37	1.46	1.58	1.73	1.94	2.24	2.73	3.70
65°	.94	.98	1.02	1.06	1.11	1.15	1.20	1.26	1.33	1.41	1.51	1.64	1.81	2.07	2.48	3.30

*Vertical shadow angle for horizontal overhangs
Horizontal shadow angle for vertical fins

it towards the shadow side. The lengths of the altered shadow coverage from overhang inclined at different angles from the normal is given in Table 2.18.

Example 2.8 It is required to find the vertical coverage for a horizontal louvre of width 1 m when it is fixed

(i) normal to the wall

(ii) inclined downwards through an angle of 30°

for a wall facing east at 29°N lat. The coverage (at 10 a.m.) is to be calculated for April 16.

Vertical shadow angle = 60°

The vertical shadow throw v_t = 1.73 m

(i) *Normal overhang:* The coverage is 1.73 x 1 m = 1.73 m

(ii) *Inclined overhang:* If the overhang is inclined say at an angle of 30°, then the coverage provided by it can be read from Table 2.18, which provided a value of 2.00 m instead of 1.73 m.

Givoni and Hoffman (1964) have examined the efficiency of the following shading devices at a latitude of 32°N (in Israel).

(i) Horizontal shading extending only above the window (H)

(ii) Horizontal shading extending along the whole facade.

(iii) Vertical shading perpendicular to the wall on both sides of the window extending only up to its top.

(iv) Vertical shading as above but extending throughout the whole height of the building.

(v) A frame of perpendicular vertical and horizontal members.

(vi) A frame whose vertical members are oblique at 45° towards the south.

A summary of the investigations for various orientations is as follows:

East and West Orientations. Adequate shading for east and west orientations can be provided by an egg-crate shading, especially if the vertical components are inclined at 45° to the south. Horizontal shading is more effective than vertical shading. Vertical shading with even infinite height provides very poor shading in summer, while cutting almost all radiation in winter.

South, South-East and South-West Orientations. While a frame shaped shading is most effective, horizontal shading is found to be more effective for this orientation also.

REFERENCES

Arumi, F. N. (1977). Finite Air Mixing Rates in Buildings and its Impact on the Energy Requirements for Heating and Cooling, *Energy and Buildings, 1*, 175-182.

ASHRAE (1982). *Applications Handbook*, Chapter 58, Solar Energy Utilisation for Heating and Cooling, *ASHRAE*, NY.

ASTM (1973). Standard solar constant and air mass zero solar spectral irradiance tables, ASTM standard, E490-73a, *Annual Book of ASTM standards*, Part 41, Philadelphia, PA.

Ballantyne, E. R. (1967). Solar Tables and Diagrams for Building Designers, Proc. CIE Conference on Sunlight in Buildings, Rotterdam, Vol. *1*, No. 20, 251-64.
Benrod, F. and Rock, J. E. (1934). A time analysis of sunshine, *Trans. Am. Illum. Eng. Soc.*, *39*, 200-218.
Brooks, C. E. P. (1951). *Climate in Everyday Life*, Philosophical Library, New York, pp. 270-272.
Coffari, E. (1977). The sun and the celestial vault, *Solar Energy Engineering*, ed. A. A. M. Saygigh, Ch. 2, Academic Press, NY.
Cooper, P. I. (1969). The absorption of Solar Radiation in Solar Stills, *Solar Energy*, *12* (3), 333-346.
Drummond, A. J. and Thekaekara, M. P. (1973) editors, Extraterrestrial Solar Spectrum, Institute of Environmental Science, Mount Prospect, Illinois.
Dryden, I. G. C. (1975). *The efficient use of energy*, IPC Science and Technology Press, Department of Energy, UK.
Erbs, D. G., S. A. Kelin and J. A. Duffie (1982). Estimation of the diffuse radiation fraction for hourly, daily and monthly averaged global radiation, *Solar Energy*, *28* (4), 293-302.
Farber, E. A. and C. A. Morrison (1977). Clear Day Design Values, ASHRAE GRP 170, *ASHRAE*, NY.
Fröhlich, C. and R. W. Brusa (1981a). Solar radiation and its variation in time, *Sol. Phy.*, *74*, 209-215.
Fröhlich, C. and C. Wehrli (1981b). Spectral distribution of solar irradiance from 25000 nm to 250 nm, World Radiation Centre, Davos, Switzerland, Private Communication.
Givoni, B. and M. E. Hoffman (1964). Effectiveness of Shading Devices, Research Report, Building Research Station, Technion, Israel Institute of Technology, Haifa.
Givoni, B. (1976). *Man, Climate and Architecture*, Elsevier, Amsterdam.
Iqbal, M. (1983). *An Introduction to Solar Radiation*, Academic Press, NY.
Johnson, F. S. (1954). The Solar Constant, *J. Meteorol.*, *11* (6), 431-439.
Kasten, F. (1966). A new table and approximate formula for relative optical air mass, *Arch. Meteorol. Geophys. Bioklimatel Ser. B. 14*, 206-223.
Kondratyev, K. Y. (1969). *Radiation in the Atmosphere*, Academic Press, NY.
Leckner, B. (1978). The spectral distribution of solar radiation at the earth's surface elements of a model, *Sol. Energy*, *20* (2), 143-150.
Lim, B. I. *et al.* (1980). Environmental Factors in the Design of Building Fenestration.
Miller, A. (1961). *Climatology*, Methuen, London.
Moon, P. (1940). Proposed Standard Solar Radiation Curves for Engineering Use, *J. Franklin Inst.*, *230*, 583-617.
NASA (1968). The solar constant and the solar spectrum measured from a research aircraft at 38000 feet NASA, Goddard Space Flight Centre, Rep. X -322 -68 -304.
Neubauer, L. W. (1968). Effect of size, shape, colour and orientation on temperature characteristics of model buildings, Paper No. 68-413, American Society of Agricultural Engineers, Utah State University, Logan, Utah.
O'Callaghan, P. W. (1978). *Buildings for energy conservation*, Pergamon Press, Oxford.
Olgyay, V. and Olgyay, A. (1957). *Solar Control and Shading Devices*, Princeton University Press, NJ.
Parmelea, C. (1954). Irradiation of vertical and horizontal surfaces by diffuse solar radiation from cloudless skies, Trans. ASHRAE, *60*, 341.
Perrin de Brichambant, Chr. (1975). "Cahiers A.F.E.D.E.S.", supplement on no. 1. Editions Europeennes Thermique et Industrie, Paris.
Sepsy *et al.* (1978). Fuel utilisation in residences, Final project report No. 137, Electric Power Research Institute, Palo Alto, CA.
Shaviv, E. (1975). A method for the design of fixed external sunshades, *Build International 8*, 121.
Spencer, J. W. (1965). Estimation of solar radiation in Australian localities on clear days, Division of Building Research, Tech. Paper No. 15, CSIRO Australia.

Spencer, J. W. (1971). Fourier Series Representation of the Position of the Sun, *Search 2* (5), 172.
Stephenson, D. G. (1967). Table of solar altitude and azimuth, intensity and solar heat gain tables, Tech. Paper No. 243, National Research Council of Canada, Ottawa.
Threlkeld, J. L. and Jordan, R. C. (1957). Direct solar radiation available on clear days, *Heat Pipe Air Cond., 29* (12), 135.

Chapter 3

BUILDING CLUSTERS AND SOLAR EXPOSURES*

So far as a single building is concerned, it is now known that a building with its longer axis oriented east-west gets most sun in winter and least sun in summer (Koenigsberger, 1975). It is also the general belief that compact building forms have better thermal performance. Markus and Morris (1980) have shown that a low surface to volume ratio results in lower heat losses from buildings. Sahu (1982) states that air-conditioned buildings in hot climates also give similar findings. O'Cathain (1981-1982) examined the implications of density and overshadowing for passive solar housing from the point of view of providing minimum shading. The study (O'Cathain, 1981) concluded that narrower frontages and single storey housing suffer less from overshadowing. Gupta (1984) studied the problem of shadowing in a cluster of buildings for non-air-conditioned buildings in hot climates. In non-air-conditioned buildings, the internal temperature is not constant but swings in response to external conditions and this makes a difference.

3.1 BUILDING FORMS

Martin and March (1972) have carried out a general anlaysis of building forms from the point of view of land utilization. They have classified buildings into three basic types, i.e. Pavilions, Streets and Courts (Fig. 3.1). "Pavilions" are isolated buildings, single or in clusters, surrounded by large open spaces. For the sake of simplicity, it is assumed that such buildings are dquare in plan. The "Street" is long building blocks arranged in parallel rows, separated by actual Streets or just open spaces. "Courts" are defined as open spaces surrounded by buildings on all sides. While Pavilions and Streets are simple enough to visualize in terms of real buildings, it is a bit difficult to see how a "Court" form, as described by Martin and March (1966) becomes a group of buildings. But if the court is seen not as a "cross" but as a square building incorporating a courtyard and surrounded by Streets or open spaces all around, it becomes a real building. Each of these three building types (Fig. 3.2) are defined by a number of variables.

---

*This chapter is based on the doctoral work of Dr V. Gupta submitted at the Indian Institute of Technology, Delhi.

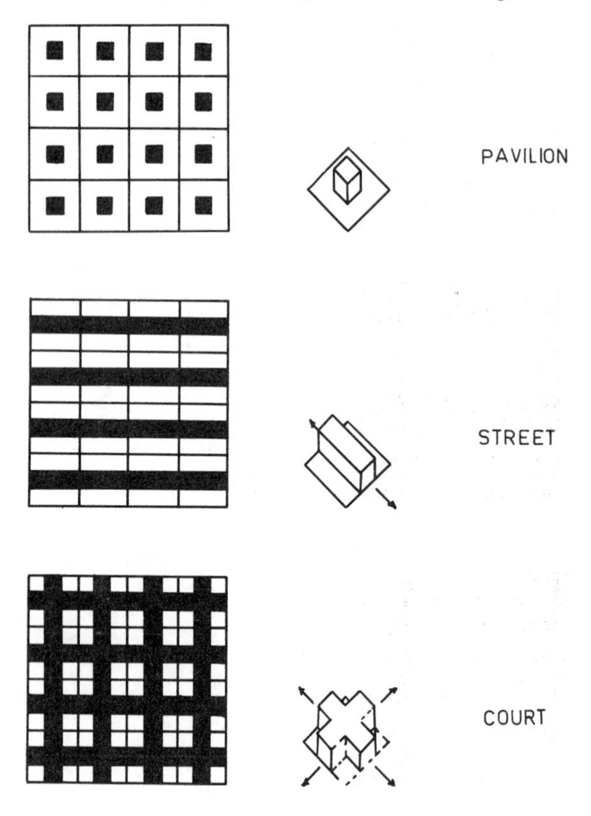

Fig. 3.1 Three different dispositions of built forms

Type		
Pavilion	-	Length, height
Street	-	Length, width, height
Court	-	Length, width, height

When considering groups of buildings (as opposed to single buildings) two other variables need to be defined. These are, the number of blocks in the cluster, and the width of the Street or the open space between the buildings. The arrangement of blocks in plan also needs to be defined. Gupta (1984) studied the configurations shown in Fig. 3.2, where the number of pavilion and Court blocks is N^2 and the number of Street blocks is N. Thus, N becomes the number of subdivisions or modules in one direction. Rather than defining the Street by its width, it is possible to express it as a fraction H_w equal to the ratio of the building height to the Street width. This has the advantage that the percentage of overshadowing of one building by another which is a function of the obstruction angle (Fig. 3.3) remains constant for different building heights with same H_w. In real situations also the distance between buildings is usually related to the height in order to obtain sufficient daylight and ventilation indoors. The different physical properties of the three building types expressed in terms of the basic variables are given in Table 3.1.

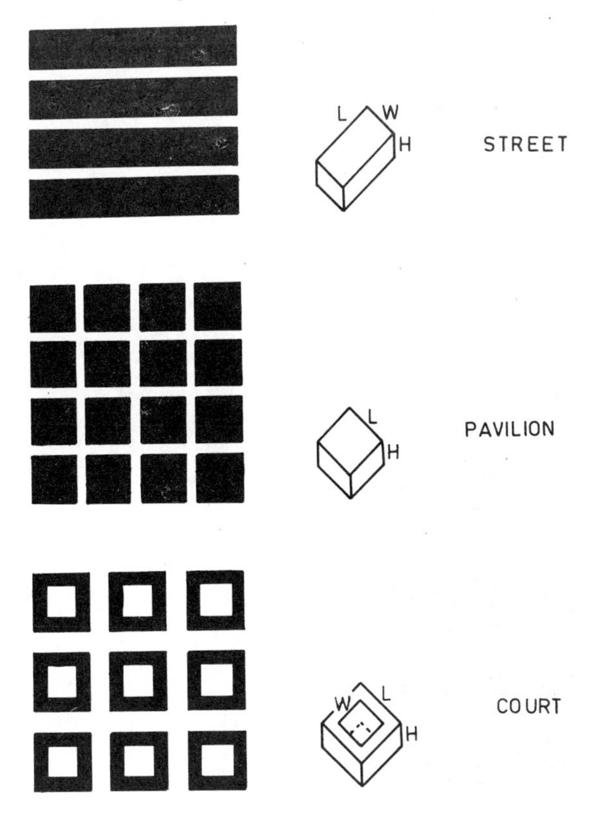

Fig. 3.2. Modified Court, Pavilion and Street forms

TABLE 3.1

Type	Variables	Volume	Surface area	Roof area
Pavilion	L,H,N,H_W	$L^2.H.N^2$	$N^2\ L(L+4H)$	$L^2.N^2$
Street	L,W,H,N,H_W	L.W.H.N	N.W.L+2H(L+W)	N.L.W.
Court	L,W,H,N,H_W	$4.H.W.N^2(L+W)$	$4N^2W(L+W)+2H(L-W)$	$4W.N^2(L+W)$

3.2 SURFACE AREAS

To compare the performance of the three building types with respect to the variables, it is necessary to assume a reference building volume and floor area and to generate the possible combinations of the variables. The reference volume needs to be sufficiently large to give various building forms for larger values of N and H. In this study, this reference volume is assumed as 800,000 m^3. Tables 3.2, 3.3 and 3.4 show the range of possible building configurations with this volume, for Pavilions, Streets and Courts respectively. Figure 3.4 shows the variations of surface areas of buildings with changes in N and H. The roof

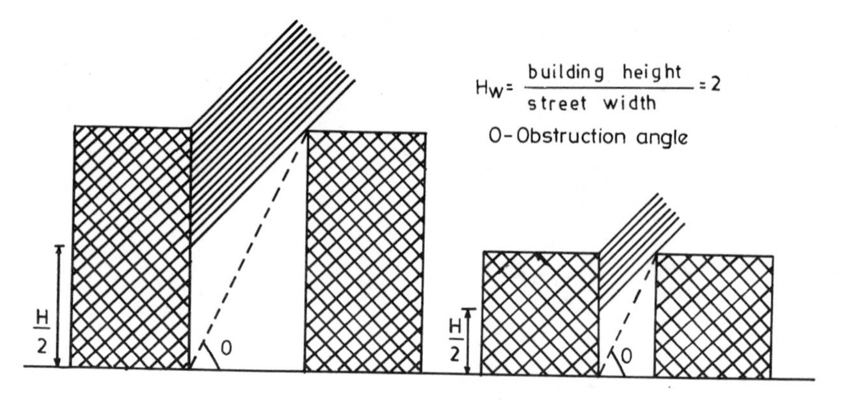

Fig. 3.3. Shading of building as a function of the Street width

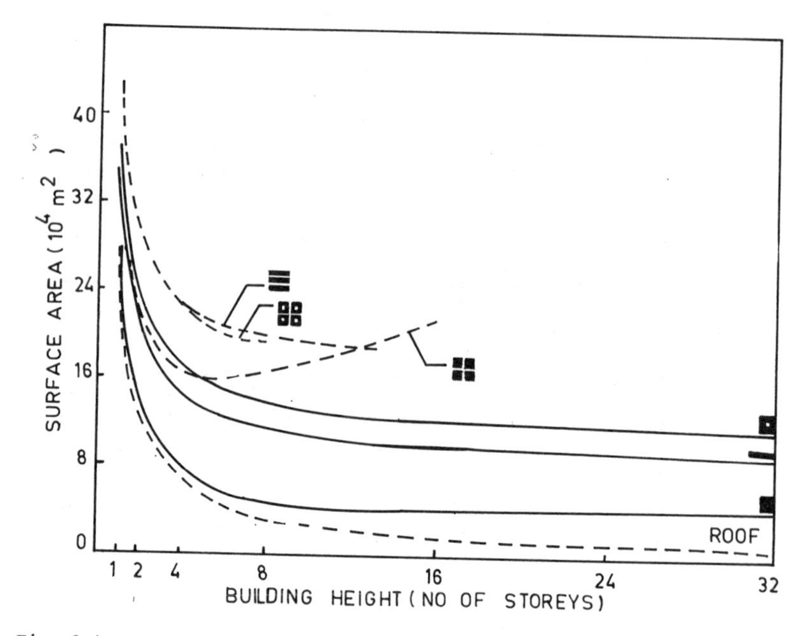

Fig. 3.4. Total surface area of a building, with a volume of 800,000 m^3, with respect to building height, for single and multiple building blocks of Pavilion, Court and Street type building forms

area of a building cluster (with a given volume) which is independent of the plan form and depends upon the building height only is shown by the chain line. The solid lines show the surface area for a single block while the broken lines show the areas for building clusters. It can be seen that the least surface area is obtained with single Pavilions and this area is not much larger than the roof area. For all three types of forms the surface area decreases sharply as height of the building is increased from one to four storeys after which, increase in height does not change the surface area very much. The situation is very different for larger values of N, and it can be seen that the surface area of the

TABLE 3.2 Pavilions

N	H	L	N	H	L
1	3.000	516.398	32	3.000	16.137
1	6.000	365.148	32	6.000	11.411
1	12.000	258.199	32	12.000	8.069
1	21.000	182.574	36	3.000	14.344
1	48.000	129.099	36	6.000	10.143
1	96.000	91.287	36	12.000	7.172
4	3.000	129.099	40	3.000	12.910
4	6.000	91.287	40	6.000	9.129
4	12.000	64.550	40	12.000	6.455
4	24.000	45.644	44	3.000	11.736
4	48.000	32.275	44	6.000	8.299
8	3.000	64.550	44	12.000	5.868
8	6.000	45.644	48	3.000	10.758
8	12.000	32.275	48	6.000	7.607
8	24.000	22.822	48	12.000	5.379
8	48.000	16.137	52	3.000	9.931
12	3.000	43.033	52	6.000	7.022
12	6.000	30.429	52	12.000	4.965
12	12.000	21.517	56	3.000	9.221
12	24.000	15.215	56	6.000	6.521
16	3.000	32.275	56	12.000	4.611
16	6.000	22.822	60	3.000	8.607
16	12.000	16.139	60	6.000	6.086
16	24.000	11.411	60	12.000	4.303
20	3.000	25.820	64	3.000	8.069
20	6.000	18.257	64	6.000	5.705
20	12.000	12.910	64	12.000	4.034
20	24.000	9.129	68	3.000	7.594
24	3.000	21.519	68	6.000	5.370
24	6.000	15.215	68	12.000	3.797
24	12.000	10.758	72	3.000	7.172
24	24.000	7.607	72	6.000	5.072
28	3.000	18.443	72	12.000	3.586
28	6.000	13.041	76	3.000	6.795
28	12.000	9.221	76	6.000	4.805
28	24.000	6.521	76	12.000	3.397

TABLE 3.3 Courts

N	H	L	W	N	H	L	W
1.000	3.000	6676.666	10.000	32.000	3.000	18.021	5.000
4.000	3.000	426.667	10.000	36.000	3.000	15.288	5.000
8.000	3.000	114.167	10.000	40.000	3.000	13.333	5.000
12.000	3.000	56.296	10.000	44.000	3.000	11.887	5.000
16.000	3.000	36.042	10.000	48.000	3.000	10.787	5.000
20.000	3.000	26.667	10.000	1.000	6.000	6671.666	5.000
24.000	3.000	21.574	10.000	4.000	6.000	421.667	5.000
1.000	6.000	3343.333	10.000	8.000	6.000	109.167	5.000
4.000	6.000	218.333	10.000	12.000	6.000	51.296	5.000
8.000	6.000	62.083	10.000	16.000	6.000	31.042	5.000
12.000	6.000	33.148	10.000	20.000	6.000	21.667	5.000
16.000	6.000	23.021	10.000	24.000	6.000	16.574	5.000
1.000	12.000	1676.667	10.000	28.000	6.000	13.503	5.000
4.000	12.000	114.167	10.000	32.000	6.000	11.510	5.000
8.000	12.000	36.042	10.000	1.000	12.000	3338.333	5.000
1.000	24.000	843.333	10.000	4.000	12.000	213.333	5.000
4.000	24.000	62.083	10.000	8.000	12.000	57.083	5.000
1.000	48.000	426.667	10.000	12.000	12.000	28.148	5.000
4.000	48.000	36.042	10.000	16.000	12.000	18.021	5.000
1.000	96.000	218.333	10.000	20.000	12.000	13.333	5.000
1.000	192.000	114.167	10.000	1.000	24.000	1671.667	5.000
1.000	3.000	1338.333	5.000	4.000	24.000	109.167	5.000
4.000	3.000	838.333	5.000	8.000	24.000	31.042	5.000
8.000	3.000	213.333	5.000	12.000	24.000	16.574	5.000
12.000	3.000	97.593	5.000	1.000	48.000	838.333	5.000
16.000	3.000	57.083	5.000	4.000	48.000	57.083	5.000
20.000	3.000	38.333	5.000	1.000	96.000	421.667	5.000
24.000	3.000	28.148	5.000	1.000	192.000	213.333	5.000
28.000	3.000	22.007	5.000	1.000	3.000	4459.444	15.000

Pavilion forms increases with increasing building height beyond four storeys. For buildings taller than four storeys, the surface area of the Street and Court does not change much with changes in H. The simplest way of describing this form is to think of it as a series of "Courts" forming an overall layout of "Streets" (Fig. 3.5). The wall surface area of this type of built form is less than that of "Courts".

The solar radiation interception properties of these building forms can now be studied in order to determine their relative efficiencies.

To obtain an estimate of the total incident radiation for a building cluster, it is therefore necessary to determine angle i, the angle of incidence, for each of

TABLE 3.4 Streets

N	H	L	W	N	H	L	W
1	3.00	26666.67	10.00	28	12.00	238.10	10.00
1	6.00	13333.33	10.00	28	24.00	119.05	10.00
1	12.00	6666.67	10.00	28	48.00	59.52	10.00
1	24.00	3333.33	10.00	32	3.00	833.33	10.00
1	48.00	1666.67	10.00	32	6.00	416.67	10.00
1	96.00	833.33	10.00	32	12.00	208.33	10.00
4	3.00	6666.67	10.00	32	24.00	104.17	10.00
4	6.00	3333.33	10.00	32	48.00	52.08	10.00
4	12.00	1666.67	10.00	36	3.00	740.74	10.00
4	24.00	833.33	10.00	36	6.00	370.37	10.00
4	48.00	416.67	10.00	36	12.00	185.19	10.00
4	96.00	208.33	10.00	36	24.00	92.59	10.00
8	3.00	3333.33	10.00	36	48.00	46.30	10.00
8	6.00	1666.67	10.00	40	3.00	666.67	10.00
8	12.00	833.33	10.00	40	6.00	333.33	10.00
8	24.00	416.67	10.00	40	12.00	166.67	10.00
8	48.00	208.33	10.00	40	24.00	83.33	10.00
8	96.00	104.17	10.00	40	48.00	41.67	10.00
12	3.00	2222.22	10.00	44	3.00	606.06	10.00
12	6.00	1111.11	10.00	44	6.00	303.03	10.00
12	12.00	555.56	10.00	44	12.00	151.52	10.00
12	24.00	277.78	10.00	44	24.00	75.76	10.00
12	48.00	138.89	10.00	48	3.00	555.56	10.00
12	96.00	69.44	10.00	48	6.00	277.78	10.00
16	3.00	1666.67	10.00	48	12.00	138.89	10.00
16	6.00	833.33	10.00	48	24.00	69.44	10.00
16	12.00	416.67	10.00	52	3.00	512.82	10.00
16	24.00	208.33	10.00	52	6.00	256.41	10.00
16	48.00	104.17	10.00	52	12.00	128.21	10.00
16	96.00	52.08	10.00	52	24.00	64.10	10.00
20	3.00	1333.33	10.00	56	3.00	476.19	10.00
20	6.00	666.67	10.00	56	6.00	238.10	10.00
20	12.00	333.33	10.00	56	12.00	119.05	10.00
20	24.00	166.67	10.00	56	24.00	59.52	10.00
20	48.00	83.33	10.00	60	3.00	444.44	10.00
20	96.00	41.67	10.00	60	6.00	222.22	10.00
24	3.00	1111.11	10.00	60	12.00	111.11	10.00
24	6.00	555.56	10.00	60	24.00	55.56	10.00

Table 3.4 (cont'd)

N	H	L	W	N	H	L	W
24	12.00	277.78	10.00	64	3.00	416.67	10.00
24	24.00	138.89	10.00	64	6.00	208.33	10.00
24	48.00	69.44	10.00	64	12.00	104.17	10.00
28	3.00	852.38	10.00	64	24.00	52.08	10.00
28	6.00	476.19	10.00	68	3.00	392.16	10.00

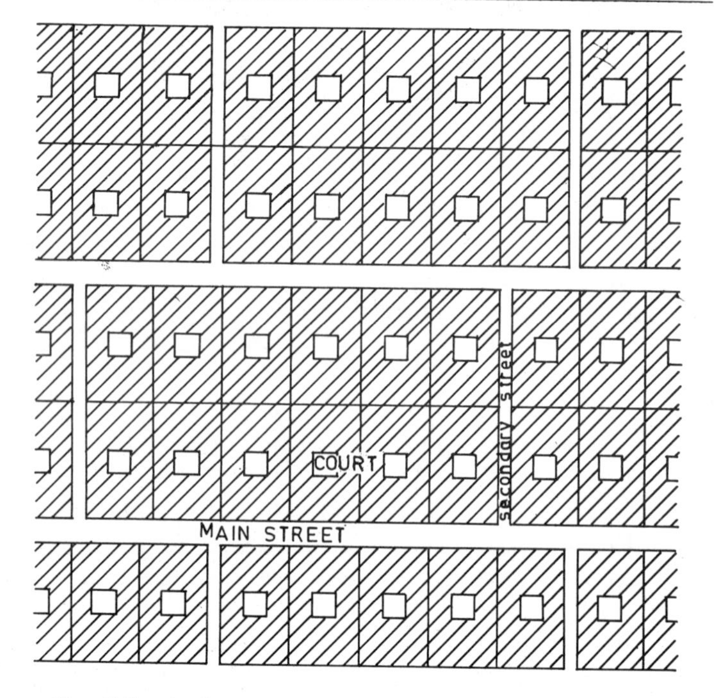

Fig. 3.5. A simple plan of taller buildings arranged in a cluster

the vertical surfaces and to calculate the area on which beam radiation is incident. The incident radiation and the angle of incidence can be calculated by the expressions given in Chapter 2.

3.3 MUTUAL SHADING OF BUILDINGS

The thermal properties of a cluster of buildings can be determined by analysing a single building in the cluster, but this is not possible with simple solar radiation analysis because of the shading of one building by another. Figure 3.6 shows a typical cluster of Courts with the shading pattern for a particular combination of solar azimuth and altitude. The exact boundary of the shaded part of each building changes continuously from sunrise to sunet. The normal

Fig. 3.6. Shade and light pattern of a cluster of Courts

graphic methods of shadow projection used by architects are too tedious to be used for this analysis. Knowles (1974) has used an elegant photographic technique for shading analysis of very complex forms. A scale model of the building is constructed and put on a helidon. The model is then rotated and photographed in such a way that the camera sees the model as the sun would see it. The total area of the building in the photograph, then represents the projected area of sun exposed parts of the building (Fig. 3.7). The advantage of this technique is that very complex forms can be easily handled, because the angle of incidence of solar radiation on individual surfaces need not be calculated separately. The projected area obtained from the photograph is automatically corrected by the factor cos i for vertical surfaces and by sin i for horizontal surfaces. This technique is most useful for a detailed study of a complex form but it becomes difficult to use for studying a large number of simpler forms.

Another technique that is becoming increasingly popular (Arumi, 1979) involves the use of a computer with graphic display and print-out facilities. The computer produces images of the building (as seen by sun) after suitable rotations along two axes. The projected area of the building form can then be calculated from the print-out or in the case of interactive systems, it can be calculated by the computer itself.

Another technique developed by Gupta (1984) involves the use of a micro-computer without graphic facilities. Essentially it consists of rotating the mathematical image of the building in the same way as done by Knowles (1974) and to obtain modified images from which the solar exposed projected area can be calculated by geometrical analysis. With a suitable programme, the computer can directly give this area for hourly intervals and daily totals of the same. For the Pavilion and Court building forms, the geometrical analysis can be greatly simplified and the shading pattern of large clusters predicted by the shading pattern of a

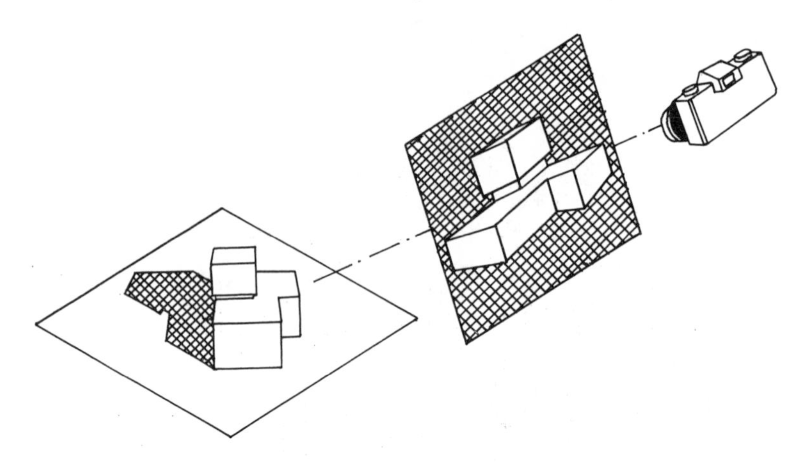

Fig. 3.7. Photographic method of calculating the sunlit area of a complex building form

2 x 2 cluster of blocks. Figure 3.8 shows a generalized solar view of a 2 x 2 cluster of cube blocks. Block A is shaded by blocks B, C and D. Block B and C are shaded only by block D on the right and left respectively. Block D is not shaded at all. Blocks A, B, C and D can represent all the blocks in a larger cluster. By further analysis of a figure it can be easily seen that the exposed area of block A is given by the parallelogram 1, 2, 3, 4 and the shaded area of blocks B and C is given by the parallelograms, 5, 6, 7, 8 and 9, 10, 11, 12. The total solar exposed area of each block can be easily calculated. It can be seen that the shaded area of block A depends upon the street width (W) and height (H) of the blocks. For a given H_W (= H/W) ratio, the percentage of shaded wall area remains constant for any size of block. Thus the shading of walls (there is no shading of roofs with blocks of equal height) depends only upon the dimensions of the Street and is independent of the length and width of the blocks.

During the course of a day other different situations can also arise but in each case the same analysis can give the solar exposed area. A slight inaccuracy

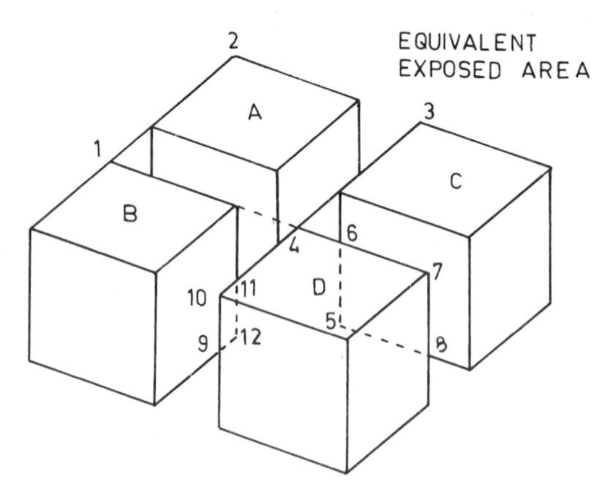

Fig. 3.8. Solar view of a 2 x 2 cube cluster

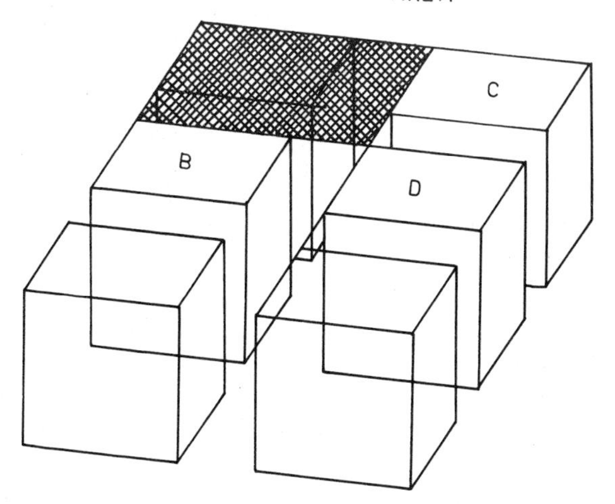

Fig. 3.9. Solar view of a 2 x 2 cube cluster for a low solar altitude

arises in the case of Fig. 3.9, where the area 1, 2, 3, 4 is somewhat less than the solar exposed area of block A. But this is not a serious error as the situation arises only with low solar altitudes when the "Solar exposure" (defined later) is small. In the case of courtyards a deduction is necessary for the floor area of the courtyard defined by the parallelogram 1, 2, 3, 4 in Fig. 3.10 which can be easily computed.

For the Street type of blocks, no separate analysis is necessary because there are only two possible situations (i) when one side is shaded and (ii) when no shading takes place (Fig. 3.11).

With this analysis it is possible to develop a computer programme that can give the solar exposed area for any of the three building types, for any given latitude, solar declination and time of the day.

3.4 COMPUTER PROGRAMME

A computer programme to evaluate the solar exposure of various building forms

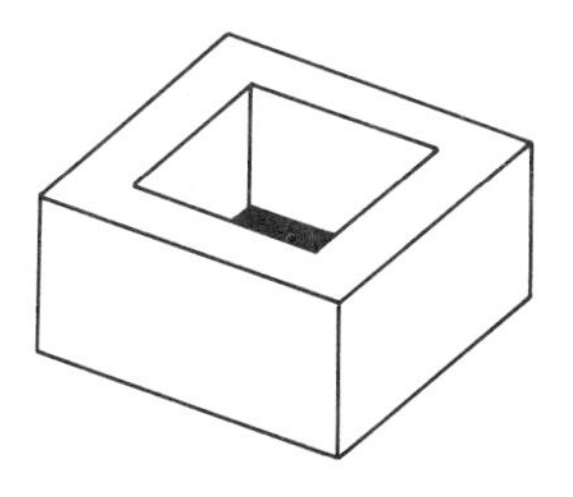

Fig. 3.10. Solar view of a Court

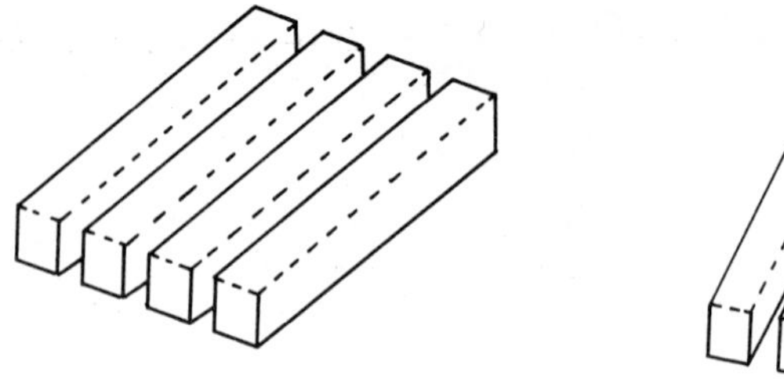

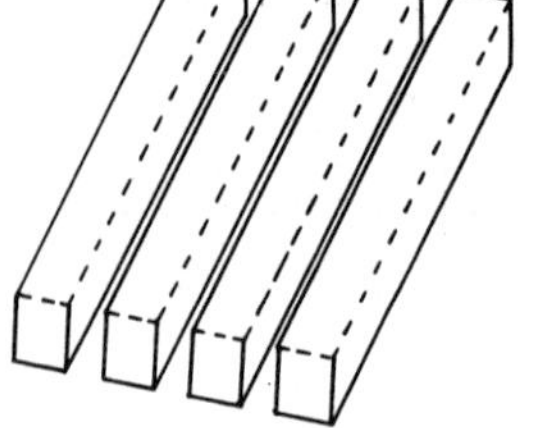

Fig. 3.11. Solar view of Street.

requires the following steps:

(i) The building form is first defined as a matrix [A] of 3 x N dimensions where N is the number of points that define the building evelope (8 for a cuboid and 16 for court form). The elements in the matrix correspond to X, Y and Z co-ordinates of the envelope. In cases where the number of blocks is more than one [A] defines only one single block.

(ii) Solar altitude and azimuth is calculated for each mid-hour from sunrise to sunset.

(iii) The building matrix [A] is rotated along X and Y axis, the angle of rotation corresponding to the solar azimuth and altitude being A_Z and A_α respectively. The rotation matrix [R] for rotation along one axis (Rooney, 1977) is

$$R = \begin{bmatrix} \cos A_Z & -\sin A_Z & 0 \\ \sin A_Z & \cos A_Z & 0 \\ 0 & 0 & 1 \end{bmatrix} \qquad (3.1)$$

and for rotation along two axes the matrix becomes,

$$R = \begin{bmatrix} \cos A_Z & 0 & -\sin A_Z \\ \sin A_Z \sin A_\alpha & \cos A_\alpha & \cos A_Z \sin A_\alpha \\ \sin A_Z \cos A_\alpha & -\sin A_\alpha & \cos A_Z \cos A_\alpha \end{bmatrix} \qquad (3.2)$$

The resulting matrix [B] is given by [B] = [R] x [A].

(iv) The X and Y co-ordinates of each element of [B] then describe the building as it would be seen by the sun. A new matrix [C] is now described by taking elements of [B] for each possible position after rotation (refer Fig. 3.12). For these eight positions, two generalized diagrams are adequate (Fig. 3.13).

Thus the new matrix [C] has a dimension 2 x 6 for "Pavilion" and "Street" forms and 2 x 12 for "Court". The elements of [C] are selected taking into consideration that either the single element or the two equal elements of [B] have the maximum value of Y co-ordinate.

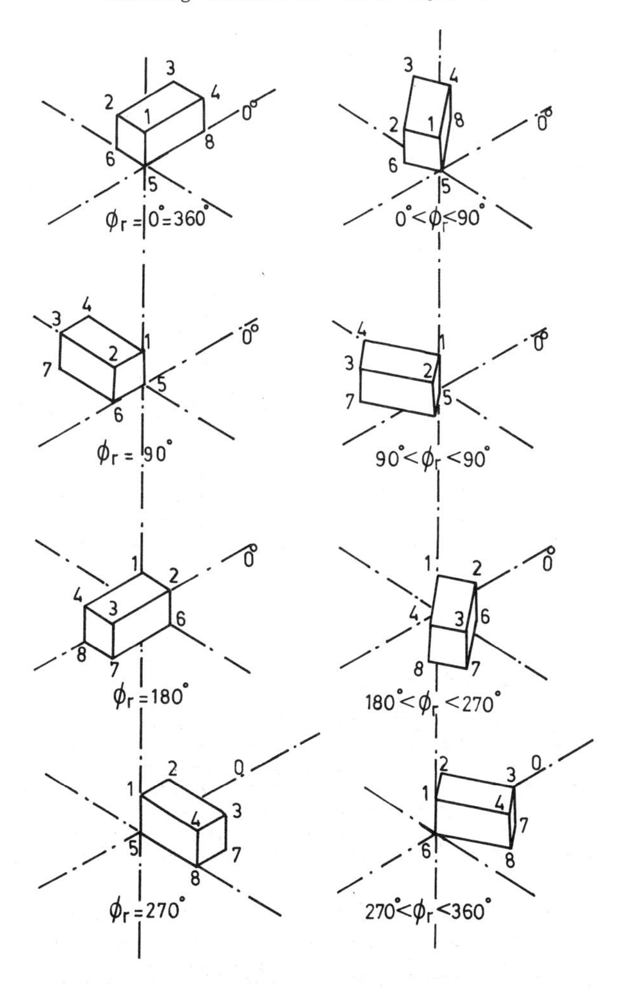

Fig. 3.12.

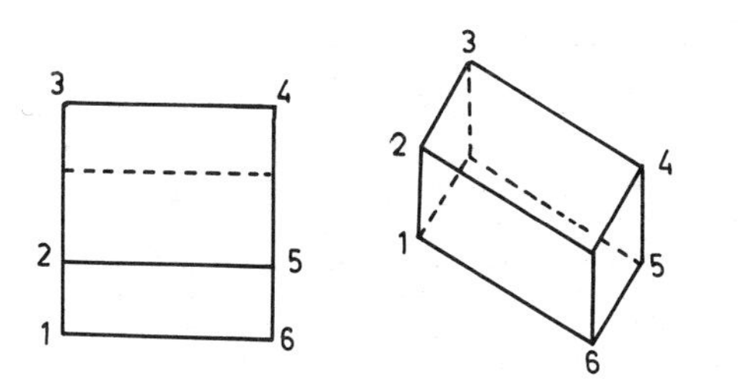

Fig. 3.13.

(v) The area of [C] represents the area of a single building block seen by the sun and it is easily calculated. A deduction is made for the visible area of the courtyard floor (Fig. 3.14) which is not part of the building.

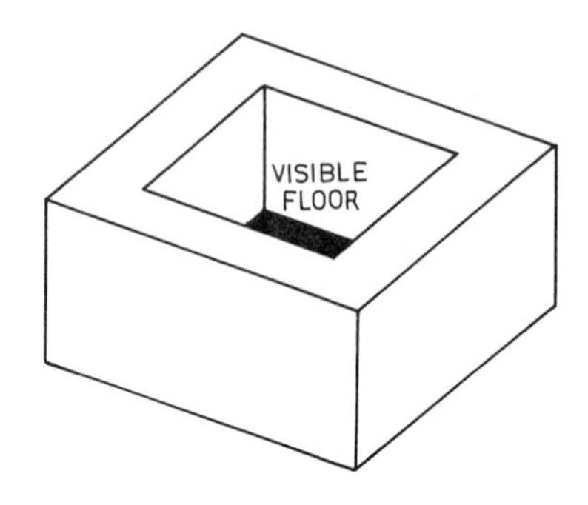

Fig. 3.14.

(vi) The other building blocks can now be described by translation of [C] to new positions [D], [E] and [F] (Fig. 3.15).

(vii) The area of [C] seen by sun is calculated from the area of parallelogram D(3), C(3), E(3), F(3). The area of [E] and [D] seen by sun is calculated by making deductions for the shaded parallelograms.

(viii) There are certain positions of [C] when there is no shading or when only one side of [C] is shaded. The condition for shading is found out as follows:

Let us examine the limiting case when the shadow of E(2) falls at the base of block [C] on the line C(5), C(6) (Fig. 3.16). From geometry, one can write

$$\psi = 180^{o} - A_{ZW} \tag{3.3}$$

$$\text{and, } \tan A_{\alpha} = H/[P, E(1)] \tag{3.4}$$

$$= \frac{H}{W/\sin\psi} = \frac{H \sin A_{ZW}}{W}$$

$$\text{Therefore } \frac{H}{W} = \frac{\tan A_{\alpha}}{\sin A_{ZW}} \tag{3.5}$$

and for shading to take place, H/W must be greater than $\tan A_{\alpha}/\sin A_{ZW}$.

(ix) The total area of the cluster can be determined from the visible areas of [C], [D], [E] and [F], from which the direct solar exposure can be calculated.

For a given combination of variables, i.e. the building cluster dimensions, latitude, declination and hour angle, the incident solar flux may be defined as

$$S_D = S_c \times A_{SD} \tag{3.6}$$

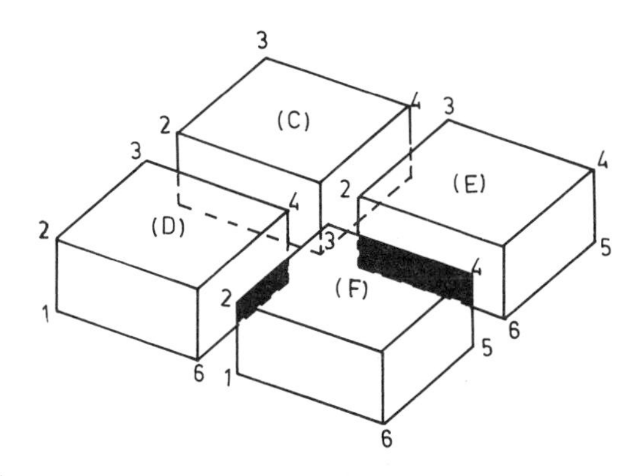

Fig. 3.15.

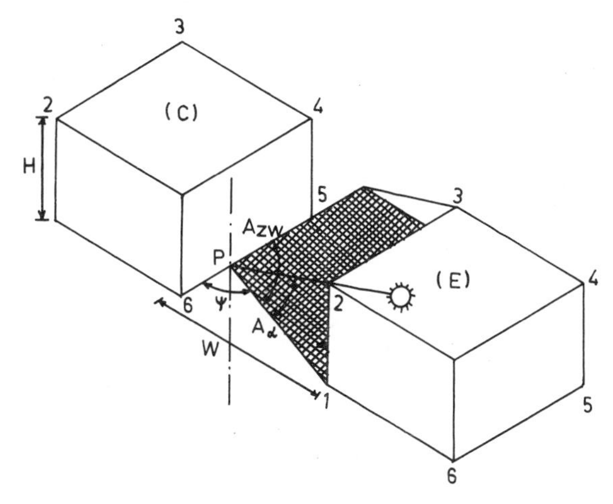

Fig. 3.16.

where A_{SD}, the effective solar exposed area for beam radiation may be calculated from

$$A_{SD} = A_S \times S_{DN}/S_c \ , \tag{3.7}$$

and the direct normal radiation on earth's surface, S_{DN}, may be estimated by the expression (Rao and Seshadari, 1961)

$$S_{DN}/S_c = 0.921/(1 + 0.3135 \text{ cosec } A_\alpha) \tag{3.8}$$

As for the direct radiation above, one can define the total diffuse radiation incident on the building

$$S_d = S_c \times A_{sd} \tag{3.9}$$

The total solar exposure is then defined by the summation of S_D and S_d i.e.

$$S_E = S_D + S_d \qquad (3.10)$$

and this has been named as the solar exposure and it is used to compare the solar properties of various building forms.

The solar exposure for a cube on a winter and summer day is shown in Fig. 3.17 for latitude 29°N.

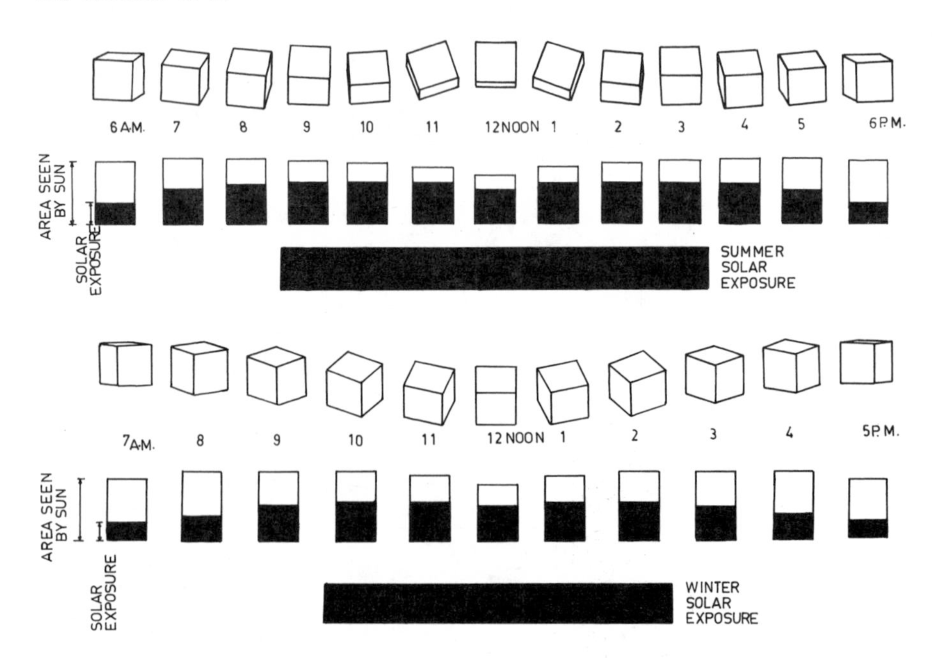

Fig. 3.17. Hourly solar views, projected area and solar exposure and total daily solar exposure for a cube for summer and winter (for 29°N latitude.)

3.5 EFFICIENCY OF BUILDING FORM

For those regions where winter heating is a necessity, the insolation efficiency of a building form can be measured (Knowles, 1974) by comparing the summer solar exposure with the winter solar exposure. But in many areas, winter heating is not a problem and efficiency has to be measured by minimizing the summer solar exposure. As solar exposure depends upon the reference building floor area (and therefore volume), the summer solar exposure on the building form compared to the summer solar exposure of equivalent area on the ground is a measure of efficiency for summer conditions.

This efficiency can be viewed as the comparative solar exposure of a given floor area on the ground and the same floor area when distributed in a certain way in building (Fig. 3.18).

Two kinds of efficiencies can be defined now:

$$E_W = \frac{\text{winter solar exposure}}{\text{summer solar exposure}} \times 100\% \qquad (3.11)$$

$$E_S = (1 - \frac{\text{building solar exposure in summer}}{\text{ground solar exposure in summer}}) \times 100\% \qquad (3.12)$$

E_S is valid for areas with predominant summer conditions. The possible range of E_W extends beyond 100% and that of E_S below zero per cent. The efficiency factors E_W and E_S are not to be confused with the normal quantitative concept of efficiency in which the range is from 0 to 100% only. These are mere qualitative indicators used for comparing the relative performance of buildings.

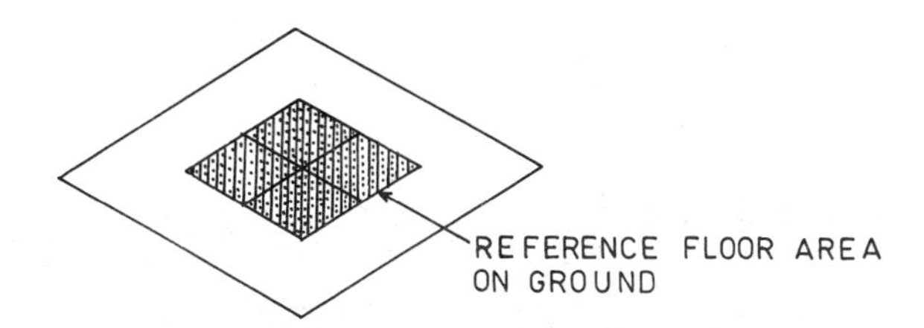

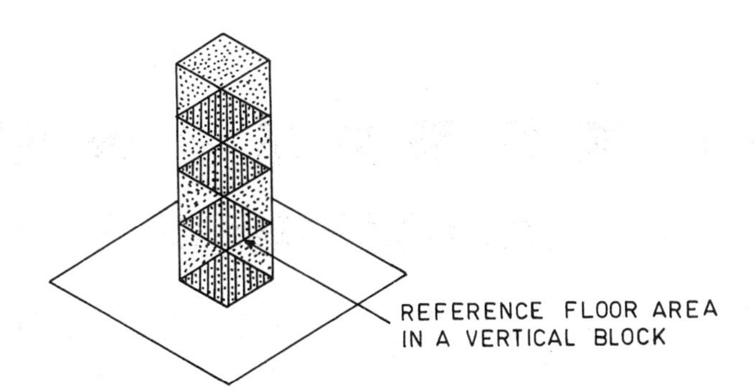

Fig. 3.18. Two different ways of arranging a given floor area: on the ground and in a four-storeyed building. The summer efficiency is given by the ratio of solar exposure for the two configurations

3.6 PERFORMANCE OF THE BUILDING FORMS

It is generally accepted (Evans, 1980; Knowles, 1974) that a high surface to volume ratio renders the building more susceptible to environmental stress resulting in a poorer thermal performance. While this is quite true of buildings in a cold climate, it is not necessarily so for a hot climate.

The change in the surface area for the reference volume (800,000 m^3) with respect to change in building height for the three types of forms was shown in Fig. 3.4. It is seen that for N = 1 the Court and Street forms have a larger surface area than that of the Pavilion, for all building heights. The surface area of all the forms decreases with increase in building height. The situation changes when the reference volume is distributed in a large number of blocks. The surface area of

Pavilions decreases up to four storeys height after which it increases and after twelve storeys it becomes greater than that of Court and Street forms. The four-storey height is thus critical for further study.

The effect of change of building height on solar exposure of single building blocks and the resulting efficiencies is shown in Fig. 3.19. Efficiency E_S (measure of building performance in hot climates) increases with building height for all the three types of forms. It is seen that after eight-storey height, the increase is negligible, while it is maximum up to four-storey height. The maximum efficiency is obtained by the Pavilion form. Efficiency E_W (measure of building

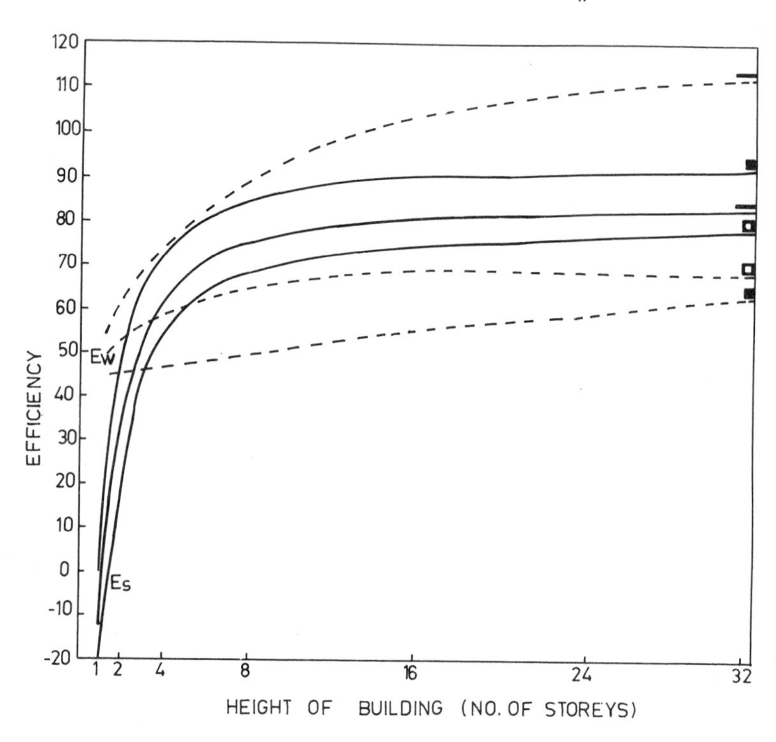

Fig. 3.19. Solar efficiency of single building block with respect to building height

performance in temperate climates) does not increase so dramatically with increasing height but for the "Street" and "Pavilion" the increase is continuous. The most efficient form is the "Street" for which the efficiency is greater than 100% after about twelve storeys. The efficiency of the Court is higher than that of the Pavilion but much less than the Street. In the case of building clusters (Fig. 3.20) there is a marginal difference between E_W for Street and Court for building height up to four storeys; but for taller buildings the Street is definately more efficient than the Court and the Pavilion. The choice of building form for a temperate climate is thus clear. The Street is the most efficient building form for such climates. For warm climates, the situation is somewhat more complicated. The Pavilion, has the highest efficiency up to eleven storeys after which the "Court" becomes more efficient. E_S for the Street always lies between E_S for the Court and Pavilion. A further analysis of these three building forms is necessary to account for changes in Street width and degree of sub-division.

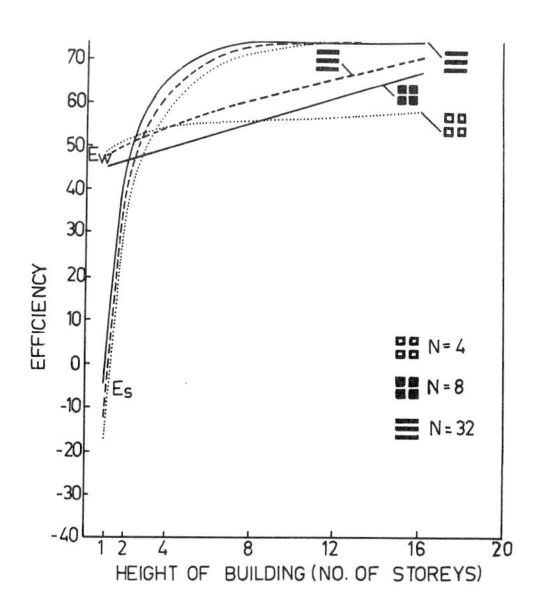

Fig. 3.20. Solar efficiency of building clusters with respect to building height

3.7 STREET WIDTH AND SUB-DIVISION

Figures 3.21, 3.22 and 3.23 show the effect of sub-division of the reference volume and Street width on efficiency for building heights of one, two and four storeys. For Street type buildings (Fig. 3.21), E_s is not affected by either change of Street width or the number of sub-divisions. However, there is a marked decrease in the E_w efficiency with decrease in Street width particularly for four-storey height. For Pavilions (Fig. 3.22), the E_s efficiency is reduced as the number of sub-divisions is increased. This drop in efficiency is partially made up by a narrower width of Street. On the other hand, the E_w efficiency increases with larger number of blocks. The Court (Fig. 3.23) exhibits unique characteristics of E_s increases with greater sub-division of reference volume and with narrower Street widths. For the Court, E_w is reduced as Street width is decreased.

3.8 ORIENTATION

In all the above calculations, the building clusters were oriented to face north-south direction. But it is possible that efficiency of building forms varies with the orientation of the facades. Figures 3.24 and 3.25 show E_w and E_s respectively, in relation to orientation. The conventional orientation of long-axis along east-west direction is represented by the building azimuth of 0° while long axis along south-north direction is taken as 90° building azimuth. The wisdom of the conventional orientation of major facades facing north-south direction is borne out by the curve for a single Street in Fig. 3.24. E_w varies from 47.5% for 90° azimuth to 75% for 0° azimuth. But some of the other forms show completely different properties. The single Pavilion is hardly affected by orientation changes, while the single Court achieves equally high efficiencies at 30° intervals. But it is more fruitful to examine the efficiency of multiple

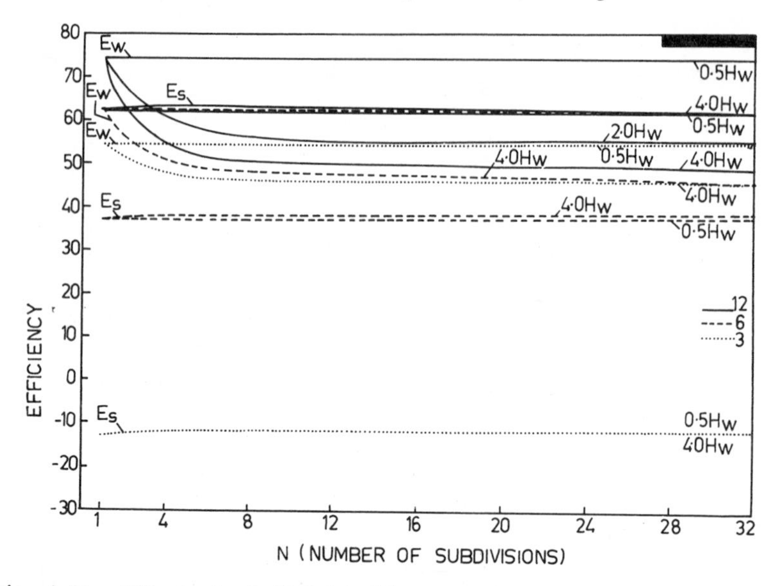

Fig. 3.21. Effect of sub-division (N) on the solar efficiency of Streets. H_w defines the ratio of building height to the Street width. The building height has been taken to be 3, 6 and 12 metres.

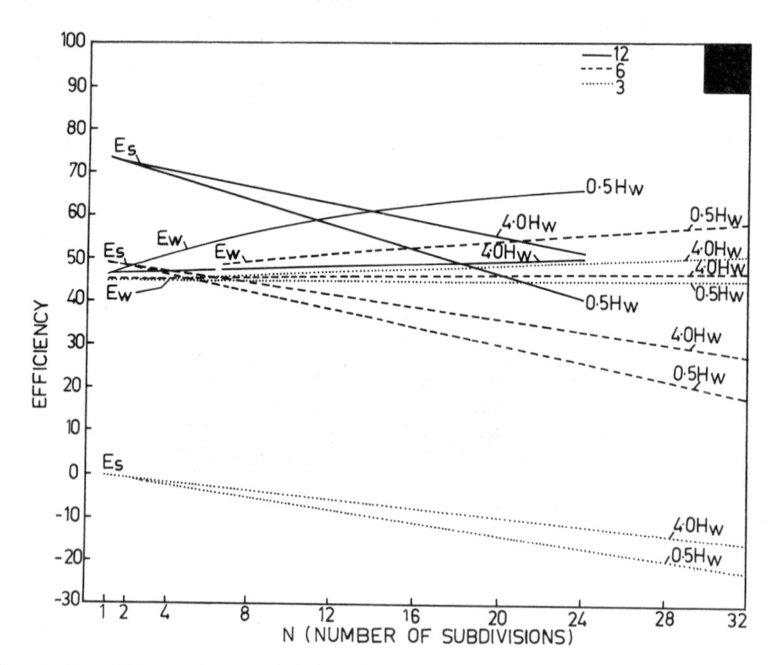

Fig. 3.22. Effect of sub-division (N) on the solar efficiency of Pavilions. H_w is the ratio of building height to Street width. The considered building heights are 3, 6 and 12 metres

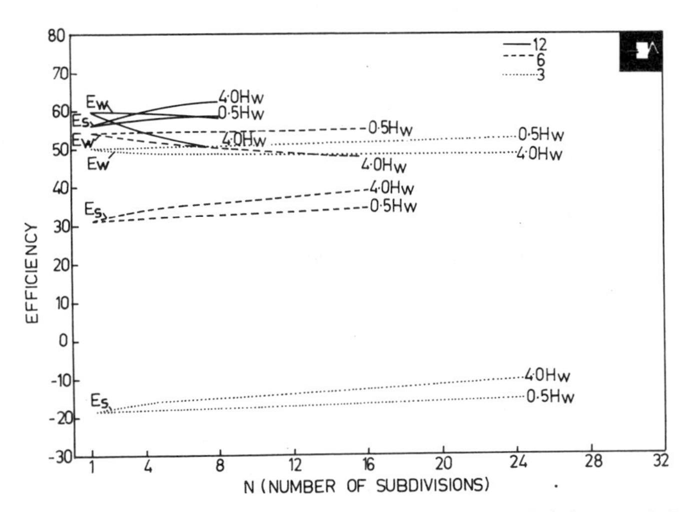

Fig. 3.23. Effect of sub-division (N) on the solar efficiency of Courts. H_w is the ratio of the building height to Street width. The considered building heights are 3, 6 and 12 metres

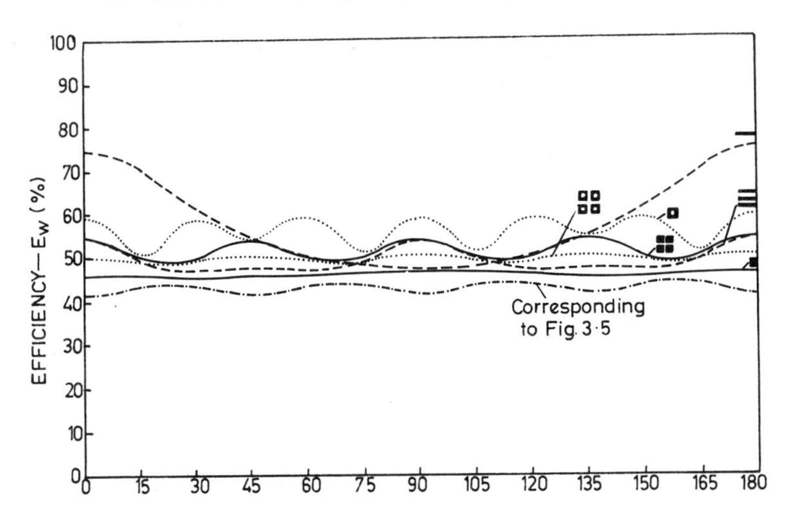

Fig. 3.24. Effect of orientation on the E_W solar efficiency of different building configurations

forms, as a single building volume of 800,000 m^3 is unrealistic. The peak efficiency E_W of multiple Streets is the same as that of multiple Pavilions while the peak efficiency of multiple Courts is lower and the form of Fig. 3.5 has the lowest efficiency. It can be seen that in all cases the efficiency at 0° azimuth is the same as at 90° azimuth.

The relative efficiency of different forms (Fig. 3.25) is totally changed for E_S. The single Pavilion is most efficient while the single Street and Court are less efficient. Amongst the multiple forms, the E_S of the form shown in Figs. 3.5 is

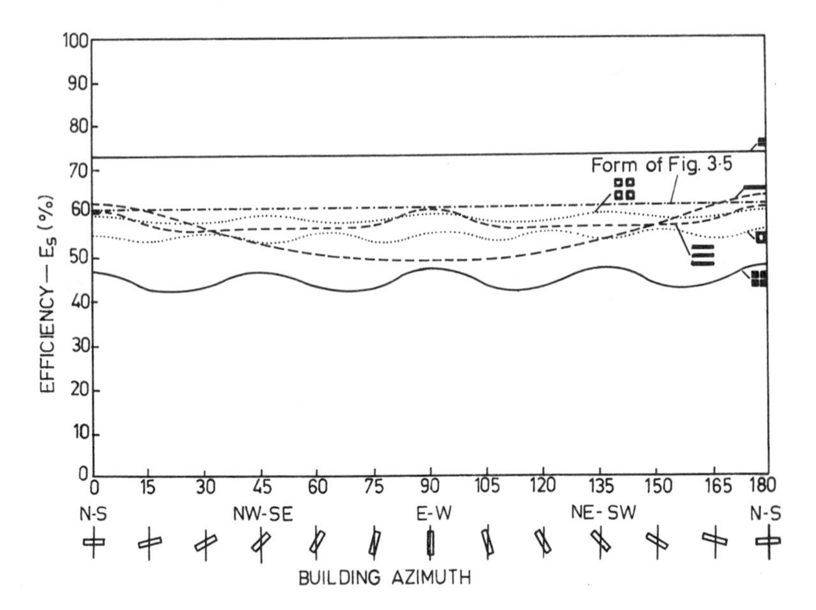

Fig. 3.25. Effect of orientation on the E_S solar efficiency of different building configurations

the highest and it is not affected by orientation. The efficiency of multiple Courts is similar to that of multiple Streets. Surprisingly the E_S of multiple Streets is the same with east-west or north-south orientation. This is because of the mutual shading of buildings which is almost zero at 0° azimuth, is maximum at 90° azimuth. If the street width is decreased (i.e. H_W is increased beyond 2), the efficiency at 90° azimuth will be even higher than with 0° azimuth.

Since the effectiveness of structural shading devices commonly used in buildings is maximum on the north and south facades and much less on east and west facades, it is possible when the relevant shading factors are taken into account, the simple efficiencies given in Figs. 3.24 and 3.25 will change.

3.9 LATITUDE

The efficiency of building forms for different latitudes is shown in Figs. 3.26 and 3.27. For warm climates the efficiency E_S of all forms is nearly independent of latitude and almost identical efficiency is achieved by the Street (single or multiple), the multiple Court and the forms of Fig. 3.5 type. The E_W efficiency (Fig. 3.27) of all forms shows a drop with the increase of latitude and at all latitudes the single Street gives the highest efficiency, followed by the single Court. For multiple forms, E_W is nearly the same for Streets and Pavilions.

3.10 CONCLUSIONS

The following conclusions may be drawn from the foregoing studies:

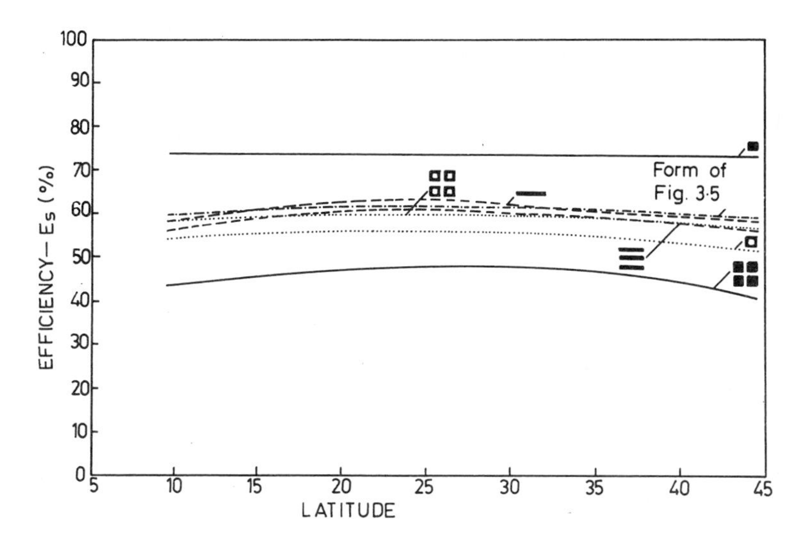

Fig. 3.26. Solar efficiency E_s of building forms at different latitudes

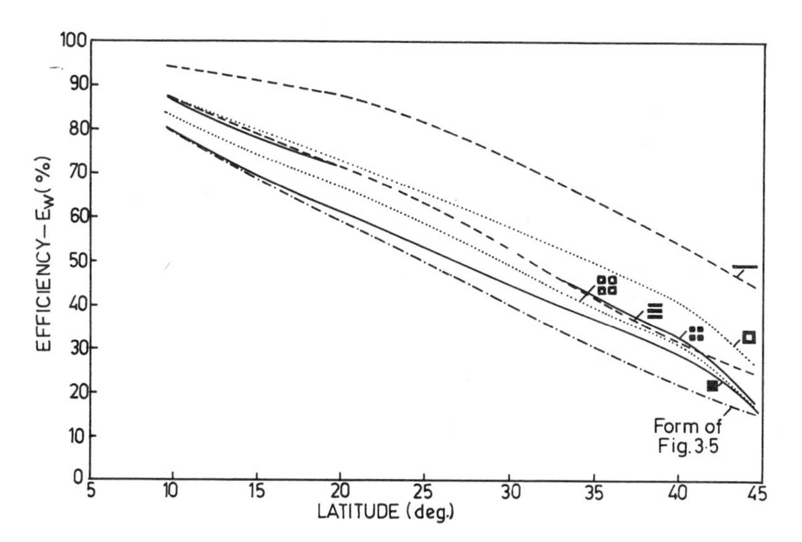

Fig. 3.27. Solar efficiency E_W of building forms at different latitudes

Warm Climates

1. The choice of building form lies between the multiple Court and Street. Pavilion type buildings give rise to low solar efficiency.

2. Orientation of building is important for the single Street type, for which it should be such that the long axis is along the east-west direction. "Courts"

can be oriented in any direction. Multiple Streets can have east-west or north-south orientation with equal efficiency.

The width of Streets running east-west is unimportant from a solar efficiency viewpoint while Streets running north-south should be as narrow as possible.

4. Unless special measures are adopted to shade the roof, buildings should not be less than four storeys high. They need not be higher than eight storeys.

5. For "Courts", a large number of blocks with smaller courtyards are preferable to fewer blocks with large courtyards.

Temperate Climates

1. The ideal form for best E_w efficiency is the Street type.

2. Depending upon the latitude, the distance between buildings should be carefully chosen to eliminate shading in winter. For 29° (north or south) latitude, a Street twice the height of buildings, ensures the condition of no shading. This is infact, applicable for higher latitude stations also.

3. An increase of building height is advantageous up to sixteen storeys after which it does not lead to significant changes in efficiency.

4. The orientation of Streets must be along the east-west direction. Any deviation from this leads to lower efficiency. For Courts, the orientation does not matter.

5. For both Courts and Streets, greater sub-division does not affect solar efficiency, but for Pavilions sub-division is not desirable.

The ultimate parameter of interest in any building form is the air temperature and the magnitude of discomfort or comfort index. This apart from solar exposure depends on various other parameters like (i) light or heavyweight structures (ii) ventilation and (iii) additional shading. In view of this, the thermal performance of various building forms has been discussed later in Chapter 7.

REFERENCES

Arumi, F. N. (1979). *Computer Aided Energy Design for Buildings. Energy Conservation through Building design*. Donald Watson, Ed., McGraw Hill, New York.
Evans, Martin (1980). *Housing, Climate and Comfort*, Architectural Press, London.
Gupta, V. (1984). *A Study of the Natural Cooling Systems of Jaisalmer*, Ph.D. Thesis, Indian Institute of Technology, Delhi, India.
Knowles, R. L. (1974). *Energy and Form: An Ecological Approach to Urban Growth*, M.I.T. Boston.
Koenigsberger, O. H. *et al.* (1975). *Manual of Tropical Building and Housing Orient*, Longman, New Delhi.
Markus, T. A. and Morris, E. N. (1980). *Buildings, Climate and Energy*, Pitman, San Francisco.
Martin, L. and March, L. (1966). *Speculation 4 in Urban Space and Structures* (1972) Cambridge University, Cambridge.
O'Cathain, C. S. (1981). A model for passive solar building density, *The International Journal of Ambient Energy*, *3* (1), 31.
O'Cathain, C. S. (1982). Exploration with a model of passive solar housing, *Energy and Buildings*, *4*, 181-183.
Rao, K. R. and Seshadari, T. N. (1961). Solar Insolation Curves, *Indian J. Met. and Geophysics*, *12* (2), 267-72.

Rooney, J. (1977). *A Survey of Representations of Spatial Rotation about a Fixed Point Environment and Planning B*, Vol. *4*, 185-210, Pion, England.
Sahu, S. (1982). Multistorey Building Envelope and Solar Heat Gains, Asian Regional Conference on Tall Buildings and Urban Habitat, Kuala Lumpur.

Chapter 4

PASSIVE CONCEPTS AND COMPONENTS

4.1 INTRODUCTION

Ancient architecture, all over the world, had many characteristics which led to thermal comfort in buildings. The buildings were shaped and different parts of the building (e.g. indoor spaces, doors, windows etc.) located and oriented to take maximum advantage of the climate, and the role of trees, vegetation and water around the building in determining the thermal comfort was well appreciated. The massive walls and clustered residences (to reduce the surface to volume ratio) for reducing the temperature swings were also commonly employed.

The Greeks appreciated the importance of south* aspect of houses as is evident from the statement of Socrates (400 *B.C.*): "Now in houses with south aspect, the sun's rays penetrate into porticos in winter, but in summer the path of the sun is right over our heads and above the roof (so there is shade). If then this is the best arrangement, we should build the south side loftier to get the winter sun and the north side lower to keep out the cold winds". Ancient Iranian architecture (Bahadori, 1978) exploited the concepts of clustering (decreasing surface to volume ratio, to reduce the thermal load), of thick walls (whose large thermal storage capacity smoothed out the temperature fluctuations), of plantations for shade and living in the basements (during extreme heat and cold). Iranians also introduced the concept of wind towers, which along with cooling by earth and water evaporation made the buildings comfortable in summer.

American Indians are also known to have used (Haskins and Stromberg, 1979) passive solar techniques as early as *A.D.* 1100 in the construction of dwellings in Chaco Canyon and Mesa Verde.

Indigenous ancient Indian forts and domes are good examples of buildings responsive to the demands of sun and climate. As an example, the hot arid climate of Rajasthan is characterized by high daytime temperatures and uncomfortable low night temperatures. The solution best suited to such a wide temperature fluctuation is to delay the entry of heat into the building by such a period that it reaches the interior when it is least bothersome (or even welcome). The inhabitants of this area achieved this desired thermal performance by using materials of high thermal capacity, such as mud and stone which absorbs the heat

---

*north in southern hemisphere.

of the sun during the day and introduce it into the dwelling during night. By clustering their dwellings together, the Rajasthanis achieved maximum volume with minimum surface area exposed to the sun while increasing the mass of the building as a whole, thus increasing the thermal time lag. Indian and other ancient architecture are representative of the intuitive approach to building design for local climatic conditions. These proven concepts have been ignored in the design of the modern buildings which mainly rely on conventional fuels for providing the desired heating or cooling. The ancient architects were, however, handicapped by the nonavailability of glass (or similar material which lets in the solar radiation and kept the cold air out) and were therefore unable to incorporate solar heating without letting the outside air in (presently called the direct gain concept). After the ready availability of glass, it was used extensively in the west. The portions of the house which admitted sunlight through the glass were hot during the sunshine hours and cold otherwise; the cold, to some extent, was countered by having a double window (one of glass and the other of wood which could be closed when there was no sunshine). However, primarily the use of glass has been in the interest of aesthetics and the needed heating and cooling effect was achieved by expenditure of conventional energy. The thoughtless use of glass in hot climates with enormous thermal discomfort or expenditure or energy for air-conditioning is a striking example of such an approach.

Conscious scientific application of solar energy for passive heating may be said to have started in 1881 when Professor E. L. Morse (1881) was granted a patent on a glazed south facing dark wall for keeping the house warm. This idea was applied by Professor Morse only to one room of his house and not followed up by either him or others for a very long time. Morse's concept was repatented by Trombe (1972, 1974) who, starting in 1972 built a series of houses at Odeillo in Pyrenees, France and made an engineering success of the idea. Hollingsworth (1947) had also employed such a wall in the MIT experimental house.

In 1947, under the sponsorship of Libbey-Owens-Ford Glass Co., a remarkable book appeared entitled *Your Solar House* (Simon, 1947). Forty-eight highly regarded architects prepared designs for direct gain solar houses, one for each state then in US. As might be expected, most of the designs featured Thermopane glass, but few if any recognized the importance of building mass as a means of providing storage. Overheating on even very cold, sunny winter days would have been a problem for most of the designs shown in this remarkable volume.

In 1952 the Kech brothers designed a 24 unit solar home development in which they used double-glazing to maintain comfortable conditions despite the biting cold of northern Illinois winters. Overheating and wide temperature swings were problems encountered in these and similar designs; openable windows or ventilating fans were generally required to maintain comfort in winter. Year-round air-conditioning was not contemplated in those days.

Hay and Yellot (1969) introduced the concept of a roof pond to store heat during the days and deliver it to the living space in the night in winter. The same system could be employed in hot weather to cool the building - using convection, radiation and evaporation to cool the water in the night. Moveable insulation is a special feature of the system.

The importance of structures (fully or partly underground) in maintaining thermal comfort had long been recognized. The pioneering work (1978) of Underground Space Centre at University of Minnesota may be mentioned in this connection. Passing air through tunnels, deep in earth provides a source of warm air in winter and cool air in summer.

With the advent of the energy crisis there was a renewed interest in those aspects of architecture which contributed to thermal comfort in a building without (or with minimum) expenditure of energy. This led to the formal recognition of the passive (or natural) heating and cooling of buildings as a distinct science. Since the sun played a dominant role in all such considerations, the science came to be known as the passive solar architecture (or in short, the passive solar). The conference and workshop on passive solar heating and cooling of buildings at Albuquerque in May 1976 marked the recognition of the maturity of the science of passive solar architecture.

There are many passive solar buildings all around the world (Stromberg and Woodall, 1977 and AIA Research Corporation, 1978). Many new buildings, which are being presently planned, bear ample testimony to the success of the passive concept.

4.2 THERMAL ENVIRONMENT WITHIN A BUILDING

The internal environment within a building results from the response of the building fabric to the changing outdoor conditions of air temperature, solar radiation, humidity, precipitation or evaporation, wind velocity and direction and the clearness of the sky. Building elements with their specific orientation, thermal conductivity, absorptivity and emissivity react differently to the outdoor conditions. In typical desert cool climates, solar radiation can be intercepted and trapped into the building for providing heating, while in warm climates the interception, absorption and inward transmission of solar radiation can be reduced to minimize the cooling load demand. A building will be judged by the occupants according to its ability to satisfy the demand of thermal comfort. To determine the state of thermal comfort inside a building space, one has to do a detailed thermal analysis which will be presented in the next couple of chapters. Before that one has to, however, discuss the various concepts which are usually involved in passive building architecture to increase or decrease heat gain into the building.

4.3 PASSIVE HEATING CONCEPTS

Various concepts used for increasing the heat gain due to solar radiation into the building are:

4.3.1 Direct Gain

This is a straightforward concept (Fig. 4.1), in which sunlight is admitted through a window or wall of glass, facing south (to admit maximum solar radiation in winter) to heat the walls, floors and objects (consequently also the air) in the room; double-glazing is used to reduce the loss of heat from the room to outside air and the windows are covered by insulation in the night for

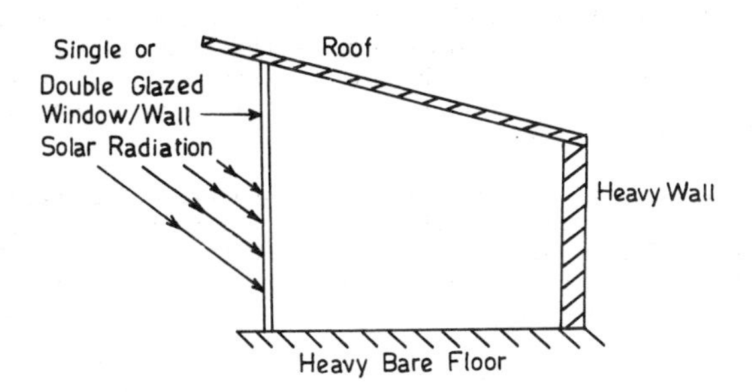

Fig. 4.1. Direct gain

the same purpose. The area of the house thus heated tends to get very hot in the day unless storage mass (in the form of bare massive floor or wall or otherwise) is provided in the room. The oscillations in the temperature of the air are large. These oscillations are reduced by providing a thermal storage media either under the floor or in the north wall. The material used for storage can either be masonry/concrete or water contained in the drum placed under the floor (Singh and Bansal, 1984). A suitable overhang provided on the south wall/window helps to keep out the summer sun to avoid undesired heating.

4.3.2 Indirect Method

In spite of the storage provided in the direct gain concept, the fluctuations in the room temperature are usually higher than tolerated by man for the desired comfort level. A more effective method for reducing the swings in the room temperature is to introduce a thermal storage wall between the direct solar radiation and the living space. The sun's energy in this concept is introduced into the room in an indirect fashion as a result of convection and longwave radiation emitted by the thermal wall which gets heated due to the absorbed energy at its surface. Based on this type of storage wall the following concepts have been forwarded for the indirect gain passive heating.

4.3.2.1 Trombe wall. A massive thermal wall of concrete or masonry usually facing south, suitably blackened and glazed greatly reduces the temperature swings in the room air. This concept was first patented by E. L. Morse in USA in 1881 (Morse, 1881) and later revived and repatented by Trombe (1972, 1974). The south facing glazed massive wall was used for collection, storage and transfer of heat to the inside building. A typical arrangement for employing this concept is illustrated in Fig. 4.2.

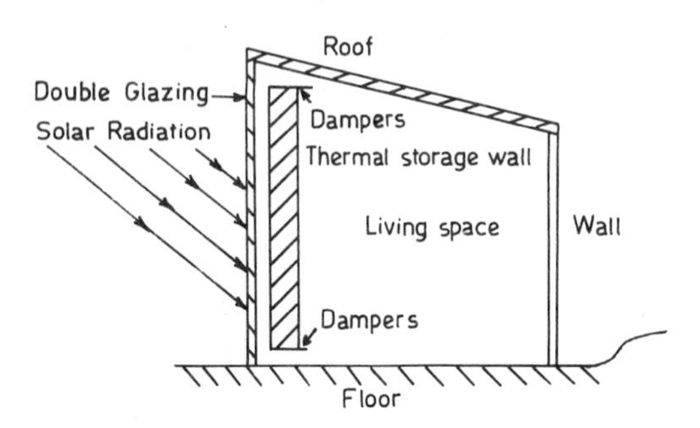

Fig. 4.2. Indirect gains: thermal storage wall

The doubly-glazed dark south wall gets heated up by the sun during the day. The air entering the space between the wall and the glass gets heated and returns to the living space through the vents provided in the massive wall. The heat input into the room can be reduced by adjusting the flow of air through dampers. The glazing over the wall is covered by insulation during off sunshine hours and the dampers closed to reduce heat losses to the ambient. The heat conducted through the wall gets transferred into the room by radiation and convection all the time.

4.3.2.2 Water wall. If the massive wall is made up of drums (Fig. 4.3), full of water stacked over each other, the wall is usually termed as a Water Wall. Water wall is more effective in reducing the temperature swings but in this case, the desired time lag between the maximum/minimum of solar radiation and the maximum/minimum of heat flux into the room is less.

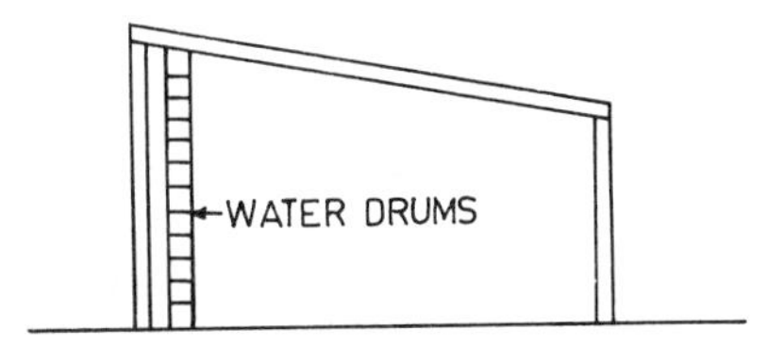

Fig. 4.3. Water wall

For a given area of south wall, the total heat input into the building through a thermal storage wall is much less (about half) than the corresponding amount for a direct gain configuration. However, the heat enters the building in the case of thermal storage wall at a much more uniform rate, reducing the temperature swing.

4.3.2.3 Transwall. The above discussed water wall or Trombe wall concepts completely block the solar radiation from the southern side thus reducing the illumination inside the living space. The concept of a Transwall, proposed by Fuchs and McClelland (1979) helps to achieve the storage effect and simultaneously provides the desired illumination in the living space. Transwall is a partially transparent thermal storage wall (Fig. 4.4) placed adjacent to a window admitting solar energy. Part of the solar energy is absorbed within the transwall and the remaining part is transmitted to the living space.

4.3.3 Sunspace (Attached Solarium/Greenhouse)

This concept proposed by Balcomb (1978) represents a marriage of the concepts of direct gain and indirect gain. The living space has a thermal storage wall on the south side; attached to the south wall is space enclosed by glass. The glass enclosure called sunspace receives heat by direct gain, while the living space receives heat by indirect gain, through the thermal storage wall (Fig. 4.5). This concept provides a pleasant sunspace, which may be used for recreation or growth of plants. Moving insulation over the walls of the sunspace improves the performance considerably.

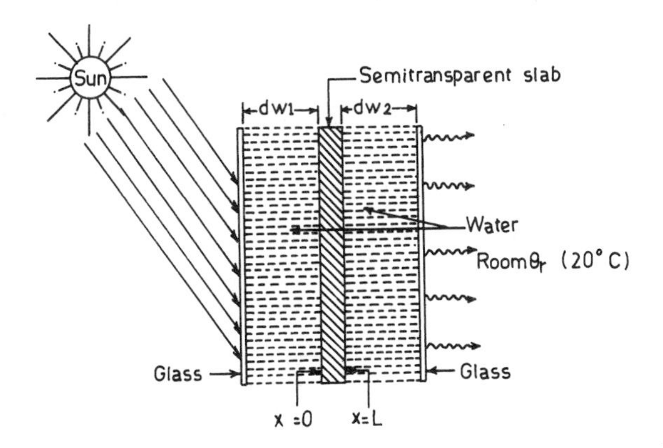

Fig. 4.4. Design of the transwall.

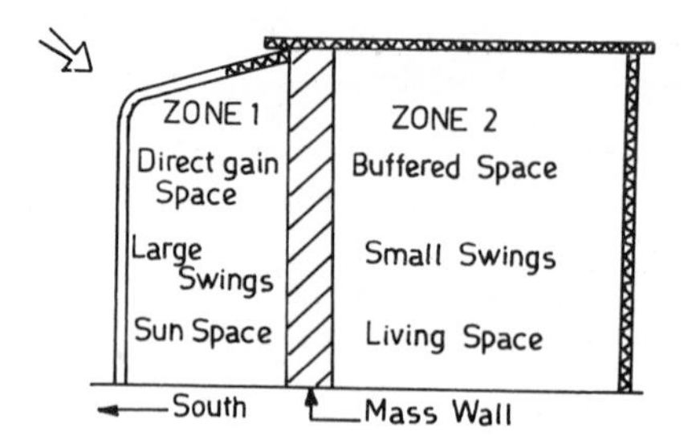

Fig. 4.5. Sunspace: solarium

4.3.4. Isolated Gain

Greater flexibility in design and operation can be obtained by isolating the building, the collector of solar energy and the storage. The most common application of this concept is the natural convective loop, of which the thermosyphoning water heater represents the simplest version. This arrangement is characterized by a flat plate collector, connected to a well insulated tank with insulation wrapped piping (Fig. 4.6). The tank is always located above the collector to induce a convective flow of fluid. The fluid from the tank can be cycled through the house to heat it.

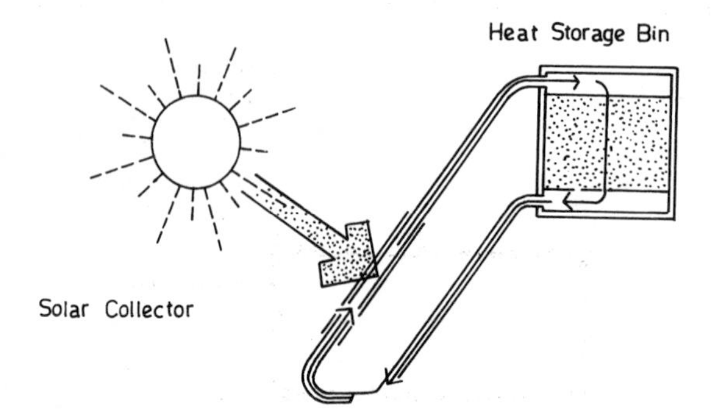

Fig. 4.6. Thermosyphoning water heater

4.4 PASSIVE COOLING CONCEPTS

Natural cooling techniques can reduce the peak cooling power demand of a building thus reducing the size of the air-conditioning equipment and the period for which it is generally required. Passive cooling encompasses a number of natural heat rejection mechanisms including ventilation, evaporation, infrared transfer to the sky, and earth contact cooling.

The first step in the design of any passive cooling system is to reduce the unnecessary thermal loads on buildings. There are usually two types of thermal load in the building, (i) exterior loads due to the climate and (ii) internal loads due to people, appliances, cooking, bathing and lights. Proper zoning of different components and local ventilation of major heat sources can reduce the overall impact of internally generated heat loads.

Internal lighting of building creates an important heat load at the commercial scale, while in the domestic sector lighting is not a major thermal energy source. There have been efforts to determine the benefits of using diffuse daylight to replace electric light fixtures for the purpose of reducing thermal load due to lighting. Depending on weather, the thermal load enters the building in three major ways.

(i) Penetration of direct beam sunlight.

(ii) Conduction of heat through walls/roof etc.

(iii) Infiltration of outside air.

4.4.1 Minimization of the Direct Beam Radiation: Shading

The entry of direct radiation through windows into the room is the source of maximum heat gain inside the building. The entry of direct solar radiation can be controlled through use of vertical, horizontal and inclined louvers, movable screens, deciduous trees and plants. As discussed in Chapter 2, the effectiveness of sunshades is not equal for all orientations of walls and therefore glazed areas should be provided only in those portions where effective protection from the sun can be provided.

Shading against direct radiation is easiest to provide on the south wall. A horizontal projection of appropriate depth will exclude the summer sun (Fig. 4.7), while still permitting sunlight into the building in winter. The east and west walls can be protected by a combination of horizontal and vertical louvers

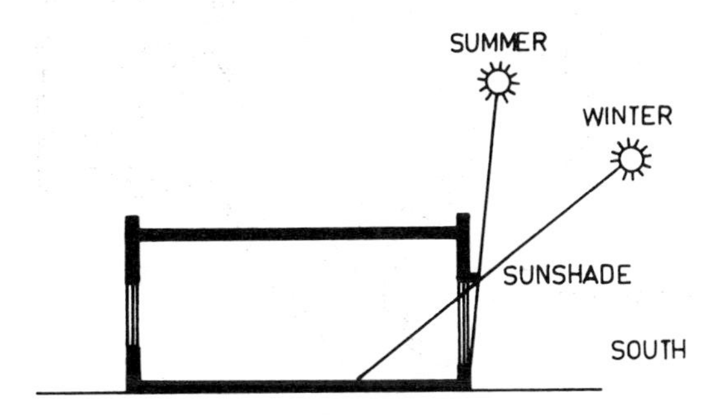

Fig. 4.7. Summer sun protection

4.4.2 Conduction of Heat through Walls/Roofs

The conduction of heat into the building through roof/walls is directly proportional to the temperature difference between the outside surface and the inside surface. In order to minimize the conducted heat, one has to therefore minimize the outside surface temperature. The table 4.1 below shows the relative heat load on surfaces of various orientations in summer and winter respectively. One of the methods to reduce outside surface temperature is, therefore, to minimize the amount of absorbed radiation by the surfaces. Various methods employed to achieve this are briefly described below.

4.4.2.1 Shading. It is evident that to reduce the heat gain into the building as a result of sun radiation, the various building elements should be provided adequate shading to minimize the amount of radiation absorbed by the outside surfaces; the roof in this regard requires maximum attention since it receives the maximum radiation in summer.

TABLE 4.1 (for latitudes 17°N to 31°N, After Seshadri et al., 1969)

	Roof	Walls			
	(%)	North (%)	South (%)	East (%)	West (%)
Summer (June 22)	48-51	6-13	0-2	19-20	19-20
Winter (December 21)	28-34	0	35-44	14-15	14-15

Surface shading can be provided as an integral part of the building element or it can be provided by separate cover. Highly textured walls have a portion of their surfaces in shade (Fig. 4.8). The radiation absorbing area of such a textured surface is less than its radiation emitting area and therefore it will be cooler than a flat surface. The increased surface area will also result in an increased coefficient of convective heat transfer which will permit the building to cool down faster at night when the ambient temperature is lower than the building temperature.

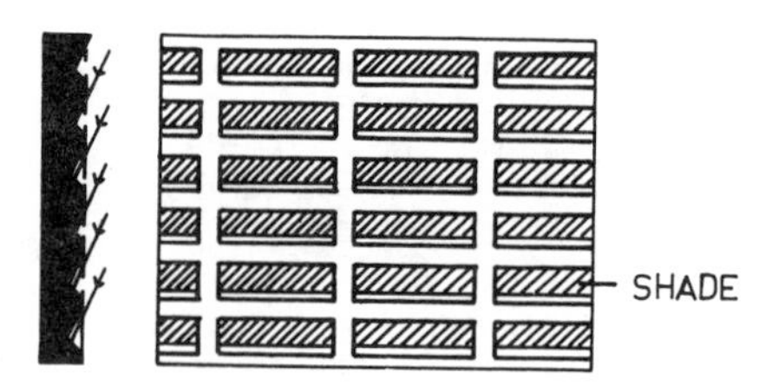

Fig. 4.8. Shading by texture

Shading provided by external means should be such that it does not interfere with the night-time cooling. This is particularly important for roof surfaces which are exposed to the cool night sky (Fig. 4.9). A cover over the roof provided by the solid concrete or galvanized iron sheets provides protection from the direct radiation but it will not permit radiation to the night sky. An alternative method is to provide a cover of deciduous plants or creepers (Fig. 4.10). Because of the evaporation from the leaf surfaces, the temperature of such a cover will be lower than the daytime air temperature and at night it may

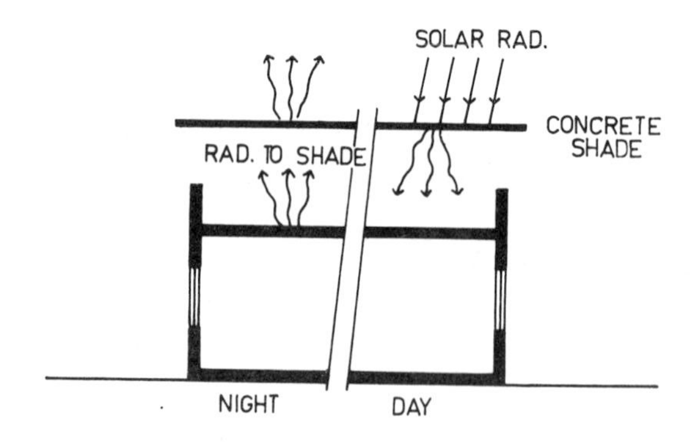

Fig. 4.9. Roof shading

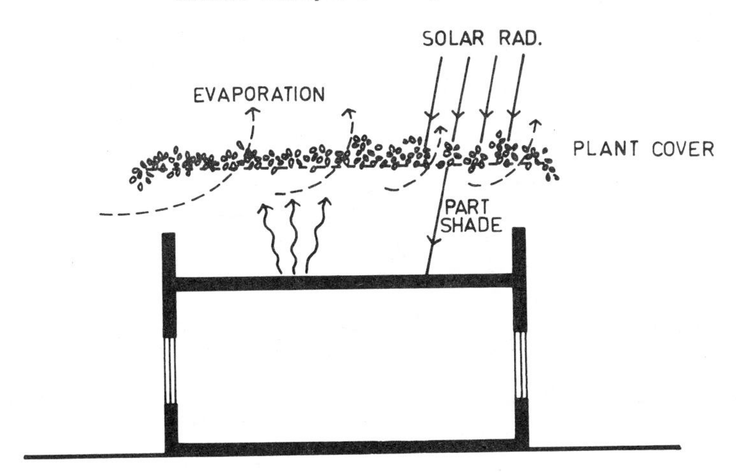

Fig. 4.10. Roof shading by vegetation

be even lower than the sky temperature.

Another shading device used in some traditional buildings is the covering of the entire roof surface area with small closely packed inverted earthen pots (Fig. 4.11). In addition to shading, this arrangement provides increased surface area for radiation emission and insulating cover of still air over the roof which impedes heat flow into the building while still permitting upward heat flow at night. Although the system of earthen pots is thermally efficient, the method suffers from practical difficulties because the roof is rendered unusable and its maintenance is difficult.

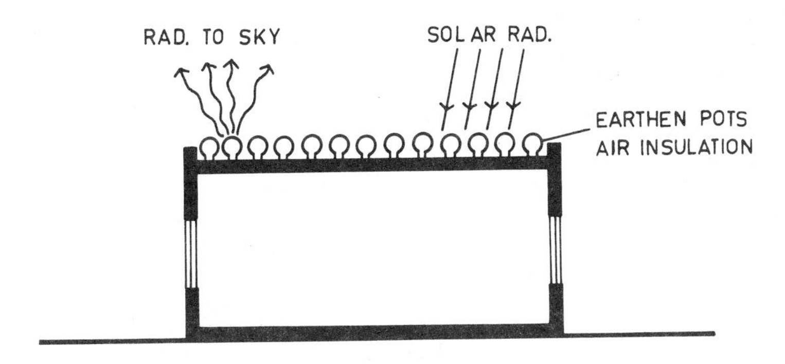

Fig. 4.11. Roof shading by pots

An effective roof shading device is a removable canvas cover (Fig. 4.12). This can be mounted close to the roof in the day time and at night it can be rolled up to permit radiative cooling. The upper surface of the canvas should be painted white to minimize the amount of absorbed radiation by the canvas and consequent conductive heat gain through it.

4.4.2.2 Paints. If the external surfaces of the building are painted with such colours which have minimum absorption of solar radiation but the emission in long

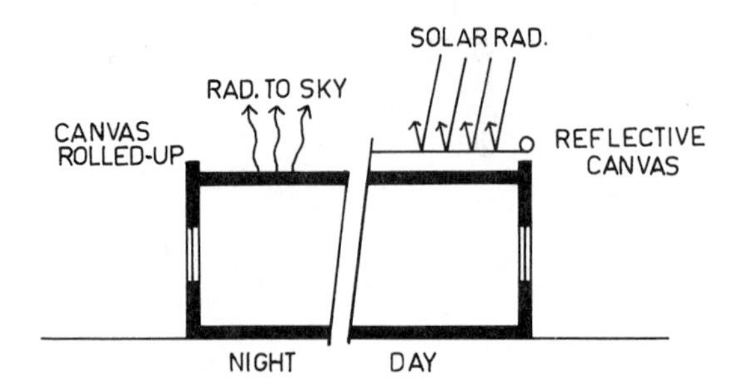

Fig. 4.12. Removable roof shade

wave region is high then the heat flux transmitted into the building is reduced considerably. In Table 4.2, the reflectivity of the surfaces for solar radiation and their emission in the long wave region are given for some common building materials.

TABLE 4.2

Material	Reflectivity (Solar Radiation)	Emissivity (low Temp.)
Aluminium foil, bright	0.95	0.05
Aluminium foil, oxidized	0.85	0.12
Polished aluminium	0.80	0.05
Aluminium paint	0.50	0.50
Galvanized steel bright	0.75	0.25
Whitewash New	0.88	0.90
White Oil Paint	0.80	0.90
Grey colour, light	0.60	0.90
Grey colour, dark	0.30	0.90
Green colour, light	0.60	0.90
Green colour, dark	0.30	0.90
Red Brick	0.40	0.90
Glass	0.08	0.90

Whitewash though has a lower reflectivity than aluminium, it will stay cooler when exposed to solar radiation because of its very high emissivity at low temperatures.

4.4.3 Evaporative Cooling

Evaporation of water takes place by the conversion of sensible heat into latent heat; a large amount of heat is therefore removed through this method. There are many ways to achieve evaporative cooling. Since evaporation occurs only at liquid air interface, it is best to create as much surface area as possible between water and the air.

Evaporative cooling methods have long been used in dry climates, where water is not scarce. Courtyard fountains provide cool spaces near massive buildings in many southern European cities, and Islamic architecture in medieval Indian buildings.

Most commonly employed evaporative cooling system is a window unit air cooler with evaporative pads, a fan and a pump. Central air cooling systems with a spray chamber and a blower are also used for larger buildings. To produce comfortable conditions both these systems require a high rate of air movement through the living space.

Many innovative evaporative coolers have been developed by CSIRO in Australia during the past two decades (Close, 1965). They typically involve evaporative cooling followed by a heat exchange phase. If the cooler (but relatively dry) output air from the heat exchanger is subjected to another stage of evaporative cooling a two-stage unit results. In many work areas excessive humidity and air movement are not desirable. For such cases, the heat exchanger has been implemented in the form of a rock bed (Fig. 4.13). It uses two rock beds set side by side and separated by an air space in which a damper is located. Water sprays are mounted close to the inner surface of each rock bed, and two fans are used. The rock beds are cooled alternatively by spraying water and letting it evaporate on the stones. While one rock bed is getting cooled, the other one (already cooled in the previous operation cycle) supplies cool air to the house. Very little moisture is thus added to the air entering the house as the rocks are almost dry before these are used to cool the incoming air in the next operating cycle. The humid air from the rock bed produced during its evaporation cycle is vented to the outside.

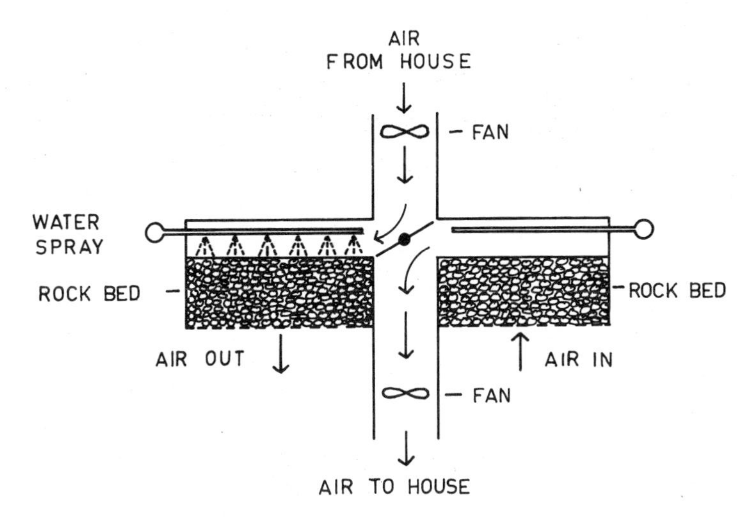

Fig. 4.13. Rock bed regenerative cooler

Evaporation from the surface of the roof can be used to provide cooling for one- or two-storeyed buildings. Continuous evaporation from a thin film of water over the roof, lowers the temperature of the roof, which in turn cools the living space below it. Solar radiation intensity and wind velocity over the roof, affect the rate of evaporation of water but not the temperature of the roof. For this method to be effective, the roof slab should be water proof and made as thin as possible.

One major problem, common to all the above-mentioned evaporative cooling systems, is the low operating efficiency during the humid part of the summer. Conventional evaporative cooling cannot be used at all in regions where the humidity remains high throughout the year. A conceivable alternative for such regions is the desiccant cooling, where the outside air is first dried by passing over a desiccant material like silica gel, and then cooled in the evaporator. Solar energy could be used for regenerating the spent desiccant material.

Desiccants were used as an integral part of the building in the Altenrich House (Dannies, 1959) which was built in Israel in the fifties. This building was oriented with its long axis along the north-south direction. The hollow east and west walls were filled with a sorbent material which permitted air flow through it. Evaporative coolers placed on top of both walls and dampers were so arranged that air could flow through each wall from top to bottom or vice versa.

During the forenoon, the sun would shine on the east wall and humid outdoor air would first dry by circulation through the west wall and cool in the evaporator. The cooled air which entered the living space from the top of the west wall would blow out through the east wall where it would carry away the moisture from the solar heated desiccant. During the afternoon, the air flow would reverse so that the regenerated desiccant in the east wall would be used for drying the air and the spent desiccant in the west wall be regenerated by solar heating. The use of desiccant for natural cooling is limited by the non-availability of materials suitable for large scale application.

4.4.4 Evaporative Air Coolers

Figure 4.14 illustrates three types of evaporative air coolers. The drip- (or desert)- type cooler is a fan unit with evaporative pads over which water is passed by a small circulating pump. The pads are normally made of aspen wood, glass fibre, metal wire or expanded paper. The spray-type cooler has evaporative pads, and a fine water spray that is thrown into the air and on to the pads by a water slinger, a centrifugal vaporizer, or spray nozzles. This type can be manufactured with or without an integral fan unit. The rotary-type cooler is a device that continuously wets and washes the evaporative pad by rotating it through a water bath and presenting it to the airstream.

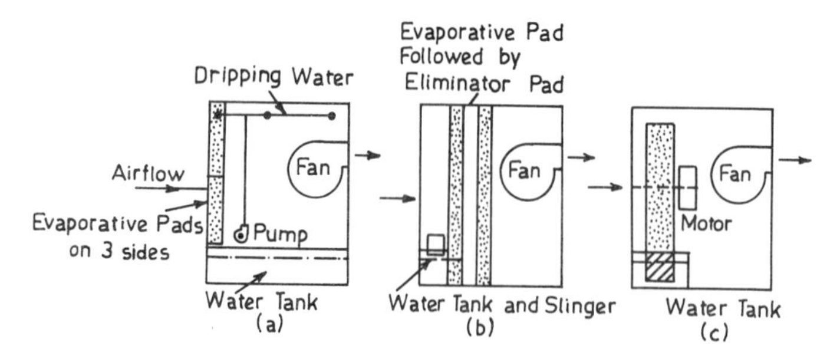

Fig. 4.14. Evaporative air coolers: (a) drip type, (b) slinger type, (c) rotary type

4.4.5 Evaporative Water Coolers

The different types are shown in Fig, 4.15 (Croome and Roberts 1981). The cooling pond is the simplest and cheapest form of water cooling and relies on natural wind effect to evaporate water from a large surface and hence cools the main body of water. Performance can be improved by the use of a spray pond where the water is sprayed several metres above the pond surface, thereby increasing the effective transfer area. Natural draught (or atmospheric) towers rely mainly on wind effect to circulate air through frames made mostly of wood and wetable material called fillers. This system has spray nozzles on top with wooden sides and fillers. The object here is to increase the wetted surface area and allow time for air/water contact. The hyperbolic tower uses the stack effect of a chimney above the packing to induce airflow up the tower in counterflow to the descending water droplets.

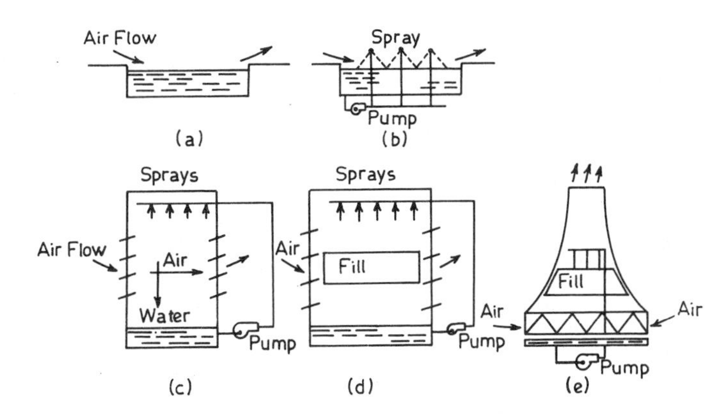

Fig. 4.15. Evaporative water equipment: (a) cooling pond, (b) spray pond, (c) spray-filled tower, (d) wood filled tower, (e) hyperbolic tower

4.4.6 Radiative Cooling

Radiative cooling, the net transfer of heat by thermal infrared radiation from a warm radiating body to a cooler heat sink, has two important applications in the comfort cooling of buildings. One is the direct radiative cooling of people by control of the mean radiant temperature of their immediate environment, and the other is the discharge of heat from a radiative heat dissipator to the cool sky. This later effect is strongest with clear night skies, but the northern sky (in the northern hemisphere) is often cool enough during the day to provide useful heat sink. A horizontal surface is the most effective radiative cooling configuration. Obstructions such as trees, walls, or clouds can significantly reduce the radiation of heat to the sky.

As the warm roof surface gets cooled by convection and radiation, a stage is reached when its surface temperature equals dry bulb temperature of the ambient air. Further cooling by radiation continues as the night sky temperature is lower than the ambient temperature. If the net heat exchange reduces the roof surface temperature to the wet bulb temperature of the surrounding air, condensation of the atmospheric moisture takes place on the roof and heat gained due to condensation limits further cooling.

If the roof surfaces are sloped towards an internal courtyard (Fig. 4.16), the cooled air sinks into the court and enters the living spaces through the low level openings. A parapet wall is raised around the roof to prevent air mixing.

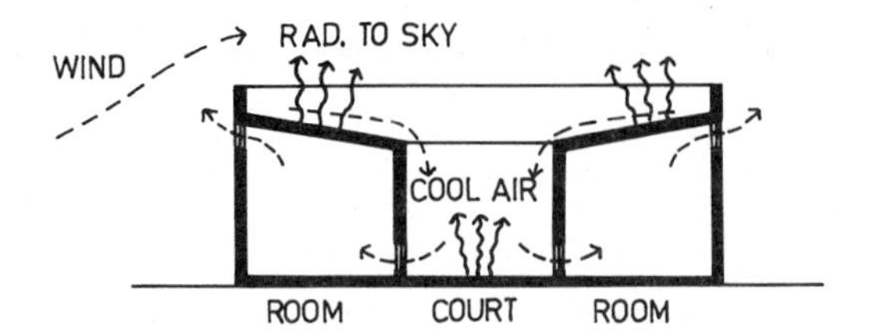

Fig. 4.16. Radiative cooling

However, this method will not work in windy conditions. The effect of wind movement and convective heat gain can be reduced by covering the roof with a sheet of polyethylene (PE) which is transparent to long wave radiation (Fig. 4.17). Inlet and outlet openings for air are provided in the roof itself. The major drawback of this method is however a short life span of the PE sheet. An alternative method is to cover the roof with white painted corrugated iron sheeting. Openings are provided in the roof for circulating air under the corrugated sheet. During the day the openings are kept closed and no air circulation takes place. At night, air is circulated under the sheet with the help of a blower and the cooled air is used in the living space. In this case the corrugated iron sheeting acts as the outer surface of the roof and cooling efficiency is limited due to convective heat gain from the outside air.

In recent years efforts have been made to develop surfaces (Catanoloth *et al.*, 1975; Harrison *et al.*, 1978) which enhance the radiative cooling effect. The surfaces have poor absorptivity in the visible region and low emissivity in the near infrared region. The data on these developments is however too limited for their large scale use.

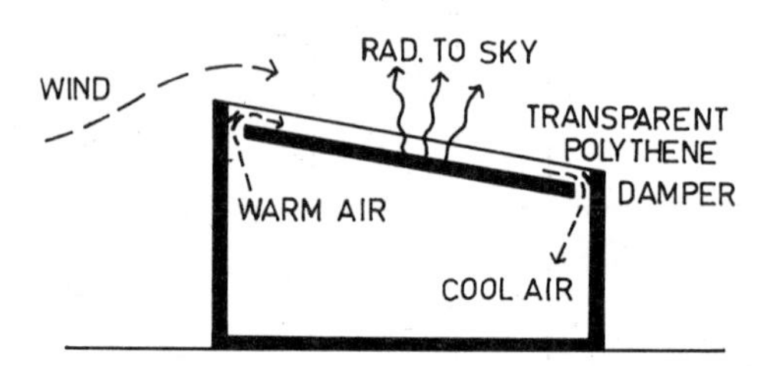

Fig. 4.17. Radiative cooling thermosyphon

4.4.7 Control of Conducted Heat by Insulation and Cavity Walls

The amount of heat coming into the room can also be decreased by insulating the various building components. The insulation is usually put over the outermost surface. The thickness of the insulation should have an optimum value from the economics point of view; any additional insulating layer has to be weighed for its thermal performance against its cost. Savings in the cooling have to be balanced by the cost of installing and additional insulation.

The roof and walls of the building having cavities also act as good insulators and inhibit the inward (or outward) transmission of heat from that component.

4.4.8 Exploitation of Wind, Water and Earth for Cooling in Hot Arid Climates

Characteristic systems described for cooling in this section are wind tower, the air vent, the cistern and the ice maker. The latter two store cooling which can be used in the summer. A wind tower (Fig. 4.18) operates in various ways, according to the time of day and the presence or absence of wind. The walls and airflow passages of the tower absorb heat during the day and release it to the cool air at night. The next day the walls are cool. When there is no wind, hot ambient air (solid Arrows in Fig. 4.18) enters the tower through the openings in the sides (Fig. 4.18, ≠ 1) and is cooled when it comes in contact with the tower. Since the cooler air is denser than the warmer air, it sinks down through the tower, creating a downdraught (Fig. 4.18, Nos. 2, 3, 5). When there is a

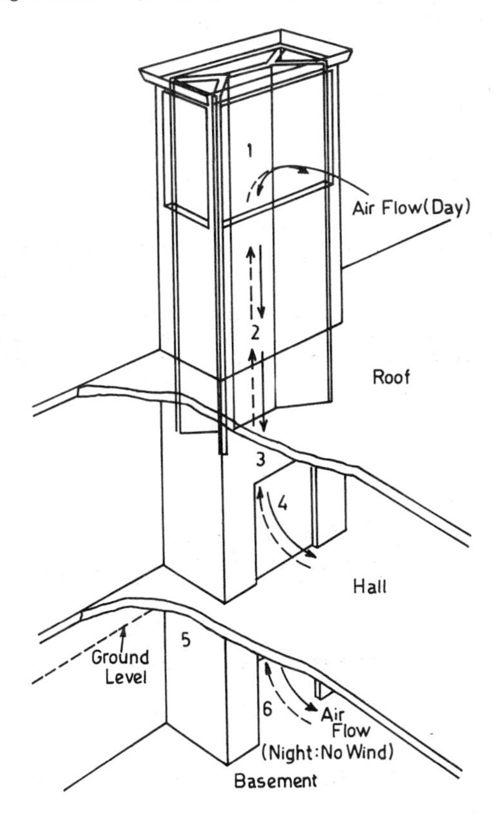

Fig. 4.18. Wind tower

wind the air is cooled more effectively and flows faster. Doors in the lower part of the tower (Fig. 4.18, ≠ 4, 6) open into the central hall and basement of the building. When these doors are open, the cooled air from the tower is pushed through the building and out of the windows and other doors, entraining room air with it. The cooled air's path of circulation depends on the arrangement of doors in the tower and the building (some of the air flowing down the windward passages of the tower is forced back through the opposite air passages and out through the leeward openings). When there is no wind at night (Fig. 4.18, broken arrows), the tower operates like a chimney. Heat that has been stored in walls during the day warms the cool night air in the tower. Since the warmer air is less denser than the cooler air, the pressure at the top of the tower is reduced, creating an updraught. Air in building is entrained up through the tower and cool night air is pulled into building through the doors and windows. When there is wind at night, air flows down the tower and through the building. Since the tower walls warm night air before it enters the building, rate of cooling can be lower.

A combination of sensible cooling in ground and evaporating cooling with the flow of air induced by the wind tower can be achieved by a configuration, shown in Fig. 4.19. The heat loss from air (on account of sensible cooling) results in a decreased air temperature but no change in the water-vapour content of the air.

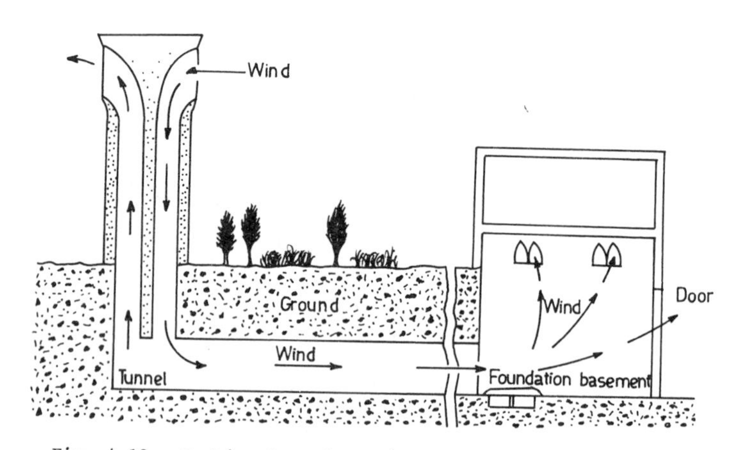

Fig. 4.19. Combination of sensible and evaporative cooling

Air in the upper part of a wind tower is sensibly cooled. When water is introduced into a system, evaporative cooling occurs. Such cooling involves a change in both the water-vapour content and the temperature of the air. When unsaturated air comes in contact with water, some water is evaporated, thus lowering the temperature of the air and increasing its water-vapour content. A wind-tower system that cools air evaporatively as well as sensibly is particularly effective. In most wind towers water in the ground seeps through to the inside of the basement wall of the tower, so that air passing over the wall is evaporatively cooled. Evaporative cooling plays an even larger part in the system shown above. The wind tower is placed about 50 metres away from the building and is connected to it by a tunnel. When the trees, shrubs and grass in the ground over the tunnel are watered, water seeps through the soil and keeps the inside surfaces of the tunnel walls damp. Thus air from the tower is evaporatively cooled as it passes through the tunnel. Pool and foundation in the basement of the building further cool the air. Whenever underground streams are present, the wind can be made to flow over them and the fountain dispensed with. Wind towers are normally employed with curved roofs, which offer a larger area for heat transfer to ambient air. Since the wind velocity over a curved roof is higher than that over a flat roof the corresponding heat transfer coefficient (surface-outside air) is also higher. Further hot air from the room gathers below the roof and thus the heat transfer from the roof to inside air gets further reduced.

Air vents (Fig. 4.20) are employed in areas where dusty winds make wind towers impractical. A typical vent is a hole cut in the apex of a domed or cylindrical roof. Openings in the protective cap over the vent direct wind across it. When air flows over a curved surface, its velocity increases and its pressure decreases at the apex of the curved roof, thereby, inducing the hot air under the roof to

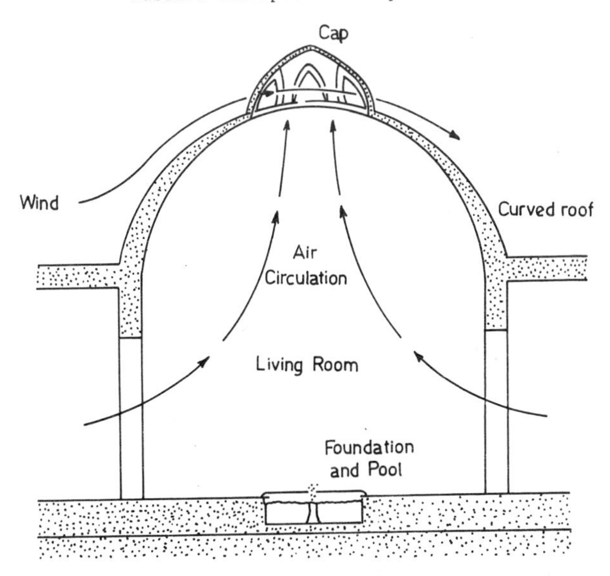

Fig. 4.20. Air vent

flow out through the vent. In this way air is kept circulating through the room under the roof. Air vents are usually placed over living-rooms, often with a pool of water directly under the vent to cool air moving through the room.

The cistern is an interesting concept to have a reservoir of cold water in the summer which can be circulated through a building to cool it. The cistern (Fig. 4.21) is filled with cold water during winter nights when the temperature is usually only a few degrees above freezing. The wind towers, which surround the cistern keep the water cold for use during the hot summer months. When the domed roof of the cistern is heated by the sun, it warms the air over the water in the cistern and increases the rate at which it evaporates. The towers maintain a draught across the surface of water as a result of which the water

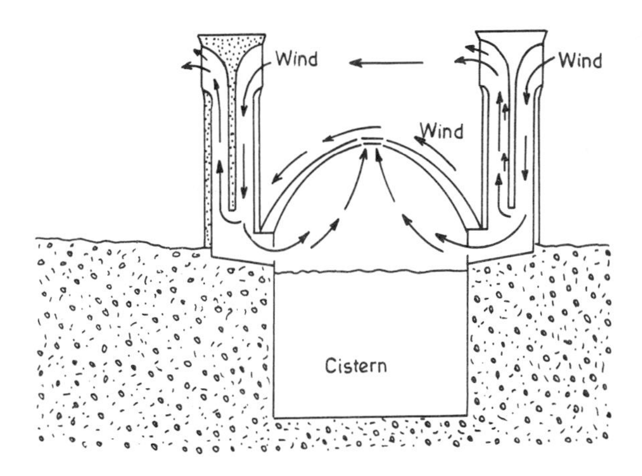

Fig. 4.21. Cistern

vapour is removed, saturation is prevented and evaporation can continue. The deeper layers of water are little warmed because the heat of air is nearly spent in evaporating the water at the surface. When there is an air vent in the roof of the cistern, wind flows down the towers across water and up through the vent (see black arrows) entraining vapour-laden air from the cistern out through the vent. When there is no air vent, wind flows down the towers and is forced back up through the air passages on the leeward side of the towers (see arrows). The updraught created in the leeward passages entrains cistern air out through the leeward tower openings. Cistern is partly buried to take advantage of the insulating properties of the ground. A domed roof as mentioned before is more easily cooled than a flat one and transmits less heat into cistern.

The ice maker is a passive cooling system that takes advantage of the near-freezing temperatures of winter nights in the desert. Several shallow ponds, 10 to 20 metres wide on a north-south axis and several hundred metres long, are filled with cold water on winter nights. A tall adobe wall on the south side of each pond and lower walls at the east and west ends shield the pond from the wind. At night water in the pond loses heat to the sky by radiation and gains heat from the ground by conduction and from the air by convection (that is, by the movement of air across the water surface). Shielding the pond from the wind reduces the heat gain by convection, so that on cloudless nights the heat loss to the sky by radiation is sufficient to freeze the water. On the following day the ice is cut up and placed in a covered storage pit 10 to 15 metres deep. The wall shades the pond during the day so that the ice does not melt before it can be cut up and stored.

4.5 PASSIVE HEATING AND COOLING

4.5.1 Roof-Pond/Skytherm

In this arrangement (Fig. 4.22), suggested by Hay and Yellot (1969), a mass of water is supported on a metallic or thin concrete roof. The water mass may be in the form of an open pond (with a transparent plastic cover in the winter) with moving insulation above it. Alternatively it may be in the form of water bags with the provision that air may be blown between the bags and the transparent cover (in winter to provide an insulating capacity) and that the bags may be flooded with additional water and the transparent cover removed (in summer).

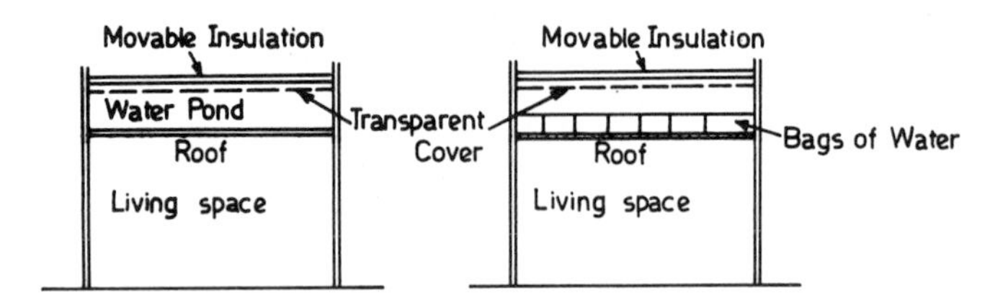

Fig. 4.22. Indirect gain: roof-pond

The operation of the pond is illustrated in Fig. 4.23. In the summer during the day the reflecting insulation keeps the solar heat away from the water, which keeps receiving heat trhough the roof from the space below (thereby cooling it). In the night the insulation is removed and water, despite cooling the living space below, gets cooled on account of losing heat by evaporation, convection and radiation. Thus the water attains its capacity to cool the living space.

In winter during the day the insulation is removed, allowing the solar energy to be absorbed by water and black surface of the roof; the living space continues receiving heat through the roof. In the night the insulation is put on the water

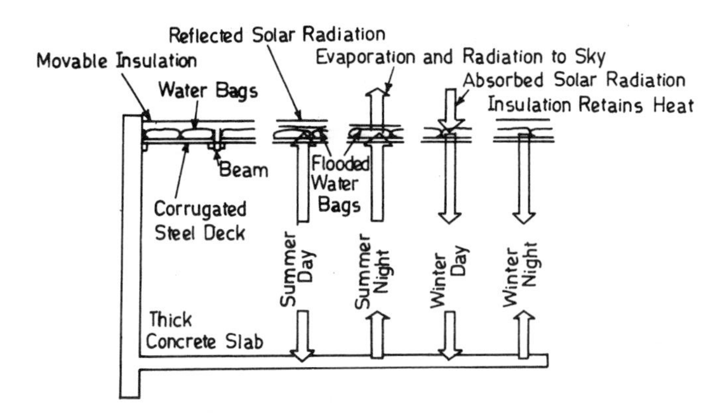

Fig. 4.23. Operation of roof-pond

mass, reducing the heat loss to the outside air. The heat loss to the inside air through the roof keeps the living space warm.

4.5.2 Roof Radiation Trap System

A system which uses roof for heating and cooling was developed by Givoni (1976a, 1976b). In this system called the "Roof Radiation Trap", solar energy for winter heating is absorbed directly at the roof's surface beneath a fixed insulation layer. Part of the heat is transferred into the occupied space by conduction across the roof, with a time lag which can be precalculated by the thermal time constant of the roof. The other part of the collected solar energy is transferred by convection, from the hot air space between the absorbing roof and the insulating layer, to a thermal storage (gravel) under the floor of the building or inside the occupied space.

The radiation trap (Fig. 4.24) consists of a vertical (or slightly inclined) glazed plane facing south, and a north-sloping insulated plane, which extends over the glazing so that in summer the glazing is shaded while in winter the sun penetrates through the glass. Between the roof and the insulation an air plenum is thus formed, where solar radiation is trapped. The floor of this radiation trap is painted black and acts as a solar energy absorbing surface. The inclined insulating layer is covered by sheets of corrugated metal, preferably painted white with a plastic paint to increase its emissivity. Thermal storage is provided partly by the roof itself and partly by a gravel storage placed under the floor of the building or inside the occupied space.

In winter, solar radiation during the day penetrates through the glazing, heats the upper surface of the roof and the air above it (by convection). Some of the absorbed energy is conducted through the roof and subsequently radiated to the living space.

The hot air above the roof is either drawn through the room or preferably for night-time storage, into a thermal storage system (Gravel) located in the basement or within the building. In this concept also a movable insulation during night time reduces heat losses through the glazing.

During summertime, a corrugated metal sheet located over the north sloping panel of the roof trap will be cooled by night outgoing longwave radiation so that its temperature can be lowered by several degrees below the temperature of ambient air. Air allowed to flow under this corrugated sheet is consequently cooled.

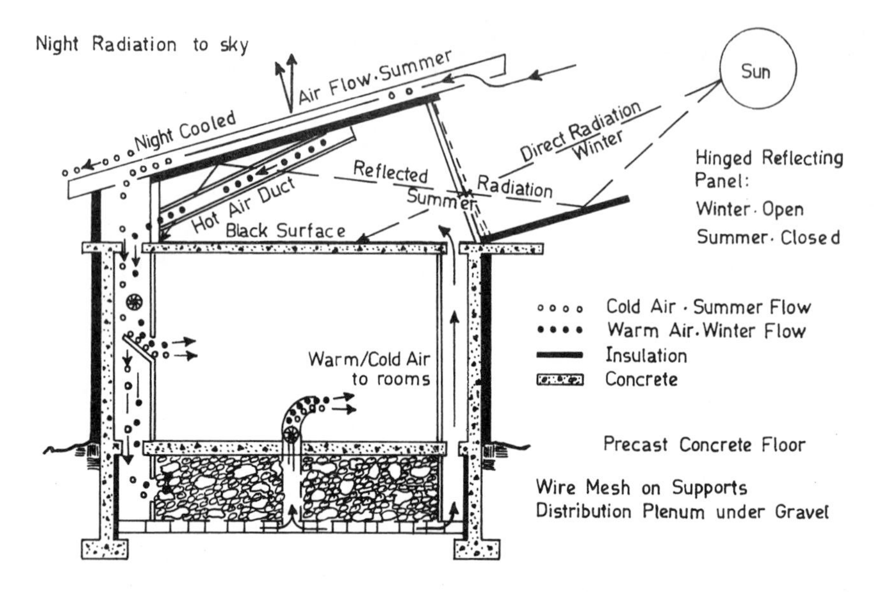

Fig. 4.24. The roof radiation trap

Experiments carried out at the Israel Institute of Technology in 1975-76 show that the air flowing under a white corrugated metal sheet exposed to night sky, was cooled by about 4-5°C below the D.B.T. of the outside air and for certain periods even to 1-2°C below W.B.T. The air may be drawn by means of a fan into a storage unit where gravel will be cooled. During the day, the air drawn through the storage may thus be cooled, and even partly dehumidified and then used to cool the rooms.

4.5.3 Vary Therm Wall

Controlling of the air movement in magnitude and direction both gives rise to wall components with varying thermal resistance. Such a system (Fig. 4.25) can be used for mild winter heating and summer cooling in mixed climates like Delhi. The external wall component of this concept is made of light material while the internal component is the usual brick (or concrete) wall. During the summer daytime, the wall provides an effective air insulation to heat flow into the room; during the night the cool ambient air comes in contact with the warm brick wall and gets heated establishing a natural flow of air. This air movement

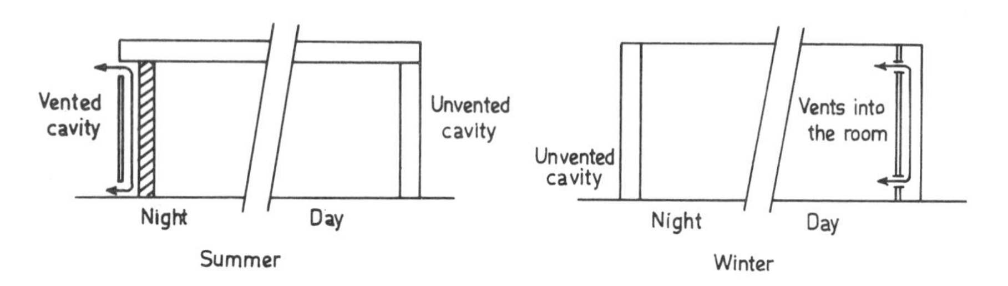

Fig. 4.25. Vary therm wall

helps in quick removal of the heat flux. For the winter operation, vents are opened during the day into the room supplying warm air of the cavity to the room, all vents are kept closed during the night time thus providing an air insulation which minimizes heat losses to the ambient.

4.5.4 Earth Sheltered or Earth Bermed Structures

The average thermal load of a building depends on the infiltration of outside air and the amount of insulation between the inside and the outside air. In an earth sheltered building the reduced infiltration of outside air and the additional thermal resistance of the surrounding earth considerably reduces the average thermal load. Further the addition of earth mass to the thermal mass of the building reduces the fluctuations in the thermal load. Hence from the point of view of thermal comfort an earth sheltered building presents a significant passive approach.

The roof of the structure should be under grass or foliage cover at a depth such that the additional load of the ground above can be supported by the structure. It is obvious that the deeper we go, the better the thermal point of view, but more expensive from the structure angle. Hence at a depth, easily obtainable the roof may be insulated by synthetic insulators in the normal manner. The additional heating, when needed may be supplied by means of direct gain, through windows near the roof, projecting above the ground. When cooling is needed one can suck outside air through ducts underground (where the temperature is low); if necessary ducts may be cooled by evaporation.

Several earth sheltered buildings, which are thermally desirable and architecturally pleasing have been described in a book, written by Underground Space Centre (1978), University of Minnesota.

Earth bermed structures such as shown in Fig. 4.26, help to block the transmission of heat from or into the building and thus helping to maintain a constant temperature.

4.5.5 Earth-Air Tunnels

The use of earth-air tunnels to heat or cool a building is an ancient concept (Bahadori, 1978) which has been revived by practioners of solar passive architecture. The temperature of the ground, a few metres below is almost constant throughout the year; hence air passed through a tunnel at a depth of a few metres will get cooled in summer and heated in winter. The magnitudes of this constant temperature depends on the nature of the ground surface, the lowest temperature resulting from a shaded and wetted surface. Table 4.3 indicates the expected temperatures at a depth of 4 metres from the surface, for various ground surface conditions. The expected temperatures from the outlet of the tunnel at this depth for various surfaces are given in the Table 4.4.

4.6 VENTILATION: REMOVAL OF HEAT FLUX

The heat flux coming into the building as a result of infiltration of the outside air and due to conduction has to be removed to attain the conditions of thermal comfort. This is usually accomplished through proper ventilation. Ventilation is an integral part of vernacular architecture in most hot and humid climates. Ventilation of living spaces is also necessary for removal of odours and gases produced by normal metabolic functions of the human body. Air movement relieves the heat stress imposed on the human body by humid conditions. During the periods when the outdoor temperature is lower than the indoor temperature, ventilation will cause cooling of the interior. When the outdoor temperature is

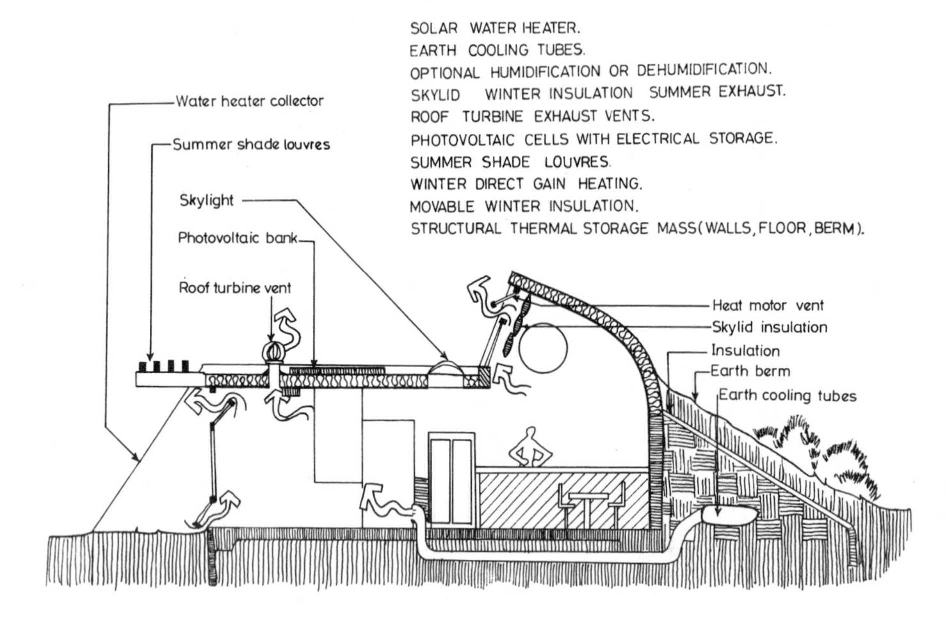

Fig. 4.26. Earth bermed structures

TABLE 4.3 Ground temperature for various surface conditions at a depth of 4 metres

Surface condition	Ground temperature
Dry sunlit	27.5°C
Dry shaded	18.5°C
Wet sunlit	21.5°C
Wet shaded	21.0°C

hotter than the indoor temperature, ventilation must be kept to a minimum unless the air is cooled before it enters the living space.

Ventilation thus serves two basic requirements namely (i) Health and (ii) Thermal comfort. The basic factors associated with each of them are shown in Fig. 4.27.

It is noted that a distinction is made between requirements for health and those for comfort. The former should be satisfied under all weather conditions, while the latter correspond to requirements in certain weather conditions only.

Natural ventilation of the building results from differential wind forces on the various building surfaces and from thermal effects due to temperature difference

TABLE 4.4 Outlet Air Temperature from the Tunnel for Various Flow Rates

(Length of the tunnel = 10 m)

(Radius of the tunnel = 0.3 m)

S. No.	Surface and tunnel characteristic	Ambient air temperature (°C)	Outlet temperature from tunnel for various flow velocities (m/s)			
			1.0	3.0	5.0	7.0
1	Dry shaded surface and wet tunnel	30	23.1	24.3	24.6	24.8
		35	23.7	25.7	26.3	26.5
		40	24.4	27.1	27.6	28.3
		45	25.4	28.6	29.6	30.0
2	Wet sunlit surface and dry tunnel	30	25.3	26.7	27.0	27.1
		35	28.2	30.2	30.7	30.9
		40	31.1	33.7	34.3	34.6
		45	34	37.3	38.0	38.3
3	Wet sunlit surface and wet tunnel	30	20.3	22.1	22.1	22.8
		35	21.0	23.5	24.1	24.6
		40	21.8	24.8	25.8	26.2
		45	22.3	25.3	27.6	27.7
4	Wet shaded surface and dry tunnel	30	24.7	26.3	26.6	26.8
		35	27.6	29.8	30.3	30.5
		40	30.5	33.3	34.0	34.2
		45	33.4	36.8	37.6	38.0

between outside air and the inside air. The easiest way to provide good ventilation is to have openable windows and vents on more than one side of each room for flow through ventilation; and open floor plan with few partitions will also facilitate air flow through the building. Factors such as surrounding landscape, location of other buildings, the building form orientation with respect to wind direction, size and proportion of window openings and arrangement of internal partitions etc. affect the air flow within the building. A low window will allow air to flow through the lower part of the room where it can effectively cool the inhabitants.

As with all passive strategies, however, ventilation does not provide comfort under all weather conditions. In hot arid climates, when the daytime air temperature becomes more than the skin temperature, two competing mechanisms are operative, namely, convective air motion tends to transfer heat from the air to the body, while at the same time it enhances the evaporative cooling process at the skin surface. The cooling effect of air motion, is, therefore, strongly dependent on the humidity of the air and on the level of a person's physical activity and clothing. Emerick (1971) suggested the following probable impact of various wind speeds on people indoors under comfortable thermal conditions.

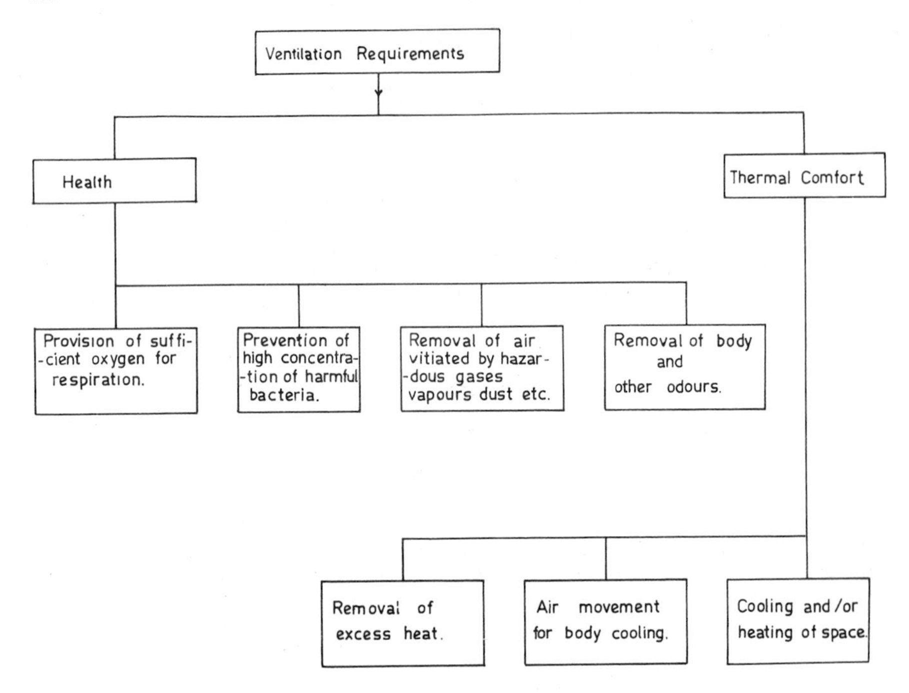

Fig. 4.27. Schematic illustration of ventilation requirements (after van Strattan, 1967)

Wind speed (m/s)	Probable impact
0.25	Unnoticed
0.25 to 0.50	Pleasant
0.50 to 1.00	Generally pleasant but causing constant awareness of air movement
1.00 to 1.50	From slightly draughty to annoyingly draughty
1.50	Requires corrective measures of work and health which are to be maintained at an acceptable level

In dry climates, when the nocturnal temperature drops to a comfortable level, thermal mass in a building can be cooled by night-time ventilation so that the daytime heat build up is dissipated. If night-time temperatures remain above the desired comfort range, ventilation will hinder the conditions of thermal comfort. On the other hand, in humid climates the dry-bulb air temperature has a tendency to remain at more moderate levels than in hot dry climates. In such cases, ventilation will usually achieve comfort, or at least reduce discomfort. Ventilation through low mass shaded structures is the classic technique for approaching comfort in the tropics.

Potential for cooling by air movement is large. As an example Fig. 4.28 shows a map in which the number of uncomfortable hours per year is plotted for zero air velocity, using typical meteorological year weather data for 28 US cities. The level of comfort was calculated using Fanger's predicted mean vote corresponding to light clothing in the shade where the radiant temperature is identical to the air temperature (Martin, 1981).

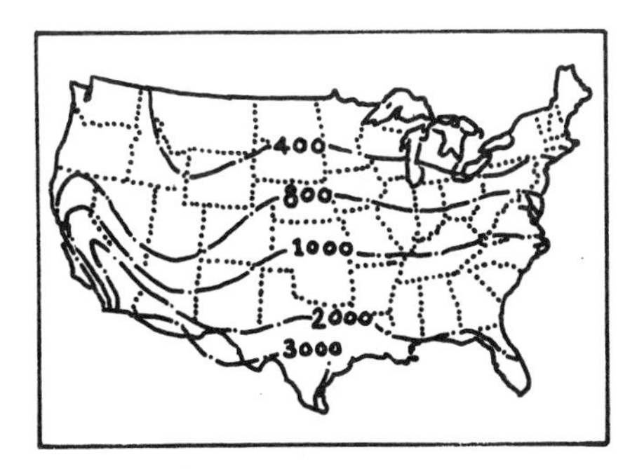

Fig. 4.28. Yearly number of hours of summer discomfort assuming no air motion

Figure 4.29 shows a similar map of discomfort hours for the case where the air velocity has been increased to a modest 0.6 m/s. Such a velocity can be easily obtained with the help of natural air motion or fans of low power consumption.

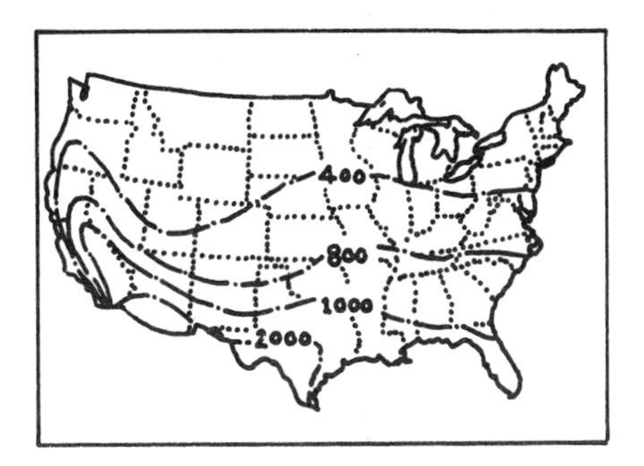

Fig. 4.29. Yearly number of hours of summer discomfort with an air velocity of 0.6 m/s

In order to trap some of the cooling potential of ventilation, a building must be thoroughly flushed with outdoor air at a rate of 13-30 air changes per hour (Martin, 1981). It may be noted that the air changes per hour due to infiltration in a well constructed building is of the order of unity. Experiments at the Florida Energy Centre in Cape Canaveral are measuring the effectiveness and characteristics of natural ventilation.

Ventilation can play a major cooling role in residences without a great deal of additional research. Factors inhibiting the increased use of this technique include security and noise control.

4.7 COMPONENTS

At present a designer of passive systems has available to him a number of options for windows (a general term for the means of solar radiation input), storage and other subsystems/components. In this section we briefly discuss some of these.

4.7.1 Windows

1. *Self-Inflating Curtain* (Shore, 1978). The curtain consists of a number of layers of thin flexible material of high reflectivity and low emissivity. Radiation (from the sun in summer or storage in winter) warms the air between the layers, causing increase in pressure and decrease in density in the upper part of the system. The pressure pushes the layers apart, causing fresh intake of air from the bottom. Thus the system of reflecting layers separated by air gaps (thus produced) provides good insulation. When insulation is not needed the air is evacuated from the sides.

2. *Window Quilt Shade* (Schnelbly, *et al.*, 1978). The quilt consists of a sandwich of five layers, as shown in Fig. 4.30. The five layers are assembled with an ultrasonic fibre welder. The quilt is enclosed in decorative polyester fibre. The effectiveness is apparent from the following figures for an airtight window:

U Value -- 5.84 $W/m^{2o}C$ without quilt

U Value -- 1.306 $W/m^{2o}C$ with quilt

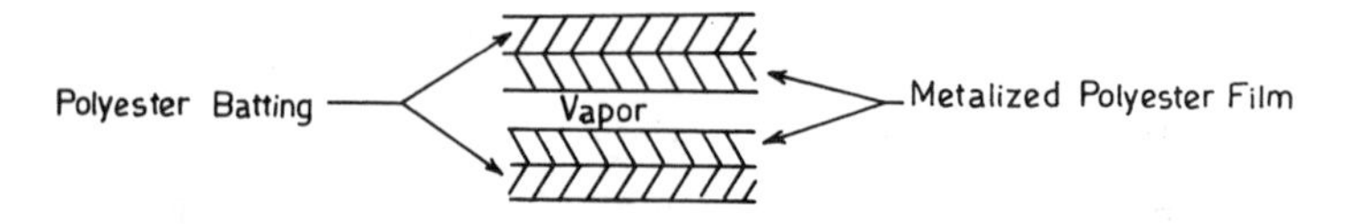

Fig. 4.30. Window quilt shade

3. *Venetian Blind Between the Glasses* (Berlad, *et al.*, 1978). The effective system to reduce the heat loss through a double-glazed window is shown in Fig. 4.31. The characteristic dimension of a unit is small and hence convective heat transfer is stopped.

4. *Transparent Heat Mirrors* (Selkowitz, 1978). One approach to reduce the heat loss from a glazed surface is to coat glass by a film which reflects, to a large extent, the infrared radiation from the surface. However, the coating also reduces the transmissivity of glass for solar radiation and a suitable compromise has to be made. In Fig. 4.32 the performance of a heat mirror (glass with a coating of emmissivity) is shown as compared to that of a multipane system.

The coating may consist of single or multiple layers of different substances, deposited by vacuum evaporation or spray (on a hot surface) techniques. Thus we see that the heat mirror gives much less heat loss and higher transmission than multipane systems.

5. *Heat Trap*. A reasonable thickness of insulating material with good transmissivity reduces the heat transfer considerably because the outer layer is at a temperature not much different from the ambient.

6. *Optical Shutter* (Chahroudi, 1978). Suntek has developed a material named Cloud Gel which changes its transmission characteristics with intensity of light and temperature. The characteristics of a shutter comprised of three layers of transparent sheets and one layer of Cloud Gel are shown in Fig. 4.33. It can be used for reducing air-conditioning loads and preventing overheating in greenhouses

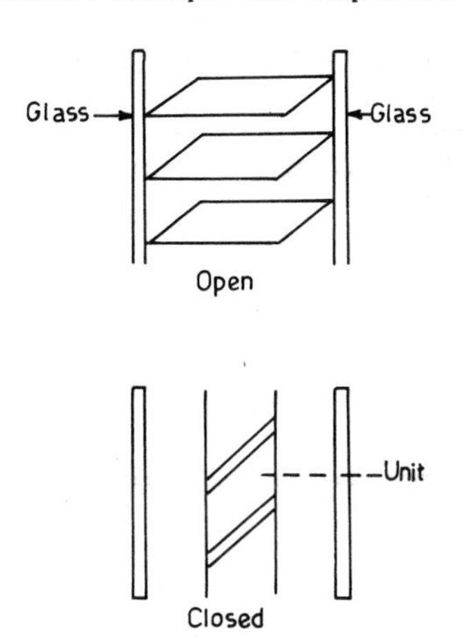

Fig. 4.31. Venetian blind between the classes

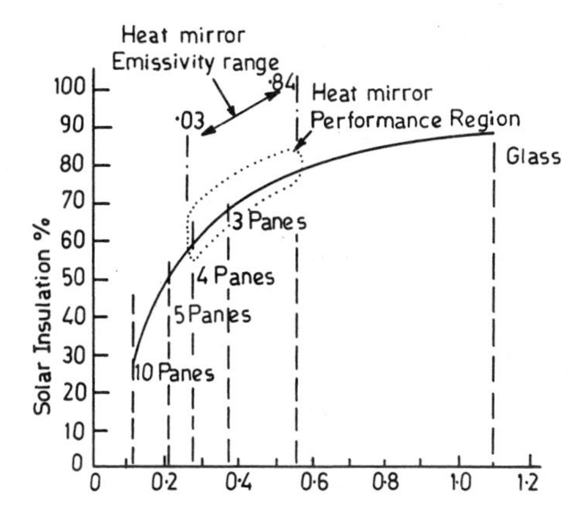

Fig. 4.32. Comparison of heat window with multipanes

and solar collector systems because of its property to be opaque at high temperatures and highlight intensities.

4.7.2 Thermal Storage

1. *Latent Heat Storage Panel* (Faunce *et al.*, 1978; Hauer *et al.*, 1978). The panel consists of an acrylic sheet, followed by a polyhedral wall, an air cavity and heavy insulation. In the air cavity, is placed, an array of tubes containing the phase change material ($Na_2SO_4.10H_2O$, melting at 32^oC).

For heating of the room the following procedure was adopted:

(a) Irradiation by sun during the day:

(b) At sunset heavy insulation was placed between the panel and outside atmosphere;

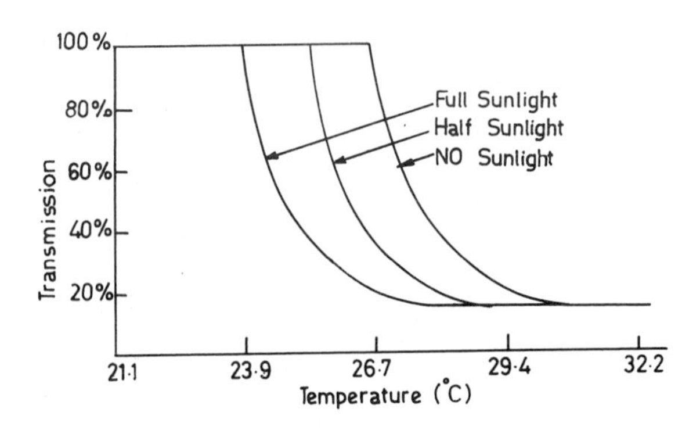

Fig. 4.33. Characteristics of optical shutter

(c) Heat was retrieved in the night by flow of air through the tubes.

On a cold day in Newark, Delaware (with a temperature of -4.4°C) 46% of the solar energy was utilized at night. With suitable materials (melting at 16°C) such panels can also be used for cooling.

2. *Water Wall* (Maloney, 1978). The first water walls consisted of drums of water stacked one over the other. Later modules of 30 mil vinyl and a height of 10' were used but abandoned because of their amenability to puncture. Smaller modules 8' x 2' x 1½" nesting on each other have indeed been used with success.

3. *Solar Window* (McClinktock *et al.*, 1978). The solar window is a transparent plastic panel with a sheet of glass on the outside to provide an insulating space. The window will thus admit solar radiation in the room as needed; when this is not desirable a black liquid is circulated through the panel and a storage unit, thus building up the storage of heat.

4. *Concrete Block Storage*. In climates where the nights are cool and the days are hot, cooling during the day can be achieved by using storage in a large concrete block on which the house sits. During the night cool air circulates through the block, cooling it and during the day the air inside the building is circulated through the block causing it to cool the building. A reverse arrangement may be utilized for heating during the night in winter. Block *et al.* (1979) have described a residence using this arrangement.

4.7.3 Miscellaneous

1. *Skylid* (Baer, 1976; Hymer, 1978). The skylid opens and closes itself in response to the sun. The driving power is provided by weight of freon shifting from one take to another through a tube. The bubble of gas in the warmer container expands, pushing the freon to the cooler tank.

By this principle, insulation filled louvers open and close in response to the sun.

2. *Beadwalls* (Harrison, 1976). Since glass is a popular enclosure material, it is necessary to insulate it during the night to prevent heat loss. The beadwall is an interesting concept for transforming double-glazing into an opaque well insulated wall. A granular insulating material from a container is blown into the cavity formed by the two glass panes. When the sun is shining the granular material is withdrawn under vacuum to leave the glazing transparent.

3. *Thermic Diode Solar Panels* (Buckley, 1976). Each panel (Fig. 4.34) is composed of two layers, a thin sensor layer facing the outside and a storage layer on the inside. Thick insulation is placed between the two layers. Both layers are filled with water and connected through a tube and a check valve which allows the flow of water in only one direction. When the sun is shining the outer layers gets heated and the buoyancy derived pressure forces open the check valve which allows hot water to flow to the storage panel. However when the sun is not shining the outer layer gets cooled, causing buoyancy derived pressure force to close the valve; hence no heat transfer takes place in the night.

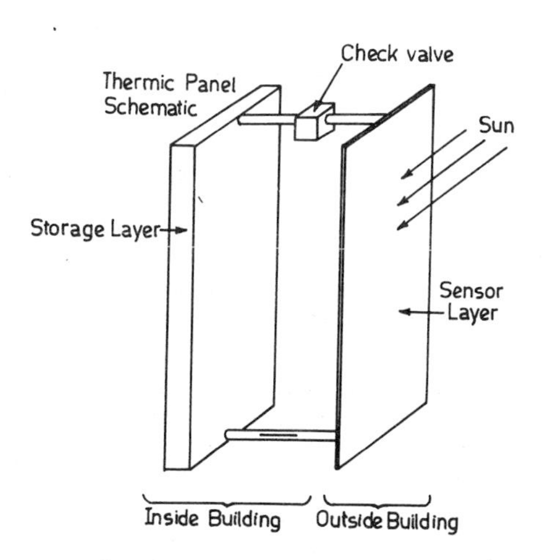

Fig. 4.34. Thermic solar diode.

For summer the check value may be set differently to allow flow from hotter storage layers to cooler sensor layers in the night and stopping it in the day.

To be effective thermic panels should be installed in the roof or in the walls. On account of the heavy weight of the panels special designing of roofs to take the load is necessary.

4. *Heat Pipes* (Grover *et al.*, 1964). Since its invention the heat pipe has found widespread application as an efficient, reliable, and passive heat transfer device. The principle of the heat pipe is shown schematically in Fig. 4.35. It consists of a sealed container (frequently a tube or pipe) which is evacuated and partially filled with a suitable working fluid such as water or a refrigerant. It transports heat via a closed cycle process of evaporation and condensation. When heat is applied to one section of the heat pipe, some of the fluid evaporates. The vapour flows to the unheated sections where it condenses and released the latent heat of condensation.

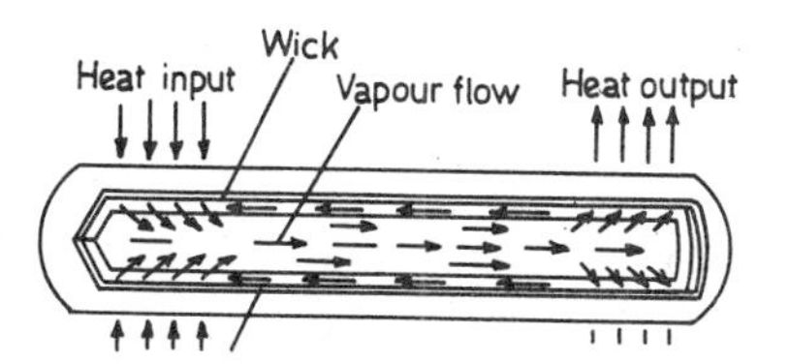

Fig. 4.35. Heat pipe

Since the latent heat of all fluids is very large, small mass transport can result in a large transport of heat. A unique feature of the heat pipe is the means of returning the condensate back to the evaporator. It is accomplished through capillary pumping in a wicklike porous structure.

5. *Energy Roof* (Pittinger *et al.*, 1978). The "Energy Roof" process differs from the conventional roof pond by moving water instead of the insulation. A metal roof ceiling supports a relatively large volume of water which is covered by a floating roof of rigid insulation and enclosed by a continuous sheet of ultraviolet-resistance Mylar film.

Water is maintained at a temperature which is in summer a few degrees cooler than the rooms below the roof; in winter, water is a few degrees warmer than the desired room temperature. Heat-transfer to or from the metal ceiling is accomplished primarily by radiation and virtually all of the ceiling area participates in the process.

In winter, the storage water is warmed during the day by causing it to flow over the upper surface of the floating insulation beneath a clear polymer film which is used to prevent evaporation. A small pump transfers water from storage to the top of the floating roof panels, the upper surface of which is faced with a thin layer of aluminium foil. The foil becomes sufficiently oxidized after a brief period of operation that, in combination with about 1 cm of water, it forms an adequate absorber for the incoming solar radiation. The temperature rise of the storage water during a clear winter day is about 5°C in Phoenix, Arizona. When a second plastic film was placed over the heat absorber supported by air from a small blower an increase of 20% was observed in water temperature.

During the hot season the operation of the Energy Roof is reversed so that no water flows across the collector during the day and the upper insulating panel reflects the unwanted solar radiation back to the sky. The circulator pump functions at night during the cooling mode when the environment temperature falls 2.5°C below the storage water temperature. Thus, when conditions are favourable for cooling by convection and radiation, water begins to flow across the roof, under the two films of ultraviolet-resistant plastic and heat is lost both by convection to the ambient air and by radiation to the sky. When further cooling is required, a roof spray system is activated.

It appears that the Energy Roof can do the whole cooling job whenever the wet bulb drops to 18.3°C at night and goes not higher than 23.9°C during the day. Within those conditions, the Energy Roof will handle dry-bulb temperatures of at least 100°F (37.8°C).

6. *Suncatcher and Cool Pool* (Hammond, 1978). Two useful architectural devices, one for natural heating and the other for natural cooling are described as follows.

The heating device, or Suncatcher, is a simple reflective roof and soffit configuration that increases the solar collection efficiency of a south-facing window while reflecting the summer sun. The cooling device, or Cool Pool, is a pool of water that is entirely shaded from the sun but open to the north sky.

The Suncatcher (Fig. 4.36) is a reflective roof and soffit combination that forms a cone that concentrates the winter sun and directs it through a clerestory window. The same configuration reflects the summer sun away. The concentrating effect also reduces the amount of glass and the movable insulation needed to cover it. By moving the collector glass on to the roof, the thermal mass can be put anywhere in the building, freezing the ground level south glass for views. This also allows row houses to be built along a north/south street and still provide them with a direct solar heat source.

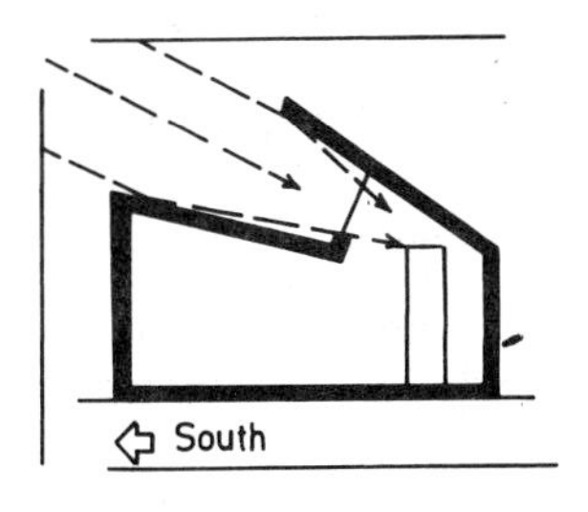

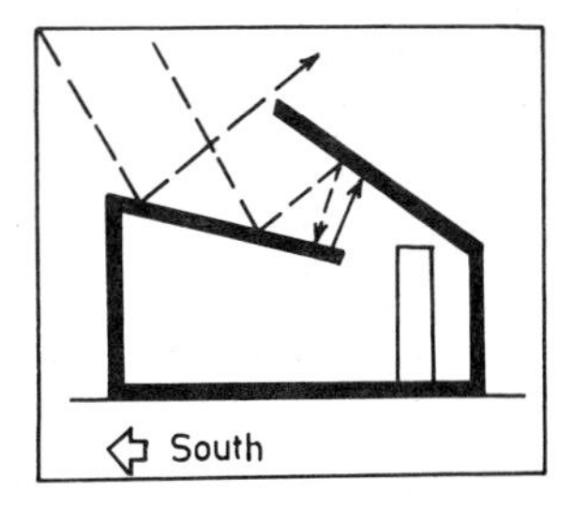

Fig. 4.36. Suncatcher

The Cool Pool: Experiments by Neubauer (1965) with different shade configurations and dry horizontal panels showed that a shade configuration open to the north sky kept the panels about 2°C lower than that of the surrounding air. This effect can be combined with the familiar roof pond to produce a very powerful cooling device that doesn't need any movable insulation. The cool bottom layer of water wicks the heat from the building below and disperses it by evaporation and radiation to the cool night sky.

In the summer of 1977 a number of different shading devices were used on pools of water of varying depths. With the best combination, temperatures could be maintained 17°C below the ambient air temperature.

REFERENCES (STATE OF ART)

General References

Conference 1 Proceedings of the Passive Solar Heating and Cooling Conference and Workshop, Albuquerque, N. M., May 18-19, 1976; edited by M. H. Keller -- Report No. LA-667-37-C (Available from NTIS*).

Conference 2 Proceedings of the Second National Passive Solar Conference, May 1978, Passive Solar State of Art -- Vol. 1/Buildings, Vol. 2/ Components Simulation and Testing, Vol. 3/Policy, Education and Economics; edited by D. Prowlor (Midatlantic Solar Energy Association, Philadelphia, Pa., USA).

Conference 3 Proceedings of the Third National Passive Solar Conference, Jan. 11-13, 1979, Vol. 3; edited by H. Miller, M. Riordan and D. Richards (Publishing Office of American Section of International Solar Energy Society, Inc., McDowell Hall, University of Delaware, Newark, Delaware, USA).

Conference 4 Proceedings of the Fourth National Passive Solar Conference, Oct. 3-5, 1979, Vol. 4; edited by G. Granta (Publishing Office of American Section of International Solar Energy Society Inc., McDowell Hall, University of Delaware, Newark, Delaware, USA).

A.I.A. Research Corporation, 1978, *A Survey of Passive Buildings*, Number HUD-PDR-287 (Available from NTIS).

Balcomb, J. D., *et al. Passive Solar Design Handbook* -- Vol. *1*/Passive Solar

*National Technical Information Service (NTIS) US Department of Commerce, 5285 Post Royal Road, Springfield, Va. 22161, USA.

Design Concepts, Vol. 2/Passive Sikar Design Analysis -- Report No. DOE/CS-0127/182 (Available from NTIS), 1980.

Haskins, D. and P. Stromberg, 1979, *Passive Solar Buildings* -- Report No. SAND79-0284 (Available from NTIS).

Mazaria, E., 1979, *The Passive Solar Energy Book* (Rodale Press, Emmans, Pa., USA).

REFERENCES

Baer, S. C. (1976) Freon Actuated Controls, Conference 1, 282.
Bahadori, M. N. (1978) Passive Cooling Systems in Iranian Architecture, *Scientific American*, 144-154.
Balcomb, J. D. (1978) State of Art in Passive Solar Heating and Cooling, 2nd National Passive Solar Conference, Philadelphia, March 16-18.
Berlad, A. L. *et al.* (1978) *Energy Transport Control in Window Systems*, Solar Energy Association, Philadelphia, USA, 326-328.
Block, D. A. *et al.* (1979) Use of Concrete Cored Slab for Passive Cooling in an Iowa Residence Conf. 4, 488-491.
Buckley, S. (1976) Thermic Diode Solar Panels: Passive and Modular, Conf. 1, 293-299.
Catanoloth, S. *et al.* (1975) The Radiative Cooling of Selective Surfaces, *Solar Energy, 17*, 81.
Chahroudi, D. (1978) Variable Transmission Solar Membrane, Conf. 2, 343-348.
Close, D. J. (1965) Rock Pile Thermal Storage for Comfort Air Conditioning, Institution of Engineers, Australia, *Mechanical and Chemical Engineering Transactions*, Vol. *1*, No. 1, pp. 11-72.
Croome, D. J., Roberts, B. M. (1981) Airconditioning and Ventilation of Buildings (Pergamon).
Dannies, J. H. (1959) Solar Air-Conditioning and Solar Refrigeration, *Solar Energy*, *3* (1), 34-39.
Emerick, Robert H. (1971) Common Factors Affecting Cooling Design, *Prog. Archit.* 97-99.
Faunce, S. F. *et al.* (1978) Application of Phase Change Materials in a Passive Solar System, Conf. 2, 475-480.
Fuchs, R. and McClelland, J. F. (1979) Passive Solar Heating Using a Transwall Structure, *Solar Energy, 23*, 123-128.
Gagge, A. P., Herrington, L. P. and Winslow, C. E. A. (1937) Thermal Interchanges Between the Human Body and its Atmospheric Environment, *Amer. J. of Hyg., 26*, 84, 102.
Givoni, B. (1976a) *Man, Climate and Architecture*, Applied Science Publishers, London.
Givoni, B., Paciuk, M. and Weiser, S. (1976b) Natural Energies for Heating and Cooling of Buildings - Analytical Survey, Research Report 017-325, Building Research Station, Technion Haifa.
Grover, G. M. *et al.* (1964) Structures of Very High Thermal Conductance, *J. Appl. Phys. 35*, 1990-91.
Gupta, Vinod (1981) Natural Cooling of Buildings, Innovative Informations Incorporated, Research Report No. 1: S1.
Hammond, J. (1978) Conf. 2, 137-140.
Harrison, D. (1976) Beadwalls, Conf. 1, 283-387.
Harrison, A. W. *et al.* (1978) Radiative Cooling of TiO_2 White Paint, *Solar Energy, 20*, 185.
Haskins, D. and Stromberg, P. (1979) Passive Solar Buildings Report, Sandia Report, SAND 79-0824.
Hauer, C. R. *et al.* (1978) Passive Solar Collector Wall Incorporating Phase Change, Conf. 2, 405-488.
Hay, H. R. and Yellot, J. I. (1969) Natural Air Conditioning with Roof Pond and Movable Insulation, *ASHRAE Trans.*, *75*, 178.
Hollingsworth, F. N. (1947) Solar Heat Test Structure at MIT, *Heating and Ventilating*, May 1947.

Hymer, R. (1978) Movable Insulation: New Developments at Zomeworks, Conf. 2, 489-492.
Maloney, T. (1978) Four Generations of Water Wall Design, Conf. 2, 489-492.
Martin, Marlo (1981) State of Art in Passive Cooling, *Sunworld*, *5*, 179.
McClinktock, M. *et al.* (1978) Solar Space Heat and Domestic Hot Water by a System Operating Both Passively and Actively, Conf. 2, 505-508.
Morse, E. L. (1881) Warming and Ventilating Apartments by Sun's Rays, US Patent 246, 626.
Neubauer (1965) *Trans. Amer. Soc. Agr. Engrs.*, *8*, 470-475.
Pithinger, A. L. *et al.* (1978) Conference 2.
Schnelbly, J. *et al.* (1978) The Window Quilt Insulating Shade, Conf. 2, 314-346.
Selkowitz, S. (1978) Transparent Heat Mirrors for Passive Solar Heating Applications, Conf. 2, 329-334.
Seshadri, T. N. *et al.* (1969) Climatilogical and Solar Data for India, C.B.R.I. Roorkee, India.
Shore, R. (1978) A Self Inflating Movable Insulation System, Conf. 2, 305-309.
Simon, M. J. (1947) *Your Solar House*, Simon and Schuster, New York.
Singh, S. and Bansal, N. I. (1984) Study of Three Different Underground Storage Systems, (Communicated).
Stromberg, R. P. and Woodall, S. O. (1977) Passive Solar Buildings: A Compilation of Data and Results, Report No. SAND-77-1204.
Trombe, F. (1972) US Patent 3, 832, 992.
Trombe, F. (1974) Maisons Solaires, *Techniques de Inginiear*, *3*, C777.
Underground Space Centre, University of Minnesota (1978), Earth Sheltered Housing Design, Van Nostrand Reinhold Co., NY.
Van Strattan (1969) *Thermal Performance of Buildings*, Elsevier NY.

Chapter 5

HEAT TRANSMISSION IN BUILDINGS AND BUILDING MATERIALS

The thermal response of a building may be determined from the heat gain and/or loss factors through various structural elements viz. walls, windows, roof and floor, the internal heat loads, and rate of ventilation. The transfer of heat between a building and its environment largely depends upon the overall conductance of the structure, the degree of exposure to weather including sunshine and the existing temperature differences between inside and outside. These factors, in turn, are closely influenced by the building materials used and their moisture contents, the type of construction, the workmanship and the geographical location of the building. Therefore, the structural heat gains or losses depend on certain properties of the elements concerned; for instance, heat gains through walls depend upon the colour of the outside surfaces, the heat storing capacity of the wall (i.e. material) and their thermal resistance or insulation properties. The processes of heating or cooling imply basically the transfer of heat by virtue of an existing temperature difference between two or more objects and can take place in three ways, namely conduction, convection and radiation. Each one of these heat-transfer modes depends on the surface temperature and because they are interdependent the calculation of any one component requires the simultaneous consideration of all the heat transfer modes.

5.1 MODES OF HEAT TRANSFER

5.1.1 Conduction

Thermal conduction is the process of heat energy transport from warmer region to colder region of a body at the molecular level. The higher the temperature difference, the larger will be the rate of heat conduction. Although conduction may occur in gases and liquids, significant conduction occurs only in solid materials and is evaluated using the steady state one dimensional Fourier equation for heat flow

$$q = KA\frac{\Delta\theta}{L} = \frac{A\Delta\theta}{R} , \qquad (5.1)$$

where $\Delta\theta$ is the temperature drop across a wall of thickness L and area A; $R(=L/K)$ ($m^2\ {}^oCW^{-1}$) being the thermal resistance opposing heat flow by conduction. All materials have some finite resistance to heat flow; those with a particularly high resistance are called thermal insulants. The reciprocal of resistance, $C_t(= 1/R)$, is termed as thermal conductance of that section. The thermal properties of many building materials are listed in Table A1 given in Appendix 3. If n layers of different materials in series comprise a composite wall, the total resistance is simply the sum of the individual resistances of the components i.e.

$$R_t = R_1 + R_2 + R_3 + \text{------} + R_n$$

or

$$C_t = 1/R_t = 1/\sum_{i=1}^{n} (L_i/K_i), \tag{5.2}$$

The higher the R-value of a material, the greater its insulating ability. The expressions of C_t for different geometries have been given in detail by Wong (1977).

5.1.2 Convection

Thermal convection is the process of heat transfer from a solid surface to fluid by the combined action of conduction and mass transport. Rates of convective heat transfer depend upon the velocity and nature of fluid flow. If the fluid movement is taking place due to buoyancy forces arising from density differences caused by temperature gradients, the mode of heat transfer is termed as free or natural convection, whereas the process in which the fluid movement is caused by some external agency, such as a pump or blower, it is termed as forced convection.

Normally, the convection losses are significant at three places in the buildings. Firstly, the convection losses occur within the walls and between the layers of glass in the skin of the building due to some air space and temperature difference existing between the two layers (Fig. 5.1a). Secondly, the convection losses occur in conjunction with the conduction losses through the skin of the building within the living space. In case of an uninsulated wall or window, the living space is generally hotter than the perimeter walls and therefore, hot air moves towards the wall (or window) and loses its heat in raising the surface temperature. This cooled air being heavier tends to sink down and runs across the floor, while the warm air at the top of the room runs in to take its place. This convection loop is rapidly set up in the room which eventually accelerates the cooling effect (Fig. 5.1b). In contrast to this, an insulated wall or window

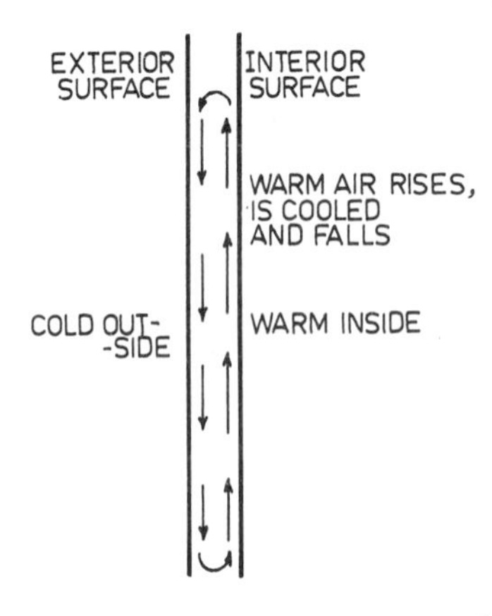

Fig. 5.1(a). Convection losses in an air space

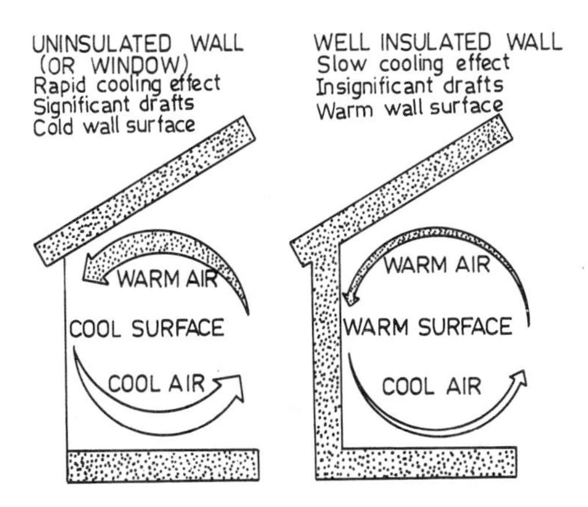

Fig. 5.1(b). Differences in comfort conditions between insulated and uninsulated walls

will be at higher temperature than the other inside walls and therefore, the cooling effect may not be significant. The third kind of convection losses occur due to air infiltration through openings in buildings (or windows) and cracks (air spaces) around doors and windows. These are primarily convection losses and are not easy to quantify accurately.

The convective heat transfer coefficient between the fluid and the boundary surface may be given by the relation

$$q = h_c A\ (\theta_{ws} - \theta_f), \tag{5.3}$$

The heat transfer coefficient, h_c, is a complex function of the geometry of the surface, flow characteristics and the physical properties of the fluid and is evaluated by a relationship between one dependent non-dimensional group, the Nusselt Number (Nu), and the other three independent non-dimensional groups, the Reynolds Number (Re), the Grashof Number (Gr) and the Prandtl Number (Pr) depending upon whether the mode of heat transfer is free or forced.

5.1.2.1 Dimensionless numbers

(a) *Nusselt Number.* The Nusselt Number is the ratio of the temperature gradient at the wall to the overall temperature drop between the wall and the edge of the boundary layer, thereby, indicating the relative conductive to convective heat transport as

$$Nu = \frac{h_c L_c}{K} \tag{5.4}$$

(b) *Reynolds Number.* The Reynolds Number is the ratio of fluid dynamic force (i.e. inertia) and the viscous drag force and characterizes the nature of the flow in forced convection:

$$Re = \frac{\rho u^2}{\mu u/L_c} = \frac{\rho u L_c}{\mu} = \frac{u L_c}{\nu}, \tag{5.5}$$

It predicts the criterion for the stability of laminar flow.

(c) *Grashof Number*. The Grashof Number is the ratio of the buoyancy force to the viscous force in the mode of natural convection.

$$Gr = g\beta_t \rho^2 L_c^3 \Delta\theta / \mu \, , \tag{5.6}$$

(d) *Prandtl Number*. The Prandtl Number represents the relation of heat transfer to fluid motion and is the ratio of momentum diffusivity to the thermal diffusivity

$$Pr = \frac{(\mu/\rho)}{(K/\rho c)} = \left(\frac{\mu c}{K}\right), \tag{5.7}$$

5.1.2.2 Free Convection. If a cold fluid comes in contact with a hot surface, the fluid in immediate contact with the surface is stationary being arrested by viscous action and therefore, gets heated by molecular conduction, becomes lighter and moves vertically due to difference in density with adjacent fluid. At steady state, a natural convective boundary layer is formed with viscous forces opposing buoyancy forces. The velocity and temperature profiles will be of the form as shown in Fig. 5.2 (Rogers and Mayhew 1974). As we move away from the surface, fluid layers move at a faster rate giving rise to the forced convective boundary layer. The velocity profile of the fluid exhibits maxima in between the stationary bulk fluid and the wall whilst the fluid temperature continuously increases until the fluid temperature acquires the temperature of the surface of the wall. The affected regions of temperature and velocity profiles due to heated surface (wall) are normally called hydrodynamic and thermal natural convective boundary layers respectively. If the flow of the fluid within the boundary layer is streamline (no mixing with adjacent layers), it is called laminar or streamline flow. Instead, the unsteady flow with turbulent eddies which activates the mixing in the fluid and leads to a higher rate of heat-transfer is called turbulent flow.

Heat transfer due to free convection has been experimentally determined for various shapes of surfaces and is correlated in terms of the dimensionless numbers as illustrated below:

$$Nu = c \, \{Gr.Pr\}^n \tag{5.8}$$

where $n = \begin{cases} 1/4 \text{ for laminar flow i.e. when } (Gr.Pr.) < 10^8 \\ 1/3 \text{ for turbulent flow i.e. when } (Gr.Pr.) > 10^8 \end{cases}$

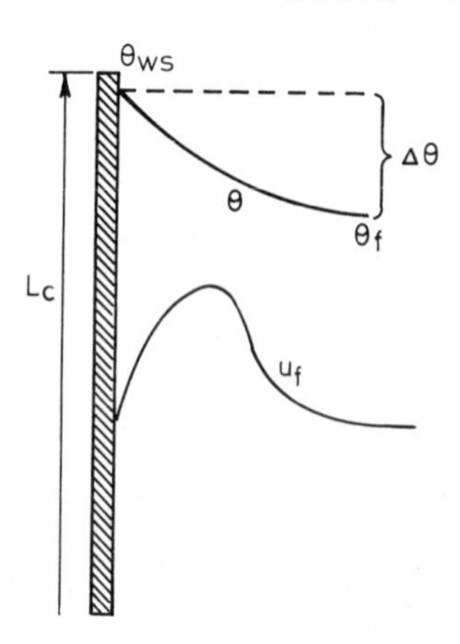

Fig. 5.2(a). Temperature and velocity profiles in natural convection

and the value of constant "c" is determined from the shape and geometry of particular surface. Some of the values of "c" with direct relevance to building geometries are illustrated in Table 5.1

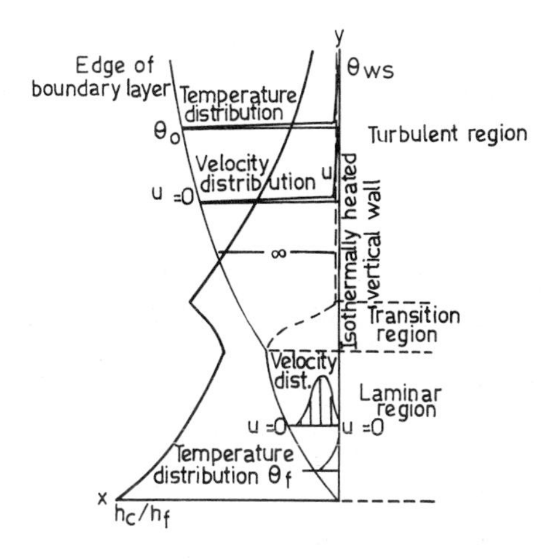

Fig. 5.2(b). Velocity and temperature distributions and the variation of the natural convective heat-transfer coefficient for fluid in contact with a vertical isothermally heated wall (O'Callaghan 1978)

TABLE 5.1 Correlations for Natural Convection

System	c		Dimension taken as L_c in (Nu) and (Gr)
	Laminar	Turbulent	
Vertical plates or cylinders	0.56	0.12	Height of plate of cylinder
Horizontal plates heat flow from top face	0.54	0.14	Length of long side
Horizontal plates heat flow from bottom face	0.25	-	Length of long side
Horizontal cylinders	0.47	0.10	Diameter of cylinder

5.1.2.3 Forced convection. When the fluid flows over a flat plate under a pressure from some external agency, a boundary layer forms adjacent to the surface. The velocity of the fluid changes from zero at the wall to the normal velocity within this layer. The local value of the heat transfer coefficient is inversely proportional to the thickness of the boundary layer and is very large at the leading edge of the plate (Fig. 5.3). In the region of laminar boundary layer, it progressively decreases until a sudden increase in heat transfer coefficient is noticed due to the transition to the turbulent boundary layer. The region between the breakdown of the laminar boundary layer and the establishment of the turbulent boundary layer is called the transition region. In this region, the heat transfer coefficient increases rapidly and thereafter, it decreases as the turbulent boundary layer grows. For flat plate with a smooth

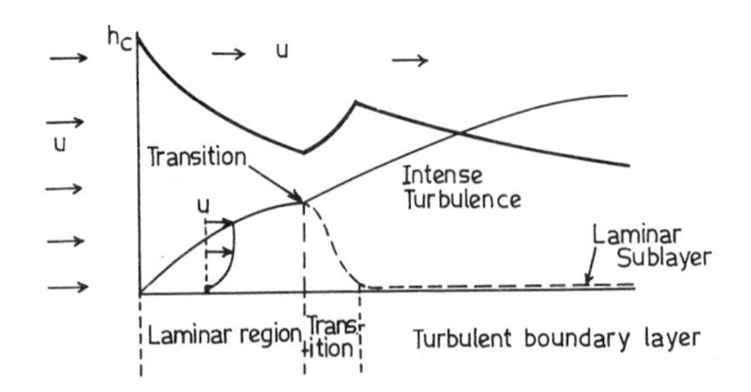

Fig. 5.3. Variation of the magnitude of the surface heat-transfer coefficient for fluid flowing in forced convection over a flat surface (O'Callaghan 1978; Rogers and Mayhew 1974)

leading edge, the turbulent boundary layer starts at Reynolds Numbers of about 0.3×10^6 to 0.5×10^6. For blunt edged plates, it starts at much smaller Reynolds Numbers (ASHRAE, 1977). In general

$$Nu \propto Re^n . Pr^m \tag{5.9}$$

The convective heat transfer coefficient, h_c, under laminar and turbulent modes can be calculated from the relations listed in Table 5.2 for various building geometries of interest.

5.1.3 Radiation

Thermal radiation is the transfer of heat from a body at higher temperature to another at lower temperature by electromagnetic waves passing through a separating medium. The solar radiation impinging on a surface is divided into three parts

$$\alpha + \tau + \zeta = 1.0$$

i.e. a part of the sunlight is absorbed to produce heat (α), a part is transmitted (τ) (if surface is not opaque) and some is reflected away due to the surface colour, specularly or diffusely, depending upon the texture of the surface. The light reflectance, for various colours (from the PPG Design-a-colour TM system) of the walls/floors is given in Table 5.3. Multiple reflections occur from a light-coloured surface, eventually, yielding multiple absorptions. For instance, a white surface which reflects 85% of the light will absorb most of the incoming radiation except 12% of the energy which will escape an off-white room with a window along one wall.

The amount of thermal radiation that leaves a surface (in the wavelength region 0.1-100 μm), is not only a function of its temperature but is also related to its surface property known as emissivity (Rogers and Mayhew, 1974). Most building materials, including glass, have high emissivity (absorption) in the far-infrared band (typically 90-95%) regardless of their colour and emit/absorb the impinging radiation equally well for a given surface temperature (see Table 5.4). The characteristics of thermal radiation are studied by the concept of black body and grey surfaces. A black body is an ideal body which absorbs all the radiation (i.e. $\alpha = 1.0$, $\tau = 0.0$, $\zeta = 0.0$) and emits maximum possible amount of radiation at all wavelengths depending upon its temperature. The total energy radiated over all wavelengths is given by the relation

$$q_b = \sigma A T^4 \tag{5.10a}$$

TABLE 5.2 Correlations for Forced Convective Heat-Transfer over Plane Surfaces (After Wong, 1977)

No.	Flow along a plane surface	Formulae		Operating conditions
1	Pure flow regime u, θ_f θ_{ws} x L	$Nu_x = 0.332\, Re_x^{1/2} Pr^{1/3}$	(local)	Laminar flow,
		$\bar{N}u_x = 0.664\, Re_x^{1/2} Pr^{1/3}$	(mean)	$Re < 5 \times 10^5$, $0 < x < L$
		$Nu_x = 0.029\, Re_x^{4/5} Pr^{1/3}$	(local)	Turbulent flow,
		$\bar{N}u_x = 0.037\, Re_x^{4/5} Pr^{1/3}$	(mean)	$Re > 5 \times 10^5$, $0 < x < L$
2	Mixed flow regime Turbulent Laminar u, θ_f s x θ_{ws} L	Mean $\bar{N}u_x$ over the distance $x < x < L$ $Nu_x = 0.037\, Pr^{1/3} (Re_x^{4/5} - C)$ where $C = 23\,500$ for $(Re)_{cr} = 5 \times 10^5$ $C = 14\,200$ for $(Re)_{cr} = 3 \times 10^5$ $C = 4\,300$ for $(Re)_{cr} = 10^5$		Laminar flow up to distance s where the critical $(Re)_{CY}$ occurs and thereafter turbulent flow to $x > s$
3	Partial wall heating u, θ_f θ_f θ_{ws} L_1 x L_2	$Nu_x = 0.332\, Re_x^{1/2} Pr^{1/3} [1 - (L_1/x)^{3/4}]^{(-1/3)}$	(local)	Laminar flow,
		$\bar{N}u_x = 0.664\, Re_x^{1/2} Pr^{1/3} [1 - (\frac{L_1}{x})^{3/4}]^{(2/3)} / (1 - \frac{L_1}{x})$	(mean)	$Re < 5 \times 10^5$, $L_1 < x < L_2$
		$Nu_x = 0.029\, Re_x^{4/5} Pr^{1/3} [1 - (\frac{L_1}{x})^{9/10}]^{(-1/9)}$	(local)	Turbulent flow,
		$\bar{N}u_x = 0.037\, Re_x^{4/5} Pr^{1/3} [1 - (\frac{L_1}{x})^{9/10}]^{8/9} / (1 - \frac{L_1}{x})$	(mean)	$Re > 5 \times 10^5$, $L_1 < x < L_2$

Since no black body practically exists in nature, the radiation from real (grey) surfaces is modified by the emissivity, ε, of the radiating grey surface as

$$q_g = \varepsilon \sigma A T^4 \tag{5.10b}$$

5.1.3.1 Radiative heat exchange. The radiative heat exchange between two surfaces at temperature T_1 and T_2 respectively, may be estimated from the relation

$$q_r = A_1\, \sigma\, F_{12} (T_1^4 - T_2^4) \tag{5.11}$$

where F_{12} is the configuration factor, ranging between 0.0-1.0, which depends upon the emissivity of each surface and relative view or geometrical shape factor F between the surfaces.

(a) *Between Building Surfaces*. The radiant heat exchange, F_{12}, between two grey surfaces at different temperatures is given by Kreith (1973) as

$$\frac{1}{A_1 F_{12}} = \left[\frac{1}{A_1} (\frac{1}{\varepsilon_1} - 1) + \frac{1}{A_2} (\frac{1}{\varepsilon_2} - 1) + \frac{1}{A_1 F} \right], \tag{5.12a}$$

Therefore, the radiation heat exchange between two parallel infinite plates may be given as

$$\frac{1}{F_{12}} = (\frac{1}{\varepsilon_1} + \frac{1}{\varepsilon_2} - 1) \tag{5.12b}$$

or for a small grey body in black surrounding,

TABLE 5.3. Light reflectance of various colours (After Johnson, 1981)

FLOORS	
Colour	Light Reflectance (%)
Atlantis	9
Really Rust	12
Slate	15
Pepper Corn	13
Slate Brown	14
Tender Taupe	20
Ultramarine Blue	11
Mint Green	22
Crimson Lips	10
Deep Red	9
WALLS	
Colour	Light Reflectance (%)
Polar Sky	72
Wisteria Blue	53
Rose Morn	76
Mission Beige	75
Old Linen	71
Vanilla Ice	83
Sunny Beige	79
Cream Supreme	81
Sun Yellow	70
Orange-Glow	62
Deep Chrome	50
Ebony Black	5
Gypsum	86

$$F_{12} = \varepsilon_1 \quad (5.12c)$$

A graphical presentation of the configuration factor, F_{12}, for different geometries of the building surfaces exchanging radiative heat is shown in Fig. 5.4. Geometrical shape factor F for various configurations has been given in Table A2 of Appendix 3.

(b) *Between Vertical Surfaces of the Building and the Ambient.* To evaluate the radiative heat exchange between a full wall of courtyard or the facade of a street, which is like a box as shown in Fig. 5.5, to the sky, one has to find out the effective area of the hemispherical sky as seen by the wall. A small elemental area dA of the full wall ABCD is assumed to be located at O' where the primed co-ordinate system X'Y'Z' is also centred. For computational convenience, the

TABLE 5.4 Average Emissivities, Absorptivities and Reflectivities for some Surfaces Common to Building (After van Straaten, 1967 and O'Callaghan, 1978)

Surface	Emissivity or absorptivity		Reflectivity
	Low-temperature	Solar radiation	Solar radiation
Aluminium, bright	0.05	0.20	0.80
Asbestos cement, new	0.95	0.60	0.40
Asbestos cement, aged	0.95	0.75	0.25
Asphalt pavement	0.95	0.90	0.10
Brass and copper, dull	0.20	0.60	0.40
Brass and copper, polished	0.02	0.30	0.70
Brick, light buff	0.90	0.60	0.40
Brick, red rough	0.90	0.70	0.30
Cement, white portland	0.90	0.40	0.60
Concrete, uncoloured	0.90	0.65	0.35
Glass	0.90	-	-
Marble, white	0.95	0.45	0.55
Paint, aluminium	0.55	0.50	0.50
Paint, white	0.90	0.30	0.70
Paint, brown, red, green	0.90	0.70	0.30
Paint, black	0.90	0.90	0.10
Paper, white	0.90	0.30	0.70
Slate, dark	0.90	0.90	0.10
Steel, galvanized, new	0.25	0.55	0.45
Steel, galvanized, weathered	0.25	0.70	0.30
Tiles, red clay	0.90	0.70	0.30
Tiles, black concrete	0.90	0.90	0.10
Tiles, uncoloured concrete	0.90	0.65	0.35

rectangular co-ordinates (X',Y',Z') have been converted into polar co-ordinates (r',θ',ϕ').

Thermal emission (dE), from the elemental area dA is given by

$$dE = \int_{\Omega'} I.\cos\beta'.d\Omega'.dA \qquad (5.13)$$

where I is the intensity of thermal emission in the directions perpendicular to the area dA, $d\Omega'$ is the elemental solid angle in the directions (r',θ',ϕ') β' is the angle between the direction (r',θ',ϕ') and the perpendicular to the area. using $\cos\beta' = \sin\theta' \, . \, \sin\phi'$

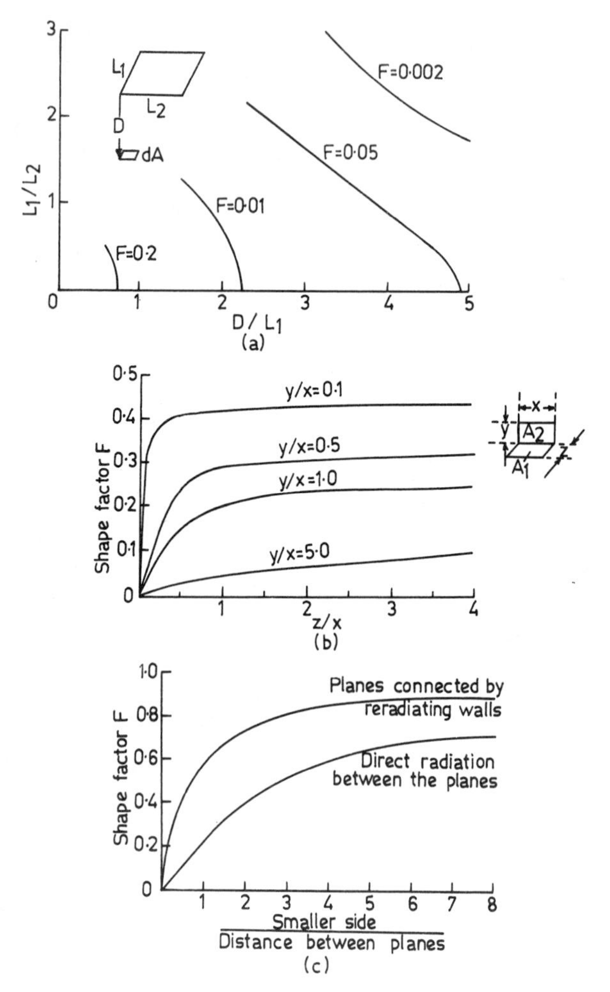

Fig. 5.4. Radiation shape factors: (a) for a surface element dA and a rectangular surface parallel to it; (b) for adjacent rectangles in perpendicular planes; (c) for equal and parallel rectangles (After Kreith, 1973)

and $d\Omega' = \sin\theta' \,.\, d\theta'.d\phi'$, equation (5.13) becomes

$$dE = I.\int_{\phi'}\int_{\theta'}\left[(\sin^2\theta')\ \sin\phi'.d\theta'.d\phi'.dA\right] \tag{5.14}$$

The integration in equation (5.14) is to be carried for that portion of the sky which is seen by the area dA'. This portion of the sky is divided into three parts (Fig. 5.6) and for each of these the relevant limits of integration are as follows:

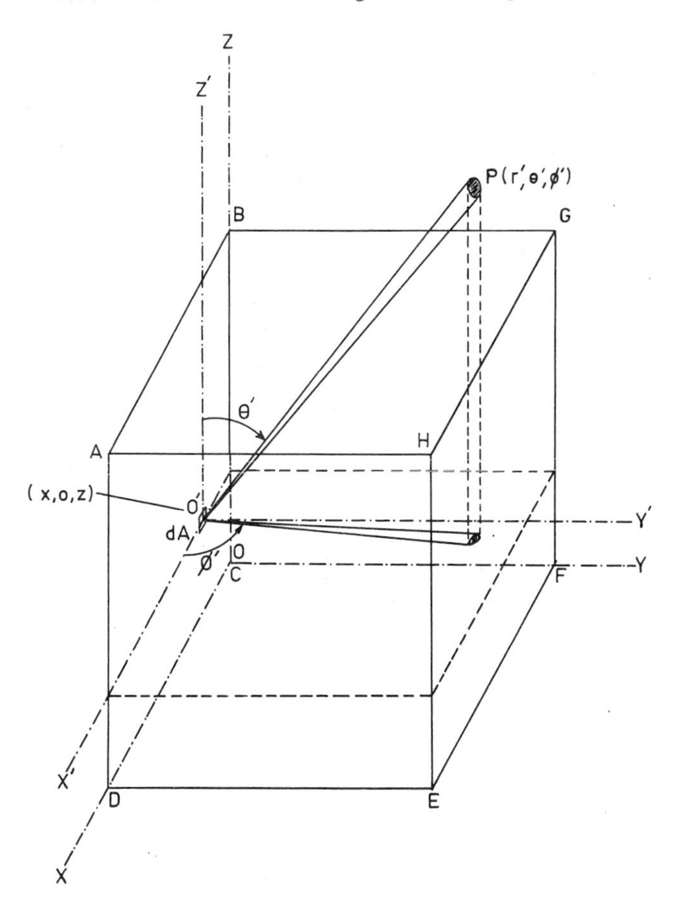

Fig. 5.5

	θ'	φ'
Part I	0° to $\pi/2-\tan^{-1}\left[(\frac{h-z}{b-x})\cos\phi'\right]$,	0 to $\tan^{-1}(\frac{d}{b-x})$
Part II	0° to $\pi/2-\tan^{-1}\left[(\frac{h-z}{d})\sin\phi'\right]$,	$\tan^{-1}(\frac{d}{b-x})$ to $(\pi/2)+\tan^{-1}(\frac{x}{d})$
Part III	0° to $\pi/2+\tan^{-}\left[(\frac{h-z}{x})\cos\phi'\right]$,	$\frac{\pi}{2}+\tan^{-}(\frac{x}{d})$ to π

where (X,O,Z) are the co-ordinates of the point O' in the unprimed system XYZ, and b, h and d are respectively the breadth, height and length of the courtyard, measured along X-axis, Z-axis and Y-axis respectively.

The sum of the three integrals yields the thermal emission to the sky from the elemental area dA in the form:

$$dE = \frac{I.dA}{2}\left[\frac{\pi}{2}\{1 - \frac{(h-z)}{\sqrt{d^2+(h-z)^2}}\} - \{\tan^{-1}(\frac{h-z}{b-x}) + \tan^{-1}(\frac{h-z}{x})\}\right.$$

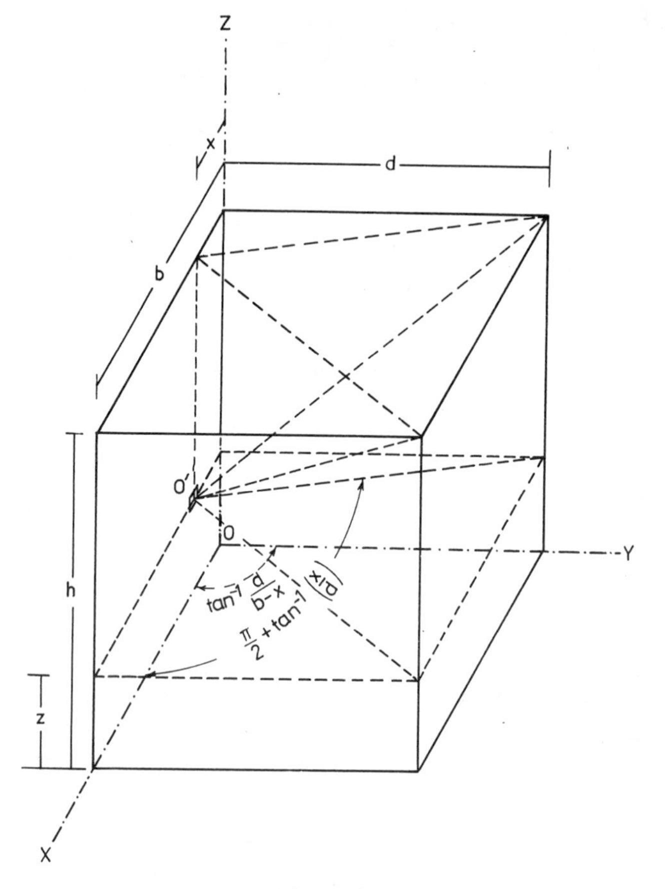

Fig. 5.6

$$+ \frac{(h-z)}{\sqrt{d^2 + (h-z)^2}} \left\{ \tan^{-1}\left(\frac{\sqrt{d^2+(h-z)^2}}{x}\right) + \tan^{-1}\left(\frac{\sqrt{d^2+(h-z)^2}}{(h-x)}\right) \right\} \Bigg], \qquad (5.15)$$

The total thermal emission, E, may, thus be evaluated by integrating (5.15) over the whole area ABCD yielding

$$E = I \Bigg[\frac{\pi b}{2} \left(h+d-\sqrt{h^2-d^2} \right) - hb\tan^{-1}\left(\frac{h}{b}\right) - db\tan^{-1}\left(\frac{d}{b}\right)$$

$$+ b\sqrt{d^2+h^2}.\tan^{-1} \frac{\sqrt{d^2+h^2}}{b} - \frac{b^2}{4} \log \left\{ \frac{(d^2+h^2+b^2)b^2}{(d^2+b^2)(h^2+b^2)} \right\}$$

$$- \frac{d^2}{4} \log \left\{ \frac{(d^2+b^2)(d^2+h^2)}{d^2(d^2+b^2+h^2)} \right\} - \frac{h^2}{4} \log \left\{ \frac{(d^2+h^2)(h^2+b^2)}{h^2(d^2+b^2+h^2)} \right\} \Bigg], \qquad (5.16a)$$

The ratio E/I gives the expression for the configuration factor F_{12}. For the special case of a square courtyard where d = b;

$$F_{12} = \left[\frac{\pi d}{2}(h+d - \sqrt{d^2+h^2}) - dh\tan^{-1}\left(\frac{h}{d}\right) - \frac{\pi}{4}d^2 + d\sqrt{d^2+h^2}\tan^{-1}\left\{\sqrt{1+\frac{h^2}{2}}\right\} - \frac{d^2}{2}\log\left\{\frac{(h^2+2d^2)}{2(h^2+d^2)}\right\} - \frac{h^2}{4}\log\left\{\frac{(d^2+h^2)}{h^2(h^2+2d^2)}\right\}\right], \qquad (5.16b)$$

For the special case street facade where b is infinite,

$$F_{12} = \frac{\pi}{2} \cdot b \cdot \left[h+d - \sqrt{d^2+h^2}\right] \qquad (5.16c)$$

If the radiant heat losses from a horizontal surface to the sky are taken as unity, the loss from a vertical surface with a particular obstruction can be expressed as

$$F_{12} = A_e \cdot \pi \qquad (5.16d)$$

where A_e is the effective area of the vertical surface for radiative heat loss and therefore A_e is given by

$$A_e = \frac{F_{12}}{\pi} \qquad (5.16e)$$

The obstruction due to surrounding vertical surfaces depends upon the ratio H_w (=H/W) (i.e. the height to width ratio of the street or courtyard) and the calculated values of A_e with respect to H_w have been presented in Fig. 5.7. It can be seen that when H_w is zero, that is the street width or courtyard width is infinite, the effective area A_e approaches 0.5. For normal street widths of H_w = 1 or 2, A_e for street walls is in the range of 0.28 to 0.18 and for courtyard walls from 0.22 to 0.14.

5.1.3.2 Radiative heat transfer to buildings. The estimation of the configuration factor, F_{12}, eventually leads to the estimation of average radiative heat transfer coefficient h_r as:

$$q_r = h_r A (T_1 - T_2) = F_{12}\sigma A (T_1^4 - T_2^4),$$

or

$$h_r = F_{12}\sigma \left[\frac{T_1^4 - T_2^4}{T_1 - T_2}\right], \qquad (5.17)$$

where surface temperatures T_1 and T_2 are expressed in degrees Kelvin.

5.2 SURFACE COEFFICIENTS

In addition to the heat-transfer by conduction, convection and radiation, other factors which influence the surface coefficient of heat-transfer are the presence of air cavity, the emissivity of the surface, its roughness, air movement and temperature relative to the surface. van Straaten (1967) has described the effect of these variables as follows.

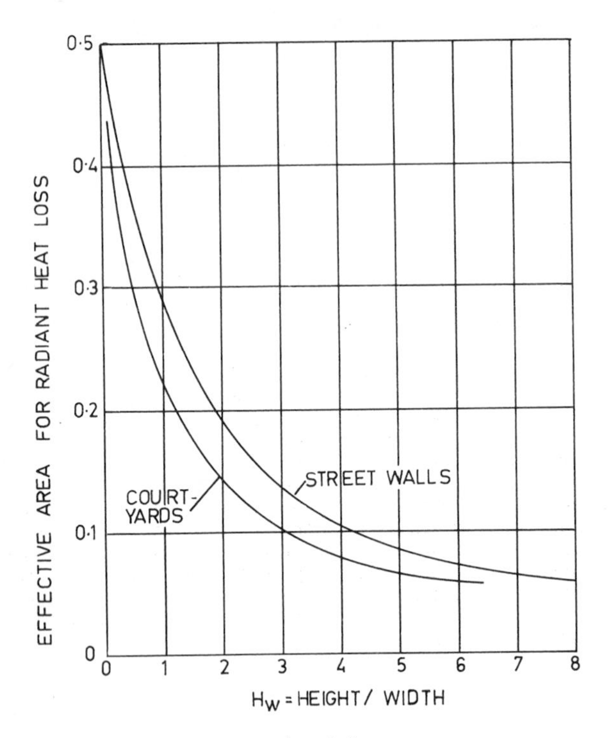

Fig. 5.7.

(i) An increased emissivity of the surface for the long wave radiation results in increased losses by radiation resulting in an increase in the surface coefficient. In a similar fashion if the emission is to an area of low temperature such as clear sky during night, the surface coefficient increases. It is due to this enhanced radiative loss that the surface temperature of a few lightweight components drops below outdoor air temperature during the night.

(ii) For higher wind speeds (higher rate of air movement), the surface coefficient is higher because the heat removed by forced convection is more than that due to natural convection.

(iii) As the temperature difference between the surface and the surrounding air increases, the surface coefficient increases because of the increase in the convective heat transfer.

(iv) A rough surface possesses a higher surface coefficient than a relatively smooth surface. This is because of the turbulent nature of the air flow over such surfaces. The roughness of the surface also increases the effective area. As an example, the surface coefficient for a corrugated surface is 20% higher than that for a flat but otherwise similar surface (Nash *et al.*, 1955)

(v) The other reasons for the differences in the values of the heat-transfer coefficients are the orientation of the components (wall, roof) and the direction of heat flow. For vertical surfaces, the coefficients are different than for the horizontal surfaces. For heat flow in the upward direction, the coefficients are greater than for the corresponding heat flow in the downward direction.

In situations where heat passes through air from the internal/external surfaces of the building as well as the air spaces existing in the skin of the wall/roof of the building, the surface heat transfer takes place by the combined action of convection and radiation as described below.

5.2.1 Air Cavity or Space

For better thermal comfort/load levelling, cavity walls and roof/ceiling combinations with an attic space in between are usually employed using construction materials such as hollow concrete block. The thermal resistance of the cavity depends upon the heat exchange between two parallel surfaces on either side of the cavity. The main factors which influence the thermal resistance of the cavity significantly are the emissivity of the two surfaces facing each other, the width of the cavity, rate of air movement and the direction of heat flow in the cavity. Nearly 60 to 65% of the heat exchange across the cavity takes place by radiation and the rest by convection currents within the air space itself. The heat transfer by conduction for air spaces thicker than 2 cm is almost negligible compared to convection and radiation. The heat exchange due to radiation can, however, be checked by using a suspended aluminium foil or shining corrugated aluminium sheet from the upper surface of the roof cavity or over ceiling joists with an air space under the foil; such an arrangement will provide low emissivity or low absorptivity to low temperature (thermal) radiation. Typical values of the thermal resistance for different conditions of an unventilated as well as a ventilated cavity in between building materials are given in Table A3 of Appendix 3.

5.2.2 Internal Surface

At the internal surface of the building, heat is transferred by convection through air and by radiation to the surrounding surfaces inside the building as

$$q_c = h_c A\,(\theta_{si} - \theta_{ai}) \tag{5.18a}$$

and

$$q_r = h_r A\,(\theta_{si} - \theta_{ai}) \tag{5.18b}$$

Therefore, the total flow of heat per unit area may be written as

$$\left(\frac{q_c+q_r}{A}\right) = (h_c + h_r)\,(\theta_{si} - \theta_{ai}) = h_{si}\,(\theta_{si} - \theta_{ai})\,,$$

$$= \left(\frac{1}{R_{si}}\right)(\theta_{si} - \theta_{ai}) \tag{5.18c}$$

where θ_{si}, θ_{ai} and R_{si} ($= 1/h_{si}$) are the inside surface temperature, air temperature and thermal resistance respectively. The recommended values of emissivity ε, h_c and h_r by IHVE Guide (1970) for internal surfaces of the building are given in Table A4 and the values of total internal surface coefficient, which depends upon the direction of heat flow, recommended for different surfaces in various countries are listed in Table A5 of Appendix 3. Subsequently, the variation of the inside surface resistance, R_{si}, with the emissivity of the surface for floor, wall and ceiling is plotted in Fig. 5.8 at mean surface temperature of 20°C. It is interesting to note that the thermal resistance is a maximum for the floor and a minimum for the roof of the building.

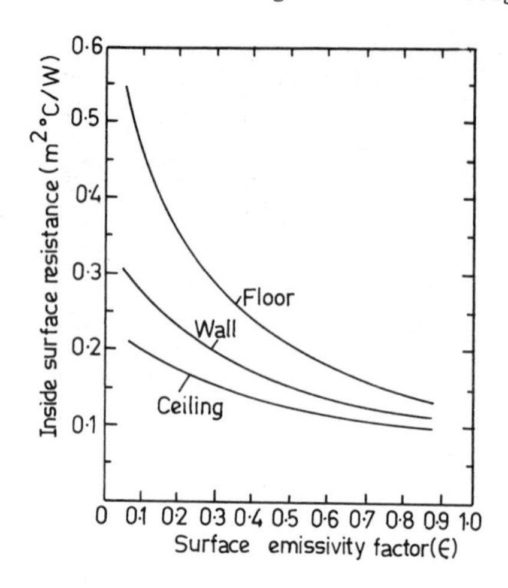

Fig. 5.8. Inside surface resistance R_{si} at mean surface temperature 20°C. (After Wong, 1977)

5.2.3 External Surface

Similar to internal surface, the thermal resistance for the external surface of a building is evaluated from the inverse of the total heat-transfer coefficient i.e. sum of the convective and radiative heat transfer coefficient

$$R_{so} = 1/(h_{co} + h_{ro}) = 1/h_{so} \tag{5.19}$$

where h_{ro} can be evaluated from Eq. (5.17), and its average value may be taken as 4.14 W/m^2 C and h_{co} being a function of air flow velocity (v), is given by (IHVE Guide A, 1970)

$$h_{co} = 5.8 + 4.1v \tag{5.20}$$

Generally, meteorological wind speed is measured at some higher level ($\sim$ 10 m) in the open which may be used to calculate the air speed at the desired level [Davenport (1967)] from the following relation

$$v = v_m \left(\frac{h}{h_m}\right)^{r_p} \quad \text{(m/s)} \tag{5.21}$$

where v and v_m are mean wind speeds at height h and h_m respectively and r_p is the roughness parameter given by

r_p = 0.15 for flat open country

= 0.29 for small town and city suburbs

= 0.45 for built-up urban centres.

Usually vertical walls receive two-thirds of wind speed so that at roof surfaces. From a field experimental study, Ito *et al.* (1972) have found that the air flow velocity near the external surface of the building is 1/3-1/5 times the wind speed when the surface is windward and 1/18-1/11 when it is leeward, and the convective heat transfer coefficient, h_{co}, is more closely related to the air velocity near the surface than to the wind speed. The surface coefficients for various wind speeds and heat flow directions in different countries are illustrated in Table 5.5.

TABLE 5.5 Outer Surface Coefficients Recommended in Various Countries (After van Straaten, 1967)

Country	Surface	Direction of heat flow	Surface coefficient ($W/m^2\ ^oC$)
America	All surfaces (15 mph wind)	All directions	34.0
	All surfaces (7.5 mph wind)	All directions	22.7
England	All wall surfaces	All directions	14.2 mean value*
	Roofs	All directions	22.7 mean value*
Holland	All surfaces	All directions	23.2
South Africa	All surfaces	All directions	19.9 mean value

*Normal conditions of exposure to wind.

As is obvious from the foregoing discussion that the rate of heat flow from an element is a complex function of wind speed and therefore, the degree of exposure of a building (relative to wind) has to be considered. Following three categories are normally selected for design purposes (IHVE Guide, 1970):

(i) Sheltered: low rise buildings of the lower floors of the buildings in urban areas (v = 1 m/s).

(ii) Normal: 4th to 8th floors of buildings in urban areas (v = 3 m/s).

(iii) Severe: Upper floors or buildings on exposed sites (v = 9 m/s).

The thermal resistance for roof and wall (of the above-mentioned three categories buildings) at different wind speeds are shown in Table 5.6.

TABLE 5.6 Values for Thermal Resistance, R_{so} (After Markus and Morris, 1980)

Surface	Exposure	Wind speed ($m\ s^{-1}$)	h_{co} ($W\ m^{-2}\,^oC^{-1}$)	R_{so} ($W^{-1}\ m^2\,^oC$)
Roof	Sheltered	1.0	9.9	0.07
	Normal	3.0	18.1	0.045
	Severe	9.0	42.7	0.02
Walls	Sheltered	0.7	8.7	0.08
	Normal	2.0	14.0	0.055
	Severe	6.0	30.4	0.03

5.3 AIR-TO-AIR HEAT TRANSMISSION: OVERALL THERMAL TRANSMITTANCE (U VALUE)

This is the rate at which heat is transmitted from air on one side of the wall or roof to the air on the other side due to difference in temperature. The heat is first transmitted from air to one surface of the element mainly by convection and radiation, then it is conducted through the fabric of the element followed by further transfer of heat from surface to the air on the other side (see Fig. 5.9). The rate of heat flow through a composite element of n layers may be given as

$$q = UA\,(\theta_{ai} - \theta_a), \tag{5.22}$$

where $(\theta_{ai} - \theta_a)$ is the air-to-air temperature difference across the element and U, the overall thermal transmittance is given by

$$U = \frac{1}{R_t} = \left(\frac{1}{R_{si} + R_{so} + R_n}\right) \tag{5.23}$$

where R_n is the total resistance per unit area of various structural components such as inner and outer leaves of masonry, the sheet of insulating materials and air spaces etc.

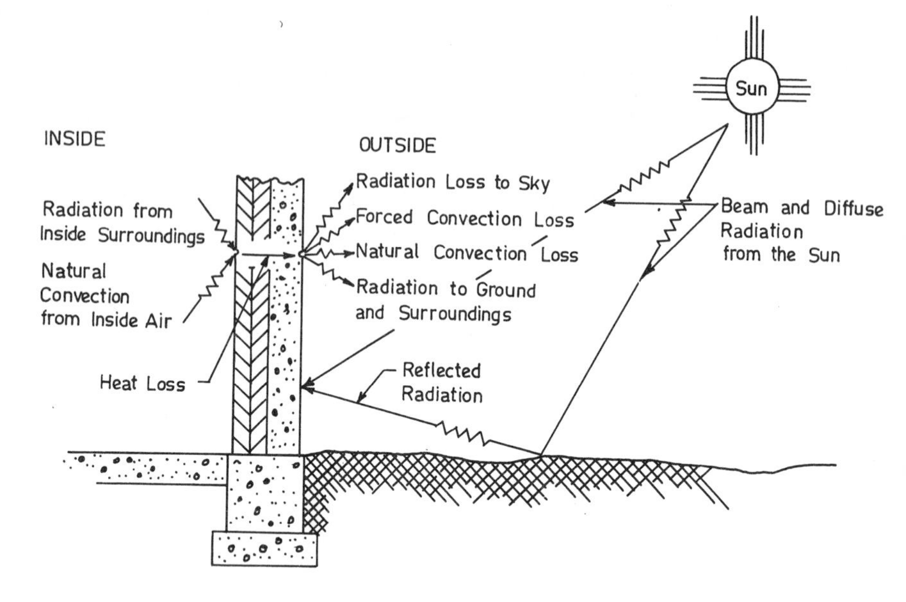

Fig. 5.9. Air-to-air heat transmission

5.3.1 Windows and Air Spaces

The total thermal resistance for heat flow through single glass may be written as

$$R_{gt} = (R_{si} + R_{so} + \frac{Lg}{Kg}), \tag{5.24}$$

For double or more glazings, the effect of air space (or cavity) and different layers of material has to be added up. The overall air-to-air transmission is depicted by the U value and for composite elements comprising different layers of materials as well as air spaces, may be written as:

$$U = \frac{1}{R_t} = \left[\frac{1}{R_{si} + R_{so} + \frac{1}{a_1} + \frac{1}{a_2} + \text{---} + \frac{1}{a_n} + \frac{L_1}{K_1} + \frac{L_2}{K_2} + \text{---} + \frac{L_n}{K_n}}\right] \tag{5.25}$$

where a_1, a_2,----a_n are the thermal conductances of n separate air spaces incorporated in the structure. The thermal resistance of the air film, that accelerates the convection phenomenon on both sides of the glass pane or in the cavity is shown in Table 5.7. The values obviously depend upon the emissivity of the surface and the flow of air. Subsequently, values of U for windows in timber frames for three types of buildings are given in Table 5.8.

5.3.2 Walls and Windows

Let us consider an element consisting of a number of units with areas A_1, A_2, --- A_n and corresponding U values U_1, U_2, --- U_n as shown in Fig. 5.10. The overall U-value of the entire element may be given (Markus and Morris, 1980) as

$$\bar{U} = \left(\frac{A_1U_1 + A_2U_2 + \text{---} + A_nU_n}{A_1 + A_2 + \text{---} + A_n}\right), \tag{5.26}$$

If A = total external wall area, A_g = window area,

A_w = solid wall area = $A-A_g$, U_g = U-value of window,

U_w = U-value of wall, r' = ratio-window to solid, hence

$$r' = \frac{A_g}{A}$$

Therefore, overall U-value for external wall

$$= (A_gU_g + A_wU_w)/(A_g + A_w)$$

$$= A_gU_g + (A - A_g)U_w/A$$

$$= r'AU_g + (A - r'A)U_w/A$$

$$= r'U_g + (1 - r')U_w \tag{5.27}$$

The effect of window exposure, room shape and colour on solar absorption is given in Table 5.9; depending on the colour of the wall, there will be an effective absorptance of the room defined in terms of the ratio of the window area to the surface area.

The standard U-values based on the effective resistances of the fabric components using standard assumptions about moisture contents of materials, rates of heat transfer to and from surfaces by radiation and convection and the presence of ventilated air spaces are given in Table A6 of Appendix 3.

5.3.3 Pitched Roof

A pitched roof (Fig. 5.11) essentially consists of three thermal resistances in

TABLE 5.7 Surface Heat Conductance and Unit Resistance for Air Films (After ASHRAE, 1972)

Position of Surface	Direction of heat flow	Surface emissivities					
		ε = 0.9		ε = 0.2		ε = 0.05	
		$W/m^2 {}^oC$	W^{-1} $m^2 {}^oC$	$W/m^2 {}^oC$	W^{-1} $m^2 {}^oC$	$W/m^2 {}^oC$	W^{-1} $m^2 {}^oC$
Still air							
Horizontal	Upward	9.26	0.11	5.2	0.194	4.3	0.232
Sloping - 45°	Upward	9.09	0.11	5.0	0.200	4.1	0.241
Vertical	Horizontal	8.29	0.12	4.2	0.238	3.4	0.298
Sloping - 45°	Downward	7.50	0.13	3.4	0.294	2.6	0.391
Horizontal	Downward	6.13	0.16	2.1	0.476	1.3	0.800
Moving air							
(any position)							
Wind is 6.7 m/s (for winter)	Any	34.0	0.029				
Wind is 3.4 m/s (for summer)	Any	22.7	0.044				

TABLE 5.8 U-values for Typical Windows (After *IHVE Guide*, 1970)

Window type	Fraction of area occupied by frame (per cent)	U-values for stated exposure ($W/m^2 °C$)		
		Sheltered	Normal	Severe
Single-glazing				
Wood frame	30	3.8	4.3	5.0
Metal frame	20	5.0	5.6	6.7
Double-glazing				
Wood frame	30	2.3	2.5	2.7
Metal frame with thermal break	20	3.0	3.2	3.5

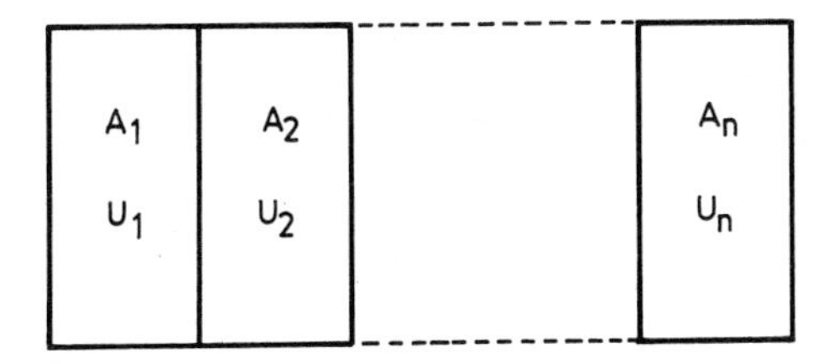

Fig. 5.10. Walls with windows

TABLE 5.9 Effective Room Absorptance vs. Wall Colour and Window Geometry for Double-Glazed Openings (After Burkhart and Jones, 1979)

Window area/ Total room surface area	Wall absorptance				
	20%	30%	50%	70%	90%
0.03	0.888	0.918	0.968	0.985	0.994
0.05	0.823	0.868	0.946	0.975	0.991
0.10	0.688	0.758	0.893	0.949	0.980
0.20	0.495	0.581	0.787	0.893	0.957

series namely, the ceiling resistance (R_c), the cavity resistance (R_v) and the slanted roof resistance (R_f). Under steady state conditions, the heat flow from room air (θ_{ai} °C) to the air space (θ_o °C) through ceiling, roof space to the ambient (θ_a °C) and from living space directly to the environment will be equal to each other and may be written as follows:

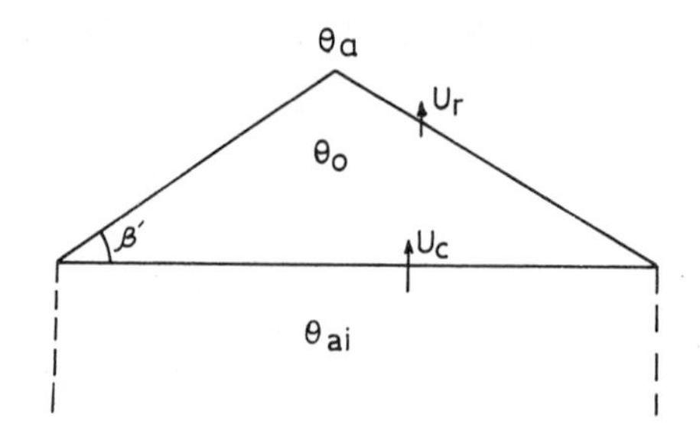

Fig. 5.11. Configuration of pitched roof

$$U_c A_c (\theta_{ai} - \theta_o) = U_r A_c (\theta_o - \theta_a) = UA_c (\theta_{ai} - \theta_a), \tag{5.28}$$

where A_c is the area of the ceiling, U_c and U_r are respectively the thermal conductances of the ceiling and roof. On solving equation (28), the overall thermal transmittance of the pitched roof system may be given by

$$U = \left(\frac{U_c U_r}{U_r + U_c \cos\beta'}\right) = \left(\frac{1}{R_c + R_r \cos\beta'}\right), \tag{5.29}$$

For $\beta' = 0$, the value of R will be simply the sum of the internal and external thermal resistances.

5.4 HEAT-TRANSFER DUE TO VENTILATION/INFILTRATION

Air enters a building due to infiltration and ventilation caused by wind flow or temperature difference existing between internal and external air. Infiltration is a natural process of leakage of air through cracks or openings round windows/doors in the building and is beyond the control of the inhabitants whereas ventilation is the flow of air through specific building openings such as windows, doors or ventilators and is within the control of the inhabitants. In both the modes, the amount of air flow depends upon the pressure gradient from outside to inside as well as the resistance offered by any opening. The wind, while striking the building at right angles, exerts pressure on the windward face and gets deflected over the top and round the sides of the building, yielding areas of suction on other faces of the building as indicated in Fig. 5.12. The pressure difference caused by the wind speed will result in the flow of air through the cracks from the windward side to the leeward side. However, the pressure difference caused by the difference in density due to the variation in indoor-outdoor temperature eventually leads to "stack effect"; the air moves from the lower level inlets to the higher level outlets in a heated building.

The two prevelent methods to estimate the air infiltration into the buildings (ASHRAE, 1972) are the following:

(i) the estimation of the leakage characteristics of the structure around windows and doors with respect to their pressure differences, normally referred to as "crack method".

(ii) the estimation based on the arbitrary number of air changes per hour in the room which is known as "air change method".

ASHRAE (1972) and IHVE (1970) have illustrated different formulae for both the methods, and simultaneously suggested the suitable ventilation rates for different

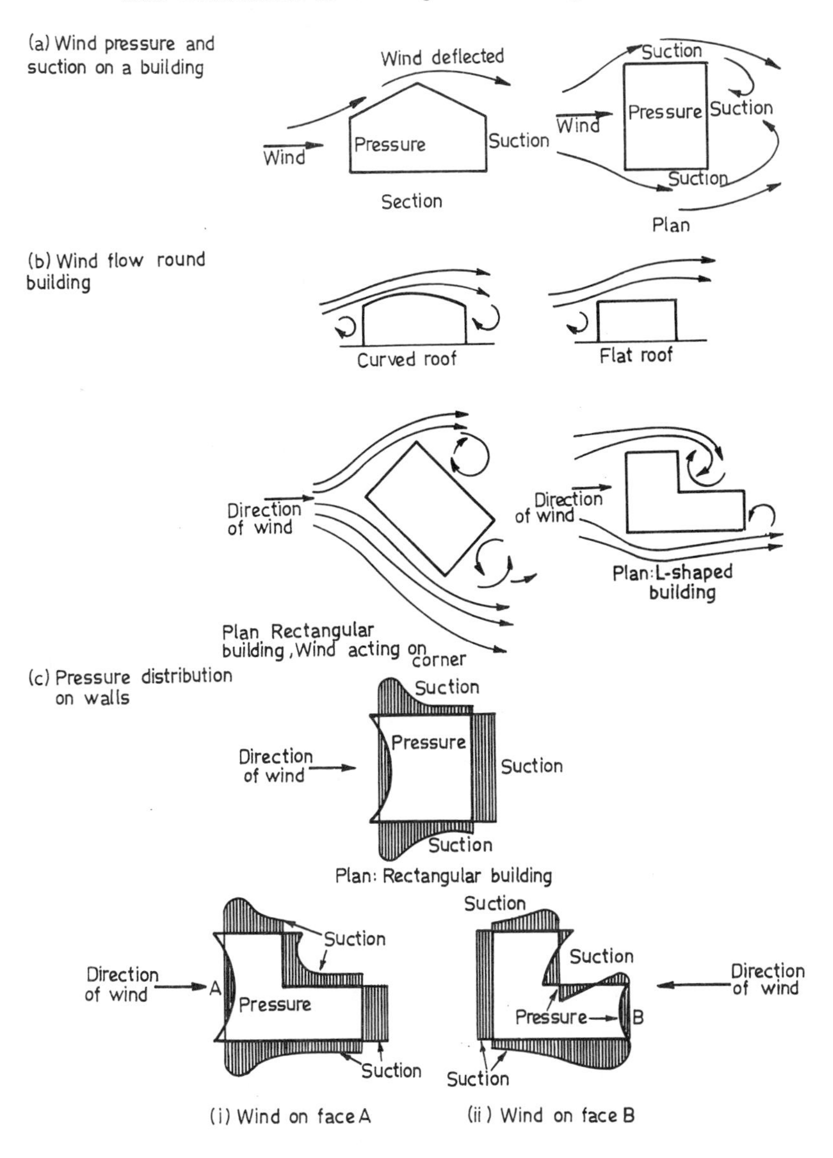

Fig. 5.12. (Contd. on next page)

orientations of the building. The appropriate relations for the rate of heat-transfer due to infiltration and ventilation, relevant to the text, are described in Table 5.10 and the ventilation rates for a domestic building are summarized in Table 5.11.

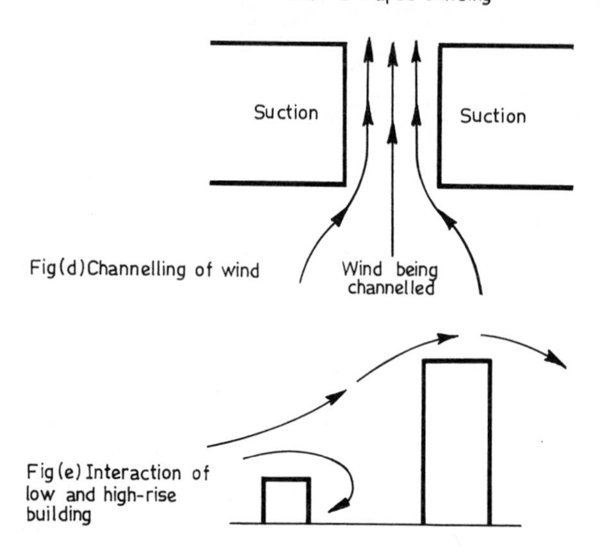

Fig. 5.12 (Contd). Wind flow directions and suction views (After Markus and Morris, 1980)

TABLE 5.10 Rates of Heat-Transfer due to Infiltration and Ventilation

	Crack Method	
	Wind Effect	Stack Effect
Pressure difference (Δp)	$1/2\ C\ \rho\ v^2$	$3462h\ (\frac{1}{\theta_a} - \frac{1}{\theta_{ai}})$
Air flow rate ($\dot{m}_v$)	$0.827\ \Sigma\ (A\ \Delta p^{\frac{1}{2}})$	$0.827\ \Sigma\ (A\Delta p^{\frac{1}{2}})$
for parallel openings $\dot{m}_v = 0.827\ (\Sigma A)\ (\Delta p)^{\frac{1}{2}}$;		
for series openings $\dot{m}_v = 0.827 \left[\frac{A_1A_2}{(A_1^2 + A_2^2)^{0.5}}\right] . (\Delta p)^{\frac{1}{2}}$		
Rate of heat-transfer (q_v)	$1200\ \dot{m}_v\Delta\theta$	$1200\ \dot{m}_v\Delta\theta$
	Air-change Method	
Air flow rate ($\dot{m}$)	NV/3600	
Rate of heat-transfer	$1/3\ NV\ (\theta_{ai} - \theta_a)$	

$\dot{m}_v$ = volumetric flow rate of air (m^3/s), C = specific heat of air (J/kg °C), ρ = density of air (kg/m^3), v = wind speed (m/s), h = height of air column (m), A = area of orifice (m^2), N = number of air changes (h^{-1}) and V = room volume (m^3).

TABLE 5.11 Natural Infiltration Rates (After ASHRAE, 1972)

Kind of room	Number of air changes per hour*
Rooms with no windows or external door	1/2
Rooms with window or external door on one side	1
Rooms with window or external door on two sides	1½
Rooms with window or external door on three sides	2
Entrance halls	2

*For rooms with weather-stripped windows use two-thirds of the values.

5.5 INTERMITTENT HEAT-TRANSFER

The internal surface temperatures of various building components vary due to intermittent heat gains from sources such as solar radiation, artificial lighting, electric equipments and occupancy or from outdoor temperature variation. Depending upon the peak values of heating loads of the buildings, for instance in summer, adequate natural ventilation (or if necessary air-conditioning) means are generally adopted to maintain the thermal comfort conditions inside the buildings particularly in temperate climates. Figure 5.13 shows that up to three air changes per hour, ventilation assists significantly in reducing overheating in summer (Croome, 1977, 1981).

Solar radiation plays a very important role in determining the rate of heat-transfer on the fabric of the building. The average heat flux entering a

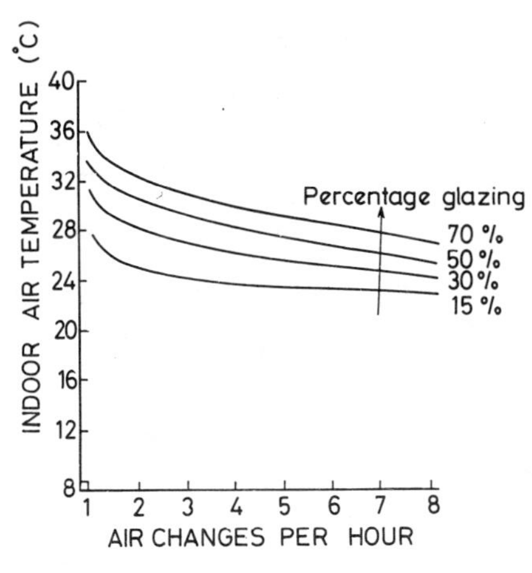

Fig. 5.13. Effect of air changes per hour on the indoor air temperature (After Croome, 1977, 1981)

building is the total mean of the solar energy admitted to the building through windows/fabric over period of 24 hours. The calculations are normally based on the concept of solair temperature, phase lag, transmission through glass windows and finally the admittance of the difference building components as described below.

5.5.1 Solair Temperature

Solair temperature, introduced by Mackey and Wright (1944), combines the effect of solar radiation, ambient air temperature and long wave radiant heat exchange with the environment. Physically solair temperature could be interpreted as that temperature of the surroundings which will produce the same heating effect as the incident radiation in conjunction with the actual external air temperature. The general formula for the solair temperature is

$$\theta_{sa} = \theta_a + \frac{1}{h_c}\left[\alpha.s - A_e\, \varepsilon\, \Delta R\right], \tag{5.30}$$

The above expression for the solair temperature can be easily derived by writing the energy balance on any wall/roof; viz.

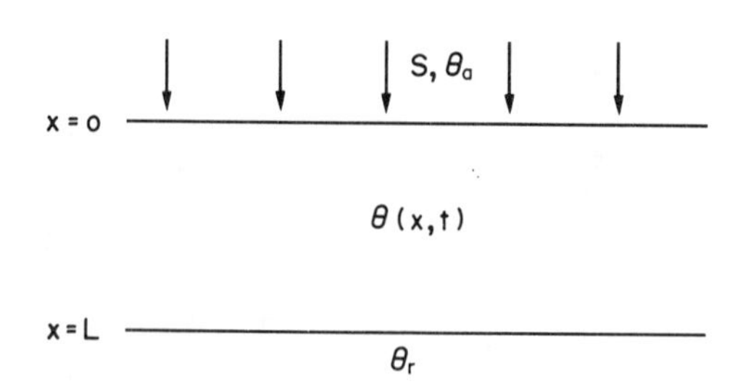

Fig. 5.14. A solid slab exposed to atmospheric conditions

Corresponding to Fig. 5.14 at x = 0

$$\alpha.s = -K\frac{\partial\theta(x,t)}{\partial x} + h_c\,(\theta(x = o,t) - \theta_a) + h_r\,(\theta(x = o,t) - \theta_{sky}), \tag{5.31}$$

The first term on the right-hand side represents the rate of heat conduction at the surface, the second convective heat losses from the outside surface to the ambient and the third radiative heat losses to the surroundings; θ_{sky} being designated as the equivalent black body sky temperature. Various relations have been proposed to estimate the sky temperature for clear sky conditions, Swinbank (1963) has forwarded the following expression

$$(\theta_{sky} + 273) = 0.0552\,(\theta_a + 273)^{1.5}, \tag{5.32a}$$

Whillier (1967) used a simple expression

$$\theta_{sky} = \theta_a - 6, \tag{5.32b}$$

whereas Brunt (1932) and Bliss (1961) related the effective sky temperature to the dew point temperature T_{dp} (in degree Kelvin) and the air temperature as

$$(\theta_{sky} + 273) = (\theta_a + 273)\left[0.8 + \frac{T_{dp} - 273}{250}\right]^{0.25} \qquad (5.32c)$$

In the overall calculation of the heat flux, it does not make much difference in choosing any one of the above equations; the range of the difference between the sky temperature and the air temperature is from 10 to 30°C for hot moist climate and cold dry climate respectively.

Equation (5.31) can be expressed in the following form

$$-K\frac{\partial\theta}{\partial x}\bigg|_{x=o} = \alpha s - h_c(\theta|_{x=o} - \theta_a) - h_r(\theta|_{x=o} - \theta_{sky}), \qquad (5.33)$$

but h_{so} (= $h_c + h_r$) is the total heat-transfer coefficient, hence, Eq. (5.33) can be written in form

$$-K\frac{\partial\theta}{\partial x}\bigg|_{x=o} = h_{so}\,(\theta_{sa} - \theta|_{x=o}), \qquad (5.33a)$$

Where the solair temperature, θ_{sa}, is defined as

$$\theta_{sa} = \frac{\alpha s}{h_{so}} + \theta_a + \frac{h_r}{h_{so}}(\theta_{sky} - \theta_a), \qquad (5.33b)$$

The concept of solair temperature allows an easy calculation of heating or cooling load through a building component as illustrated in the example below.

Example 5.1. Calculate the rate of heat flow through a south-facing concrete wall with mean incident solar radiation of 225 W/m^2, ambient air temperature 10°C, wall thickness 30 cm, wall conductivity = 0.72 W/m°C, mean room temperature, θ_r = 20°C, h_c = 8.7 W/m^2°C, hr = 3.8 W/m^2°C, α = 0.6, and R_{si} = 0.125 W^{-1} m^2 °C.

Solution: Corresponding to Fig. 5.14, the rate of heat flow may be given by Eq. (5.33a) as

$$\dot{Q} = -K\frac{\partial\theta}{\partial x}\bigg|_{x=o} = h_{so}\,(\theta_{sa} - \theta|_{x=o}),$$

Also in steady state

$$\dot{Q} = \frac{K}{L}(\theta|_{x=o} - \theta|_{x=L}),$$

and

$$\dot{Q} = h_{si}\,(\theta|_{x=L} - \theta_r),$$

Adding the three, one gets

$$\dot{Q} = U(\theta_{sa} - \theta_r),$$

where

$$\left[\frac{1}{U} = \left(\frac{1}{h_{so}} + \frac{L}{K} + \frac{1}{h_{si}}\right)\right],$$

here $h_{so} = h_c + h_r$ = 12.5 W/m^2 °C and $h_{si} = 1/R_{si}$

$$U^{-1} = \frac{1}{12.5} + \frac{0.3}{0.72} + 0.125 = 0.623$$

$$U = 1.609\ W/m^2\ ^oC$$

Using equations (5.32b) and (5.33b), the solair temperature will be

$$\theta_{sa} = \frac{0.6 \times 225}{12.5} + 10 + \frac{3.8}{12.5}(4 - 10)$$

$$= 18.98^oC$$

Therefore, the rate of heat flow

$$\dot{Q} = 1.609 \times (18.98 - 20)$$

$$= - 1.65\ W/m^2$$

The -ve sign means that the heat is lost from the living space to the environment and heat has to be added at the above rate to maintain the space at 20°C.

The same calculation, if performed, for an ambient temperature of 40°C yields the rate of heat flow as 46.6 W/m^2, measuring that heat has to be removed at this rate to maintain the space at 20°C.

5.5.2 Decrement Factor (d_f)

The effect of thermal capacity in a building-component is to delay the attainment of steady state conditions after the heat is absorbed at the external surface. The temperature change at the exposed surface is propagated through the material as a temperature wave, which gradually loses its amplitude as it progresses. For a sinusoidal change of temperature of the exposed surface, the temperature at any point will also be sinusoidal with the same frequency but lags behind the variation at the surface temperature more and more as the distance from the surface is increased creating simultaneously more attenuation of the temperature wave. The ratio between the temperature-amplitudes of the internal and external surfaces is known as the decrement factor. Time lag and the decrement factors both are connected. For building elements of massive construction, the time lag is larger and the decrement factor smaller than for a lightweight element of the same U value. To calculate the decrement factor (of a multilayered slab) the order of layering is not important. Following Milbank and Harrington-Lynn (1974), the decrement factor along with other thermal factors, for various constructions and materials is given in Table A7 of Appendix 3.

5.5.3 Phase Lag

The phase lag is the time delay between the impact of the diurnal variation of temperature and radiation on the external surface, and the resultant temperature variation on the internal surface. The heat storing capacity of building components, plays an important role in determining the phase lag bringing marked difference between the time dependent thermal performance of heavyweight and lightweight structures under such circumstances. Thus, the heat gain through a 200 mm thick brick wall under typical summer conditions in warm arid countries can be about half of that through a similar lightweight wall of a comparable U value as shown in Fig. 5.15.

The time lag for a homogeneous material subject to temperature fluctuations with a 24-hour period, is given approximately by the formula

$$\psi = 1.38\ L\sqrt{\frac{1}{\mathcal{D}}} \tag{5.34}$$

where ψ is the time lag (in seconds) L is the thickness (m) and $\mathcal{D}$ (= $K/\rho C$) is the thermal diffusivity (m^2/s)

More exact calculations of the time lag can be performed by the admittance and the Fourier series methods. Range of values of time lag for homogeneous building materials is given in Table 5.12 along with their diffusivity values. Difference in the density and moisture content also affect the time lag. The situation in the multilayer constructions is more complex since the resistance of one layer affects the rate of heat-transfer to the adjacent layer. The time lag is not the sum of the time lags of the individual layers but it will depend on the properties of the layers and the sequence in whicn these are arranged.

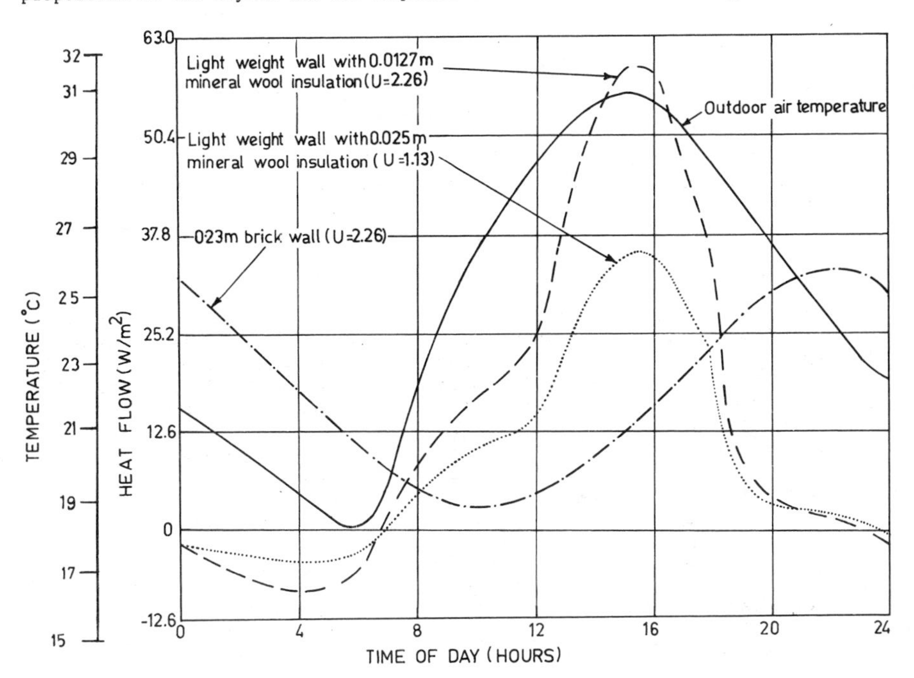

Fig. 5.15. Comparison of heat flows through lightweight and heavyweight walls facing west on a typical summer's day for Pretoria (After van Straaten, 1967)

Considerable savings in the material can be achieved by the judicious use of combinations of light insulating materials with low capacity, and heavy high heat capacity materials, to achieve long time lags without using extreme thicknesses, which would otherwise be required for the case of a single homogeneous material. The time lag through multilayered structure can be calculated using the time constant concept, defined as the time taken for a layer of wall or roof to increase in temperature by a certain proportion of an instantaneous change in external temperature. It is defined by the ratio

$$\text{Time constant}_{\tau} = \frac{\text{Volumetric heat capacity of the material layer}}{\text{Rate of heat-transfer into (or out of) the material per degree temperature difference}}$$

TABLE 5.12 Time lag for homogeneous materials (After Evans, 1980)

Material		Diffusivity ($\times 10^{-6}$ m^2/s)	Time lag in hours — Thickness of materials (mm)					
			25	50	100	150	200	300
Dense concrete	min	0.768	-	1.1	2.5	3.8	4.9	7.9
	max	1.03	-	1.5	3.0	4.4	6.1	9.2
Brick	min	0.56	-	-	2.3	-	5.5	8.5
	max	0.66	-	-	3.2	-	5.5	10
Wood	min	0.125	0.4	1.3	3.0	-	-	-
	max	0.162	0.5	1.7	3.5	-	-	-
Fibre insulating board	average	-	0.27	0.77	2.7	5.0	-	-
Concrete with foamed slag aggregate	average	-	-	-	3.25	-	8	-
Stone	average	1.86	-	-	-	-	5.5	8.0
Stabilized soil	average	-	-	-	2.4	4.0	5.2	8.1

$$\tau = \frac{Q_v}{U} \quad \text{(in seconds)}.$$

For a multilayered structure

$$\frac{Q_v}{U} = \frac{Q_{v1}}{U_1} + \frac{Q_{v2}}{U_2} + \frac{Q_{v3}}{U_3} + \text{---}$$

$$\text{(1st layer)} \quad \frac{Q_{v1}}{U_1} = (R_{so} + \frac{L_1}{2K_1})\,(L_1\rho_1 C_1) \tag{5.35}$$

$$\text{(2nd layer)} \quad \frac{Q_{v2}}{U_2} = (R_{so} + \frac{L_1}{K_1} + \frac{L_2}{2K_2})\,(L_2\rho_2 C_2)$$

$$\text{(3rd layer)} \quad \frac{Q_{v3}}{U_3} = (R_{so} + \frac{L_1}{K_1} + \frac{L_2}{K_2} + \frac{L_3}{2K_3})\,(L_3\rho_3 C_3)$$

where K_j, L_j, C_j and ρ_j are respectively the conductivity (W/m°C), thickness (m), specific heat (J/kg°C) and density (kg/m³) of the corresponding (j = 1 to 3) layers.

The Q_v/U value for thin, low heat capacity layers such as thin metal sheet, plastic film, aluminium foil and air space is negligible but the resistance of these layers will affect the heat flow into subsequent layers. For example, the time constant of a multilayer construction can be increased by using lightweight insulating layers outside high heat capacity layers. The external insulating layer, due to its low capacity, gets heated quickly. The rate of heat-transfer across this layer will however be low so that the heavy layer will heat up more slowly resulting in a longer time constant. When the position of the layer is reversed, the heavy outer layer has a higher conductivity so that the heat will be rapidly transferred to the inner insulating layer which will heat up quickly since its heat capacity is low. The result is that the time constant of this combination will be very similar to the time constant of the heavy layer alone. The value of time constant for difference types of walls and roof, as calculated by Raychaudhury and Chaudhury (1961), are summarized in Table 5.13.

TABLE 5.13 Q_v/U Values of Walls and Roofs; Order of Layers given from the Exterior Inward (After Raychaudhury and Chaudhury, 1961)

Component	Description	Q_v/U (in hours)
Wall	38.1 mm plaster, 228.6 mm brickwork, 38.1 mm plaster	18.2
Wall	12.7 mm plaster, 114.3 mm brickwork, 12.7 mm plaster	6.7
Wall	12.7 mm plaster, 114.3 mm brickwork, 50.8 mm air space 114.3 mm brickwork, 12.7 mm plaster	25.2
Wall	25.4 mm plaster over wire mesh, 25.4 mm expanded plastic, 114.3 mm brickwork, 12.7 mm plaster	50.2
Wall	12.7 mm plaster, 114.3 mm brickwork, 25.4 expanded plastic, 25.4 mm plaster over wire mesh	17.1
Roof	12.7 mm plaster, 165.1 mm reinforced concrete, 12.7 mm plaster	10.7
Roof	12.7 mm cement concrete, 50.8 mm vericulite concrete, 114.3 mm reinforced concrete	55.2

5.5.4 Admittance

The admittance of a surface is the rate at which the surface absorbs or emits heat from or to the air when the air temperature is different from the temperature of the surface. The factors determining the admittance are the thermal conductivity and the thermal capacity of the surface layers of walls and roofs. For slabs less than 75 mm thick, the admittance approximates to the U value of the structure, while for thicknesses above 200 mm, admittances tend to a constant value. In comparison to lightweight materials, dense constructions have higher admittances i.e. they absorb more energy for a given temperature swing, and it is this characteristic which leads to descriptions such as lightweight or heavyweight. Since the admittance also depends on thermal conductivity, thin plastic nylon film has, however, the same value of Y ($\sim$ 6) as that of window glass.

The room admittance is a function of the admittance, Y, of the individual elements defined as the reciprocal of the thermal impedence of an element to cyclic heat flow (Milbank and Harrington-Lynn, 1977) from the environmental temperature point and it has the same units as U value i.e.

$$Y = \tilde{q}/\tilde{\theta} \quad W\ m^{-2}K^{-1} \tag{5.36}$$

where $\tilde{q}$ is the heat input per unit area = $\tilde{Q}/A$ and $\tilde{\theta}$ = temperature swing. In the case of composite units e.g. a concrete slab with insulation on the surface receiving the variable heat input, it is the layer of insulation which will influence the value of the admittance. Admittance values of a few building elements are given in Table A7 of Appendix 3.

5.5.5 Heat-Transfer Through Glass Windows

In case of the radiant energy striking a surface, it has been stated that it would be transmitted, reflected or absorbed. The unique property of the glass and some transparent plastics, which is responsible for their specific thermal effect is the differential transparency to shortwave and longwave radiation, in the spectral range 300 to 2800 nm; its spectral distribution is shown in Fig. 5.16.

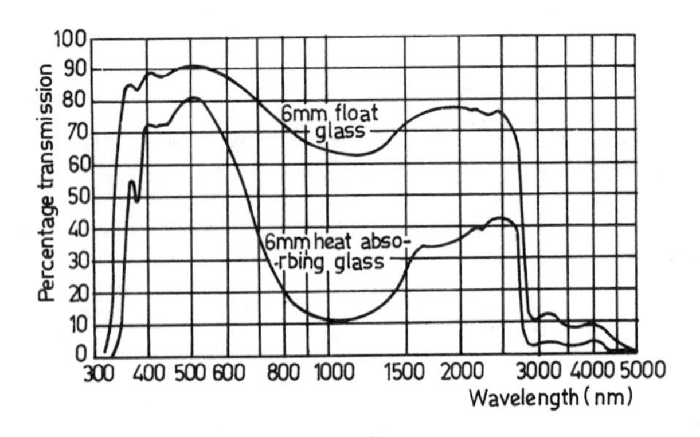

Fig. 5.16. The spectral transmission characteristics of glass (after Turner, 1969)

Glass is opaque to longwave emission and it is this phenomenon which gives rise to the greenhouse effect; some lightweight membranes like plastic films do transmit parts of longwave infrared spectrum. The basic difference between different glasses lie in their transmission characteristics. While clear glass transmits the major portion of radiation in the visible and infrared part of the spectrum, the heat absorbing glass reduces light transmission more than it reduces heat transmission. Grey and coloured (anti-glare) glasses absorb more of the visible part of the spectrum and they may be grey or coloured according to the fraction of the visible light mostly absorbed. Heat absorbing glass is also characterized by the high absorption of the infrared portion of the solar spectrum, while transmitting most of the visible light. The increased selective infrared absorption is due to a higher content of iron oxide among the ingredients of the glass. In consequence of this absorption, the surface temperature of the glazing is higher than for normal transparent glazing.

Solar heat gain through heat absorbing glass comprises of two parts: the first part is the direct transmission of the visible shortwave and infrared radiation and the second is the inward heat flow by convection and longwave radiation.

Heat reflecting glasses are obtained by depositing very fine semitransparent metallic coatings in surface of the glass which reflects selectively a greater proportion of the infrared radiation. As the coating is sensitive to mechanical damage, reflecting glasses require protection either by double-glazing with an air space or by lamination.

The energy transmitted through the glass depends upon the transmissivity of the glass. The properties of a few, commonly used glazing materials are given in Table 5.14.

The absorbed energy by the glass results in an increase in the temperature of the glass and a proportion of this absorbed energy will be retransmitted. A method of determining the retransmission coefficient is given below (Louden, 1968).

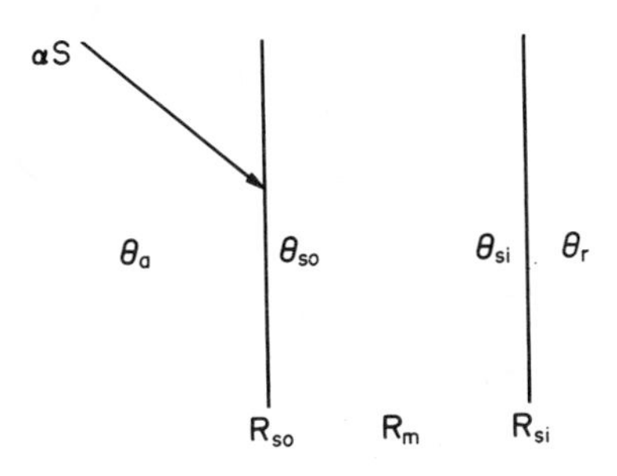

Fig. 5.17. Transmission of light through glass

For steady state conditions (Fig. 5.17), the heat flow per unit area may be defined as

$$Q/A = \alpha S - (\theta_{so} - \theta_a)/R_{so} \quad (5.37a)$$

$$Q/A = (\theta_{so} - \theta_{si})/R_m \quad (5.37b)$$

$$Q/A = (\theta_{si} - \theta_r)R_{si} \quad (5.37c)$$

Adding the above three equations, one gets

$$Q/A = U\alpha S\ R_{so} + U\ (\theta_a - \theta_r) \quad (5.37d)$$

where U has already been defined in section 5.3.1 as

$$U^{-1} = R_{so} + R_m + R_{si} \quad (5.37e)$$

and R_m is the resistance offered by the material to heat flow.

The factor $UR_{so} = f_r$ is called the retransmission factor defined physically as

$$f_r = \frac{\text{Thermal resistance from outside to point where heat is generated}}{\text{Total thermal resistance of the structure}}$$

Knowing the amount of energy absorbed, and hence retransmitted, the total radiation being transmitted through the glass can be obtained. Figure 5.18 shows typical transmission, absorption and reflection curves for different types of glass.

TABLE 5.14 Properties of Various Glazing Materials

Material	Typical trade name*	Representative thickness (mm)	Solar⁺ transmission (%)	Infra-red transmission	Maximum recommended size (m)	Weight (kg/m^2)	Flammability	Maximum operating temp. (°C)	Estimated lifetime for solar use (Years)
Clear glass (float lime)		3	85	Low	0.86 x 1.93	7.95	Low	204	Over 30
		5	81	Low	0.91 x 2.43	12.2	Low	204	Over 30
Low-iron glass	ASG Lo-Iron°	3	87	Low	0.86 x 1.93	5.03	Low	204	Over 30
		5	85	Low	0.91 x 2.43	12.2	Low	204	Over 30
Water-white	ASG Sunadex°	3	91	Low	0.86 x 1.93	7.86	Low	204	Over 30
		5	90	Low	0.91 x 2.43	11.8	Low	204	Over 30
Stabilized fiber glass reinforced polyester (F.R.P.)	Kalwall sunlite°	1	87	Low	0.60 (short dimension)	1.46	Low	121	10 - 15
		1.5	85	Low	0.76 (short dimension)	2.2	Low	121	10 - 15

⁺Solar transmission factor = $(\tau + \frac{\alpha}{2})$

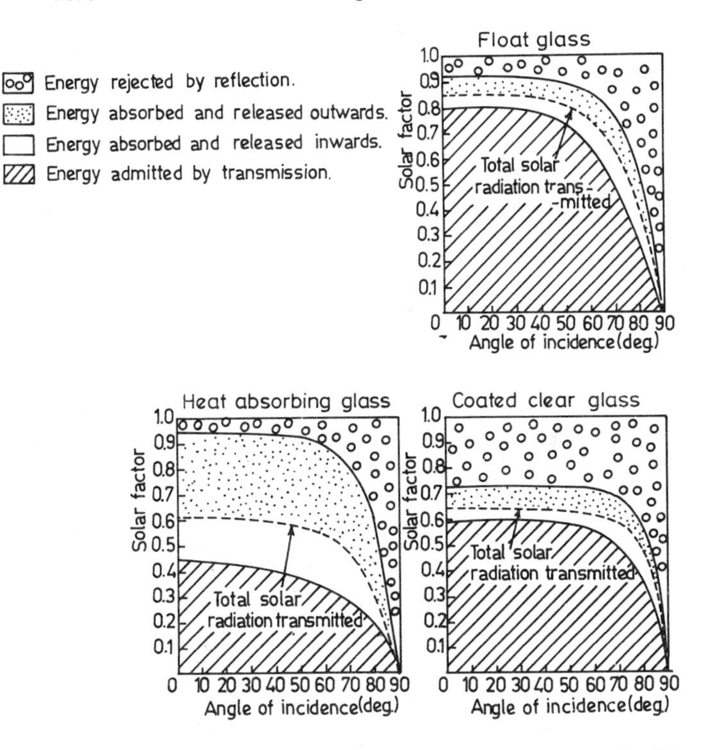

Fig. 5.18. Glass transmission and absorption vs. angle of incidence (After Turner, 1969)

An exact determination of total radiation transmitted through a window is made by using solar charts introduced in Chapter 2 and by the corresponding values of the reflection, transmission coefficients of the glass used in the windows. However, when the mean daily transmitted radiation is required or less accurate value than that obtained either by measurements or calculations, it is more convenient to use an overall solar gain factor for the glass (Louden, 1968).

Solar gain factor, f_s, depends on the angle of incidence of the radiation and additionally it takes account for interreflection between different window surfaces in the window system. Typical values (Louden, 1968) for various types of glazing and sun controls averaged over August for orientations south of east-west are given in Table 5.15; they can be used without much error for other months also.

Example 5.2. Assuming the heat loss by ventilation to be zero, calculate the average temperature of a room (typically 8 m x 6 m x 3 m) having an external wall on the long axis and a south-facing single-glazed window (typically 3 m x 2 m) for the following conditions: average ambient temperature $\bar{\theta}_a = 15^oC$, mean global irradiance on window $\bar{S}_g$ = 150 W/m^2. mean solair temperature $\bar{\theta}_{sa}$ = 22°C, heat transmission coefficient of the wall U_w = 0.7 W/m^2 ^{o}C; of window U_g = 5.6 W/m^2 ^{o}C.

Solution: Solar heat gain = $f_s . \bar{S}_g . A_g$

where A_g = area of glass = 6 m^2

and f_s = Solar gain factor = 0.76 (from Table 5.15)

TABLE 5.15 Solar Gain Factors (f_s) for Various Types of Glazings and Shading (Strictly for UK only, Approximately Correct World Wide) (After *CIBS Guide*, 1975)

Position of shading and type of sun protection		Solar gain factors* (f_s) for the following types of glazing	
Shading	Type of sun protection	Single	Double
None	None	0.76	0.64
	Lightly heat absorbing glass	0.51	0.38
	Densely heat absorbing glass	0.39	0.25
	Lacquer coated glass, grey	0.56	-
	Heat reflecting glass, gold (sealed unit when double)	0.26	0.25
Internal	Dark green open weave plastic blind	0.62	0.56
	White venetian blind	0.46	0.46
	White cotton curtain	0.41	0.40
	Cream holland linen blind	0.30	0.33
Mid-pane	White venetian blind	-	0.28
External	Dark green open weave plastic blind	0.22	0.17
	Canvas roller blind	0.14	0.11
	White louvered sunbreaker, blades at 45°	0.14	0.11
	Dark green miniature louvered blind	0.13	0.10

*All glazing clear except where stated otherwise. Factors are typical values only and variations will occur due to density of blind weave, reflectivity and cleanliness of protection.

∴ Solar heat gain = 0.76 x 150 x 6 = 684 W

Calculation of heat losses:

(i) Fabric heat loss through window $= U_g A_g (\bar{\theta}_r - \bar{\theta}_a)$

$= 5.6 \times 6 \times (\bar{\theta}_r - 15)$

$= 33.6 (\bar{\theta}_r - 15)$

(ii) Fabric heat loss due to opaque portion of external wall $= U_w A_w (\bar{\theta}_r - \bar{\theta}_{sa})$

A_w = wall area $= 8 \times 3 - A_g = 24 - 6 = 18 m^2$

∴ $U_w A_w (\bar{\theta}_r - \bar{\theta}_{sa}) = 0.7 \times 18 \times (\bar{\theta}_r - 22) = 12.6 (\bar{\theta}_r - 22)$

Now equating the heat gain to heat loss, we have

Solar heat gain = Heat losses due to (window + external wall).

∴ $684 = 33.6 (\bar{\theta}_r - 15) + 12.6 (\bar{\theta}_r - 22)$

Therefore, mean room air temperature is

$$\therefore \quad \bar{\theta}_r = 31.7^{\circ}C$$

If the conditions are not steady, such as usually the case with changing ambient conditions and incident solar radiation, one has to use an alternating solar gain factor, the values of which are given in Table 5.16.

TABLE 5.16 Alternating Solar Gain Factors ($\tilde{f}_s$) for Various Types of Glazing and Shading and for Heavyweight and Lightweight Buildings (Strictly Accurate for UK Only; Approximately Correct World World) (from *CIBS Guide*, 1975)

Position of Shading and type of sun protection		Alternating solar gain factors* ($\tilde{f}_s$) for the following building and window types			
		Heavyweight building		Lightweight building	
Shading	Type of sun protection	Single	Double	Single	Double
None	None	0.42	0.39	0.65	0.56
	Lightly heat absorbing glass	0.36	0.27	0.47	0.35
	Densely heat absorbing glass	0.32	0.21	0.37	0.24
	Lacquer coated glass, grey	0.37	-	0.50	-
	Heat reflecting glass, gold (sealed unit when double)	0.21	0.14	0.25	0.20
Internal	Dark green open weave plastic blind	0.55	0.53	0.61	0.57
	White venetian blind	0.42	0.44	0.45	0.46
	White cotton curtain	0.27	0.31	0.35	0.37
	Cream holland linen blind	0.24	0.30	0.27	0.32
Mid-pane	White ventian blind	-	0.24	-	0.27
External	Dark green open weave plastic blind	0.16	0.13	0.22	0.17
	Canvas roller blind	0.10	0.08	0.13	0.10
	White louvered sunbreaker, blades at 45°	0.08	0.06	0.11	0.08
	Dark green miniature louvered blind	0.08	0.06	0.10	0.07

*All glazing clear except where stated otherwise. Factors are typical values only and variations will occur due to density of blind weave, reflectivity and cleanliness of protection.

A gain of solar heat is desirable in cold winter conditions. In warm climates, the heat gains through windows should be reduced to a minimum value. Methods adopted to reduce heat gains in the desired months include orientation of the building and shading devices discussed earlier in Chapter 2. The relative performance of various devices to control solar heat gains is determined by a factor "shading coefficient". This factor is defined as the ratio of solar heat gain factor through the window under consideration to solar heat gain factor for ordinary class. Values of shading coefficients for various types of windows are given in Table 5.17. It is observed that double-glazing with ordinary glass is relatively ineffective as a means of reducing heat gains; it is only 10% more effective than the ordinary single-glazing. Heat absorbing glass reduces the heat gain by about 50%, while the glass with heat reflective films is even more efficient and reduces the daylight factor considerably which is a function of window size also as shown in Fig. 5.19. The use of heat absorbing glass combined with ordinary clear glass on the inside results in a reduction of solar heat gain of the order of 40%. A combination of heat reflecting glass outside the ordinary glass inside with a well insulated air space in between yields a reduction of just over 80%. The advantage of double-glazing of this type is that the internal glass protects the occupants from the high temperature of the outer glass and thus almost eliminates the effect of directional radiation. The acoustic properties of double glazing are described in Croome (1977).

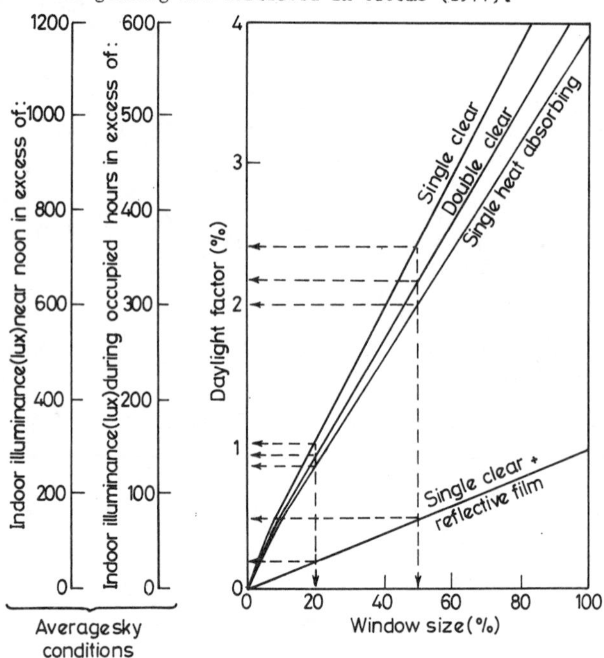

Fig. 5.19. Variation of daylight factor and indoor illuminance (in August) with window size for four types of glazing (after Wise, 1973)

The effectiveness of internal shading devices in the form of venetian and roller blinds and draperies depends on the extent to which short wave or solar radiation is reflected back through the glass. Consequently these should not be used in combination with a heat absorbing glass. This may be seen in Fig. 5.20 wherein the effect of various solar shading devices on the indoor temperature in lightweight and heavyweight buildings has been shown.

Various forms of glass shadings may be reasonably effective, but there is little doubt that external shading in the form of louvers, overhangs or in the form of

TABLE 5.17. Shading Coefficients for Different Combinations of Glasses and Various Shading Devices so Adjusted as to Exclude Direct Sun Penetration except where otherwise Stated (After van Straaten, 1967)

Method of shading	Description	Shading coefficient
Double-glazing	Ordinary clear glass (solar transmittance 0.86) both sides	0.90
Double	Heat-absorbing glass (solar transmittance 0.46) outside and	0.56
	Solarshield UV 393 I.R. 2/20 outside and ordinary clear glass inside. Air space ventilated to outside	0.17
Internal shading	Ordinary glass with ventian blind:	
	light-coloured	0.55
	medium-coloured	0.64
	Ordinary glass with opaque roller shade: white	0.25
	dark-coloured	0.59
	Ordinary glass with curtains or draperies: light, closed weave	0.44
	dark, closed weave	0.62
Between glass shading	Ordinary glass both sides and venetian blind in between:	
	white	0.33
	medium-coloured	0.36
External shading	Ordinary glass shaded with miniature louvers (Koolshade) dark-coloured:	
	17 louvers/in.	0.49-0.13*
	23 louvers/in.	0.41-0.09*
	Ordinary glass shaded completely with awnings, louvers, etc., and allowing free air movement	0.20

*Shading coefficient decreases linearly from maximum value at normal incidence to constant lower value at all angles of incidence greater than 30°

reflecting glass is the most effective method of controlling solar heat gains through windows. The quality of daylight and maintenance requirements are other factors which are important in deciding the type of glass shading to be used.

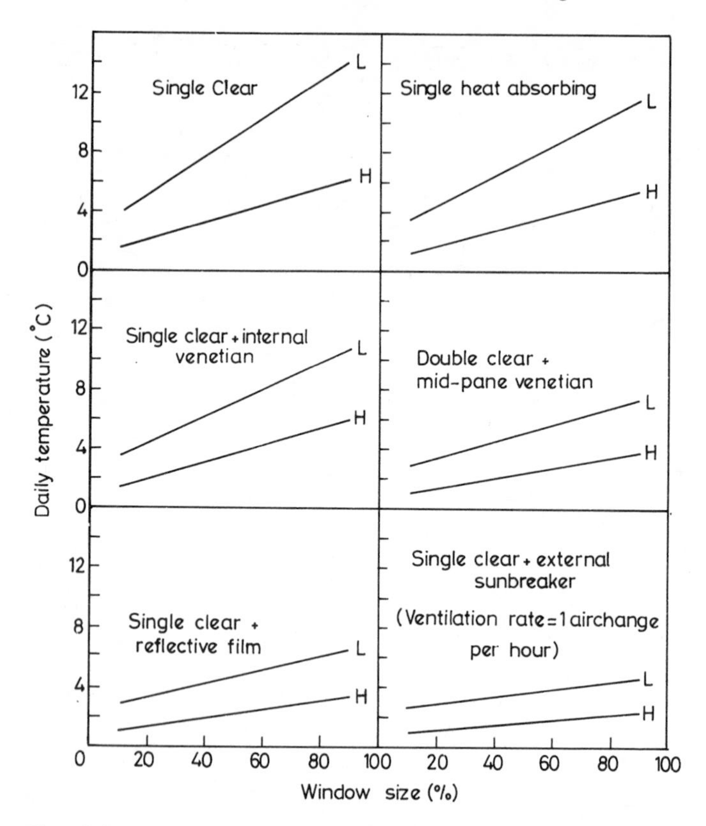

Fig. 5.20. Solar control schemes and their effect on indoor temperature in lightweight (L) and heavyweight (H) buildings (after Wise, 1973)

REFERENCES

Anon (1975) Standard U-Values, Building Research Establishment Digest, No. 108, HMSO, London.

ASHRAE *Handbook of Fundamentals* (1972) American Society of Heating, Refrigerating and Air-Conditioning Engineers.

ASHRAE *Handbook of Fundamentals* (1977) American Society of Heating, Refrigerating and Air-Conditioning Engineers.

Bliss, R. W. (1961) Atmospheric Radiation Near the Surface of the Ground, *Solar Energy, 5*, 103.

Brunt, D. (1932) Notes on Radiation in the Atmosphere, *Quart. J. Roy. Meteorol. Soc., 58*, 389-420.

Burkhart, J. F. and Jones, R. E. (1979) The Effective Absorptance of Direct Gain Rooms, Proc. 4th National Passive Solar Conference, Kansas City, International Solar Energy Society, October.

CIBS Guide (1975) Section A8 Summertime Temperatures in Buildings.

Croome, D. J. (1977) *Noise Buildings and People*, Pergamon Press Ltd., Oxford.

Croome, D. J., Roberts, B. M. (1981) Airconditioning and Ventilation of Buildings, Pergamon.

Davenport, A. G. (1967) The Dependence of Wind Loads on Meteorological Parameters, Proc. Symposium, *Wind Effects on Buildings and Structures*, Vol. *1*, Ottawa.

Evans, M. (1980) *Housing, Climate and Comfort*, John Wiley and Sons, Inc., New York.

Ito, N. Kimura, K. and J. Oka (1972) A Field Experiment Study on the Convective Heat Transfer Coefficient on Exterior Surface of a Building, ASHRAE Semiannual Meeting, New Orleans, La., Jan. 23-27, 184-191.
Johnson, T. E. (1981) *Solar Architecture: The direct gain approach*, McGraw-Hill Book Company, New York.
Kreith, F. (1973) *Principles of Heat Transfer*, Harper & Row Publishers, New York.
Louden, A. G. (1968) Summer Time Temperature in Buildings, Building Research Station Current Paper, 47/68.
Mackey, C. O. and L. T. Wright (1944) *Periodic Heat Flow - Homogeneous Walls or Roofs*, ASHVE Trans., Vol. *50*, p. 293.
Markus, T. A. and E. N. Morris (1980) *Buildings, Climate and Energy*, Pitman Publishing Limited, London, WC2B 5PB
Milbank, N. O. and J. Harrington-Lynn (1974) Thermal Response and the Admittance Procedure, Building Research Establishment, a chapter in *Energy, Heating and Thermal Comfort*, The Construction Press, Vol. *4*, New York.
Nash, G. D., J. Comrie and H. F. Broughton (1955) *The Thermal Insulation of Buildings*, HMSO, London.
O'Callaghan, P. W. (1978) *Building for Energy Conservation*, Pergamon Press, Oxford.
Raychaudhury, B. C. and N. K. D. Chaudhury (1961) Thermal Performance of dwellings in the tropics, *Indian Construction News*, 38-42.
Rogers, G. F. C., Mayhew, Y. R. (1974) Engineering Thermodynamics Work and Heat Transfer, Longman (page 575 discussion on emissivity and pages 529-554 concerning convection.
Swinbank, W. C. (1963) Longwave Radiation from Clear Skies, *Quart, J. Roy. Meteor. Soc., 89.*
Turner, D. P. (1969) *Windows and Environment*, Pilkington Bros., St. Helins.
Van Straaten, J. F. (1967) *Thermal Performance of Buildings*, Elsevier Publishing Co., New Y'rk.
Whillier, A. (1967) Low Temperature Engineering Applications of Solar Energy, New Y⅞rk, ASHRAE 1967, "Design Factor Influencing Solar Collectors".
Wise, A. F. E. (1973) Building Research Establishment Current Paper CP 6/73.
Wong, H. Y. (1977) *Heat Transfer for Engineerings*, Longman Inc., New York.

Chapter 6

MATHEMATICAL MODELS FOR PASSIVE SOLAR HOUSES

In recent years various mathematical techniques have been developed for sizing and optimizing solar passive houses. Various models developed so far can be divided into three basic categories.

(a) *Approximate methods*: These methods are used to find out the average energy requirements of a building for heating and/or cooling purposes, and are helpful during the planning stage of the project. The methods like the degree day and steady state methods are the examples of this kind. The degree day method is the one commonly practised by the architects.

(b) *Correlation methods*: In this case, the thermal relationship of a building is expressed in terms of a correlation coefficient, expressing the solar energy fraction with the heating requirements. Various parameters, like, the sizing of the storage mass and orientation of the building can be considered in these methods.

(c) *Analytical methods*: Various methods based on the solution of heat conduction equation with appropriate boundary conditions have been forwarded. These methods have been described briefly in this chapter. For complex structures one has to resort to finite element methods for solving heat conduction equation necessitating the use of big computers. However, in case the solutions are assumed to be periodic, it is possible to find explicit expressions for various parameters of interest. This allows the use of mini or micro-computers to perform the necessary numerical computation. Analytical methods make it possible to consider the effect of various parameters and construction variations over the thermal performance of the buildings.

From a study of the results of simulation models and correlation methods, various rules of thumb have been developed for the passive systems, these are described in Chapter 9. In order to evaluate building performance various computer programmes have also been developed; a summary of these programmes, their availability and scope for use has been described in the Appendix 4.

6.1 APPROXIMATE METHODS

6.1.1 Heat Losses and Degree Day

Residences require supplementary heat only when the outside temperature drops

∿ 5°C below the base value of the indoor temperature because the heat provided by occupancy, electrical usage and solar heat entering through the opaque and transparent portions (windows) of the walls usually compensates the heat losses up to a temperature difference of 5°C. Heat losses from the buildings can be expressed as

$$\dot{Q} = UA\ (\theta_b - \theta_a), \tag{6.1}$$

where θ_b is referred to as the temperature base for heat loss calculations. The base temperature commonly accepted in USA is 18°C, which was empirically determined nearly 50 years ago by US utilities to correlate fuel demand with the weather.

Integration of Eq. (6.1) over an entire day yields the total daily heat losses from a building i.e.

$$Q_T = UA\ (\theta_b - \bar{\theta}_a), \tag{6.2}$$

where $\bar{\theta}_a$ is the average daily temperature. The term over the entire day period, is evaluated only when it is positive i.e. when the average outside ambient temperature is less than θ_b. The number of degree days per month are thus defined as

$$DD = (\theta_b - \bar{\theta}_a)\ n'_d\ , \tag{6.3}$$

n'_d being the number of days in a month for which the heating is required. Monthly space heating load $\dot{Q}_m$ is thus calculated from the formula

$$\dot{Q}_m = U.A.DD. \tag{6.4}$$

The product of building overall energy loss coefficient and the area, UA, can be determined in several ways. For structures in which a record of conventional fuel requirements has been kept, UA can be calculated as the amount of energy required to heat the building for a given period (considering both the heating value of the fuel and the furnace efficiency) divided by the total number of degree days during that period; therefore

$$UA = (N_F \times H_F \times \eta_F)/DD, \tag{6.5}$$

For new structures UA has to be calculated from the details of the building construction using methods that would be described in the subsequent sections. The various methods allow to calculate the heating/cooling load for the required temperature difference $(\theta_b - \theta_a)$. UA is then calculated by

$$UA = \frac{\text{Design heating load}}{\text{Design temperature difference}}$$

Example 6.1 A residence at Leh has (UA) = 550 W/°C. If the degree days for January are 800, calculate the total space heating load for January.

Space heating load

= 550 x 800 x 24 x 3600 J

= 38.02 GJ.

Example 6.2 Find the overall heat loss coefficient of a building if the number of degree days in the month of January are 800 and 100 litres of oil has to be used to keep the space comfortable. (Heating value of various fuels are given in Table 6.1). Using Eq. (6.5)

$$UA = \frac{100 \times 43000 \times 0.8}{800 \times 24} \frac{kJ}{h^oC}$$

$$= 179.2 \frac{kJ}{h^oC} = 49.8 \ W/^oC$$

6.1.2 Steady State Method

Degree day method can not be used to calculate the heating or cooling load if the indoor conditions are not constant with time, as happens in the case of intermittent heating. For calculating the average heating/cooling load one can take the average of the inside and the ambient conditions making assumptions about the duration of the heating season and outdoor conditions.

TABLE 6.1 Heating Values of Various Combustibles and their Conversion Efficiencies

Fuel	Heating value (kJ/kg)	Efficiency of device (%)
Coal coke	29000	70
Anthracite	32000	70
Lignite	20000	70
Wood	15000	60
Straw	14000-16000	60
Gasoline	43000	80
Kerosene	42000	80
Methanol	24000	80
Methane (Natural gas)	50000	80
Butane	50000	80
Biogas (60% methane)	20000	80
Electricity	-	95

In addition, one assumes the system in steady state implying that the temperature at various positions does not change with time. For the desired indoor temperature θ_b, the heat flux into the building is calculated from the knowledge of overall heat loss coefficient described in Chapter 5. The method of calculation is illustrated below by considering some of the common passive heating and cooling concepts.

6.1.2.1 Simple glazed wall (Trombe wall without vents). For a simple glazed wall of thickness L and thermal conductivity K (Fig. 6.1), the average heat flux coming into a room to be maintained at a temperature, θ_r, can be easily calculated by the following expression

$$\dot{Q} = \alpha\tau . \bar{s} - h_o(\theta_{x=o} - \theta_a), \tag{6.6}$$

where h_o is the overall heat loss coefficient from the wall's outside surface to the ambient through glazing. In the steady state, the amount of heat flux conducted and subsequently convected and radiated into the space remains $\dot{Q}$ and can be written as

$$\dot{Q} = \frac{K}{L}(\theta_{x=o} - \theta_{x=L}), \tag{6.7}$$

$$\dot{Q} = h_{si}(\theta_{x=L} - \theta_b), \tag{6.8}$$

Adding Eqs. (6.6) to (6.8) yields

$$\dot{Q} = U\left|\frac{\alpha\tau\bar{s}}{h_o} + (\theta_a - \theta_b)\right|,$$

$$= U(\theta_{sa} - \theta_b), \tag{6.9}$$

where U is given by

$$\frac{1}{U} = \frac{1}{h_o} + \frac{L}{K} + \frac{1}{h_{si}}, \tag{6.10}$$

6.1.2.2 Water wall (WW). If θ_p is the temperature of the metallic surface of the drums, the heat balance equations corresponding to Fig. 6.2, for the steady state are written as

$$\dot{Q} = (\alpha\tau)\bar{s} - h_o(\theta_p - \theta_a), \tag{6.11}$$

$$\dot{Q} = h_1'(\theta_p - \theta_w), \tag{6.12}$$

$$\dot{Q} = h_2'(\theta_w - \theta_{si}), \tag{6.13}$$

$$\dot{Q} = h_{si}(\theta_{si} - \theta_b), \tag{6.14}$$

Adding Eqs. (6.11) to (6.14) yields

$$\dot{Q} = U(\theta_{sa} - \theta_b), \tag{6.15}$$

with

$$\frac{1}{U} = \left(\frac{1}{h_o} + \frac{1}{h_1'} + \frac{1}{h_2'} + \frac{1}{h_{si}}\right), \tag{6.16}$$

6.1.2.3 Solarium. For the system shown in the Fig. 6.3, there is no direct heat-transfer between the absorbing surface and the ambient. Net gain of heat flux is expressed as

$$\dot{Q} = (\alpha\tau)\bar{s} - h_o(\theta_{ss} - \theta_a), \tag{6.17}$$

$$\dot{Q} = \frac{K}{L}(\theta_{x=o} - \theta_{x=L}), \tag{6.18}$$

$$\dot{Q} = h_{si}(\theta_{x=L} - \theta_b), \tag{6.19}$$

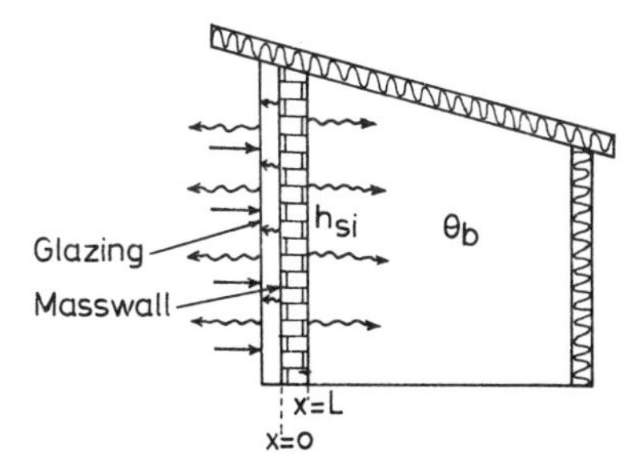

Fig. 6.1. Trombe wall

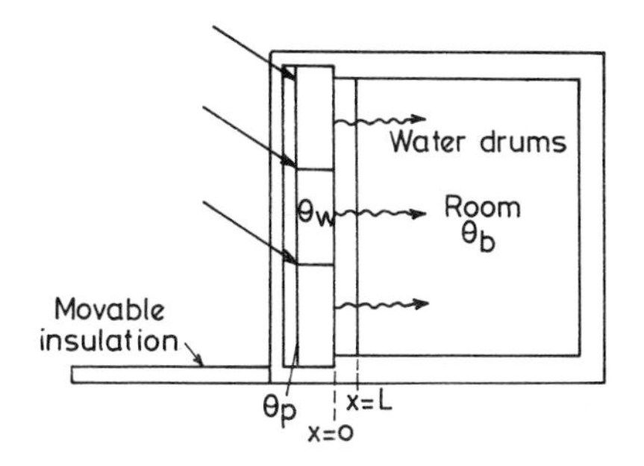

Fig. 6.2. Water wall

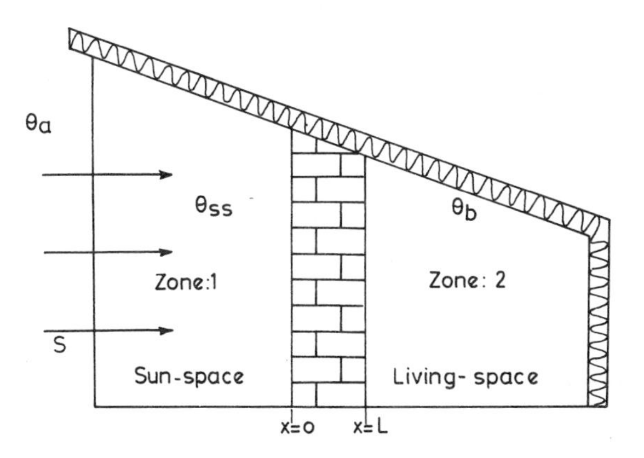

Fig. 6.3. Solarium

The unknown air temperature of zone 1, θ_{ss}, is related to the ambient air temperature through the equation

$$h_o\,(\theta_{ss} - \theta_a) = h_{TS}\,(\theta_{x=o} - \theta_{ss}), \tag{6.20}$$

yielding

$$\theta_{ss} = h_m \left\{ \frac{1}{h_o} \theta_{x=o} + \frac{1}{h_{TS}} \theta_a \right\}, \tag{6.21}$$

where

$$\frac{1}{h_m} = \left(\frac{1}{h_o} + \frac{1}{h_{TS}} \right),$$

and h_{TS} is the heat transfer coefficient between the wall and the sunspace. Substituting Eq. (6.21) into Eq. (6.17) and adding the resultant equation to Eqs. (6.18) and (6.19), one gets

$$\dot{Q} = U \left[\frac{(\alpha\tau)\bar{S}}{h_m} + \frac{h_o}{h_m} \left(1 - \frac{h_m}{h_{TS}} \right) \theta_a - \theta_b \right], \tag{6.22}$$

where

$$\frac{1}{U} = \left(\frac{1}{h_m} + \frac{L}{K} + \frac{1}{h_{si}} \right), \tag{6.23}$$

For passive cooling techniques such as placement of an insulation, air cavities or evaporative cooling, one directly calculates the heat flux by the formula

$$\dot{Q} = U\,(\theta_e - \theta_b), \tag{6.24}$$

where, θ_e, represents the effective temperature, which for a wet surface has been defined in section 8.1.4 of Chapter 8. For a dry surface θ_e automatically assumes the familiar form of solair temperature of the surface.

Example 6.3 Consider a wall facing south. Let α of wall's surface be 0.8 and transmittance of the glazing be 0.71. Mean solar radiation on south face from 6 to 18 hours is 300 W/m^2. If the average ambient temperature is 14.5°C, calculate the mean heat flux into a room maintained at 18°C temperature for (a) Trombe wall of 300 mm thick concrete without vents (b) Water wall (c) Solarium, assuming that the heat-transfer takes place only from the south and all other components of the building are perfectly insulating. The external wall heat transfer coefficient from a glazed surface is 5.0 W/m^2°C.

Solution Corresponding to Fig. 6.1 and the formula (6.9), the heat flux $\dot{Q}$ for Trombe wall is calculated as

$$\dot{Q} = \left(\frac{1}{5.0} + \frac{0.3}{0.72} + \frac{1}{8.0} \right)^{-1} \left(\frac{0.8 \times 0.71 \times 300}{5} + 14.5 - 18 \right)$$

$$= 1.348 \times 30.58$$

$$= 41.23 \text{ W/m}^2$$

Using expression (6.15), the heat flux for water wall comes out to be

$$\dot{Q} = \left(\frac{1}{5} + \frac{1}{206} + \frac{1}{206} + \frac{1}{8} \right)^{-1} \times 30.58$$

$$= 91.3 \text{ W/m}^2$$

For a solarium using expression (6.22)

$$\dot{Q} = U\left[\frac{0.71 \times 0.8 \times 300}{h_m} + \frac{h_o}{h_m}\left(1 - \frac{hm}{8.0}\right) \times 14.5 - 18\right]$$

$$h_m = \left(\frac{1}{5} + \frac{1}{8}\right)^{-1} = 3.076$$

$$U = \left(\frac{1}{5} + \frac{1}{8} + \frac{0.3}{0.72} + \frac{1}{8}\right)^{-1}$$

$$= 1.154$$

$$\dot{Q} = 1.154\left[\frac{0.71 \times 0.8 \times 300}{3.076} + \frac{5}{3.076}\left(1 - \frac{3.076}{8.0}\right) \times 14.5 - 18\right]$$

$$= 1.154\ (55.39 + 14.51 - 18)$$

$$= 59.89\ \mathrm{W/m^2}$$

Example 6.4 For the above three cases and for an average ambient temperature of 10°C, calculate the night losses for the above three systems if the inside temperature (θ_b) is maintained at 18°C and also calculate the net heat flux into the room.

Solution The radiation during night-time being zero, the heat losses in the three cases are given by the expressions (6.9), (6.15) and (6.22) with $\bar{S} = 0$

For Trombe wall (TW) losses $= -\ 1.348 \times 8\ \mathrm{W/m^2}$

$= -\ 10.784\ \mathrm{W/m^2}$

For Water wall (WW) losses $= -\ 2.987 \times 8\ \mathrm{W/m^2}$

$= -\ 23.9\ \mathrm{W/m^2}$

For Solarium losses $= 1.154\left[\frac{5}{3.1}\ (1 - \frac{3.1}{8.0}) \times 10 - 18\right]$

$= -\ 9.37\ \mathrm{W/m^2}$

The net heat flux for all the three cases, therefore, becomes

Trombe wall (TW) $= 30.4\ \mathrm{W/m^2}$

Water wall (WW) $= 67.4\ \mathrm{W/m^2}$

Solarium $= 50.5\ \mathrm{W/m^2}$

Example 6.5 If the glazing in all the above three systems is covered with a 5 cm thick movable night insulation (NI) with K = 0.025 W/m°C then calculate the net heat flux into the room.

Solution: For night time insulation the effective heat loss coefficient from room to the ambient changes thus reducing the heat losses.

For Trombe wall (TW) losses $= \left(\frac{1}{5} + \frac{0.3}{0.72} + \frac{1}{8} + \frac{.05}{.025}\right)^{-1} \times (-8)$

$= -\ 2.918\ \mathrm{W/m^2}$

For Water wall (WW) losses $= -\ 3.42\ \mathrm{W/m^2}$

For Solarium $h_o = 0.45\ W/m^2$

$h_m = 0.43\ W/m^2$

$$\text{losses} = \left(\frac{1}{5} + \frac{1}{8} + \frac{.05}{.025} + \frac{0.3}{0.72} + \frac{1}{8}\right)^{-1} \times$$

$$\times \left[\frac{.45}{.43}\left(1 - \frac{.43}{8.0}\right) \times 10 - 18\right]$$

$$= - 2.82\ W/m^2$$

The net heat flux into the room for all the three cases, therefore, becomes

Trombe wall = $38.3\ W/m^2$

Water wall = $87.9\ W/m^2$

Solarium = $57.1\ W/m^2$

6.2 CORRELATION METHODS

In these methods, the desired result is expressed in terms of one or more correlating parameters. F-chart technique developed at the University of Wisconsin for active solar systems is an example of a correlation technique (Beckmann *et al.*, 1977). Researchers at the Los Alamos National Laboratory independently developed the solar-load-ratio (SLR) method (Anon, 1976) and this has been applied extensively to passive solar systems. Another method developed for passive systems is the non-utilizability method (Monsen *et al.*, 1981) for passive systems.

These methods have two things in common. First, they use monthly weather data to predict the monthly performance. A month has been found to be a convenient long enough time, for averaging out the statistical variations, and the weather statistic can be assumed to be basically stationary. The prediction of monthly performance leads to relatively high standard errors (typically ± 8%) but annual performance is predicted with a standard error of ± 3% only. This is considered to be adequate for design purposes.

A second common feature of these methods is that the correlations are derived using data developed from hour by hour computer simulation. In case of F chart, the TRNSYS (1979) was used and for passive SLR correlations, the PASOLE (McFarland, 1978) was used. The correlation techniques are therefore second generation analytical procedures, intended to give reasonable good correspondence with simulation analyses. Intrinsically the results are no better than those obtained by simulation techniques; the calculations can, however, be performed using a hand calculator.

Usually two correlation methods, namely (i) the solar load ratio design method and (ii) the load collector ratio method are used for the calculations.

6.2.1 Solar Load Ratio

This method, devised by Balcomb and McFarland (1978), is essentially used to estimate the thermal performance of passive heating systems with collector storage wall. Many detailed hour by hour simulations of four different systems were performed for 29 locations and the results were correlated in terms of the ratio of the absorbed solar energy to loads. The systems simulated by Balcomb and McFarland included a solid storage wall and a water wall each with and without a night insulation. The basic parameters of the systems are given in Table 6.2 below.

TABLE 6.2 Parameters and Characteristics of Systems Simulated in the Development of the SLR Design Method. (After Balcomb and McFarland, 1978)

Storage capacity	$0.92\ MJ/m^2\ ^oC$
Room temperature range	$19\text{-}24\ ^oC$
Night insulation resistance (where used)	$1.6\ m^2\ ^oC/W$
Night insulation time	5 p.m. to 8 a.m.
Wall to room conductance	$5.68\ W/m^2\ ^oC$
Storage wall thermal conductivity	$K = 1.73\ W/m\ ^oC$
Storage wall thermal capacitance	$\rho_c = 2.0\ MJ/m^3\ ^oC$
Double-glazed (Normal Transmittance 0.747)	$\frac{L}{K} = 2.012$
Building thermal mass other than storage wall negligible	
Storage wall has vents with backdraught dampers	

The solar load ratio is defined as the ratio of monthly solar energy absorbed on the storage wall surface to the monthly building load including the losses through the wall in the absence of solar gains. Average daily solar energy $\bar{S}$ absorbed by the wall is calculated from the incident radiation on a particular orientation, as described in Chapter 2. Multiplying $\bar{S}$ by the area, A_r, of the receiver and n_d', the number of days in the month gives the mean monthly solar gain. Loads are calculated by the standard techniques (degree day or direct calculations) for all other components but for the storage wall. Since, in the absence of solar gains, the storage wall loses energy, a load corresponding to that is added to the total loss due to other building components. This yields the total losses (or the total heating load) of the building. The values of heat loss coefficient, U_w, for the water wall and the concrete wall as used by Balcomb and McFarland (1978) are given in the Table 6.3. Loads are calculated by using the formula

$$\text{LOAD} = (UA + A_r U_w)\ DD \tag{6.25}$$

The solar load ratio $(A_r.\bar{S}.n_d'/\text{LOAD})$ is calculated for each month and correlated with the solar heating fraction defined as

$$SHF = 1 - \frac{\text{Heating requirements of a solar house}}{\text{Heating requirements of a standard house}}$$

$$= \frac{\text{LOAD} - \text{Auxiliary}}{\text{LOAD}} = \frac{Q_s}{\text{LOAD}} \tag{6.26}$$

TABLE 6.3 Average Loss Coefficients for Collector Storage Walls for Calculating Loads in the SLR Method

Type of wall	Heat loss coefficient through double-glazed wall. (W/m^2 ^oC)	
	No night insulation	With R9 night insulation (R = 1.6 m^2 $^oC/W$)
Water wall	1.87	1.02
450 mm concrete wall	1.25	0.68

SHF depends on the solar passive system in use and it is determined from Fig. 6.4 for the particular system. The monthly required auxiliary energy is (1-SHF) LOAD. The annual auxiliary energy requirement is then the sum of the monthly auxiliary requirements. Figure 6.4 was developed from the results of a system described by the parameters of the Table 6.2 and using the weather data of 29 locations. The curves were fitted to the data so as to minimize errors in annual solar heating fraction. It may be noted that the results of simulation can vary substantially for individual months. Only annual results from the SLR method should be considered significant.

Example 6.6 A residential building in Leh (latitude 34°N) has a concrete collector - storage wall on its southern side having an area of 20 m^2 with no night insulation. The design parameters are the same as those of Table 6.2. The building (esclusive of the collector-storage wall) has a UA of 240 W/°C (including effects of infiltration and internal heat generation). The average absorbed radiation and degree days are shown in Table 6.3 below. Estimate the amount of auxiliary energy required to heat this building. If there was no concrete storage walls on its south face, the UA value of the house becomes 300 W/°C; estimate the auxiliary energy requirements in this case.

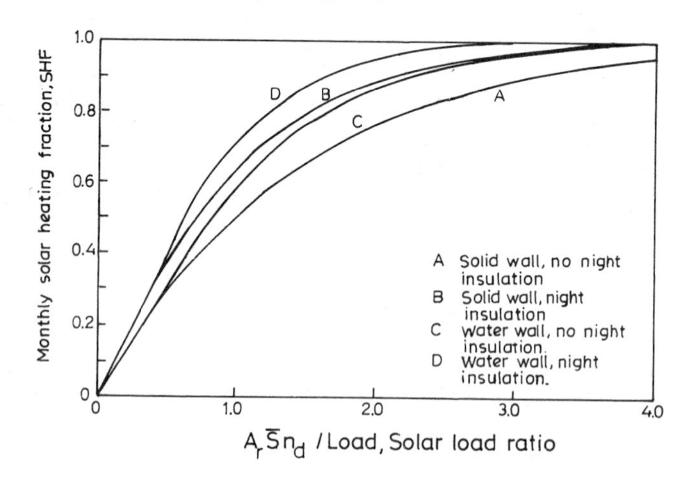

Fig. 6.4. Monthly solar heating fraction versus solar load ratio for buildings with south-facing collector-storage wall system. (After Balcomb and McFarland, 1978)

Solution We shall show here the detailed calculations for the months of December only.

For 45 cm thick concrete wall $U_w = 1.25$ and hence the load is

$$\text{LOAD} = (240 + 20 \times 1.25)\ 450 \times 24 \times 10^{-3}\ \text{kWh}$$

$$= 2.86 \times 10^3\ \text{kWh}$$

Absorbed solar radiation is

$$\text{Ar}\ \bar{S}\ n'_d = 20 \times 3.28 \times 31 = 2.034 \times 10^3\ \text{kWh}$$

The S.L.R. is

$$\text{S.L.R.} = \left\{\frac{2.034 \times 10^3}{2.86 \times 10^3}\right\} = 0.71$$

From Fig. 6.4, using curve A, the months SHF is 0.37. The auxiliary required for December is then $(1-0.37) \times 2.86 \times 10^3 = 1.8 \times 10^3$ kWh

For ordinary walls

$$\text{LOAD} = 300 \times 450 \times 24 \times 10^{-3}\ \text{kWh}$$

$$= 3.24 \times 10^3\ \text{kWh}$$

and hence

$$\text{SLR} = \frac{2.034}{3.24} = 0.63$$

From Fig. 6.4, SHF = 0.26. The auxiliary required would therefore be $(1-0.26) \times 3.24 \times 10^3 = 2.398 \times 10^3$ kWh.

Similarly, the calculations showing the auxiliary requirements of ordinary walls for other months may be carried out.

The method of correlation actually consists of a basic assumption that the monthly solar heating fraction can be expressed as a unique function of SLR, independent of either location or time of year. Using the results of hour by hour computer simulation analyses for 29 different cities scattered throughout US, southern Canada and 3 foreign locations, a plot was made (Balcomb and McFarland, 1978) between the monthly solar heating fraction and monthly solar load ratio. Based on this data, the following function forms were chosen for the relationship between SHF and SLR

$$\text{SHF} = a_1.(\text{SLR}),\ \text{SLR} < R' \tag{6.27}$$

$$\text{SHF} = a_2 - a_3.\exp\ (-a_4.\ \text{SLR}),\ \text{SLR} > R'; \tag{6.28}$$

at SLR = R', SHF = SLR. The values of least square coefficients and standard deviation of annual SHF for the mass storage wall systems are given in Table 6.4.

6.2.2 Load Collector Ratio

In this method also, one first calculate the building loss coefficient UA including the infiltration losses but excluding the passive thermal storage wall. After this the building Load Collector Ratio (LCR), defined (Balcomb and McFarland, 1978) as

TABLE 6.3. Various Head Load Parameters in Different Months at Leh, India (ϕ = 34°N)

Month	$\bar{S}$ (kWh/m^2 day)	DD(°C-days)	Load (kWh) (x 10^3)	$A_r S\ n'_d$ (kwh)	SLR	SHF	Aux. (x 10^3)
January	3.34	500	3.180	2.07	0.65	0.36	2.035
February	3.44	350	2.226	1.926	0.86	0.46	1.202
March	3.35	200	1.272	2.077	1.63	0.68	0.407
April	3.05	100	0.636	1.83	2.88	0.88	0.22
May	2.63	0	-	1.63	-		
June	2.31	0	-	1.39	-		
July	2.44	0	-	1.51	-		
August	2.84	0	-	1.76	-		
September	3.22	50	.318	1.93	6.07	0.95	0.016
October	3.42	150	.954	2.12	2.22	0.80	0.191
November	3.42	280	1.781	2.05	1.15	0.55	0.80
December	3.28	450	2.860	2.034	0.71	0.37	1.8

TABLE 6.4 Values of the Correlation Coefficients between SHF and SLR. (After Balcomb and McFarland, 1978) (System of Table 6.2)

Case	R'	a_1	a_2	a_3	a_4	σ
WW	0.8	0.5995	1.0149	1.26	1.0701	0.028
WWNI	0.7	0.7642	1.0102	1.4027	1.5461	0.026
TW	0.1	0.452	1.0137	1.0392	0.7047	0.024
TWNI	0.5	0.7197	1.0074	1.1192	1.0948	0.023

$$LCR = \frac{\text{Building loss coefficient (kwh/DD)}}{\text{Solar wall collector area (m}^2\text{)}} ,$$

is calculated. The solar wall collector area includes the net glazed area only. Corresponding to this value of the LCR, the values of SHF are read from the tables provided by Balcomb and McFarland (1978) for various cities and various passive systems, namely the water wall (WW), water wall with night insulation (WWNI), Trombe wall (TW), Trombe wall with night insulation (TWNI), Direct gain (DG) and Direct gain with night insulation (DGNI). The values of SHF, corresponding to different values of LCR and latitude are given in Table 6.5. Milborn (1980) has also given values of LCR for various solar heating fractions in the context of North European climates. These values are slightly different from the earlier ones, and are given in Table 6.6.

Based on computer simulations, the following correlations have been suggested for solar heating fraction (SHF) and the load collector ratio (LCR). The monthly performance was found to satisfy the following correlation

$$SHF = (1 - F(X))E' \tag{6.29}$$

where

$$E' = 1 + G/LCR \tag{6.30}$$

and

$$F(X) = B - C \exp(-D.X), \quad X > R' \tag{6.31}$$

or

$$F(X) = A.X, \quad X < R' \tag{6.32}$$

The above equation requires the additional condition that the maximum value of F is 1. The quantity X is the generalized solar load ratio:

$$X = (S/DD - LCRs\ H)/(LCR.E')$$

The parameter LCRs is the load collector ratio of just the glazed area of the vertical wall. H is a dimensionless parameter determined in the correlation process. Parameter S is the solar radiation absorbed in the building per month per unit of projected area. Values for the correlation coefficients A,B,C,D,G and H for various passive systems identified in the Appendix 5 are given in Table 6.7. For direct gain H = 0 and for all other systems B = 1, G = 0 and R' = - 9. For more details the reader is referred to a report by Balcomb *et al.* (1982).

6.2.3 Thermal Time Constant Method

Boundary conditions change constantly for most of the commonly encountered thermal structures. For this reason, the steady methods described above provide only the approximate solutions. On the other hand the exact analysis of heat

TABLE 6.5 Performance Parameters for Passive Solar Heating Systems using Thermal Storage Walls (Load Collector Ratio in $KJ/DD\text{-}m^2$ for Particular Values of Solar Heating Fraction). (After Balcomb and McFarland, 1978)

Latitude (DD)	System	Solar Heating Fraction								
		0.1	0.2	0.3	0.4	0.5	0.6	0.7	0.8	0.9
26°N	WW	21461	10730	7099	5182	3958	3080	2387	1795	1224
	WWNI	28540	14280	9486	6977	5406	4264	3366	2591	1836
333DD	TW	19910	10322	6610	4610	3366	2509	1856	1346	898
	TWNI	27132	13546	8874	6426	4855	3733	2856	2122	1448
28°N	WW	23399	11689	7630	5549	4284	3386	2632	1999	1408
	WWNI	31008	15504	10200	7446	5773	4631	3713	2876	2081
379DD	TW	21604	11179	7160	4998	3652	2734	2040	1489	1000
	TWNI	29437	14627	9527	6916	5263	4060	3101	2326	1632
30°N	WW	14912	6793	4325	3101	2366	1836	1407	1040	714
	WWNI	20400	9323	5957	4304	3305	2632	2081	1612	1142
688DD	TW	13505	6650	4121	2836	2040	1489	1102	796	510
	TWNI	19237	8874	5630	4019	3019	2305	1754	1306	898
32°N	WW	7364	3386	2122	1510	1326	857	632	449	286
	WWNI	10690	4998	3142	2264	1734	1367	1081	816	571
1213DD	TW	6936	3407	2101	1408	1000	714	510	347	204
	TWNI	10200	4774	3019	2142	1612	1224	918	673	449
34°N	WW	6140	2774	1693	1183	877	632	469	306	163
	WWNI	9139	4223	2632	1856	1408	1102	857	653	449
1645DD	TW	5834	2815	1693	1122	775	551	367	245	143
	TWNI	8792	4039	2509	1775	1305	979	734	530	347
36°N	WW	3652	1714	1061	734	530	367	245	143	-
	WWNI	5875	2836	1816	1306	979	755	592	428	286
3669DD	TW	3733	1816	1102	734	490	326	224	122	-
	TWNI	5773	2774	1754	1244	918	694	510	367	245
38°N	WW	4366	2020	1244	877	632	469	326	204	-
	WWNI	6834	3264	2060	1142	1102	857	673	510	347
2770DD	TW	4366	2122	1285	836	571	265	265	163	-
	TWNI	6671	3142	1979	1408	1040	775	592	428	286
40°N	WW	3998	1836	1142	796	571	408	286	163	-
	WWNI	6385	2978	1920	1367	1040	816	632	469	306
3069DD	TW	4019	1958	1183	775	530	367	245	143	-
	TWNI	6181	2917	1856	1387	734	734	551	388	265
42°N	WW	2795	1224	714	469	306	184	-	-	-
	WWNI	4916	2244	1387	979	734	551	428	306	184
3130DD	TW	2958	1367	796	490	306	184	102	-	-
	TWNI	4855	2203	1367	959	694	510	367	265	163
44°N	WW	1632	612	306	-	-	-	-	-	-
	WWNI	3488	1530	938	632	469	347	245	163	82
4594DD	TW	1918	836	428	224	-	-	-	-	-
	TWNI	3509	1571	938	632	449	326	224	143	82

TABLE 6.5 (cont'd)

Latitude (DD)	System	Solar Heating Fraction 0.1	0.2	0.3	0.4	0.5	0.6	0.7	0.8	0.9
46°N	WW	1958	796	428	224	-	-	-	-	-
	WWNI	3856	1734	1061	734	530	388	286	184	102
4933DD	TW	2203	979	530	306	143	-	-	-	-
	TWNI	3856	1754	1061	734	510	367	265	163	102
48°N	WW	3040	1285	694	408	204	-	-	-	-
	WWNI	5202	2366	1428	959	673	469	347	224	122
3697DD	TW	3080	1387	775	449	65	122	-	-	-
	TWNI	5120	2326	1387	918	653	449	326	204	102
50°N	WW	1510	551	-	-	-	-	-	-	-
	WWNI	3305	1489	898	592	408	286	184	102	-
5933DD	TW	1795	755	367	143	-	-	-	-	-
	TWNI	3346	1510	898	592	408	286	184	102	-
54°N	WW	1897	694	-	-	-	-	-	-	-
	WWNI	3754	1693	979	632	408	265	163	82	-
5704DD	TW	2081	857	408	-	-	-	-	-	-
	TWNI	3754	1693	979	632	408	286	184	102	-

WW Water wall

WWNI Water wall with night insulation (R = 1.6 m^2°C/W)

TW Trombe wall

TWNI Trombe wall with night insulation (R = 1.6 m^2°C/W)

TABLE 6.6 Load Collector Ratio for Various Passive Concepts as a Function of Solar Heating Fraction (For North European Climates, Milborn, 1980)

System	Solar Heating Fraction (SHF) 0.1	0.2	0.3	0.4	0.5	0.6	0.7	0.8	0.9
WW	2400	980	531	306	163	-	-	-	-
WWNI	4105	1838	1123	776	551	408	285	204	142
TW	2328	960	510	306	163	81	-	-	-
TWNI	3880	1736	1041	714	510	368	265	184	102
DG	1879	-	-	-	-	-	-	-	-
DGNI	4187	1817	1082	714	490	327	224	142	61

DG Direct gain

NI Night insulation (R = 1.6 m^2°C/W)

transmission in transient conditions is complex. However, an approximate estimation of the time dependent response of thermal structures can be made by assuming that the temperature at any instant is constant throughout the structure.

If this temperature changes by a small amount $d\theta$ in time dt, the change in internal energy is equal to the net heat flow rate across the boundary, i.e.

$$\rho c\ v\ d\theta = h'A\ (\theta - \theta_a)\ dt \tag{6.34}$$

yielding the solution

$$\left[\frac{\theta(t) - \theta_a}{\theta(t{=}o) - \theta_a}\right] = \exp\left\{-t\left(\frac{h'A}{\rho cV}\right)\right\}, \tag{6.35}$$

where $(\rho cV/h'A)$ is called the time constant of the system. Its value is indicative of the time taken for the difference between the system temperature and that of the surroundings to change to 36.8% of the initial temperature difference.

If the parameter $(h'L/K)$, known as Biot modulus (Bi), is less than 0.1; the error introduced by the assumption that the temperature is uniform at any instant of time is less than 5%. The time constant for a multilayered structure as defined in section 5.2.2 is

$$\frac{\text{heat stored}}{\text{heat transmitted}} = \Sigma\left(\frac{\dot{Q}}{U}\right)_i = \left(R_{so} + \frac{L_1}{K_1} + \frac{L_2}{K_2} + \text{---} + \frac{L_i}{K_i}\right)\ (L_i\rho_i c_i), \tag{6.36}$$

Raychoudhury *et al.* (1964) found a nearly exponential relation between thermal time constant and thermal damping D, time constant defined as

$$D = \left(\frac{\theta_{sa} - \theta_b}{\theta_{sa}}\right), \tag{6.37}$$

According to Raychoudhuri and Chaudhuri (1961), Parson expressed the measured time lag t_{24} between a harmonic temperature variation at the outer surface of a wall and the corresponding variation at the inner surface (exposed to constant internal air temperature) as

$$t_{24} = 1.18 + \frac{2\pi}{24}\left(\frac{\dot{Q}}{U}\right), \tag{6.38}$$

An interesting result of the study by Raychoudhury is that the value of the time constant depends significantly on the position of the insulating layer in the element of the building. For example a 110 mm thick brick wall shows a thermal damping of 35%, an addition of outer cladding of 25 mm thick thermocole insulation increases the damping to 86%, whereas the same insulation when placed over the inside surface gives damping of only 70%.

6.3 ANALYTICAL METHODS

6.3.1 Thermal Circuit Analysis

Heat conduction problems can be solved by drawing an analogy between the heat-transfer due to temperature difference and the flow of electric current due to potential difference. The first step in predicting the thermal behaviour of a building by this analysis is, therefore, to make a thermal circuit diagram using corresponding electrical symbols. In the equivalent a thermal circuit, it may be

TABLE 6.7 Passive System Data

Type	Generalized Solar Load Ratio Correlation Constants (After Balcomb *et al.*, 1982)								
	A	B	C	D	R'	G/20.4	H	LCRs/20.4	SHF*
WW A1	0.0000	1.0000	.9172	.4841	-9.0000	0.00	1.17	13.0	.465
WW A2	0.0000	1.0000	.9833	.7603	-9.0000	0.00	.92	13.0	.609
WW A3	0.0000	1.0000	1.0171	.8852	-9.0000	0.00	.85	13.0	.660
WW A4	0.0000	1.0000	1.0395	.9570	-9.0000	o.00	.81	13.0	.687
WW A5	0.0000	1.0000	1.0604	1.0387	-9.0000	o.00	.78	13.0	.714
WW A6	0.0000	1.0000	1.0735	1.0827	-9.0000	0.00	.76	13.0	.728
WW B1	0.0000	1.0000	.9754	.5518	-9.0000	0.00	.92	22.0	.481
WW B2	0.0000	1.0000	1.0487	1.0851	-9.0000	0.00	.78	9.2	.713
WW B3	0.0000	1.0000	1.0673	1.0087	-9.0000	0.00	.95	8.9	.794
WW B4	0.0000	1.0000	1.1028	1.1811	-9.0000	0.00	.74	5.8	.821
WW B5	0.0000	1.0000	1.1146	1.2771	-9.0000	0.00	.56	4.5	.816
WW C1	0.0000	1.0000	1.0667	1.0437	-9.0000	0.00	.62	12.0	.815
WW C2	0.0000	1.0000	1.0846	1.1482	-9.0000	0.00	.59	8.7	.806
WW C3	0.0000	1.0000	1.1419	1.1756	-9.0000	0.00	.28	5.5	.885
WW C4	0.0000	1.0000	1.1401	1.2378	-9.0000	0.00	.23	4.3	.860
TW A1	0.0000	1.0000	.9194	.4601	-9.0000	o.00	1.11	13.0	.460
TW A2	0.0000	1.0000	.9680	.6318	-9.0000	0.00	.92	13.0	.557
TW A3	0.0000	1.0000	.9964	.7123	-9.0000	0.00	.85	13.0	.596
TW A4	0.0000	1.0000	1.0190	.7332	-9.0000	0.00	.79	13.0	.608
TW B1	0.0000	1.0000	.9364	.4777	-9.0000	0.00	1.01	13.0	.477
TW B2	0.0000	1.0000	.9821	.6020	-9.0000	0.00	.85	13.0	.548
TW B3	0.0000	1.0000	.9980	.6191	-9.0000	0.00	.80	13.0	.559
TW B4	0.0000	1.0000	.9981	.5615	-9.0000	0.00	.76	13.0	.534
TW C1	0.0000	1.0000	.9558	.4709	-9.0000	0.00	.89	13.0	.481
TW C2	0.0000	1.0000	.9788	.4964	-9.0000	0.00	.79	13.0	.501
TW C3	0.0000	1.0000	.9760	.4519	-9.0000	0.00	.76	13.0	.479
TW C4	0.0000	1.0000	.9588	.3612	-9.0000	0.00	.73	13.0	.430
TW D1	0.0000	1.0000	.9842	.4418	-9.0000	0.00	.89	22.0	.424
TW D2	0.0000	1.0000	1.0150	.8994	-9.0000	0.00	.80	9.2	.657
TW D3	0.0000	1.0000	1.0346	.7810	-9.0000	0.00	1.08	8.9	.707
TW D4	0.0000	1.0000	1.0606	.9770	-9.0000	0.00	.85	5.8	.762
TW D5	0.0000	1.0000	1.0721	1.0718	-9.0000	0.00	.61	4.5	.765
TW E1	0.0000	1.0000	1.0345	.8753	-9.0000	0.00	.68	12.0	.760
TW E2	0.0000	1.0000	1.0476	1.0050	-9.0000	0.00	.66	8.7	.765
TW E3	0.0000	1.0000	1.0919	1.0739	-9.0000	0.00	.61	5.5	.855
TW E4	0.0000	1.0000	1.0971	1.1429	-9.0000	0.00	.47	4.3	.835
TW F1	0.0000	1.0000	.9430	.4744	-9.0000	0.00	1.09	13.0	.458
TW F2	0.0000	1.0000	.9900	.6053	-9.0000	0.00	.93	13.0	.532
TW F3	0.0000	1.0000	1.0189	.6502	-9.0000	0.00	.86	13.0	.555
TW F4	0.0000	1.0000	1.0419	.6258	-9.0000	0.00	.80	13.0	.543
TW G1	0.0000	1.0000	.9693	.4714	-9.0000	0.00	1.01	13.0	.455
TW G2	0.0000	1.0000	1.0133	.5462	-9.0000	0.00	.88	13.0	.497
TW G3	0.0000	1.0000	1.0325	.5269	-9.0000	0.00	.82	13.0	.487
TW C4	0.0000	1.0000	1.0401	.4400	-9.0000	0.00	.77	13.0	.435
TW H1	0.0000	1.0000	1.0002	.4356	-9.0000	0.00	.93	13.0	.428
TW H2	0.0000	1.0000	1.0280	.4151	-9.0000	0.00	.83	13.0	.414
TW H3	0.0000	1.0000	1.0327	.3522	-9.0000	0.00	.78	13.0	.372
TW H4	0.0000	1.0000	1.0287	.2600	-9.0000	0.00	.74	13.0	.300
TW I1	0.0000	1.0000	.9974	.4036	-9.0000	0.00	.91	22.0	.387
TW I2	0.0000	1.0000	1.0386	.8313	-9.0000	0.00	.80	9.2	.622
TW I3	0.0000	1.0000	1.0514	.6886	-9.0000	0.00	1.01	8.9	.666
TW I4	0.0000	1.0000	1.0781	.8952	-9.0000	0.00	.82	5.8	.731

TABLE 6.7 (cont'd)

Type	Generalized Solar Load Ratio Correlation Constants (After Balcomb *et al.*, 1982)							
	A	B	C	C	R'	G	H	LCR_s	SHF*
TW I5	0.0000	1.0000	1.0902	1.0284	-9.0000	0.00	.65	4.5	.745
TW J1	0.0000	1.0000	1.0537	.8228	-9.0000	0.00	.65	12.0	.739
TW J2	0.0000	1.0000	1.0677	.9313	-9.0000	0.00	.62	8.7	.739
TW J3	0.0000	1.0000	1.1153	.9831	-9.0000	0.00	.44	5.5	.834
TW J4	0.0000	1.0000	1.1154	1.0607	-9.0000	0.00	.38	4.3	.813
DG A1	.5650	1.0090	1.0440	.7175	.3931	9.36	0.00	0.0	.406
DG A2	.5906	1.0060	1.0650	.8099	.4681	5.28	0.00	0.0	.535
DG A3	.5442	.9715	1.1300	.9273	.7086	2.64	0.00	0.0	.681
DG B1	.5739	.9948	1.2510	1.0610	.7905	9.60	0.00	0.0	.480
DG B2	.6180	1.0000	1.2760	1.1560	.7528	5.52	0.00	0.0	.600
DG B3	.5601	.9839	1.3520	1.1510	.8879	2.38	0.00	0.0	.731
DG C1	.6344	.9887	1.5270	1.4380	.8632	9.60	0.00	0.0	.591
DG C2	.6736	.9994	1.4000	1.3940	.7604	5.28	0.00	0.0	.686
DG C3	.6182	.9859	1.5660	1.4370	.8990	2.40	0.00	0.0	.802
SS A1	0.0000	1.0000	.9587	.4770	-9.0000	0.00	.83	18.6	.628
SS A2	0.0000	1.0000	.9982	.6614	-9.0000	0.00	.77	10.4	.774
SS A3	0.0000	1.0000	.9552	.4230	-9.0000	0.00	.83	23.6	.574
SS A4	0.0000	1.0000	.9956	.6277	-9.0000	0.00	.80	12.4	.757
SS A5	0.0000	1.0000	.9300	.4041	-9.0000	0.00	.96	18.6	.572
SS A6	0.0000	1.0000	.9981	.6660	-9.0000	0.00	.86	10.4	.769
SS A7	0.0000	1.0000	.9219	.3225	-9.0000	0.00	.96	23.6	.494
SS A8	0.0000	1.0000	.9922	.6173	-9.0000	0.00	.90	12.4	.743
SS B1	0.0000	1.0000	.9683	.4954	-9.0000	0.00	.84	16.3	.562
SS B2	0.0000	1.0000	1.0029	.6802	-9.0000	0.00	.74	8.5	.723
SS B3	0.0000	1.0000	.9689	.4685	-9.0000	0.00	.82	19.3	.525
SS B4	0.0000	1.0000	1.0029	.6641	-9.0000	0.00	.76	9.7	.706
SS B5	0.0000	1.0000	.9408	.3866	-9.0000	0.00	.97	16.3	.485
SS B6	0.0000	1.0000	1.0068	.6778	-9.0000	0.00	.84	8.5	.712
SS B7	0.0000	1.0000	.9395	.3363	-9.0000	0.00	.95	19.3	.430
SS B8	0.0000	1.0000	1.0047	.6469	-9.0000	0.00	.87	9.7	.687
SS C1	0.0000	1.0000	1.0087	.7683	-9.0000	0.00	.76	16.3	.593
SS C2	0.0000	1.0000	1.0412	.9281	-9.0000	0.00	.78	10.0	.713
SS C3	0.0000	1.0000	.9699	.5106	-9.0000	0.00	.79	16.3	.478
SS C4	0.0000	1.0000	1.0152	.7523	-9.0000	0.00	.81	10.0	.646
SS D1	0.0000	1.0000	.9889	.6643	-9.0000	0.00	.84	17.8	.727
SS D2	0.0000	1.0000	1.0493	.8753	-9.0000	0.00	.70	9.9	.853
SS D3	0.0000	1.0000	.9570	.5285	-9.0000	0.00	.90	17.8	.660
SS D4	0.0000	1.0000	1.0356	.8142	-9.0000	0.00	.73	9.9	.835
SS E1	0.0000	1.0000	.9968	.7004	-9.0000	0.00	.77	19.6	.659
SS E2	0.0000	1.0000	1.0468	.9054	-9.0000	0.00	.76	10.8	.797
SS E3	0.0000	1.0000	.9565	.4827	-9.0000	0.00	.81	19.6	.556
SS E4	0.0000	1.0000	1.0214	.7694	-9.0000	0.00	.79	10.8	.751

*SHF for Denver, Colorado: LCR = 227 kJ/DD-m^2, θ_b = 18.3°C. This information is provided primarily to provide a checkpoint so that users can verify that the correlation constants are entered and used correctly.

noted (following Gupta and Raychoudhury, 1963) that

(i) thermal potentials are temperatures in °C

(ii) thermal currents are heat flows in W/m^2

(iii) resistance (inverse of the heat-transfer) opposes heat flow with a corresponding temperature drop.

(iv) capacitances store heat and

(v) thermal sources generate heat and thermal sinks receive heat.

The circuit diagram representing the mode of heat transport has four main parts

(a) Thermal conduction paths through the structural elements which are represented as parallel lumped RC networks.

(b) Radiation exchange between indoor surfaces of elements which is represented as a resistance network.

(c) Convective exchange between the interior surfaces and the indoor air which is represented by a resistance network grounded through a capacitor substituting the capacity of the enclosed air.

(d) Thermal inputs which are represented by time variable temperature and current sources connected to the network through suitable boundary resistances.

The main assumptions and limitations, generally implicit in circuit representation, are as follows:

(i) Thermal energy transfer through building elements is unidirectional and perpendicular to long dimensions.

(ii) Different elements as walls, roof etc. are parallel paths of heat-transfer to the enclosed space.

(iii) All surfaces are isothermal planes and the temperature is equal to the actual temperatures measured at the geometric centre of the surface.

(iv) Moisture transfer, if any, is not accounted for.

(v) Continously distributed resistance-capacitance along the heat conduction paths in actual physical system is replaced by a finite number of discrete R-C lumps in the thermal circuit; the two systems tending to be equal as the sites of the lump become smaller and their number approaches infinity.

(vi) Air in the enclosed building space is uniformly heated and is at uniform temperature at any instant.

(vii) Net radiation exchange with enclosed air mass is neglected.

(viii) All surfaces are grey radiating diffusely and without secondary inter-reflections.

Circuit Parameters. For each branch of the circuit, the thermal admittances for different circuit parameters are evaluated as per formula given below

Conduction admittance ($W/^{o}C$)

Resistivity branch $$Y_c = \frac{KA}{L}, \tag{6.39}$$

Capacity branch $$Y_o = in\omega A\,(\rho Lc) \tag{6.40}$$

Convective admittance $Y'_c = h_c A$ (6.41)

Radiative admittance (from surface area A_1 to surface area A_2) $Y'_R = A_1 F_{12} F h_r$ (6.42)

where $F_{12} = \varepsilon_1 \varepsilon_2$

Circuit Equations. For each nodal point, the amount of heat flux through it depends on the admittance of the circuit and the temperature difference across it viz.

$$\dot{Q} = Y.\Delta\theta, \tag{6.43}$$

Circuit equations are set up for each of the nodal points of the circuit corresponding to the unknown interior temperatures and junction points of R-C lumps. The thermal inputs to the circuits, namely, the ambient air temperature and the solar radiation, and the output at any point are assumed to be periodic and expressed by Fourier series i.e.

$$\theta = a_o + \sum_{n=1}^{\infty} (a_n \cos n\omega t + b_n \sin n\omega t), \tag{6.44a}$$

$$= a_o + \sum_{n=1}^{\infty} \sqrt{a_n^2 + b_n^2} \cos (n\omega t - \psi), \tag{6.44b}$$

where, ω the frequency usually corresponds to a daily cycle of 24 hours. A general form of the equation at point x is

$$\sum_{m=1}^{M} y_{mx} \theta_m - \left(\sum_{m=1}^{M} Y_{mx} \right) \theta_x = 0, \tag{6.45}$$

where Y_{mx} are the admittances of circuit branches through x; M is the number of circuit branches through x, and θ_m is the known or unknown temperature at the other end of branch mx.

Example 6.7 As an example of the above-mentioned procedure we consider a single room masonry structure of 200 mm wall thickness. The interior dimensions of the room are 3.47 m x 2.9 m x 3.23 m and those of lobby are 1.2 m x 2.9 m x 2.9 m. The room is supposed to be exposed from all the sides. Only north wall is shaded. Roof is flat allowing it to be considered as horizontal. Structural specifications of the test unit components and thermo-physical properties of materials used in construction are given in Tables 6.8 and 6.9.

The external surface temperature for the roof, east, west and south walls have been considered as the input potentials for respective building elements. These are given in Table 6.10a. The harmonic coefficients derived by Fourier analysis of the data, using equations (6.44a) for various surfaces are given in Table 6.10b. The equivalent thermal circuit for the enclosure is shown in Fig. 6.5. Thermal admittance of each building element is then identified. The symbol Y_{EW}, for example, refers to admittance from east to west wall. Values of admittances, with reference to Fig. 6.5, and calculated from Eqs. (6.39) to (6.42) are given in the Table 6.11. The values of the surface convective heat-transfer coefficients (h_c) as given by Nottage and Parmelee (1954) are: for floor = 6.246 W/m^2°C, for ceiling 2.27 W/m^2°C and for walls 3.97 W/m^2°C. The radiative heat-transfer coefficient has been calculated corresponding to a temperature of 38°C i.e. h_r = 6.9 W/m^2°C. The shape factors as calculated by Gupta and Chaudhuri (1963) are given in Table 6.12. The calculated values satisfy the

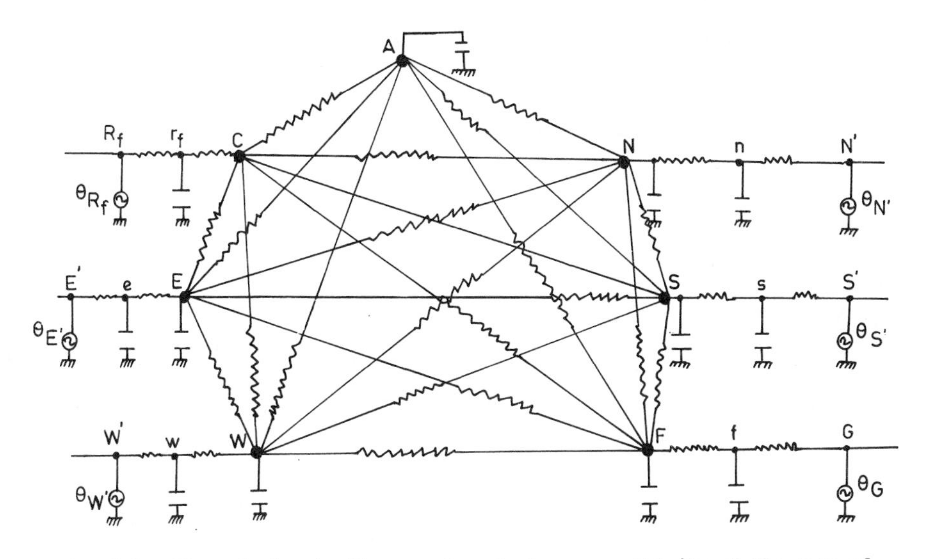

Fig. 6.5. Thermal circuit diagram of a single room. Details of the type of admittances between two nodes are given in Table 6.11. (After Gupta and Raychoudhuri, 1963)

TABLE 6.8 Structural Specifications of the Test Unit Selected in the Example (After Gupta and Raychoudhuri, 1963)

Elements	Specifications (outwards-inwards)		Overall thickness in mm	Interior surface area in m^2	Symbol in thermal circuit Fig. 6.5
	Material	Thickness in mm			
Roof	Cement concrete	165	178	9.995	$R_f r_f C$
	Lime cement plaster	13			
East wall	Lime cement plaster	13			
	Brick work	220	246	11.184	E'e E
West wall	Lime cement plaster	13			W'w W
North wall (interior)	Lime cement plaster	13			N'n N
South wall (interior)	Brick work	220	246	9.30	s's S
	Lime cement plaster	13			
Floor	Lime concrete	76	114	9.995	G f F
	Cement concrete	38			
Enclosed air volume	3.47 x 2.9 x 3.23 m^3				A

TABLE 6.9 Thermophysical Properties of Materials

Property	Units	Brick	Cement concrete	Lime cement plaster	Lime concrete
Thermal conductivity (K)	W/m ^oC	0.969	1.991	1.16	0.969
Density (ρ)	kg/m^3	1762	2322	1762	1762
Specific heat (C)	kJ/kg^oC	0.84	0.92	0.92	0.84

TABLE 6.10a Thermal Inputs - Measured Temperatures (^oC)

Time h	Shade	Roof θ_{Rf}	east wall $\theta_{E'}$	west wall $\theta_{W'}$	south wall $\theta_{S'}$	Exposed ground θ_G
11	36.7	45.9	51.0	40.1	39.3	46.4
12	38.6	48.7	49.0	41.9	44.1	47.9
13	40.0	50.6	46.5	45.2	45.3	48.7
14	40.2	50.8	45.6	49.0	45.6	47.9
15	40.9	50.8	45.0	53.4	45.3	46.05
16	40.9	49.0	44.05	54.7	44.7	43.8
17	40.7	46.7	43.05	54.1	43.3	40.6
18	40.0	44.2	41.7	49.4	41.9	38.7
19	38.9	42.2	40.2	45.8	40.2	36.8
20	37.5	40.7	38.5	41.2	38.6	35.4
21	36.4	39.6	37.6	39.6	37.7	34.7
22	35.4	38.3	36.6	38.1	36.7	34.05
23	35.0	37.5	36.0	37.3	36.2	33.6
0	34.1	36.7	35.4	36.6	35.6	33.5
1	33.6	35.8	34.8	35.7	34.9	33.2
2	32.1	35.1	34.05	35	34.2	32.2
3	31.1	34.6	33.5	34.6	33.8	30.7
4	30.6	33.4	32.9	33.4	33.05	31.9
5	29.8	32.6	32.05	32.9	32.3	31.5
6	29.6	32.4	32.05	32.6	32.2	31.8
7	29.4	33.1	34.8	32.9	32.9	32.9
8	30.6	34.7	40.8	34.1	34.2	35.2
9	32.2	37.6	46.4	35.7	36.1	38.6
10	33.3	40.7	48.3	37.3	38.0	42.1

TABLE 6.10b Harmonic Coefficients of External Surface Temperatures

Coefficients		Roof θ_{Rf}	East wall $\theta_{E'}$	West wall $\theta_{W'}$	South wall $\theta_{S'}$	Exposed ground θ_G
Steady state mean level	a_o	40.5	40.0	40.4	37.98	37.87
1st harmonic	a_1	3.73	5.87	1.84	2.32	5.78
	b_1	7.4	4.51	8.92	5.61	5.20
2nd harmonic	a_2	8.45	2.84	-2.35	0.239	1.835
	b_2	2.78	0.216	2.313	1.879	2.346

reciprocity theorem ($A_1F_{12} = A_2F_{21}$) and the condition that the summation of shape factors for a sequence of surfaces, surrounding as emitting surfaces, should be unity.

Assuming that cooling conditions prevail and heat is flowing from outdoors to indoors, heat balance equations are set up using Eqs. (6.43) and (6.45) at all surfaces, air and lumping point. These equations are given in the matrix form in Table 6.13. The symmetric pattern of matrix is significant and is of valuable help in setting up the equations, checking for mistakes and solving them. The notation refers to summation of admittances of all branches through the point to which the equation applies. The set of equations obtained for various harmonics can be solved by usual techniques. The solutions obtained for mean values (n = 0) and two harmonic coefficients for the internal surface temperatures are given in Table 6.14. From these solutions, it is easy to synthesize the temperature variations.

6.3.2 The Admittance Procedure of the Matrices

The admittance procedure developed by the UK Building Research Establishment (Petherbridge, 1974; Milbank and Harrington-Lynn, 1974) is a conceptually simple method used in the calculation of room temperature of nonconditioned buildings for many years. By writing the simple energy balance equations, average room temperature is calculated from the formula

$$\bar{\theta}_r = \left(\frac{\bar{\dot{Q}}}{\Sigma AU + C_v}\right), \tag{6.46}$$

where $\bar{\dot{Q}}$ is given, in general, by the expression

$$\bar{\dot{Q}} = f_s \, \bar{S} \, A_g + U_w \, A \, \bar{\theta}_{sa} + (C_v + U_g \, A_g) \, \bar{\theta}_a, \tag{6.47}$$

where A is the opaque wall area and C_v, the ventilation conductance, which may be taken from Table 5.10 for small ventilation rates. However, for large ventilation rates, C_v should be calculated from the expression (Markus and Morris, 1980)

$$\frac{1}{C_v} = \frac{3}{NV} + \frac{1}{4.8 \, \Sigma A} \tag{6.48}$$

TABLE 6.11 Admittances (W/°C) for Circuit Parameters

Test Unit elements	Conduction admittances				Convective admittances		Radiative** admittances	
	Resistive		Capacitive*					
	Parameter	Value	Parameter	Value	Parameter	Value	Parameter	Value
Ceiling (C)	$Y_{R_f r_f}$	223.6	$Y_{r_f o}$	138.2n	Y_{CA}	22.7	Y_{CE}	12.99
							Y_{CW}	12.99
	$Y_{r_f C}$	202.4	Y_{Co}	133.3n			Y_{CN}	10.25
							Y_{CS}	10.25
							Y_{CF}	11.43
East Wall (E)	$Y_{E'e}$	89.7	Y_{eo}	149.5n	Y_{EA}	44.5	Y_{EC}	12.99
							Y_{EW}	15.87
	Y_{eE}	89.7	Y_{Eo}	149.5n			Y_{EN}	12.46
							Y_{ES}	12.46
							Y_{EF}	12.99
West Wall (W)	$Y_{W'w}$	89.7	Y_{wo}	149.5n	Y_{WA}	44.5	Y_{WC}	12.99
							Y_{WE}	15.87
	$Y_{w'W}$	89.7	Y_{Wo}	149.5n			Y_{WN}	12.46
							Y_{WS}	12.46
							Y_{WF}	12.99

TABLE 6.11 (cont'd)

North Wall (N)							Y_{NC}	10.25
	$Y_{N'n}$	74.6	Y_{no}	124.4n			Y_{NE}	12.46
					Y_{NA}	37.0	Y_{NW}	12.46
	Y_{nN}	74.6	Y_{No}	124.4n			Y_{NS}	2.65
							Y_{NF}	3.00
South Wall (S)							Y_{SC}	10.25
	$Y_{S's}$	74.6	Y_{so}	124.4n			Y_{SE}	12.46
					Y_{SA}	37.0	Y_{SW}	12.46
	Y_{sS}	74.6	$Y_{\ o}$	124.4n			Y_{SN}	2.65
							Y_{SE}	10.25
Floor (F)							Y_{FC}	11.43
	Y_{Gf}	169.9	Y_{fo}	61.4n			Y_{EE}	12.99
					Y_{FA}	62.4	Y_{FW}	12.99
	Y_{fF}	258.3	Y_{FO}	79.5n			Y_{FN}	3.00
							Y_{FS}	3.00
Air (A)			Y_{Ao}	2.38n				

* capacitive parameters - n order of harmonic

**radiative parameters $Y_{EW} = - Y_{WE}$

TABLE 6.12 Shape Factors for Indoor Radiation Exchange

F_{CE}	0.222	F_{EC}	0.198	F_{WC}	0.198
F_{CW}	0.222	F_{EW}	0.242	F_{WE}	0.242
F_{CN}	0.175	F_{EN}	0.190	F_{WN}	0.190
F_{CS}	0.175	F_{ES}	0.190	F_{WS}	0.190
F_{CF}	0.195	F_{EF}	0.198	F_{WF}	0.198
Total	0.989	Total	1.018	Total	1.018
F_{NC}	0.188	F_{SC}	0.188	F_{FC}	0.195
F_{NE}	0.222	F_{SE}	0.222	F_{FE}	0.222
F_{NW}	0.222	F_{SW}	0.222	F_{FW}	0.222
F_{NS}	0.1666	F_{SN}	0.166	F_{FN}	0.175
F_{NF}	0.188	F_{SF}	0.188	F_{FS}	0.175
Total	0.985	Total	0.986	Total	0.989

where ΣA is the total surface area of the room. The deviations of the room temperature from its mean value at any point are found from the relation

$$\tilde{\theta}_r(t) = \frac{\tilde{\dot{Q}}(t)}{\Sigma AY + C_v}, \tag{6.49}$$

where

$$\tilde{\dot{Q}}(t) = \tilde{\dot{Q}}_s(t) + \tilde{\dot{Q}}_g(t) + \tilde{\dot{Q}}_{gc} + \tilde{\dot{Q}}_v, \tag{6.50}$$

The alternating heat gain $\tilde{Q}_s$ through solid surface is given by

$$\tilde{Q}_s(t) = f_s\, A\, U\, d_f\, \{\theta_{sa}(t - \psi) - \bar{\theta}_{sa}\}, \tag{6.51}$$

Similarly the heat gain $\tilde{\dot{Q}}_g$ through glazed area, convective heat gain/loss $\tilde{\dot{Q}}_{gc}$ through glazed area and the gain/loss Q_v through ventilation are expressed as

$$\tilde{\dot{Q}}_g(t) = f_s\, A_g\, \{S(t) - \bar{S}\}, \tag{6.52}$$

$$\tilde{\dot{Q}}_{gc}(t) = U_g\, A_g\, \{\theta_a(t) - \bar{\theta}_a\}, \tag{6.53}$$

$$\tilde{Q}_v(t) = C_v\, \{\theta_a(t) - \bar{\theta}_a\}, \tag{6.54}$$

The absolute value of the room temperature is obtained by adding the mean value of the room temperature $\bar{\theta}_r$ and the fluctuating term $\tilde{\theta}_r(t)$.

Example 6.8 A room is 5 m x 4 m x 3 m in size with one wall (5 m x 3 m) facing

TABLE 6.13 Circuit Equations in the Form of Matrix of Coefficients

Departure at input point	θ_{r_f}	θ_e	θ_w	θ_n	θ_s	θ_f	θ_c	$\theta_{E'}$	θ_w	θ_N	θ_s	θ_F	θ_A	
r_f	$-Y_{r_f}$						Y_{r_fC}							$-Y_{R_fr_f}\theta_{R_f}$
e		$-Y_e$						Y_{eE}						$-Y_{E'e}\theta_{\varepsilon'}$
w			$-Y_w$						Y_{wW}					$-Y_{W'w}\theta_{w'}$
n				$-Y_n$						$Y_{nN}+Y_{N'n}$				0
s					$-Y_s$						Y_{sS}			$-Y_{S's}\theta_{s'}$
f						$-Y_f$						Y_{fF}		$-Y_{G\,f}\,\theta_G$
C	Y_{f_fC}						$-Y_C$	Y_{CE}	Y_{CW}	Y_{CN}	Y_{CS}	Y_{CP}	Y_{CA}	0
Y_E		Y_{eE}					Y_{EC}	$-Y_E$	Y_{EW}	Y_{EN}	Y_{ES}	Y_{EF}	Y_{EA}	0
Y_W			Y_{wW}				Y_{WC}	Y_{WE}	$-Y_W$	Y_{WN}	Y_{WS}	Y_{WF}	Y_{WA}	0
N				$Y_{nN}+N'n$			Y_{NC}	Y_{NE}	Y_{NW}	$-Y_N$	Y_{NS}	Y_{NF}	Y_{NA}	0
S					Y_{sS}		Y_{SC}	Y_{SE}	Y_{SW}	Y_{SN}	$-Y_S$	Y_{SF}	Y_{SA}	0
F						Y_{fF}	Y_{FC}	Y_{FE}	Y_{FW}	Y_{FN}	Y_{FS}	$-Y_F$	Y_{FA}	0
A							Y_{CA}	Y_{EA}	Y_{WA}	Y_{NA}	Y_{SA}	Y_{FA}	$-Y_A$	0

Note: (1) Each temperature is in vector from $\theta_n = \vec{a}_n - i\vec{b}_n = a_n \cos n\omega t - ib_n \sin n\omega t$.

(2) North wall is an adiabatic surface.

(3) All elements are represented by double L-lump R-C networks.

TABLE 6.14 Solutions of Thermal Circuit Equations

Building element coefficients		Ceiling	Int. East wall	Int. West wall	Int. North wall	Int. South wall	Floor	Indoor air
Steady state mean levels, a_o (°C)		39.9	39.5	39.6	39.3	38.9	38.7	39.25
First harmonic (°C)	a_1	-19.9	19.2	-19.7	-18.6	19.3	-17.9	-18.9
	b_1	-14.95	16.97	-17.4	-17.86	-17.4	-14.4	-16.4
2nd harmonic (°C)	a_2	-18.46	-17.95	-17.91	-17.85	-17.95	-18.17	-18.03
	b_2	-17.58	-17.7	-18.0	-17.82	-17.86	-16.88	-17.58

south and having a single-glazed window (2.5 m x 2 m). If the mean external ambient temperature is 10°C, mean global radiation of south wall 200 W/m^2, mean solair temperature is 18°C, U value of the wall is 0.7 W m^{-2} K^{-1} and U value of window is 4.5 Wm^{-2} K^{-1}, calculate the mean temperature inside the room. Assume 2 air changes per hour and there are no gains or losses from other components of the room except the south wall.

Solution Solar gains = $f_s . \bar{S}.Ag$

$$- 0.76 \times 200 \times 5$$

$$= 760 \text{ W}$$

Ventilation heat loss coefficient $C_v = 0.33 \times 2 \times 60$

$$= 39.6 \text{ W/°C}$$

$$U_wA = 0.7 (15 - 5) = 7.0 \text{ W/°C}$$

$$U_gA_g = 5.6 \times 5 = 28.0 \text{ W/°C}$$

Hence $\bar{Q} = 760 + 7.0 \times 18 + (39.6 + 28) \times 10$

$$= 1562 \text{ W}$$

$$\bar{\theta}_r = \frac{1562}{(15 \times 0.7 + 5 \times 5.6) + 39.6} = \frac{1562}{78.1} = 20.°\text{C}$$

Example 6.9 For the same data as above in Example 6.8, find out the fluctuation in the room temperature at 13.00 hours when the peak external air temperature is 20°C, global solar radiation is 600 W/m^2 and the time lag ψ is 4 hours. At 9.00 hours the solair temperature is 27°C and the decrement factor d_f is 0.25.

Solution Alternating solar gain factor for single glazing $\tilde{f}_s$ = 0.42 (from Table 5.17). Alternating solar gain

$$\tilde{\dot{Q}}(t) = \tilde{\dot{Q}}_s(t) + \tilde{\dot{Q}}_g(t) + \tilde{\dot{Q}}_{gc}(t) + \tilde{Q}_v(t),$$

Now

$$\tilde{\dot{Q}}_s(t) = 0.42 \times 15 \times 0.7 \times 0.25\ (27 - 18)$$

$$= 9.923\ \text{W}$$

$$\tilde{\dot{Q}}_g(t) = 0.42 \times 5\ (600 - 200)$$

$$= 840\ \text{W}$$

$$\tilde{\dot{Q}}_{gc}(t) = 28.0\ (20-10) = 280\text{W}$$

$$\tilde{\dot{Q}}_v(t) = 39.6 \times 10 = 396\ \text{W}$$

Room temperature swing is given by

$$\tilde{\theta}_r(t) = \left(\frac{\tilde{\dot{Q}}}{\Sigma AY + C_v}\right)$$

The admittances are given as

Element	A(m^2)	Y*($Wm^{-2}{}^oC^{-1}$)	AY(W^oC^{-1})
South wall	10	3.0	30.0
South window	5	5.6	18.0
Ceiling	20	5.8	116
Floor	20	3.1	62
Internal walls	39	2.55	99.45
			$\Sigma AY = 340.45\ W^oC^{-1}$

Hence $$\tilde{\theta}_r = \frac{9.923 + 840 + 280 + 396}{340.45 + 39.6}$$

$$= 4.01^oC$$

Hence the peak room temperature is

$$\theta_r = \bar{\theta}_r + \tilde{\theta}_r$$

$$= 20.9 + 4.0$$

$$= 24.9^oC$$

Multilayered structures Milbank and Harrington-Lynn (1974) forwarded the following matrix expression for linking temperature and heat flow cycles in multilayer walls:

$$\begin{bmatrix} \theta_o \\ \dot{q}_o \end{bmatrix} = \begin{bmatrix} 1 & R_{so} \\ 0 & 1 \end{bmatrix} \begin{bmatrix} E & F \\ G & H \end{bmatrix} \begin{bmatrix} 1 & R_{si} \\ 0 & 1 \end{bmatrix} \begin{bmatrix} \theta_j \\ q_j \end{bmatrix}, \qquad (6.55)$$

*adapted from Markus and Morris (1980)

where

$$\begin{bmatrix} E & G \\ F & H \end{bmatrix} = \begin{bmatrix} A_1 & B_1 \\ D_1 & A_1 \end{bmatrix} \begin{bmatrix} A_2 & B_2 \\ D_2 & A_2 \end{bmatrix} ---- \begin{bmatrix} A_n & B_n \\ D_n & A_n \end{bmatrix}, \tag{6.56}$$

for a wall of n layers

where θ_o, θ_j = temperatures at surfaces 0 or j

$\dot{q}_o, \dot{q}_j$ = heat flow at surfaces 0 or j

R_{so}, R_{si} = resistances of surfaces 0 or j

$$A = \cosh\,(1 + i)\psi$$

$$B = \frac{R}{(1+i)\psi} \text{ Sinh } (1+i)\psi, \quad D = \frac{(1+i)\psi}{R} \text{ Sinh } (1+i)\psi$$

$$\psi = \left(\frac{\omega L^2}{2D}\right)^{0.5},$$

$$\omega = \frac{2\pi}{24 \times 3600} s^{-1}, \quad R = \frac{L}{k}$$

A multiplication of the matrices in Eq. (6.55) yields

$$\begin{bmatrix} \theta_o \\ q_o \end{bmatrix} = \begin{bmatrix} a & b \\ c & d \end{bmatrix} \begin{bmatrix} \theta_j \\ q_j \end{bmatrix}, \tag{6.57}$$

where

$$a = E + R_{so} F,$$

$$b = ER_{si} + R_{si} R_{so} G + R_{so} H + F,$$

$$c = G,$$

$$d = GR_{si} + H,$$

and hence

$$\theta_o = a\,\theta_j + B\,q_j, \tag{6.58}$$

$$q_o = c\theta_j + dq_j, \tag{6.59}$$

The admittance of the system from θ_o to θ_j is given

$$Y_{oj} = \frac{q_o}{\theta_o} = \frac{d}{b} = \left[\frac{GR_{si} + H}{ER_{si} + R_{si} R_{so} G + H R_{so} + F}\right] = r + is, \tag{6.60}$$

The admittance

$$Y = |Y_{oj}| = (r^2 + s^2)^{\frac{1}{2}}, \tag{6.61}$$

with a time lead ψ_Y given by

$$\psi_Y = \tan^{-1} \frac{s}{r} \quad \text{radians.} \tag{6.62}$$

The decrement factor d_f from θ_o to θ_j, by definition, is given by

$$d_f = \frac{\tilde{U}_{oj}}{U} = \frac{\text{Alternating U value}}{\text{U value}}, \tag{6.63}$$

where

$$\tilde{U}_{oj} = \frac{q_j}{\theta_o} = \left[\frac{1}{ER_{si} + R_{si}\, R_{so}\, G + R_{so}\, H + F}\right] \tag{6.64}$$

$$= x + iy ,$$

$$|\tilde{U}_{oj}| = (x^2 + y^2)^{\frac{1}{2}},$$

$$d_f = \frac{|\tilde{U}_{oj}|}{U} = \frac{(x^2 + y^2)^{\frac{1}{2}}}{U} , \tag{6.65}$$

The time lag ψ_f is given by

$$\psi_f = \tan^{-1} (y/x) \quad \text{radians.}$$

The admittance procedure as described above is able to take in to account only the fundamental frequency of periodic variation i.e. w in Eq. (6.56). Consideration of higher frequencies lead to slight modification of the procedure discussed in appendix 6.

6.3.3 Response Factor Method

In order to study the performance of the passive buildings, Goldstein (1978) derived an analytical model which is based on the response factor method. The heat balance on the surfaces, exposed to environmental parameters, were used along with the solutions of the diffusion equation to derive response factors for surface temperature as a function of solar flux and the ambient temperature. The procedure for using this model is illustrated for a direct gain system below.

In a direct gain system, the solar energy enters into the room and strikes a surface (e.g. floor), absorbing a fraction of it and reflecting the rest of it. The reflected energy, in turn, gets absorbed by other surfaces. If α_j is the fraction of solar energy absorbed by the surface j, its temperature, θ_{sj}, is determined by the following heat balance equation

$$h_j\, (\theta_{sj} - \theta_r) - K_j \left.\frac{\partial\theta_j(x,t)}{\theta x}\right|_{x=o} = \alpha_j\, S_{tj}, \tag{6.66}$$

where S_{tj} is the total amount of solar radiation entering the building and getting reflected on to the jth component. Eq. (6.66) sets heat losses from the surface (left-hand side) equal to heat gains (right-hand side). The equation assumes that the surface transfers the heat directly to the room air, rather than being in radiative contact with other surfaces. This results in substantial simplification of the computational effort (Goldstein, 1978).

Assuming one dimensional heat flow, the heat-transfer across the jth element of a building is given by

$$K_j \frac{\partial^2\theta_j(x,t)}{\partial x^2} = (\rho c)_j \frac{\partial\theta_j(x,t)}{\partial t} , \tag{6.67}$$

Equations (6.66) and (6.67) described heat flows at the inside surface of a material and in its interior; at the outside surface, the material can be assumed to be in contact with solair temperature θ_{sa} by a pure resistance, (later as described by Eq. 6.80) which can be described by a heat-transfer coefficient h_{so}.

Such a description allows the solution for surface temperature in terms of the driving forces of solar gain and ambient temperature. If the amplitude of solair temperature and solar radiation can be written in simple form at a steady harmonic frequency, the results can be expressed as

$$\theta_{sj} = (h_j \, \theta_r + \alpha_j \, S_{tj}) \, R_{1j} + \theta_{sa} \, R_{2j}, \tag{6.68a}$$

where R_{1j} and R_{2j} are frequency dependent response functions defined as

$$R_{1j} = \left[\frac{\text{Cosh}\ (\alpha_d L) + \dfrac{1}{R_{so} \, K \, \alpha_d} \text{Sinh}\ (\alpha_d L)}{\dfrac{1}{R_{si}} + \dfrac{1}{R_{so}} \text{Cosh}\ (\alpha_d L) + \alpha_d K + \dfrac{1}{R_{si} \, R_{so} \, K \, \alpha_d} \text{Sinh}\ (\alpha_d L)} \right] \tag{6.68b}$$

$$R_{2j} = \left[\frac{(1/R_{so})}{\left(\dfrac{1}{R_{si}} + \dfrac{1}{R_{so}}\right) \text{Cosh}\ (\alpha_d L) + \left(\alpha_d K + \dfrac{1}{R_{si} \, R_{so} \, {}_d K}\right) \text{Sinh}\ (\alpha_d L)} \right], \tag{6.68c}$$

The response functions, defined above, contain almost all the information required to describe the behaviour of the material. The function R_{1j} describing the response of the material to sunlight is more interesting and is plotted as a function of frequency for a sample material in Fig. 6.6. The value of R_{1j} is maximum at low frequencies and decreases steadily with increasing ω. This means that the surface temperature has a larger response to solar radiation at low frequencies and a smaller response at higher frequencies. A good response function for a passive solar house would begin to decrease at frequencies comparable to weather changes and would be very small at diurnal frequency to damp out day to night changes in solar gain.

The results for the surface temperatures can be combined into an expression for the room temperature using a heat balance for the room air. This is given by

$$\sum_{j=1}^{N} \hat{h}_j \, (\theta_r - \theta_{sj}) + \hat{U}_q \, (\theta_r - \theta_a) = H + \alpha_r S, \tag{6.69}$$

where

$$\hat{h}_j = h_j \, A_j$$

and H = heater output (in case the auxiliary is present).

α_r = fraction of solar radiation absorbed directly into the room air or surfaces of light objects

$\hat{U}_q$ = the sum if U-values times area for all pure conductances (e.g. windows) plus the loss rate due to infiltration.

Equations (6.68a) and (6.69) are used to derive the room temperature given as

$$\theta_r \, A(\omega) = S \, B(\omega) + \theta_a \, C(\omega) + H, \tag{6.70a}$$

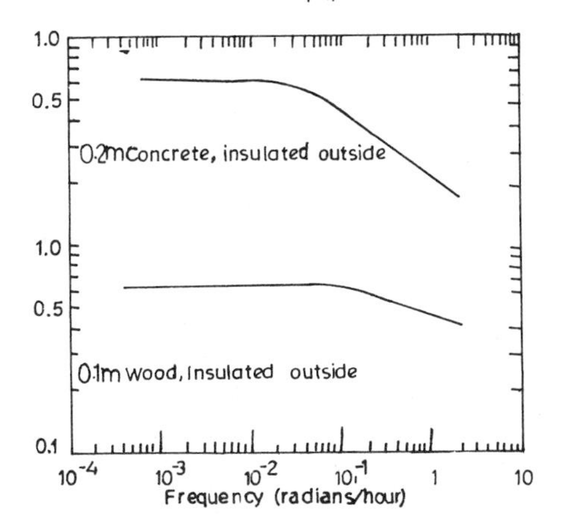

Fig. 6.6. Response functions as a function of frequency. Log $|R_1|$ is plotted vs. log ω for two materials with insulation R-8 ($1.41\ m^2\ {}^oC/W$) on the outside. (After Goldstein, 1978)

where $A(\omega)$, $B(\omega)$ and $C(\omega)$ are building response functions described as

$$A(\omega) = \sum_{j=1}^{N} \hat{h}_j\ (1 - h_j\ R_{1j}) + \hat{U}_q, \tag{6.70b}$$

$$B(\omega) = \sum_{j=1}^{N} \alpha_j\ h_j\ R_{1j} + \alpha_r, \tag{6.70c}$$

$$C(\omega) = \sum_{j=1}^{N} \hat{h}_j\ R_{2j} + \hat{U}_q, \tag{6.70d}$$

The physical interpretation of building transfer functions is as follows. $A(\omega)$ and $C(\omega)$ are analogous to conventional design heat losses. In fact, for a frequency of $\omega = 0$, the steady state term, $\hat{h}_j\ (1 - h_j\ R_{1j})$ is equal to $U_j\ A_j$, where U_j is the conventional U value of the material. For this case, $h_j\ R_{2j}$ is also equal to $U_j\ A_j$. Thus $A(\omega)$ and $C(\omega)$ are the conventional design heat losses for steady state results, and are simply frequency dependent heat transfer coefficients for $\omega \neq 0$. $B(\omega)$ describes the reduction in room temperature fluctuations due to the location of thermal mass, since this is the only term which uses the radiation balance fractions α_j.

The response of the building to incident solar radiation is given by $B(\omega)/A(\omega)$. A good passive solar house will have large B/A for low frequencies (e.g. annual variations) but small for daily frequencies. B/A will decrease with frequency for any building since B is composed of response functions, which decrease with frequency and A is made of function $\{\hat{h}_j\ (1 - h_j\ R_{1j})\}$ which increases with the frequency. Typically $A(\omega)$ begins to increase significantly before $B(\omega)$ decreases much, so the long term storage should be relatively insensitive to the location where sunlight falls within the building.

The results obtained from the response function method described above and the experimental results on a direct gain cell and Trombe wall experimental cell are given in Figs. 6.7 and 6.8.

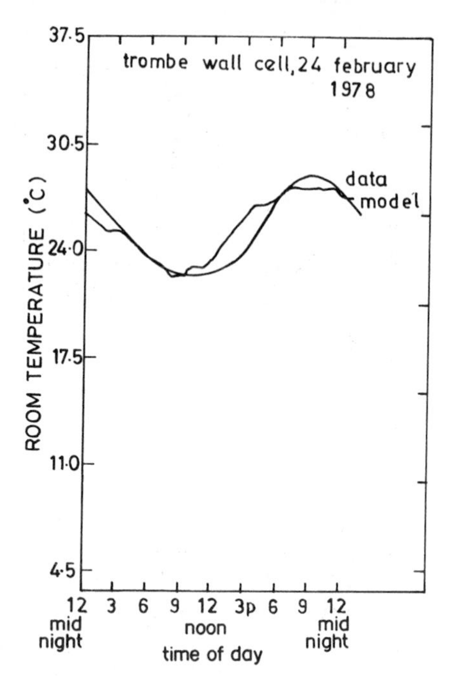

Fig. 6.7. Predicted room temperature and observed data as a function of time of day for the LASL Trombe wall cell, 24 February, 1978. (After Goldstein, 1978)

6.3.4 Finite Difference Representation

The problem of heat-transfer through a solid is studied by the solution of the three dimensional heat conduction equation viz.

$$\nabla(K.\nabla\theta) = \rho c \frac{\partial\theta(x,y,z,t)}{\partial t}, \tag{6.71}$$

The infinitesimal element $\nabla(K.\nabla\theta)$ at a certain place represents the residual heat as a result of conduction due to the temperature gradient in the three directions, each direction being characterized by the thermal conductivity K. This residual heat produces a variation of heat content in time i.e. $\rho c\ \partial\theta/\partial t$ representing the storing capacity of the material. If the heat flow can be assumed to be one dimensional, Eq. (6.71) reduces to the form given by Eq. (6.67) i.e. the Fourier conduction equation

$$\frac{\partial^2\theta}{\partial x^2} = \frac{1}{D} \cdot \frac{\partial\theta}{\partial t}, \tag{6.72}$$

In the form of a difference equation, it can be written for two consecutive homogeneous internal slabs in a wall as follows:

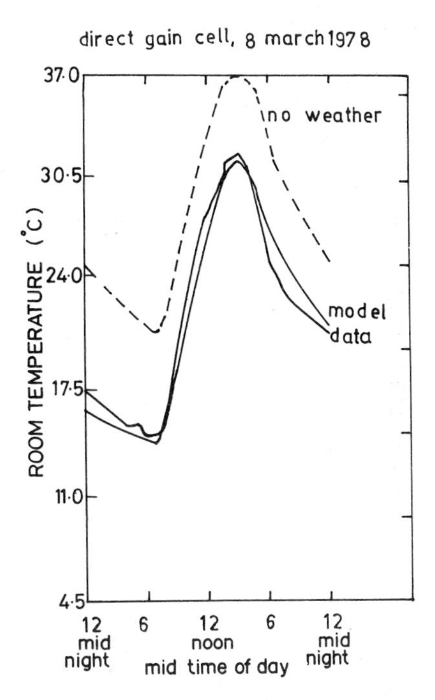

Fig. 6.8. Predicted room temperature and observed data for the direct gain cell as a function of time of day on March 8, 1978. The curve labelled "model" was calculated taking into account the weather variations for the previous two weeks. The curve labelled "no weather" shows the predictions of the model for the case where all days before March 8 were assumed to have the same weather. (After Goldstein, 1978)

$$\frac{\text{Area}_{N+1}}{R_{N+1}}\left\{\theta^M_{N+1}-\theta^M_N\right\}-\frac{\text{Area}_N}{R_N}\left\{\theta^M_N-\theta^M_{N-1}\right\}$$
$$=\left(\text{Area}_N\,\rho_N\,C_N\,\frac{d_N}{2}+\text{Area}_{N+1}\,\rho_{N+1}\,C_{N+1}\,\frac{d_{N+1}}{2}\right)\left(\frac{\theta^{M+1}_N-\theta^M_N}{\Delta\theta}\right) \tag{6.73}$$

where $(\theta^{M+1}_N - \theta^M_N)$ is the variation of temperature at the interface N in the interval $\Delta\theta$ between time M and time M+1, as a consequence of the temperature gradient at the interfaces N-1, N and N+1 at a distance d_N between them. R_N denotes the thermal resistance between inter-surfaces (N-1) and N.

The solution of this equation for θ^{M+1}_N at the interface N is of the form

$$\theta^{M+1}_N = A_N\,\theta^M_{N+1} + B_N\,\theta^M_N + C_N\,\theta^M_{N-1}, \tag{6.74}$$

where

$$A_N = \frac{\text{Area}_{N+1}}{R_{N+1}}\cdot\frac{\Delta\theta}{D_N}$$

$$C_N = \frac{Area_N}{R_N} \cdot \frac{\Delta\theta}{D_N}$$

$$B_N = 1 - (A_N - C_N)$$

$$D_N = Area_N \rho_N C_N \frac{d_N}{2}$$

R_N is the thermal resistance between the intersurfaces (N-1) and N, then $R_N = d_N/K_N$. If the surface N is in contact with air, R_N is the thermal resistance due to convection or radiation or both. If N is an internal surface, the effects due to internal heat gain, air infiltration and that of mechanical heating or cooling can be incorporated. If N is an external surface, the effects due to longwave radiation exchange with the sky and the incident radiation at the surface can be incorporated.

The use of finite difference representation necessitates the use of large digital computers. Buchberg *et al.* (1964) used this technique for evaluating the changing values of the internal temperatures for slab of each wall all over the structure (Hoffmann, 1976).

6.3.5 The Periodic Solutions

As illustrated in section 6.3.1, the variations of solar intensity and the ambient temperature can be expressed as a Fourier series with a frequency ω, viz.

$$f(t) = A_o + \text{Re} \sum_{n=1}^{\infty} A_n \exp\{i(n\omega t - \psi_n)\}, \tag{6.75a}$$

where

$$A_n = \sqrt{a_n^2 + b_n^2}$$

$$\psi_n = \tan^{-}\left(\frac{b_n}{a_n}\right), \tag{6.75b}$$

Alternatively, $f(t)$ can also be represented as

$$f(t) = \sum_{n=-}^{+\infty} C_n \exp(in\omega t), \tag{6.75c}$$

where

$$C_n = \left(\frac{a_n + ib_n}{2}\right), \quad n < 0,$$

$$= \left(\frac{a_n - ib_n}{2}\right), \quad n > 0 \tag{6.75d}$$

Since the building is exposed to solar radiation and the ambient temperature, varying periodically, for a number of years, the temperature inside the room shows variations corresponding to the variations of the input functions (solar insolation and ambient temperature), and the solution of the heat conduction equation (6.72) can be expressed as*

$$\theta(x,t) = \theta_o(x) + \text{Re} \sum_n \theta_n \exp(in\omega t), \tag{6.76}$$

Substitution of Eq. (6.76) into Eq. (6.72) and equating the coefficients of equal powers of exp($in\omega t$), one gets

$$\frac{d^2\theta_o}{dx^2} = 0 , \tag{6.77a}$$

$$\frac{d^2\theta_n}{dx^2} = \left(\frac{in\omega\rho c}{K}\right), \tag{6.77b}$$

Solutions of Eqs. (6.77) and their substitution into Eq. (6.76) yields

$$\theta(x,t) = (A_o + A_1 x) + R_c \sum_n \left|\lambda_n \exp(\alpha_n x) + \lambda_n^1 \exp(-\alpha_n x)\right| \exp(in\omega t), \tag{6.78}$$

where

$$\alpha_n = (1 + i)\left(\frac{n\omega\rho c}{2K}\right)^{\frac{1}{2}}, \tag{6.79}$$

and the constants A_o, A_1, λ_n and λ_n^1 have to be determined from the appropriate boundary conditions for the system to be studied.

6.3.5.1 A simple building element i.e. roof/wall. For a building element, such as wall, with heat flow being assumed to be one-dimensional, and internal environment to be constant at base temperature θ_b, the boundary conditions are usually written as

$$-K \left.\frac{\partial\theta}{\partial x}\right|_{x=o} = h_{so}(\theta_{sa} - \theta_{x=o}), \tag{6.80}$$

$$-K \left.\frac{\partial\theta}{\partial x}\right|_{x=L} = h_{si}(\theta_{x=L} - \theta_b), \tag{6.81}$$

Solair temperature may also be expressed in terms of the Fourier series, i.e.

$$\theta_{sa}(t) = \bar{\theta}_{sa} + \text{Re} \sum_{n=1}^{\infty} \theta_{sn} \exp\{i(n\omega t - \psi_n)\}, \tag{6.82}$$

Substitution of Eqs. (6.78) and (6.82) into the boundary conditions (6.80) and (6.81) yields simple algebraic equations for the unknown constants. The solution of these equations leads to explicit expressions for various desired parameters. The expression for heat flux entering the room through the roof/wall comes out to be (Mackey and Wright, 1946)

$$Q = h_{si}(\theta_{x=L} - \theta_b)$$

---

*If the transverse dimensions of the building element are much greater than the characteristic length $L_c = (K/\omega\rho c)^{\frac{1}{2}}$, the assumption of one-dimensional heat flow leads to only marginal inaccuracies.

$$= U.(\bar{\theta}_{sa} - \theta_b) + \text{Re} \sum_{n=1}^{\infty} h_{si}\, d_f\, \theta_{sn} \exp\{i(n\omega t - \psi_n)\}, \qquad (6.83)$$

where in analogy of Eq. (6.10), U may be expressed as

$$\frac{1}{U} = \left[\sum_{j=1}^{m} \left(\frac{L}{K}\right)_j + \frac{1}{h_{so}} + \frac{1}{h_{si}}\right], \qquad (6.84)$$

Generally, a small number of harmonics (usually six) are sufficient for a good convergence of the series. According to Dreyfus (1963), the decrement in wave amplitude for the nth harmonic may be calculated from

$$d_f = \exp\left(-\sqrt{\frac{\pi}{t}\, A^2 B}\right), \qquad (6.85)$$

and the time lag ψ_f which depends on the thermo-physical properties of the building element, is given by

$$\psi_f = \frac{1}{2}\sqrt{\frac{\pi}{t}\, A^2 B}\,, \qquad (6.86)$$

where A^2B is a kind of thermal time constant, t ($= \frac{24}{n}$ h) is the period for the harmonic n, and A and B are given by

$$A = \left(\frac{L}{K}\right)_i + \left(\frac{L}{K}\right)_{m_1} + \left(\frac{L}{K}\right)_{m_2} + \text{---} + \left(\frac{L}{K}\right)_c$$

$$B = \frac{1.1}{A}\left[\sum_{j=m_1,m_2} \left(\frac{L_j}{K_j}\right)(K\rho C)_j - \left(\frac{L}{K}\right)_o (K\rho C)_o\right]$$

$$+ \frac{(K\rho C)_o}{A}\left[\left(\frac{L}{K}\right)_o - 0.1\left(\frac{L}{K}\right)_i - 0.1\left(\frac{L}{K}\right)_{m_1} \text{--- etc.}\right], \qquad (6.87)$$

The indices i, m_1, m_2 .... and O represent the innermost, intermediate and outermost layers respectively. If the second term in computation of B is negative, it can be neglected.

Example 6.10 The roof of a building is made of 30 cm thick concrete slab covered with 5 cm thick insulation of corkboard. Its location is 29°N latitude. The typical variation of solair temperature (on 26 May, 1978 at New Delhi) is given in the table below. The space inside the building is kept at a constant temperature of 20°C. Determine the amount of heat flux entering into the room throughout the day.

Time (hr)	Solair temp. {θ_{sa} (°C)}
1	27.7
2	27.7
3	27.2
4	27.7
5	28.2
6	30.3

Time (hr)	Solair temp. {θ_{sa} (°C)}
7	35.6
8	42.7
9	48.7
10	55.6
11	60.5
12	63.3
13	64.2
14	64.0
15	60.9
16	56.1
17	53.2
18	43.9
19	37.5
20	34.2
21	32.2
22	31.0
23	31.0
24	29.2

Solution Corresponding to the above data, the mean solair temperature and the other harmonic coefficients (using Eq. 6.75) are obtained as

harmonic	0	1	2	3	4	5	6
amplitude	42.28	18.49	4.59	0.54	0.66	0.34	0.12
phase (radians)	-	3.4999	0.3674	0.399	4.391	3.891	0.891

From Eq. (6.87)

$$A = \left(\frac{L}{K}\right)_{conc.} + \left(\frac{L}{K}\right)_{ins.}$$

$$= \left(\frac{0.3}{1.4}\right) + \left(\frac{0.05}{0.028}\right) \simeq 2.0 \text{ W}^{-1} \text{ m}^2 \text{ °C}$$

$$B = \frac{1.1}{A}\left[\left(\frac{L}{K}\right).\ K\rho C_{\ conc.} - \left(\frac{L}{K}\right).\ K\rho C_{\ ins}\right]$$

$$+ \frac{(K\rho C)_{ins}}{A}\left[\left(\frac{L}{K}\right)_{ins.} - 0.1\left(\frac{L}{K}\right)_{conc.}\right]$$

$$= \frac{1.1}{2}\left[(0.3 \times 1858 \times 795) - (0.05 \times 2049 \times 115)\right]$$

$$+ \left(\frac{0.028 \times 115 \times 2049}{2}\right)\left[\left(\frac{0.05}{0.028}\right) - \frac{0.1 \times 0.3}{1.4}\right]$$

$$= 233063 \left(\frac{J}{m^2\ {}^oC} \cdot \frac{W}{m^2\ {}^oC} \right),$$

The decrement factor $d_f = \exp \left(- \sqrt{\frac{\pi}{t}\, A^2 B} \right)$

For first harmonic $t = \frac{24}{n} = \frac{24}{1}\ h$

$$\therefore\ d_f = \exp \left(- \sqrt{\frac{\pi \times 4 \times 233063}{24 \times 3600}} \right)$$

$$= 0.003$$

$$\psi_1 = \frac{1}{2} \sqrt{\frac{\pi}{t}\, A^2 B}$$

$$= 2.9 \text{ radians}$$

2.9 radians correspond to a phase lag of $(12 \times 2.9/\pi) \simeq 11$ hours.

$$U = \left(\frac{1}{h_{so}} + \left(\frac{L}{K}\right)_{ins} + \left(\frac{1}{K}\right)_{conc.} + \frac{1}{h_{si}} \right)^{-1}$$

$$= \left(\frac{1}{20.0} + \frac{0.05}{0.028} + \frac{0.3}{1.4} + \frac{1}{8} \right)^{-1}$$

$$= 0.459\ W/m^2\ {}^oC$$

Hence using only the first harmonic the heat flux is given by

$$\dot{Q} = 0.459\ (42.28 - 20)$$

$$+ 8.0 \times .003 \times 18.49 \cos \frac{2\pi}{24} t - 3.499 - 2.9$$

Maximum of solair occurs at 13 hours. Hence the maximum of heat flux occurs at 24 hours because the phase difference is 11 hours. Therefore

$$\dot{Q} = 10.23 + 0.44 \cos (0.1158)$$

$$\simeq 10.67\ W/m^2$$

6.3.5.2 Passive heating concepts with constant internal environment. The passive heating concepts that are usually employed in solar buildings are (i) Trombe wall, (ii) water wall, and (iii) solarium. Using numerical simulation analysis and correlation techniques, the performance of these concepts is well characterized (Balcomb *et al.*, 1977, Wray *et al.*, 1979, Balcomb *et al.*, 1982). Utzinger *et al.* (1980) developed a one-dimensional thermal circuit network model to determine the effects of air flow rate on the auxiliary heating energy consumption in collector storage walls. The results of Utzinger *et al.* (1980) show that the assumption of one-dimensional heat flow through the mass wall in a Trombe wall system along with the exponentially varying air temperature profile in the space between glazing and the absorber surface, yields results which are in excellent agreement with the corresponding results of a two-dimensional model. The study of Utzinger *et al.* is based on numerical solution of the corresponding heat transfer equations. A one-dimensional heat transfer model has, therefore, been used in the following analysis here also.

The use of periodic solutions of the heat conduction equation leads to close-form solutions for the heat flux coming into the room, though the periodic theory is able to predict only the typical (average) behaviour of the system. Moreover, from the design point of view the average behaviour is the most desirable aspect. The use of periodic theory implies that the meteorological parameters have a consistent time dependence over a period of 48 hours; the response function methods indicate that the variation of meteorological parameters, before 48 hours or longer, does not affect the thermal performance significantly. The analysis can be used to predict the mean daily behaviour throughout the year if mean daily values of the meteorological parameters are Fourier analysed over a period of 365 days. The simplicity and elegance of the periodic theory has to be weighed against its inability to predict the effect of large scale fluctuations in the weather on a daily time scale.

The schematics of three types of passive heating systems viz. Trombe wall (a) without and (b) with vents (natural or forced air circulation), (c) water wall and (d) solarium have already been illustrated in Figs. 6.1-6.3. To apply the periodic solution method, we shall now consider them separately. In each case it will be assumed that the room is maintained at the balance point temperature θ_b.

(a) *Trombe wall without vents*. Trombe wall is essentially a thick wall, with the outer surface blackened and glazed. The storage mass may be concrete, adobe, stone or composites of brick, block and sand. Solar radiation is absorbed by the blackened surface and is stored as sensible heat in the wall. The temperature distribution in the wall is governed by Eq. (6.78) subject to the boundary conditions given by Eqs. (6.80) and (6.81). It may, however, be noted that the external heat transfer coefficient h_{so} in this case has to be replaced by an overall heat transfer coefficient from a glazed surface given by

$$\frac{1}{h_o} = \left(\frac{1}{C_c} + \frac{Lg}{K_g} + \frac{1}{h_{so}}\right), \tag{6.88}$$

C_c, being the conductance of air gap between the absorbing surface and glazing.

I. In the case when the glazing is not insulated during off-sunshine hours, the expression for heat flux takes the form

$$\dot{Q} = U(\bar{\theta}_{sa} - \theta_b) + K\, h_{si}\, h_{so}\ \mathrm{Re}\ \Sigma\ \theta_{sn}\, U_n\ \exp\ (in\omega t) \tag{6.89}$$

where $U_n = \left[(h_o\, h_{si} + k^2\alpha_n^2)\ \sin h\ \alpha_n L + (h_o A\, h_{si})\ k\alpha_n\ \cos h\ \alpha_n L\right]^{-1}$,

α_n defined by Eq. 6.79, corresponds to the physical properties of the thermal wall.

II. In case the entire glazing is insulated during off-sunshine hours, the top heat-transfer coefficient h_o instead of being a constant value, becomes time dependent. As a reasonable approximation, h_o can be assumed to be a rectangular function of time with a constant value during sunshine hours and another value during off-sunshine hours (Fig. 6.9). One can still assume a following type of Fourier series expansion for h_o

$$h_o(t) = H_o + \sum_{\substack{n=-\infty \\ n\neq o}}^{+\infty} H_n\ \exp\ (in\omega t), \tag{6.90}$$

Since the Fourier series expansions with term n = - 6 to n = + 6 have been found to be good representations of the functions, only terms corresponding to n = - 6 to n = + 6 are retained in the above expansion for the purpose of numerical computation.

In this case of time varying overall heat-transfer coefficient from the absorbing

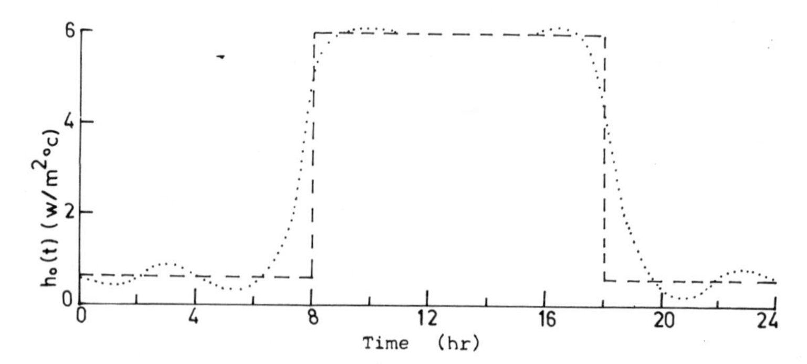

Fig. 6.9. Hourly variation of overall heat-transfer coefficient. Dashed curve corresponds to the data, dotted curve to that calculated from Fourier coefficients

surface to the ambient, it is desirable to express all the functions by a Fourier series by the form (6.90) and retaining terms corresponding to n = - 6 to n = + 6. For example θ_{sa} is now expressed as

$$\theta_{sa} = b_o + \sum_{\substack{n=-\infty \\ n\neq o}}^{+\infty} b_n \exp(in\omega t), \tag{6.91a}$$

and

$$\theta_T(x,t) = (A_o + A_1 x) + \sum_{\substack{n=-6 \\ n\neq o}}^{+6} \{\lambda_n \exp(\alpha_n x) + \lambda_n^1 \exp(-\alpha_n x)\} \exp(in\omega t), \tag{6.91b}$$

Naturally $b_o = \overline{\theta}_{sa}$

Substituting Eqs. (6.90) and (6.91b) into the boundary conditions (6.80) and (6.81) yields two sets of equations for λ_n, λ'_n, A_o and A_1 with

$$A_1 = \frac{h_{si}(\theta_b - A_o)}{(h_{si}L + K)}, \tag{6.92}$$

$$\lambda'_n = -\left\{\frac{(h_{si} + K\alpha_n)\exp(\alpha_n L)}{(h_{si} - K\alpha_n)\exp(-\alpha_n L)}\right\}\lambda_n, \tag{6.93}$$

$$(A_o H_o - K A_1) = \sum_{m=1}^{6} H_{-m} b_m + b_{-m} H_m, \tag{6.94}$$

$$\left[(H_o - K\alpha_n)\lambda_n + \sum_m \sum_n H_m \lambda_n + (H_o + K\alpha_n)\lambda'_n + \sum_m \sum_n H_m \lambda'_n + A_o H_n\right] = b_o H_n \tag{6.95}$$

Eliminations λ'_n from Eqs. (6.93) and (6.95), one obtains the following equations for the unknown constants

$$[M] \times [N] = [R], \tag{6.96}$$

$[M]$ is a square matrix having the following elements

$$m_{ij} = U_{i-j} - Y_{i-(\ +1)}\ \ _{ij}, \quad \text{for } i = 1 \text{ to } (2\ell+1)$$
$$j = 1 \text{ to } (i+\ell)\ \{\text{for } 1 \leqslant i \leqslant (\ell+1)$$
$$j = (i-\ell) \text{ to } (2\ell+1)\ \{\text{for } (\ell+2) \leqslant i \leqslant (2\ell+1)\} \tag{6.97}$$

otherwise = 0

and $[N]$ and $[R]$ are column vectors defined as

$$n_{j1} = C_j - (\ell+1), \quad \text{for } 1 < j < (2\ell+1) \tag{6.98}$$

$$r_{j1} = p_{j1} + U_b\ \theta_b\ \delta_j(\ell+1) \tag{6.99}$$

where

$$U_{i-j} = H_{i-j}\left[1 + X_{j-(\ell+1)}\right], \quad \text{for all } i \text{ and } j$$
$$(\text{except } i = (\ell+1) \text{ and } j = (\ell+1))$$
$$= H_o + U_b; \text{ for } i = (\ell+1) \text{ and } j = (\ell+1)$$

$$Y_{i-(\ell+1)} = K\ \alpha_{i-(\ell+1)}\left[1 - X_{i-(\ell+1)}\right], \quad \text{for } i \neq (\ell+1)$$
$$= 0; \text{ for } i = (\ell+1)$$

$$p_{j1} = \sum_{r=0}^{\ell} H_{-(\ell-r)}\ b_{-(r-j+1)} + \sum_{\substack{r=1\\ r>j}}^{j-1} H_r\ b_{-(r-j+\ell+1)}, \quad \text{for } 1 \leqslant j \leqslant j(\ell+1)$$

$$= \sum_{r=0}^{2\ell+1} H_{r+j-(2\ell+1)}\ b_{(\ell-r)} + \sum_{r=1}^{\ell} H_r\ b_{-(r-j+\ell+1)}, \quad \text{for } (\ell+2) \leqslant j \leqslant (2\ell+1)$$

$$X_n = \left(\frac{K\alpha_n + h_{si}}{K\alpha_n - h_{si}}\right) \exp\ (2\alpha_n L), \tag{6.100}$$

δ_{ij} is Kronecker's delta

and $\ell = 6$.

C's are the unknown constants λ_n's with $C_o = A_o$.

Equation (6.96) can be solved for [N] by matrix inversion techniques. After some algebra the expression for $\dot{Q}$ comes out to be

$$\dot{Q} = \left[(C_o - \theta_b) U_b + 2 K h_{si} \sum_{\substack{n=-6 \\ n \neq o}}^{+6} \left(\frac{C_n \alpha_n}{K\alpha_n - h_{si}} \right) \exp(\alpha_n L) \exp(in\omega t) \right], \tag{6.101a}$$

where

$$\frac{1}{U_b} = \left(\frac{L}{K} + \frac{1}{h_{si}} \right), \tag{6.101b}$$

(b) *Trombe wall with vents.* Vents at upper and lower ends of Trombe wall provide good facilities for air circulation from the sunspace (the space between the glazing and the wall) to the living space by natural/forced circulation. Air sweeping over the blackened surface of the wall carries heat along with it into the living space. The heat flux entering the living space due to air circulation (in addition to diffusion of heat through the wall) can be estimated from the following analysis.

The energy balance at the absorbing surface (x = 0) of Trombe wall (Fig. 6.10) can be written as

$$\alpha\tau . S(t) = h_1 \left(\theta_T \Big|_{x=o} - \bar{\theta}_{ss} \right) + h_r \left(\theta_T \Big|_{x=o} - \theta_{g\ell} \right) - K \frac{\partial \theta_T}{\partial x} \Big|_{x=o}, \tag{6.102}$$

while at the inside surface (x = L), it is given by Eq. (6.81); with $\theta(x,t) = \theta_T(x,t)$; $\theta_T(x,t)$ being chosen to represent the temperature distribution in the thermal storage wall. The temperature rise of air in an elementary element $b.\Delta y$ of sunspace (refer to Fig. 6.1b) can be written as

$$\dot{m}_a C_a \left(\frac{\partial \theta_{ss}}{\partial y} \right) \Delta y = \left[h_1 \left(\theta_T \Big|_{x=o} - \theta_{ss} \right) - h_2 \left(\theta_{ss} - \theta_{g\ell} \right) \right] b\Delta y, \tag{6.103}$$

where C_a is the specific heat of air (J/kg°C)

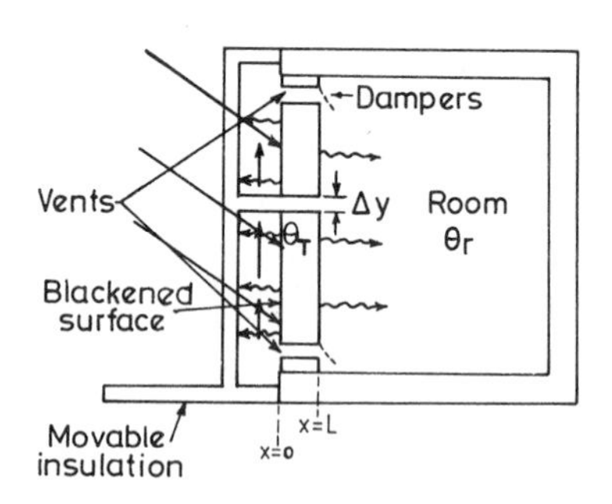

Fig. 6.10. Trombe wall with vents

Therefore, the instantaneous average temperature $\bar{\theta}_{ss}$ of air in sunspace and the temperature rise $(\theta_{out} - \theta_b)$ of air mass coming out of sunspace can be written as

$$\bar{\theta}_{ss} = \left(h_1 \theta_T \Big|_{x=o} + h_w \theta_{g\ell} \right) g_1 + \theta_b g_2, \tag{6.104}$$

and

$$(\theta_{\theta ut} - \theta_b) = \left[h_1 \, \theta_T\Big|_{x=o} + h_2 \, \theta_{g\ell} - (h_1 + h_2) \, \theta_b\right] \left(\frac{g_2 A}{\dot{m}_a C_a}\right), \tag{6.105a}$$

where

$$g_2 = \frac{\dot{m}_a C_a}{A(h_1 + h_2)} \left[1 - \exp\left\{- \frac{(h_1 + h_2) \, A}{\dot{m}_a C_a}\right\}\right], \tag{6.105b}$$

$$g_1 = \left(\frac{1 - g_2}{h_1 + h_2}\right), \text{ and } A = b.L \text{ being the area of the mass wall.} \tag{6.105c}$$

The glass temperature $\theta_{g\ell}$ can be obtained from the equation

$$h_r \, (\theta_T\Big|_{x=o} - \theta_{g\ell}) + h_2 \, (\bar{\theta}_{ss} - \theta_{g\ell}) = h_3 \, (\theta_{g\ell} - \theta_a), \tag{6.106a}$$

where

$$\frac{1}{h_3} = \left(\frac{L_g}{K_g} + \frac{1}{h_{so}}\right), \tag{6.106b}$$

It may be noted that in the above equations, the heat capacity of glass and its absorptivity for solar radiation are assumed to be negligibly small. The terms on the left-hand side of Eq. (6.106a) represent the radiative and convective heat gain by the glazing's inside surface from the mass wall and air in sunspace respectively, while the term on the right-hand side denote the heat losses from the glazing to the ambient.

The heat flux entering through this Trombe wall with vents is a summation of two terms i.e.

$$\dot{Q} = h_{si} \, (\theta_T\Big|_{x=L} - \theta_b) + \frac{\dot{m}_a C_a}{A} \, (\theta_{\theta ut} - \theta_b), \tag{6.107}$$

the first term on the RHS is corresponding to the heat flux conducted through the opaque portion of the wall and the second term is the heat energy brought by the flowing air in space between the room and the sunspace through vents.

In Trombe wall, the air circulation is generally, maintained by natural convection. The mass flow rate $\dot{m}_a$ thus created can be determined by analysing the stack effect. In the simplified case of a ventilated Trombe wall (Fig. 6.1), a control volume is set up inside the stack in which the heated air has an average density (ρ_h) smaller than that of the air outside (or inside the room, $\rho_{\theta ut}$). This density difference produces the driving force for an upward velocity v inside the stack. A steady state force balance in vertical direction including pressure forces, resistance to friction and weight is solved for the friction force yielding (Heidt, 1983)

$$F' = gLA \, (\rho_{\theta ut} - \rho_r), \tag{6.108}$$

On the other hand, the force F' can be expressed in terms of a pressure loss coefficient ρ_ℓ as defined by

$$F' = \rho_\ell \cdot \frac{1}{2} \rho_h v^2 A, \tag{6.109}$$

which on substitution in Eq. (6.108) gives

$$v = \left[2gL\ (\rho_{\theta ut} - \rho_h)/\rho_\ell\ \rho_h\right]^{\frac{1}{2}}, \tag{6.110}$$

From perfect gas equation, this can be written in terms of temperature rather than density difference as:

$$v = (2gL.\Delta\theta/\rho_\ell \cdot \theta_{\theta ut})^{0.5}, \tag{6.111}$$

The temperature difference $\Delta\theta$ is calculated from Eq. (6.105a) which shows that it depends upon the temperatures of the surface and the heat transfer coefficients which again depend upon the air velocity v. An iterative approach has to be adopted therefore to calculate v and hence $\dot{m}_a = \bar{\rho}\ A\ v$, where $\bar{\rho} = \frac{1}{2}\ (\rho_{out} + \rho_h)$.

The glass temperature $\theta_{g\ell}$ can be eliminated from the Eqs. (6.104) and (6.106a). Substitution of the resulting expression for in Eq. (6.102) and subsequently substituting expressions for $\theta_T(x,t)$ and the Fourier series representation of the solar radiation and the ambient temperature, one obtains explicit expressions for the heat fluxes.

I. If the glazing is not insulated during off-sunshine hours

$$\dot{Q}_w = \left[U_b\ (B - \theta_b) + K\ h_{si}\ \mathrm{Re} \sum_{n=o}^{\infty} \frac{\alpha_n}{D_n}\ (q_{1n} + U_1\ \theta_{gn})\ \exp\ (in\omega t)\right], \tag{6.112}$$

and

$$\dot{Q}_A = \Big[g_2\ \{h_1 B + h_2\ \theta_{go} - (h_1 + h_2)\ \theta_b\}$$

$$+ g_2.\mathrm{Re} \sum_{n=o}^{\infty} \frac{1}{D_n}\ \{h_1\ (K\alpha_n\ \mathrm{Cosh}\ \alpha_n L + h_{si}\ \mathrm{Sinh}\ \alpha_n L)\ (q_{1n} + U_1\ \theta_{gn})$$

$$- h_2\ \theta_{gn}\}\ \exp\ (in\omega t)\Big], \tag{6.113}$$

where the different constants are defined as

$$B = \left[\frac{U_3\ (q_{10} + U_b\ \theta_b) + U_1\ (g_2 h_2 \theta_b + h_3\ \bar{\theta}_a)}{U_3\ (U_2 + U_b) - U_1^2}\right], \tag{6.114a}$$

$$\theta_{go} = \left\{\frac{U_1\ (q_{10} + U_b\ \theta_b) + (U_2 + U_b)\ (g_2\ h_2\ \theta_b + h_3\ \bar{\theta}_a)}{U_3\ (U_2 + U_b) - U_1^2}\right\}, \tag{6.114b}$$

$$\theta_{gn} = \left[\frac{D_n\ h_3\ \theta_{an} + U_1\ q_{1n}\ (K\alpha_n\ \mathrm{Cosh}\ \alpha_n L + h_{si}\ \mathrm{Sinh}\ \alpha_n L)}{D_n\ U_3 - U_1^2\ (K\alpha_n\ \mathrm{Cosh}\ \alpha_n L + h_{si}\ \mathrm{Sinh}\ \alpha_n L)}\right] \tag{6.114c}$$

$$D_n = \left[(U_2 + h_{si})\ K\alpha_n\ \mathrm{Cosh}\ \alpha_n L + (U_2\ h_{si} + K^2\alpha_n^2)\ \mathrm{Sinh}\ \alpha_n L\right], \tag{6.114d}$$

$$U_1 = (h_1\ h_2\ g_1 + h_r),$$

$$U_2 = h_1 (1 - g_1 h_1) + h_r,$$

$$U_3 = h_3 + h_r + h_2 (1 - g_1 h_2), \tag{6.114e}$$

and $\bar{\theta}_a$, θ_{an}, q_{10} and q_{1n} are the Fourier coefficients defined by

$$\theta_a = \bar{\theta}_a + \text{Re} \sum_{n=1}^{\infty} \theta_{an} \exp (in\omega t), \tag{6.115a}$$

$$(\alpha\tau.S(t) + h_1 g_2 \theta_b) = q_{10} + \text{Re} \sum_{n=1}^{\infty} q_{1n} \exp (in\omega t), \tag{6.115b}$$

II. In case, the glazing is insulated during off-sunshine hours, the heat flux can be obtained in the same manner as discussed in the case of Trombe wall (without vents). In this case, the heat transfer coefficient from the inside surface of the glazing to the ambient becomes time dependent and hence, as indicated earlierr can be expanded as

$$h_3(t) = H'_o + \sum_{\substack{n=-6 \\ n\neq o}}^{+6} H'_n \exp (in\omega t), \tag{6.116}$$

An expression for the heat flux entering into the living space can thus be derived. It may be noted that Eqs. (6.112) and (6.133) still remain as the expression for heat fluxes in this case as well, except that the constants B, θ_{go}, and θ_{gn} are re-defined as

$$B = \left(\frac{q_{10} + U_b \theta_b + U_1 \theta_{go}}{U_2 + U_b} \right), \tag{6.117}$$

and θ_{gn} (n = - 6 to + 6 including n = 0) are given by a matrix equation similar to Eq. (6.96). The matrix elements of $|M|$ are given by an expression similar to expression (6.97) with following definitions for U's and Y's

$$U_{i-j} = H'_{i-j} + \{h_r - h_2 (g_1 h_2 - 1)\} \delta_{ij}, \tag{6.118}$$

$$Y_{i-(\ell+1)} = \left[\frac{U_1^2 \{K\alpha_{i-(\ell+1)} \text{Cosh} (\alpha_{i-\ell-1}L) + h_{si} \text{Sinh} (\alpha_{i-\ell-1}L)\}}{\{(U_2+h_{si})K\alpha_{i-\ell-1}\text{Cosh}(\alpha_{i-\ell-1}L)+(U_2h_{si}+K^2\alpha^2_{i-\ell-1})\text{Sinh}(\alpha_{i-\ell-1}L)\}} \right] \tag{6.119a}$$

$$= U_1/(U_2 + U_b),$$

$$\text{for } i = 1 + 1 \tag{6.119b}$$

$|N|$ is a column vector generated by θ_{gn} for n = - 6 to n = + 6 while $[R]$ is a column vector whose elements are defined as

$$r_{j1} = \left[p_{j1} + \frac{Y_j}{U_1} q'_{1n} + g_2 h_2 \theta_b + \frac{U_1 (q_{10} + U_b \theta_b)}{U_2 + U_b} \delta_{j(\ +1)} \right], \tag{6.120a}$$

where

$$P_{j1} = \sum_{r=o}^{\ell} H'_{-(\ell-r)} \, a'_{-(r-j+1)} + \sum_{\substack{r=1 \\ r>j}}^{j-1} H'_r \, a'_{-(r-j+\ell+1)}; \quad 1 \leqslant j \leqslant (\ell+1)$$

$$= \sum_{r=o}^{2\ell+1} H'_{r+j-(2\ell+1)} \, a'_{(\ell-r)} + \sum_{r=1}^{\ell} H'_r \, a'_{-(r-j+\ell+1)}; \quad (\ell+2) \leqslant j \leqslant (2\ell+1) \tag{6.120b}$$

and q'_{1n} and a'_n are the Fourier coefficients defined as

$$\alpha\tau.S(t) + g_2 \, h_1 \, \theta_b = q_{10} + \sum_{\substack{n=-6 \\ n\neq o}}^{+6} q'_{1n} \exp(in\omega t), \tag{6.121a}$$

and

$$\theta_a = a'_o + \sum_{\substack{n=-6 \\ n\neq o}}^{+6} a'_n \exp(in\omega t), \tag{6.121b}$$

Naturally $a'_o = \bar{\theta}_a$.

(c) *Water wall.* Water wall is based on the same principle as Trombe wall except that it employs water as the storage material. It consists of containers (metallic) filled with water and is kept south facing. One surface of it is blackened and glazed while other surfaces can either be in direct contact with the living space or be separated from it by a thin concrete wall or insulating layer. In this case, for generality, we assume a thin concrete wall in close contact with the water wall (Fig. 6.2).

The energy balance at the absorbing surface can be written as

$$\alpha\tau.S(t) = h'_1 (\theta_p - \theta_w) + h_o (\theta_p - \theta_a), \tag{6.122}$$

The energy balance of water mass is

$$M_w \frac{d\theta_w}{dt} = h'_1 (\theta_p - \theta_w) - h'_2 (\theta_w - \theta_I|_{x=o}), \tag{6.123}$$

The boundary conditions at surfaces x = 0 and x = L can be written as

$$- K \left.\frac{\partial\theta_T}{\partial x}\right|_{x=o} = h'_2 (\theta_w - \theta_T|_{x=o}), \tag{6.124}$$

and

$$- K \left.\frac{\partial\theta_T}{\partial x}\right|_{x=L} = h_{si} (\theta_T|_{x=L} - \theta_b), \tag{6.125}$$

Equations (6.122) and (6.123) can be combined to eliminate θ_p so that one can write

$$M_w \frac{d\theta_w}{dt} = h_{eff} (\theta_{sa} - \theta_w) - h_2' (\theta_w - \theta_T|_{x=o}), \tag{6.126}$$

where

$$\frac{1}{h_{eff}} = \left(\frac{1}{h_1'} + \frac{1}{h_o}\right), \tag{6.127}$$

Assuming periodic variations as before, Eqs. (6.124) to (6.126) can be used to determine the unknown constants in the periodic solution for $\theta_T(x,t)$ and hence the heat flux coming into the room.

I. In the case when the water wall remains exposed even in off-sunshine hours, the heat flux is given by (Nayak *et al.*, 1983)

$$\dot{Q} = U (a_o - \theta_b) + K h_2' h_{si} \operatorname{Re} \sum_{n=1}^{\infty} \frac{\alpha_n \theta_{wn}}{D_{wn}} \exp (in\omega t), \tag{6.128}$$

where

$$\frac{1}{U} = \left(\frac{1}{h_o} + \frac{1}{h_1'} + \frac{1}{h_2'} + \frac{L}{K} + \frac{1}{h_{si}}\right),$$

and θ_{wn} and D_{wn} are defined as follows

$$\theta_{wn} = h_{eff}.a_n/[(in\omega M_w + h_{eff}) D_{wn}$$

$$+ K h_2 \alpha_n (K\alpha_n \operatorname{Sinh} \alpha_n L + h_{si} \operatorname{Cosh} \alpha_n L)], \tag{6.129}$$

and

$$D_{wn} = \{K \alpha_n (h_2' + h_{si}) \operatorname{Cosh} \alpha_n L + h_2' h_{si} K^2 \alpha_n^2 \operatorname{Sinh} \alpha_n L\}, \tag{6.130}$$

II. For the case when water wall is insulated during off-sunshine hours, the expression for heat flux changes to

$$\dot{Q} = \left[U_T (\bar{\theta}_w - \theta_b) + h_2' h_{si} K \sum_{\substack{n=-6 \\ n\neq o}}^{+6} \frac{\alpha_n \theta_{wn}}{D_{wn}} \exp (in\omega t)\right], \tag{6.131}$$

where

$$\frac{1}{U_T} = \left(\frac{1}{h_2'} + \frac{L}{K} + \frac{1}{h_{si}}\right), \tag{6.132}$$

and $\bar{\theta}_w$, θ_{wn} and D_{wn} are defined by a matrix equation similar to Eq. (6.96), with the matrix elements m's same as given by Eq. (6.97) with following expressions for U and Y.

$$U_{i-j} = H_{i-j}, \tag{6.133}$$

$$Y_{i-(\ell+1)} = [\{i-(\ell+1)\} \omega M_w \sqrt{-1}$$

$$+ \frac{\{K\,\alpha_{i-(\ell+1)}\,\mathrm{Sinh}\,(\alpha_{i-(\ell+1)}L) + h_{si}\,\mathrm{Cosh}\,(\alpha_{i-(\ell+1)}L)\}K\alpha_{i-(\ell+1)}h_2'}{\{K\,\alpha_{i-(\ell+1)}\,\mathrm{Cosh}\,(\alpha_{i-(\ell+1)}L) + (h_2'\,h_{si} + K^2\alpha_{i-(\ell+1)}^2)\mathrm{Sinh}(\alpha_{i-(\ell+1)}L)\}}\Bigg] \tag{6.134}$$

$$= U_T, \quad \text{for } i = \ell+1,$$

$[N]$ is a column vector generated by θ_{wn} for n = - 6 to n = + 6 while $[R]$ is a column vector whose elements are defined as

$$r_{j1} = p_{j1} + U_T\,\theta_b\,\delta_{j(\ell+1)}, \tag{6.135}$$

where p_{j1} has expression similar to that defined in Eq. (6.100).

(d) *Conservatory.*+ This integrated system (Fig. 6.3) greatly reduces temperature swings in the living space and also provides a comfortable environment for sunbath and other activities in the sun in winter. Now we consider a complete heat balance through all the components.

Solar radiation through the south facing wall and the roof glazing, is incident on the mass wall, ground, east and west walls; these sun faces are suitably blackened for maximum absorption of solar radiation. The temperature distribution in each of these is given by Eq. (6.78) subject to the appropriate boundary conditions which corresponding to Fig. 6.11 are

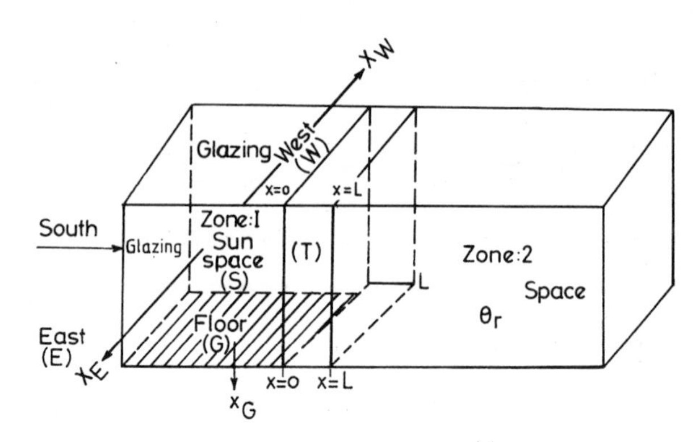

Fig. 6.11. Schematic of the conservatory

(i) *Thermal wall.* At the surface, exposed to zone 1, of thermal wall, energy balance is

$$a.S_T(t) = -K\left.\frac{\partial\theta_T}{\partial x}\right|_{x=o} + h_{TS}\left(\theta_T\big|_{x=o} - \theta_{ss}(t)\right), \tag{6.136}$$

while at the sun face exposed to zone 2, it is

$$-K\left.\frac{\partial\theta_T}{\partial x}\right|_{x=L} = h_{si}\left(\theta_T\big|_{x=L} - \theta_b\right), \tag{6.137}$$

(ii) *Floor of zone 1.* At the floor the energy balance can be written as

+Conservatory or greenhouse referred to as solarium in some parts of the world.

$$\alpha_G . S_G(t) = - K_G \frac{\partial \theta_G}{\partial x_G}\bigg|_{x_G=o} + h_{GS}' \{\theta_G\big|_{x_G=o} - \theta_{ss}(t)\}, \tag{6.138}$$

while the other condition is

$$\theta_G \text{ is finite as } x_G \to \infty, \tag{6.139}$$

(iii) *East wall.* Energy balance at the inside surface of east wall is

$$\alpha_w . S_w(t) = - K_w \frac{\partial \theta_E}{\partial x_E}\bigg|_{x_E=o} + h_{ws} \,|\theta_E\big|_{x_E=o} - \theta_{ss}(t)|, \tag{6.140}$$

while at the outside surface is

$$\alpha_c . S_E(t) - K_w \frac{\partial \theta_E}{\partial x_E}\bigg|_{x_E=d_E} - h_{wa}' \{\theta_E\big|_{x_E=d_E} - \theta_a(t)\}, \tag{6.141}$$

(iv) *West wall.* At the inside surface the boundary condition is

$$\alpha_w \; S_E(t) = - K_w \frac{\partial \theta_w}{\partial x_w}\bigg|_{X_w=o} + h_{ws}' \{\theta_w\big|_{x_w=o} - \theta_{ss}(t)\}, \tag{6.142}$$

while for the outside surface is

$$\alpha_c . S_w(t) - K_w \frac{\partial \theta_w}{\partial x_w}\bigg|_{x_w=d_E} = h_{wa}' \{\theta_w\big|_{x_w=d_E} - \theta_a(t) , \tag{6.143}$$

(v) *Isothermal mass.* The energy balance of isothermal mass (if any) in zone 1 can be written as

$$M_B \frac{d\theta_B(t)}{dt} = A_B \, h_{SB}' \{\theta_{ss}(t) - \theta_B(t)\}, \tag{6.144}$$

(vi) *Air mass.* Energy balance of the air column of zone 1 can be written as

$$\begin{aligned} M_A \frac{\partial \theta_{ss}(t)}{\partial t} = {} & A_a \, h_{TS}' \{\theta_T\big|_{x=o} - \theta_{ss}(t)\} + A_G \, h_{GS}' \{\theta_G\big|_{x_G=o} - \theta_{ss}(t)\} \\ & + A_w \, h_{ws}' \{\theta_E\big|_{x_E=o} + \theta_w\big|_{x_w=o} - 2\theta_{ss}(t)\} \\ & - A_B \, h_{SB}' \{\theta_{ss}(t) - \theta_B(t)\} \\ & - U_a \, (A_a + A_G)' \{\theta_{ss}(t) - \theta_a(t)\}, \end{aligned} \tag{6.145}$$

It may be noted that the area of mass wall is the same as that of the glazed south wall of solarium; the area of roof glass also is about the same as that of the floor of the sunspace. In the above equations, the subscripts on the heat-transfer coefficient denote the exchange of heat amongst the two components. For example h_{TS} means the convective heat-transfer coefficient between mass wall and air in the sun space.

Substitution of the periodic solutions $\theta(x,t)$ for various walls and the Fourier series representation for various temperatures in the above equations (6.136) to (6.145), gives equations for the unknown constants.

I. In the case when the glazing is left exposed, even in the off-sunshine hours, then heat flux entering into the living space can be written as

$$\dot{Q} = h_{si} \left| (A_o + A_1 L - \theta_b) + \mathrm{Re} \sum_{n=1}^{6} \{\lambda_n \exp(\alpha_n L) + \lambda'_n \exp(-\alpha_n L)\} \exp(in\omega t) \right| \tag{6.146}$$

The expressions for A_o, A_1, λ_n and λ'_n are

$$A_o = \frac{(K + h_{si} L)(h_{TS}\bar{\theta}_{ss} + \alpha_T \bar{S}_T) + K h_{si} \theta_b}{(h_{TS}(K + h_{si} L) + K h_{si}}, \tag{6.147}$$

$$A_1 = \frac{h_{TS} h_{si} \theta_b - h_{si}(h_{TS}\bar{\theta}_{ss} + \alpha_T \bar{S}_T)}{K h_{si} + h_{TS}(K + h_{si} L)} \tag{6.148}$$

$$\lambda_n = \frac{\left(1 - \frac{K\alpha_n}{h_{si}}\right)\left(\theta_{ssn} + \frac{\alpha_T S_{Tn}}{h_{TS}}\right) \exp(-\alpha_n L)}{2\left\{\left(1 + \frac{K^2 \alpha_n^2}{h_{TA} h_{si}}\right) \mathrm{Sinh}\, \alpha_n L + K\alpha_n \left(\frac{1}{h_{si}} + \frac{1}{h_{TS}}\right) \mathrm{Cosh}\, \alpha_n L\right\}} \tag{6.149}$$

$$\lambda'_n = \frac{\left(1 + \frac{K\alpha_n}{h_{si}}\right)\left(\theta_{ssn} + \frac{\alpha_T S_{Tn}}{h_{TS}}\right) \exp(\alpha_n L)}{2\left\{\left(1 + \frac{K^2 \alpha_n^2}{h_{TA} h_{si}}\right) \mathrm{Sinh}\, \alpha_n L + K\alpha_n \left(\frac{1}{h_{si}} + \frac{1}{h_{TS}}\right) \mathrm{Cosh}\, \alpha_n L\right\}} \tag{6.150}$$

In case, when glass wall and roof of solarium is not insulated from outside during off-sunshine hours, the olar intensity may be represented by the expression

$$S_k = \bar{S}_k + \mathrm{Re} \sum_{n=1}^{6} S_{kn} \exp(im\omega t), \tag{6.151}$$

The Fourier coefficients, $\bar{\theta}_{ss}$ and θ_{ssn} may, therefore, be given as

$$\begin{aligned} \bar{\theta}_{ss} = \Big[&\{A_G \alpha_G \bar{S}_G + A_T U_R \theta_b + A_W U_W \{2\bar{\theta}_a + \frac{\alpha_c}{h_{wa}}(\bar{S}_E + \bar{S}_W)\} \\ &+ U_a (A_T + A_G)\bar{\theta}_a + A_T \left(1 - \frac{U_R}{h_{TS}}\right) \alpha_T \bar{S}_T + A_W \left(1 - \frac{U_W}{h_{WS}}\right) \times \\ &\times \alpha_W (\bar{S}_E + \bar{S}_W)\} / \{A_T U_R + 2 A_W U_W + U_a (A_T + A_G)\} \Big] \end{aligned} \tag{6.152}$$

and

$$\theta_{ssn} = \left[\{A_T \frac{P_1}{R_T} \alpha_T S_{Tn} + A_w \frac{P_2}{R_w} \alpha_w (S_{En} + S_{wn}) + \frac{A_G \alpha_G}{R_a} S_{Gn} + \frac{2 A_w K_w \alpha_{nw}}{R_w} + Ua (A_T + A_G) \theta_{an}\} / \{in\omega M_A + P_3 + P_4 + \frac{A_G \alpha_G K_G}{R_G} + Ua (A_T + A_G) + i \left(\frac{A_B h_{AB} n\omega M_B}{A_B h_{AB} + in\omega M_B} \right) \} \right] \tag{6.153}$$

where

$$\frac{1}{U_R} = \left(\frac{1}{h_{TS}} + \frac{L}{K} + \frac{1}{h_{si}} \right),$$

$$\frac{1}{U_w} = \left(\frac{1}{h_{wa}} + \frac{1}{h_{ws}} + \frac{d_E}{K_w} \right),$$

$$P_1 = \left\{ \text{Sinh } (\alpha_n L) + \frac{K.\alpha_n}{h_{si}} \text{Cosh } (\alpha_n L) \right\}$$

$$R_T = \left\{ \left(1 + \frac{K^2 \alpha_n^2}{h_{TS} h_{si}} \right) \text{Sinh } (\alpha_n L) + K\alpha_n \left(\frac{1}{h_{TS}} + \frac{1}{h_{si}} \right) \text{Cosh } (\alpha_n L) \right\},$$

$$P_2 = \text{Sinh } (\alpha_{nw} d_E) + \frac{K_w \alpha_{nw}}{h_{wa}} \text{Cosh } (\alpha_{nw} d_E),$$

$$R_w = \left(1 + \frac{K_w^2 \alpha_{nw}^2}{h_{ws} h_{wa}} \right) \text{Sinh } (\alpha_{nw} d_E) + K_w \alpha_{nw} \left(\frac{1}{h_{ws}} + \frac{1}{h_{wa}} \right) \text{Cosh } (\alpha_{nw} d_E),$$

$$P_3 = A_T h_{TS} \left(1 - \frac{P_1}{R_T} \right),$$

$$P_4 = 2 A_w h_{ws} \left(1 - \frac{P_2}{R_w} \right), \tag{6.154}$$

II. In case when the glass wall and roof is covered with insulation during the off-sunshine hours the heat flux is still given by Eq. (6.146). However, the overall heat-transfer coefficient between air in zone 1 and ambient now becomes time dependent and can be expressed as (Sodha *et al.*, 1982)

$$U_a(t) = H_o'' + \sum_{\substack{n=-6 \\ n \neq o}}^{+6} H_n' \exp (in\omega t), \tag{6.155}$$

The Fourier series of solar intensity in this case is defined as

$$S_k(t) = \bar{S}_k + \sum_{\substack{n=-6 \\ n\neq o}}^{+6} S_{kn} \exp(in\omega t), \tag{6.156}$$

θ_{ssn} (for n = - 6 to n = + 6 including zero) are evaluated by a matrix equation similar to Eq. (6.96). The matrix elements of $[M]$ are given by an expression similar to (6.97) with the following definitions for U's and Y's.

$$Y_n = [(X_1 - 1) A_T h_{TS} + A_G h_{GS} (X_6 - 1) + A_w h_{ws} (X_3 - 2)$$

$$+ A_B h_{BS} (X_8 - 1) - in\omega M_A]/(A_T + A_G); \quad \text{for } n \neq o$$

$$= \left\{ \frac{(Z_1 - 1) A_T h_{TS} + 2 (Z_s - 1) A_w h_{ws}}{(A_T + A_G)} \right\}; \quad \text{for } n = o \tag{6.157}$$

$[N]$ and $[R]$ are column vectors defined as

$$n_{j1} = \theta_{j-(\ell+1)}; \quad \text{for } 1 < j < (2\ell+1)$$

$$r_{j1} = V_{j-(\ell+1)} + T_{j-(\ell+1)} a'_{j-(\ell+1)} + p_{j1} \delta_{j(\ell+1)}, \tag{6.158}$$

where

$$p_{j1} = \sum_{r=o}^{\ell} H''_{-(\ell-r)} a'_{-(r-j+1)} + \sum_{r=1}^{j-1} H''_r a'_{-(r-j+\ell+1)}$$

$$\text{for } 1 < j < (+1)$$

$$= \sum_{r=o}^{2\ell+1} H''_{r+j-(2\ell+1)} a'_{(\ell-r)} + \sum_{r=1}^{\ell} H''_r a''_{-(r-j+\ell+1)}$$

$$\text{for } (\ell+2) < j < (2\ell+1) \tag{6.159}$$

The V's and T's are given by

$$V_n = \frac{X_2 A_T h_{TS} + x_7 A_G h_{GS} + X_5 A_w h_{ws}}{(A_G + A_T)}$$

$$T_n = A_w h_{ws} X_4/(A_G + A_T), \tag{6.160}$$

where the X's are defined as

$$X_1 = \left[\frac{\text{Sinh } \alpha_n L + \left(\frac{K\alpha_n}{h_{si}}\right) \text{Cosh } \alpha_n L}{\left(1 + \frac{K^2 \alpha_n^2}{h_{TS} h_{si}}\right) \text{Sinh } \alpha_n L + K\alpha_n \left(\frac{1}{h_{si}} + \frac{1}{h_{TS}}\right) \text{Cosh } \alpha_n L} \right],$$

$$X_2 = \alpha_T S_{Tn} X_1/h_{TS},$$

$$X_3 = \left[\frac{2\,\{\text{Sinh}\,(\alpha_{nw}\,d_E) + \left(\frac{K_w\,\alpha_{nw}}{h_{wa}}\right)\text{Cosh}\,(\alpha_{nw}\,d_E)\}}{\left\{\left(1 + \frac{K_w^2\,\alpha_{nw}^2}{h_{wa}\,h_{ws}}\right)\text{Sinh}\,(\alpha_{nw}d_E) + K_w\alpha_{nw}\left(\frac{1}{h_{wa}} + \frac{1}{h_{ws}}\right)\text{Cosh}\,(\alpha_{nw}d_E)\right\}}\right],$$

$$X_4 = \left(\frac{K_w\,\alpha_{nw}}{h_{ws}}\right) X_3\,\{\text{Sinh}\,(\alpha_{nw}\,d_E) + \left(\frac{K_w\,\alpha_{nw}}{h_{wa}}\right)\text{Cosh}\,(\alpha_{nw}\,d_E)\},$$

$$X_5 = \frac{1}{2}\left\{\frac{\alpha_w X_3}{h_{ws}} + \frac{\alpha_c X_4}{h_{wa}}\right\}(S_{En} + S_{wn}),$$

$$X_6 = h_{GS}/(K_G\,\alpha_{nG} + h_{GS}),$$

$$X_7 = \alpha_G\,S_{Gn}/(K_G\,\alpha_{nG} + h_{GS}).$$

$$X_8 = A_B\,h_{SB}/(A_B\,h_{SB} + in\omega M_B), \tag{6.161}$$

while Z's are defined as

$$Z_o = \left[A_T h_{TS} Z_2 + A_G \alpha_G \bar{S}_G + A_w h_{ws}\,\{(Z_6 + Z_9)\,\bar{\theta}_a + Z_6' + Z_9'\}\right]/(A_T + A_G),$$

$$Z_1 = (K + h_{si}L)\,h_{TS}/D_1,$$

$$Z_2 = \{(K + h_{si}L)\,\alpha_T\,\bar{S}_T + Kh_{si}\theta_b\}/D_1,$$

$$Z_5 = (K_w + h_{wa}\,d_E)\,h_{ws}/D_2,$$

$$Z_6 = K_w\,h_{wa}/D_2,$$

$$Z_6' = \{(K_w + h_{wa}\,d_E)\,\alpha_w\,\bar{S}_w + K_w\,\alpha_c\,\bar{S}_E\}/D_2,$$

$$Z_9 = Z_6,$$

$$Z_9' = \{(K_w + h_{wa}\,d_E)\,\alpha_w\,\bar{S}_E + K_w\,\alpha_c\,\bar{S}_w\}/D_2,$$

$$D_1 = \{K\,h_{si} + h_{Ts}\,(K + h_{si}L)\},$$

$$D_2 = \{K_w\,h_{wa} + h_{ws}\,(K_w + h_{wa}\,d_E)\}, \tag{6.162}$$

where

$$\alpha_n = (1 + i)\left(\frac{n\omega\rho c}{2K}\right)^{\frac{1}{2}},$$

$$\alpha_{nW} = (1 + i)\left(\frac{n\omega\ \rho_W\ C_W}{2\ K_W}\right)^{\frac{1}{2}}$$

$$\alpha_{nG} = (1 + i)\left(\frac{n\omega\ \rho_G\ C_G}{2\ K_G}\right)^{\frac{1}{2}},$$

and $i = \sqrt{-1}$, (6.163)

6.3.5.3 Passive cooling concepts with constant internal environment. Evaporative cooling is one of the most effective ways of reducing the heat gains into the building. The techniques to evaluate various ways of evaporation have been described in detail in Chapter 8. Other ways of reducing heat gains are the shading of roof, which can be incorporated by means of plants, removable canvas or inverted earthen pots as described in detail in Chapter 4.

In North India a normal roof is a three-layered structure (Fig. 6.12) consisting of concrete, mud phuska* and brick tiles. In rural India, certain vegetable plants are grown by people in pots and these are allowed to creep on wire mesh erected at a level of about 2.5 m above the roof, so that the whole structure will act as a roof shading device. The normal plants usually grown are Tecoma grand-flora, Impea plamata, Quisqualis indica, Antigonum Leptopus etc.

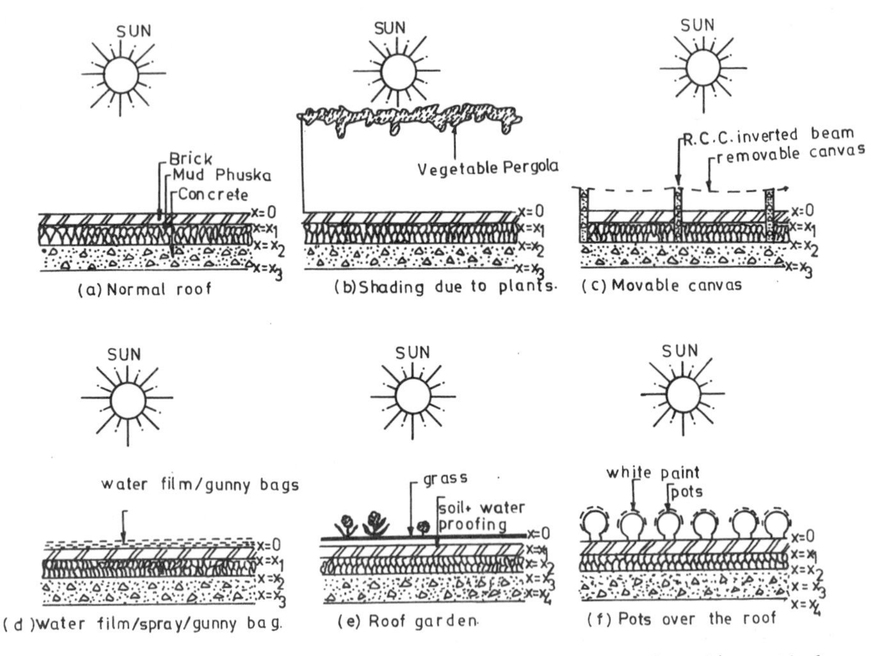

Fig. 6.12. Configurations of normal roof and other natural cooling methods. (dimensions of the roof = x_1 = 0.05 m, $x_2 - x_1$ = 0.10 m, $x_3 - x_2$ = 0.15 m)

*Mud phuska is a mud mortar mixed with hay (proportion of hay is 35 kg/m^3 of earth).

Figures (6.12b) to (6.12e) schematically illustrate the four concepts for natural cooling viz. (i) roof with sunshading of plants, (ii) roof with a movable insulation, (iii) roof garden and (iv) roof with inverted earthen pots. Fig. (6.12d) illustrates evaporative cooling described in detail in chapter 8.

To evaluate the performance of these concepts by periodic solution method, one assumes an expression for the temperature distribution of the same form as given in Eq. (6.79) in each of the regions, i.e.

$$\theta_j(x,t) = \left[(A_j x + B_j) + \mathrm{Re} \sum_{n=1}^{\infty} \{\lambda nj \exp(\alpha_{nj}\, x) + \lambda'_{nj} \exp(-\alpha_{nj} x)\} \times \exp(in\omega t)\right], \tag{6.164}$$

where j corresponds to the region. The unknown constants A_j, B_j, λ_{nj} and λ'_{nj}'s are determined by appropriate energy balance equations at various interfaces. Corresponding to Fig. (6.1), at the surface x = 0 and $x = x_3$, the boundary conditions essentially remain the same as given by Eqs. (6.80) and (6.81) with minor modifications to incorporate the effects of shading on the roof, i.e. solair temperature θ_{sa} in this case is defined as

$$\theta_{sa}(t) = \left\{\frac{1}{h_{so}} (\alpha .\ f_{sh}.S(t)) + \theta_a(t)\right\}, \tag{6.165}$$

where f_{sh} is the shading factor. f_{sh} = 1 means no shading and f_{sh} = 0.0 means 100% shading. At the interfaces i.e. at $x = x_1$ and $x = x_2$, the temperature and heat flux must be continuous, i.e.

$$\theta_1\ (x,t)\Big|_{x=x_1} = \theta_2\ (x,t)\Big|_{x=x_1}, \tag{6.166}$$

$$K_1 \frac{\partial \theta_1}{\partial x}\Big|_{x=x_1} = K_2 \frac{\partial \theta_2}{\partial x}\Big|_{x=x_1}, \tag{6.167}$$

$$\theta_2\ (x,t)\Big|_{x=x_2} = \theta_3\ (x,t)\Big|_{x=x_2}, \tag{6.168}$$

and

$$K_2 \frac{\partial \theta_2}{\partial x}\Big|_{x=x_2} = K_3 \frac{\partial \theta_3}{\partial x}\Big|_{x=x_2}, \tag{6.169}$$

Using boundary conditions and periodic series representation of the ambient temperature and solar radiation along with the temperature distributions (6.164), one can find explicit expressions for the heat flux into the room as shown in the case of passive heating concepts.

For the case of a movable canvas over the roof, h_{so} (in Eq. (6.165)) becomes time-dependent and in this case as before, the summation in the periodic series should be taken from n = - 6 to n = + 6. The analysis is done exactly in the same fashion as illustrated in the case of passive heating concepts i.e. Trombe wall without vents.

In the case of earthen pots over the roof, these are kept inverted and their outer surfaces painted white. The pots will shade almost the entire roof. Depending upon the place of consideration and the day of the year, the roof will have very small unshaded area to receive solar radiation and that too only for a few hours. The whole arrangement can therefore be considered equivalent to shading of the roof. Only the heat-transfer coefficients in this case get changed because the air is now trapped inside the pot inhibiting large air movement. The exact expressions for the heat flux are given by Nayak and others (1982).

6.4 UNCONDITIONED BUILDINGS

The basic physical system representing a building is shown in Fig. 6.13. The south-facing wall is assumed to have an ordinary glass window desired for ventilation. The energy balance equation for the room air temperature $\theta_r(t)$ is written as

$$M_A \frac{d\theta_r(t)}{dt} = \dot{Q}_I + \dot{Q}_{window} - \dot{Q}_G - \dot{Q}_I - \dot{Q}_V, \tag{6.170}$$

$\dot{Q}_T$, the rate at which the heat enters the building through various walls/roof is given by

$$\dot{Q}_T = \sum_j h_{sij} (\theta_{sij} - \theta_r) A_j \tag{6.171}$$

where j refers to the particular component, A_j the area, h_{sij} the heat transfer coefficient between the internal surface (at temperature θ_{sij}) and the room air.

For calculating the net heat gain through the window, an assumption is made that the heat capacity and the absorptivity of window glass is negligible and Q_{window} is written as

$$Q_{window} = \alpha_a \tau_g A_w S(t) - U_w A_w (\theta_{si} - \theta_a), \tag{6.172}$$

$\dot{Q}_G$ and $\dot{Q}_I$, the heat loss rate to the ground and to the isothermal mass, are calculated by a formulae like Eq. (6.171).

The heat losses due to infiltration or ventilation are expressed by the formula

$$\dot{Q}_V = M_A N \{\theta_r(t) - \theta_a(t)\}, \tag{6.173}$$

In many calculations N is taken to be a constant, the value of which is equivalent to some average value of the air changes. To take into account the variable ventilation rates, N is made time dependent and then the

$$\dot{Q}_v = M_A N(t) \{\theta_r(t) - \theta_a(t)\}, \tag{6.174}$$

Assuming periodic solutions θ_{sij}, the internal surface temperatures for various walls and roofs can be obtained. Substituting all quantities in Eq. (6.170), and assuming periodic solutions for $\theta_r(t)$, one can find explicit expressions for $\bar{\theta}_r$ and θ_{rn}, the Fourier coefficients of the room temperature. In Eq. (6.174), $N(t)$ can also be expressed as a Fourier series and this multiplies by the Fourier series for θ_r and θ_a retaining only the desired number of finite terms (Chandra, 1982).

Based on various mathematical models described above various studies have been performed to characterize the performance of various passive heating and cooling concepts. These have been presented in a summary form in the next chapter.

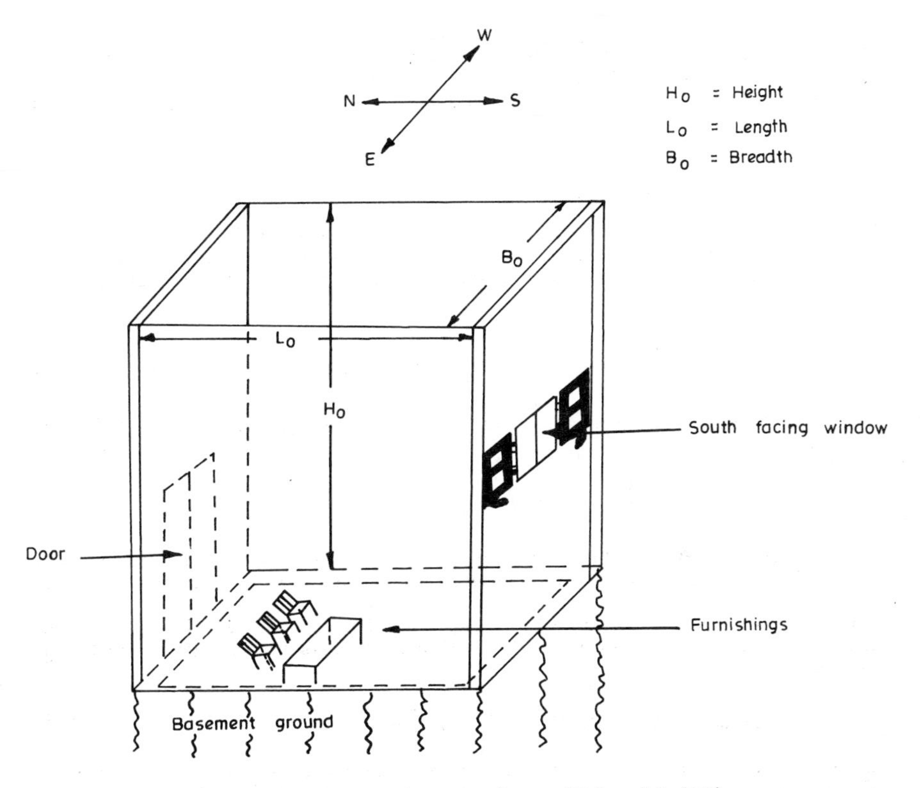

Fig. 6.13. Sketch of a non-air-conditioned building

REFERENCES

Anon (1976) *ERDA's Pacific Region Solar Heating Handbook*, Stock No. 060-000-0024-7, US Government Book Store.

Balcomb, J. D., McFarland, R. D. and Moore, S. W. (1977) Simulation Analysis of of Passive Solar Heated Buildings - Comparison with Test Room Results, LA-UR-77-939, Los Alamos Scientific Laboratory, University of California, Los Alamos, New Mexico.

Balcomb, J. D. and McFarland, R. D. (1978) A Simple Empirical Method for Estimating the Performance of a Passive Solar Heated Building of a Thermal Storage Wall Type, Proc. 2nd National Passive Solar Conference, *2*(2), 377.

Balcomb, J. D., Jones, R. W., McFarland, R. D. and Wray, W. O. (1982) Performance Analysis of Passive Solar Heated Buildings of the Solar Load Ratio Method, LA-UR-82-670.

Beckmann, W. A., Klein, S. A. and Duffie, J. A. (1977) *Solar Heating Design by the F-Chart Method*, Wiley-Interscience, New York.

Buchberg, H., Bussel, B. and Reisman, R. (1964) On the Determination of Optical Stress Enclosures, *Int. J. Biometeor.*, *8*, 103-111.

Chandra, S. (1982) Solar thermal modeling of buildings and space conditioning systems, Ph.D. Thesis, IIT Delhi.

Dreyfus, J. (1963) Transmission de la chaleur en regime periodique a' travers les murs de habitants, *Rev. Gen. de Therm.*, *22*, 1155-1162.

Goldstein, D. B. (1980) Modelling Passive Solar Buildings with Hand Calculations, Conf. 4, Proc. IVth National Passive Solar Conference, Oct. 3-5, 1979, Vol. *4*, ed. G. Granta, American Section of ISES, Univ. of Delaware, Delaware, USA.

Goldstein, D. B. (1978) Some models of solar passive building performance, LBL-7811, Lawrance Berkley Laboratory.
Gupta, C. L. and Raychaudhury, B. C. (1963) Thermal Circuit Analysis of Unconditioned Buildings, Symposium on Analogies in Engineering, May 21, 1963, 43rd Annual Convention Bangalore India.
Hoffmann, M. E. (1976) Prediction of indoor temperature - the influence of thermophysical parameters of building elements on internal thermal configurations, Chapter 19 of *Man, Climate and Architecture,* by Givoni, B., Applied Science Publishers Ltd., London.
Heidt, F. D. (1983) Physical principles of passive solar heating systems, Solar World Congress, 14-19 Aug., Perth, Australia.
Mackey, C. O. and Wright, L. T. (1946) Periodic Heat Flow Composite Walls or Roofs, *Heating, Piping and Air-Conditioning, 18*, 107-110.
Markus, T. A. and Morris, E. N. (1980) *Buildings, Climate and Energy,* Pitman, London.
McFarland, R. D. (1978) PASOLE, A General Simulation Program for Passive Solar Energy, Los Alamos National Laboratory Report LA-7433-MS.
Milborn, G. (1980) Berechnungsverfahren Passive Solar Systeme, Vorlesung University of Insburck, Institut für Bauphysik, West Germany.
Milbank, N. O. and Harrington-Lynn, J. (1974) Thermal Response and the Admittance Procedure, *The Bldg. Serv. Eng. (J. IHVE) 42,* 38-51.
Monsen, W. A., Klein, S. A. and Beckman, W. A. (1981) The Un-utilizability Design Method for Collector Storage Walls, Proceedings of the 1981 Annual Meeting of the AS ISES, May 26-30, Philadelphia PA, USA.
Nayak, J. K., Srivastava, A., Singh, U. and Sodha, M. S. (1982) The Relative Performance of Different Approaches to Passive Cooling of Roofs, *Building and Environment, 17*(2), 143.
Nayak, J. K., Sodha, M. S. and Bansal, N. K. (1983) Analysis of Passive Heating Concepts, *Solar Energy, 30*(1), 51-69.
Nottage, H. B. and Parmelee, G. V. (1954) Circuit Analysis Applied to Load Estimating, *ASHRAE Trans., 60*, 59.
Petherbridge, P. (1974) Limiting Temperatures in Naturally Ventilated Buildings in Warm Climates, BRE, CP 7174.
Raychoudhury, B. C. and Chaudhury, N. K. D. (1961) Thermal Performance of Dwellings in Tropics, *Indian Construction News*, 38-47.
Sodha, M. S., Nayak, J. K., Bansal, N. K. and Goyal, I. C. (1982) Thermal Performance of a Solarium with a Removable Insulation, *Building and Environment, 17*(1), 23-32.
Utzinger, D. M., Klein, S. A. and Mitchell, J. W. (1980) The Effect of Air Flow Rate in Collector-Storage Walls, *Solar Energy 25,* 511.
Wray, W. O., Balcomb, J. D. and McFarland, R. D. (1979) A Semi-empirical Method for Estimating the Performance of Direct Gain Passive Solar Heated Buildings, LA-UR-74-1176, Los Alamos Scientific Lab., University of California, Los Alamos, New Mexico.
TRNSYS (1979) Version 10.1 Solar Energy Laboratory, University of Wisconsin, Madison.

Chapter 7

PERFORMANCE OF PASSIVE HEATING AND COOLING CONCEPTS

Architecture as a discipline is concerned with the social, philosophical and aesthetic implications of the built environment as well as its economic and technical characteristics. In contrast to the energy intensive approaches which overlooked these factors, and has resulted in complex active systems, the solar passive concepts have proved to be simple and economical. As mentioned earlier, these concepts encompass the methods collecting, storing and distributing the energy by natural means.

Various passive concepts for heating/cooling of the buildings, and their adaptability in a typical climate have already been discussed in Chapter 4. However, to evaluate the suitability of these concepts for a specific location, an idication of their performance is required and the present chapter aims to cover this aspect. To appreciate the performance of many concepts, for which experimental data is not available, mathematical models discussed in the last chapter are to be used.

7.1 PERFORMANCE OF PASSIVE HEATING CONCEPTS

Careful design of the building incorporating one or more concepts of passive heating can provide substantial heating of the building space. Storage of heat becomes another important factor for extending the duration of heating to the off-sunshine periods. The performance of various concepts (which have already been described in Chapter 4) used for heating, is discussed below.

7.1.1 Direct Gain Concept

The rigorous investigations on direct gain and indirect gain concepts were carried out at Los Alamos Scientific Laboratory. Wray and Balcomb (1979) have reported the simulation results for direct gain buildings in Albuquerque, USA. Their investigations included the variation of design parameters viz. the amount of glazing (NGL), the resistance of movable insulation (R_n), and the acceptable fluctuations in the indoor temperature (ΔT) about the desired room temperature value of 20°C. The performance of resulting eight configurations (the variations being in the amount of glazings) is shown in Fig. 7.1. The percentage of the fraction which can be met by using solar energy of the total energy consumption gets considerably affected by the presence of movable insulation. A comparison of the configurations 1 and 3 shows that in the absence of insulation, the amount of glazing has a pronounced effect on performance while in the presence of movable insulation, the increased amount of glazing improves the performance only marginally. It may be mentioned here that in order to restrict the variations in room temperature to a desired range the rest of the energy-fraction has to be provided by the conventional energy-sources. If the acceptable temperature variations are increased, the solar fraction also increases as seen from the results of configurations 7 and 8.

Placement of storage inside the enclosure heated by direct gain concept affects the performance of the system. For Los Alamos climate Balcomb *et al.* (1977) have theoretically found that storage materials placed directly in the sunlight could be very effective in capturing heat for later use in direct gain situation, while storage material placed out of the direct sunlight and illuminated only by diffuse reflection from room furnishings would be very inefficient.

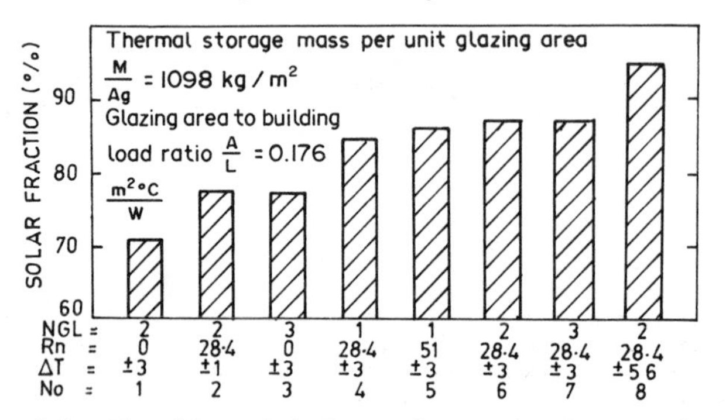

Fig. 7.1. The effect of design option combinations on direct gain-performance in Albuquerque, USA. (After Wray and Balcomb, 1979)

In an example of direct gain studies, Mazaria, *et al.* (1977) showed the importance of diffuse illumination of distributed storage. Using a quasi - periodic solution it was shown that for a clear winter day in Portland, Oak Ridge, 400 mm thick north storage wall of concrete, illuminated through a clear southwall window, produced a room air temperature varying from -4.5°C to 1.8°C. On the other hand, with 100 mm concrete in all walls and roofs, illuminated by sunlight coming through a diffusing south window, the room temperature variation was only 5°C. In another calculation, Mazaria *et al.* (1977) showed that the performance of storage wall in a direct gain situation was strongly dependent on the thermal conductivity of the storage material. Water used as the storage medium provided least daily variation in room temperature while successive larger variations were observed with magnesite brick, concrete, common brick and adobe. Adobe has the lowest conductivity of these five materials and shows a room temperature swing roughly twice as larger as that of water (Parry, 1977).

Wray (1981) has discussed the design and analysis of direct gain solar heated buildings in much detail. He has applied the monthly solar load ratio method (SLR) to analyse the performance of heavy mass direct gain buildings. The amount of thermal storage mass was set equal to 0.92 MJ/m^2°C of glazing nad it was found to be adequate for preventing excessive overheating in most circumstances. The thermal storage mass was provided on the floor. For a typical building with parameters given in Table 7.1 the solar saving fraction calculated as a function of the ratio of mass surface to glazing surface area (Am/Ag = 2, 3, 6, 10), and for various thickness is plotted in Figs. 7.2 and 7.3, for no night insulation and for night insulation respectively. The reference design conditions (thickness = 152 mm, Am/Ag = 3) are indicated by solid points.

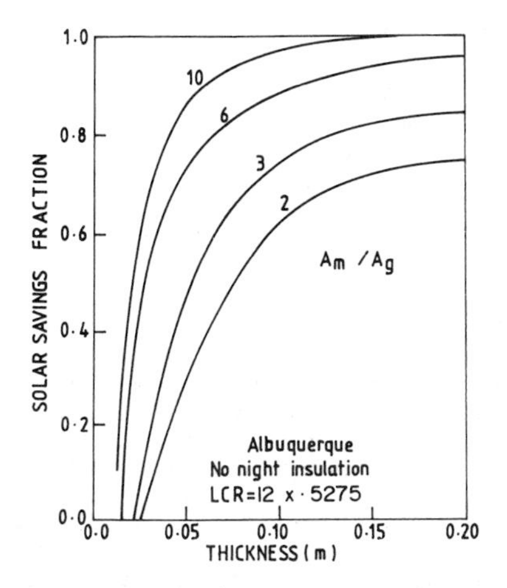

Fig. 7.2. Solar savings fraction vs. mass thickness, Albuquerque, no night insulation. (After Wray, 1981)

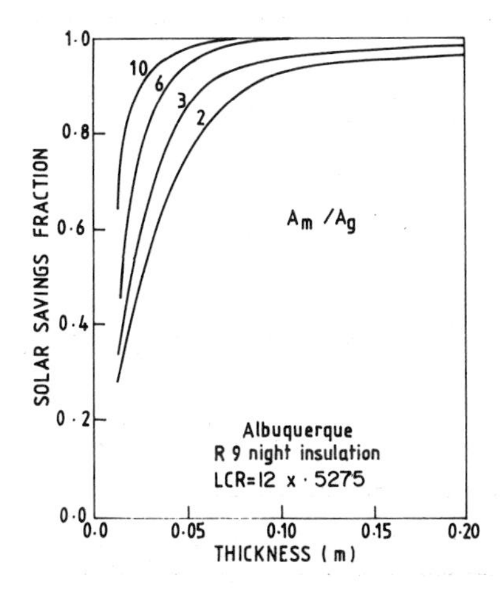

Fig. 7.3. Solar savings fraction vs. mass thickness, Albuquerque, R9 night insulation. (After Wray, 1981)

The following conclusions may be drawn from the figures:

- Performance variations for mass thickness between 102 and 204 mm are small. This is true irrespective of the location, configuration and mass surface area.
- The range of mass thicknesses between 51 mm and 102 mm can be called a transition region. In this region, a reduction in the mass thickness severely affects the performance.
- For mass thickness below 51 mm, the performance falls off more rapidly than in the transition region. Under most conditions the mass thicknesses should not be less than 51 mm, particularly, if one is striving for a high SHF.

TABLE 7.1 Parameters of the Reference Building

(i)	Thermal Storage Capacity = 0.92 MJ/^{o}C m^2 glazing The building has no other thermal mass.
(ii)	Double-glazing having normal thermal transmittance = 0.747
(iii)	Spacing between the glazing = 12.7 mm
(iv)	Room temperature control range = 18.3 - 23.9^{o}C
(v)	Night insulation is R9 with K = 1.6 W/m^{o}C
(vi)	Duration of night insulation = 5.30 p.m. - 7.30 a.m.
(vii)	Masonry properties K = 1.7307 W/m^{o}C ρ = 2403 kg/m^3 C = 837 J/kg^{o}C
(viii)	Infrared emittance of the solid surface = 0.9
(ix)	Internal heat-generation = 0.0
(x)	Non-mass absorption factor = 0.2
(xi)	Mass of masonry = 152.4 mm
(xii)	Mass area is 3 times the glazing area.

The thermophysical parameters (ρ, C, K) and the thickness (d) of the storage mass affects the performance considerably. The effect of thermal capacity has already been described above. The other combination of thermophysical properties, i.e. ρCK affects the performance, the typical curve being shown in Fig. 7.4. LCR of 12, 48 and 120 for configurations with and without night insulation are included.

It is seen that the performance of the direct gain is insensitive to variations in ρCK, between 2320 and 4640 and only moderately sensitive between 1160 and 2320. However, if the value of ρCK drops much below 1160, performance begins to erode seriously.

The above observation about ρCK value does not necessarily imply that materials with ρCK values less than 1160 should not be used for thermal storage mass in direct gain buildings. Adobe, for example, has ρCK value of approximately 800, but nevertheless adobe has been successfully used as thermal storage medium in many passive solar homes in the south west. The ratio of mass to glazing area in these buildings, however has to be much greater than the designed reference value of 3. This conclusion essentially means, that as a general rule, the storage materials having relatively small values of ρCK will necessitate the use of larger mass surface areas.

7.1.2 Indirect Gain Concept

In direct gain concepts, the temperature fluctuations of the order of 6 to 7°C over the day can always be anticipated. This is an intrinsic character of direct gain buildings (Balcomb, 1977). A mass storage wall interposed between a solar

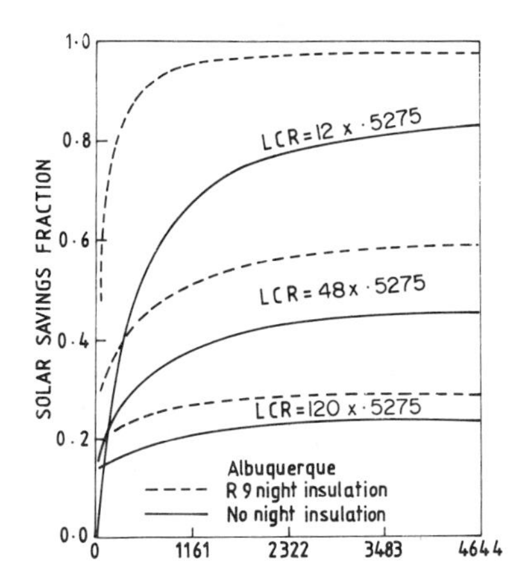

Fig. 7.4. Solar savings fraction vs. ρck, Albuquerque. (After Wray, 1981)

heated space and a living space can be used to decrease the magnitude of temperature fluctuations tremendously. The storage material can be water in containers, or masonry walls such as concrete, brick, stone work or adobe. The outer storage surface is black, or dark, to promote absorption of solar radiation. When equipped with upper and lower vents to enhance thermal circulation of hot air from the glazed wall space into the room, these walls are termed Trombe walls. Mazaria *et al.* (1977) augmented the results of Balcomb *et al.* (1977) about placement of storage by showing that the performance of a storage wall in an indirect-gain concept depends strongly on the thermal-conductivity of the storage material. Water, therefore, having a high effective conductivity because of convection, reduces daily variations in the room temperature more effectively than the solid storage of concrete of gravel for example.

Based on the periodic solutions of the heat conduction equation and analysis of the passive heating concepts presented in the last chapter (section 6.3.4), the thermal performance of Trombe wall concept has been illustrated in Fig. 7.5. The corresponding set of parameters is given in Table 7.2. The figure shows the hourly variation of the heat flux into the living space through the Trombe wall. The living space has been assumed to be maintained at a temperature of 20°C. It is seen that the performance of Trombe wall significantly depends on the fact whether (i) the vents are open, and/or (ii) the night insulation is used. The use of vents in Trombe wall is useful only when a quick removal of heat flux is required and the building is used only during the daytime. For round the clock use, a simple glazed wall, without vents, is desirable from the point of view of thermal load levelling.* In addition to the reduction of heat losses during off-sunshine hours, another notable feature of the movable insulation is that it helps in better load levelling of the heat flux.

The above set of calculations correspond to harsh North American winter climate of Boulder, Colorado (latitude 40.02°N). For the same data the performance of water wall is given in Table 7.3. The general conclusion of the results is that irrespective of the storage capacity of water wall, the maximum of the heat flux

*Thermal load levelling refers to a process of decreasing the maxima and increasing the minima of the heat flux entering the room.

$\dot{Q}(t)$ always occurs at 5 p.m. With more capacity only the relative fluctuations about the mean get decreased. If a concrete wall is provided between the water wall and the living space, the maxima of $\dot{Q}(t)$ occurs at a later time than at 5 p.m., the exact shift depending upon the thickness of the concrete wall. For a 0.22 m thick concrete wall behind the water wall, the desirable phase lag of 12 hours between the maxima of $\dot{Q}(t)$ and the solair temperature is obtained.

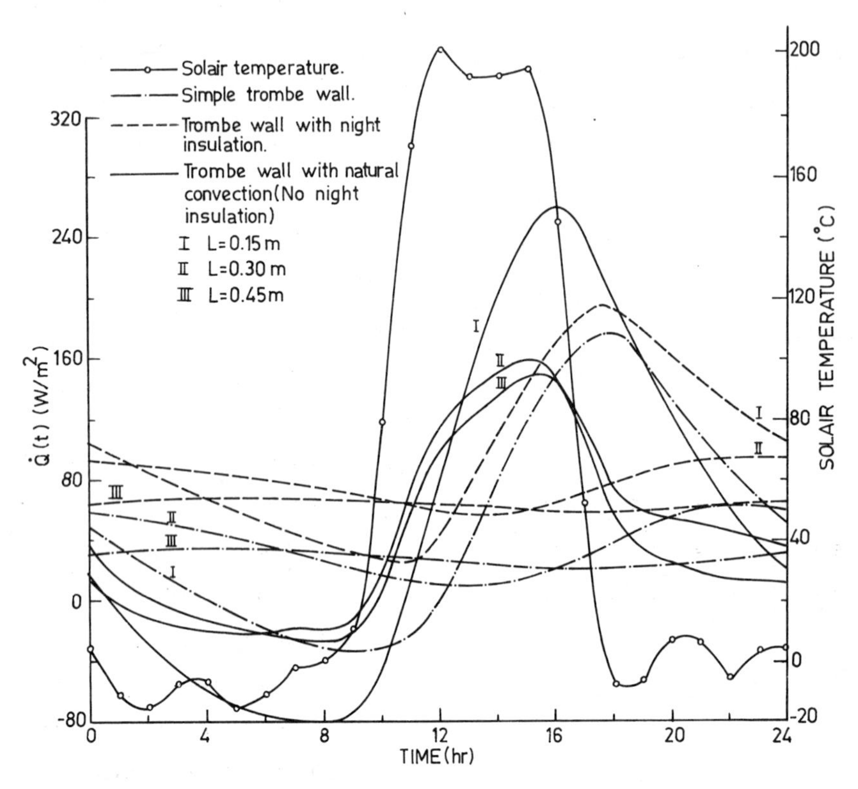

Fig. 7.5. Solair temperature and hourly variation of heat flux $\dot{Q}$ into the room for a Trombe wall system. (After Nayak *et al.*, 1983)

Perry (1977) has examined the suitability of Trombe wall and the water wall for various climatic zones of USA (Table 7.4). The thickness of concrete and water-walls was taken to be 0.45 m and 0.22 respectively. Evidently, the water wall performed more satisfactorily in comparison to the Trombe wall. It is also clear that in all the climates it is not advisable to leave vents opened continuously because the performance gets affected severely, especially in cold climates. The parametric calculations made by Perry (1977) were also presented in terms of annual solar heating fraction (SHF) for various magnitudes of thermal storage in the walls. An optimum is seen to occur for the Trombe wall beyond which the value of SHF starts decreasing (Fig. 7.6). Most of the benefits of storage, however, are obtained at a value of 340 kJ/°C per m^2 of glazed area. For a water wall the improvement obtained with a double-glazing is considerable as shown in Figs. 7.7 and 7.8. In fact, a single-glazed wall without night insulation can hardly be considered a viable option for passive solar heating because only 30% heating can be achieved even with maximum storage. The night insulation

TABLE 7.2 Parameters Corresponding to the Fig. 7.5.
(Aftet Nayak *et al.*, 1983)

Trombe wall	Free convection (for air)	Forced convection (corresponding to an air velocity) = 1.0 m/s
C_c (air) = 6.6 W/m^2 ^{o}C	h_1 = 3.2 W/m^2 ^{o}C	$h_1 = h_2$ = 29.4 W/m^2 ^{o}C
h_{si} = 8.29 W/m^2 ^{o}C	h_2 = 2.9 W/m^2 ^{o}C	h_r = 4.6 W/m^2 ^{o}C
α = 0.9	h_r = 4.9 W/m^2 ^{o}C	$\dot{m}_a$ = 4.2 Kg/s
	$\dot{m}_a$ = 0.104 Kg/s	
	C = 1008 J/Kg ^{o}C	
	h_{so} = 21.96 W/m^2 ^{o}C	

Water wall	Solarium
Concrete thickness	h_{wa} = 24.4 W/m^2 ^{o}C
	h_{ws} = 8.29 W/m^2 ^{o}C
	h_{si} = 8.29 W/m^2 ^{o}C
0.0 m $h_1' = h_2'$ = 206.5 W/m^2 ^{o}C	h_{TS} = 8.29 W/m^2 ^{o}C
0.15 m $h_1' = h_2'$ = 180.8 W/m^2 ^{o}C	= 0.6 (Ground)
0.30 m $h_1' = h_2'$ = 166.6 W/m^2 ^{o}C	= 0.9 (Trombe wall)
α = 0.9	= 0.9 (Water wall)
	= 0.6 (Concrete)
	U_a = 6.0 W/m^2 ^{o}C (without insulation)
	= 0.61 W/m^2 ^{o}C (with insulation)
	M_B = 0.0 Kg

Note: α corresponds to the absorptivity of the surface and subscripts G, T, W and C correspond to the components shown in brackets.

considerably improves the storage wall performance. A conclusion drawn from the performance figures is that night insulation is more effective for single-glazing than for double-glazing. A strategy of placing night insulation based on observed conditions rather than a time clock would result in only a small increase in the performance ($\approx$ 2%) (Balcomb, 1981).

It can be appreciated from Fig. 7.8 that the effect of varying the glass area is reverse of the effect of varying the building thermal load.

For a Trombe wall, the important parameters affecting the performance are, therefore, found to be wall's thickness and wall's thermal conductivity. The performance is shown in Fig. 7.9. The net annual thermal contribution for three different thicknesses is not markedly different; 30 cm thick wall is in fact best of the three, giving an annual solar heating fraction of 68%. This compares well with a figure of 73% for the same storage capacity of water wall.

TABLE 7.3 Thermal Flux into the Living Space in a Water Wall System for Various Parameters. (After Nayak *et al.*, 1983)

Water depth	Concrete thickness = 0.0 m									
	Q_{max}		Q_{min}		Q		Maxima (hr)		Minima (hr)	
	I	II	I	II	I	II	I	II	I	II
0.10	229.3	262.8	-21.9	55.0	85.9	146.5	17	16	9	9
0.30	134.2	187.3	42.4	113.9	85.9	149.1	17	17	9	10
0.50	114.5	172.0	59.1	127.5	85.9	149.3	17	17	9	10
1.00	100.0	186.5	72.3	138.2	85.9	149.3	27	17	9	10
	Concrete thickness = 0.15 m									
0.10	80.3	121.3	19.27	79.5	50.2	101.4	21	20	12	12
0.30	61.3	109.4	37.7	93.6	50.2	102.2	21	20	12	12
0.50	57	106.6	42.4	96.9	50.2	102.2	21	20	12	12
1.00	53.6	104.4	46.2	99.5	50.2	102.2	21	20	12	12
	Concrete thickness = 0.30 m									
0.10	43.6	82.98	27.4	73.4	36.2	78.9	2	2	15	16
0.30	39.0	81.0	32.7	77.4	36.2	79.5	3	2	16	16
0.50	37.95	80.5	34.07	78.1	36.2	79.5	3	2	16	16
1.00	37.1	80.0	35.11	78.8	36.2	79.5	3	2	16	16

I Without Night Insulation

II With Night Insulation

7.1.3 Transwall

As described earlier, the transwall is a thermal storage wall which partially absorbs and partially transmits the incident solar radiation into the living space. It differs from the conventional storage walls because most of the incident radiation is absorbed at its centre, in contrast to at the outer-most surface. The room gets heated and illuminated directly, and a self-imposed control of the transmitted fraction of the solar radiation alleviates the disadvantages of direct gain concept.

Fuchs and McClelland (1979) were the first to propose and investigate the performance of transwall placed adjacent to a window admitting solar radiation. The studied transwall structure consisted of a water-column contained between two vertical transparent glass plates, and a semitransparent glass plate placed vertically in the middle of water column. The 0.152 m thick transwall provided a heat capacity equivalent to a 0.33 m thick concrete wall. It was seen that

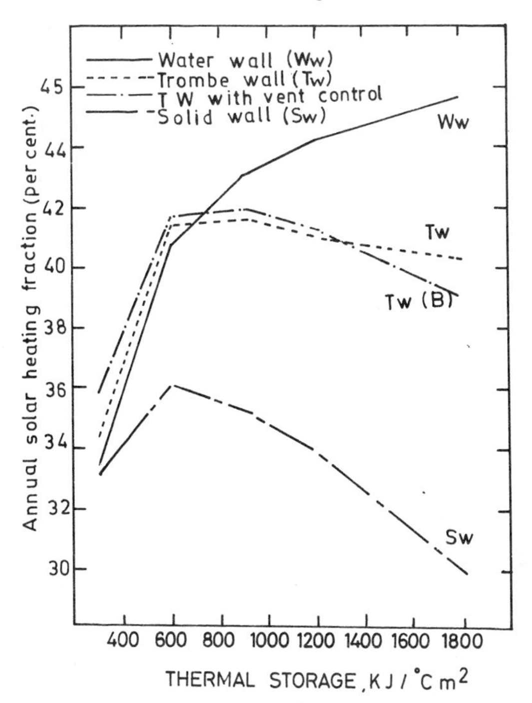

Fig. 7.6. Effect of storage mass and wall type on the performance of a passive system in Madison, WI, USA (After Perry, 1977)

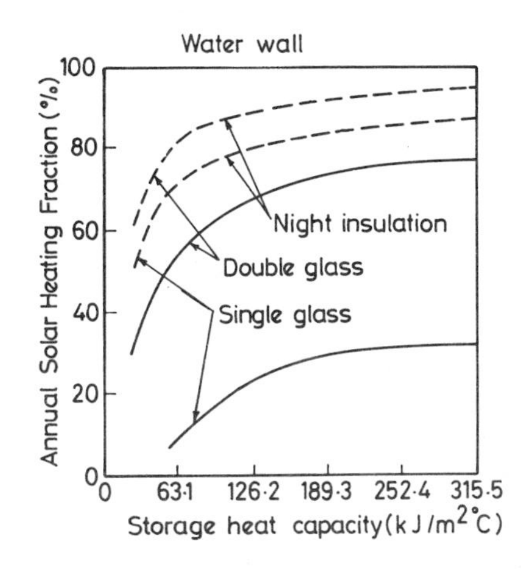

Fig. 7.7. Effect of storage heat capacity for the four different cases of single and double-glazing, with and without night insulation. (After Balcomb, 1981)

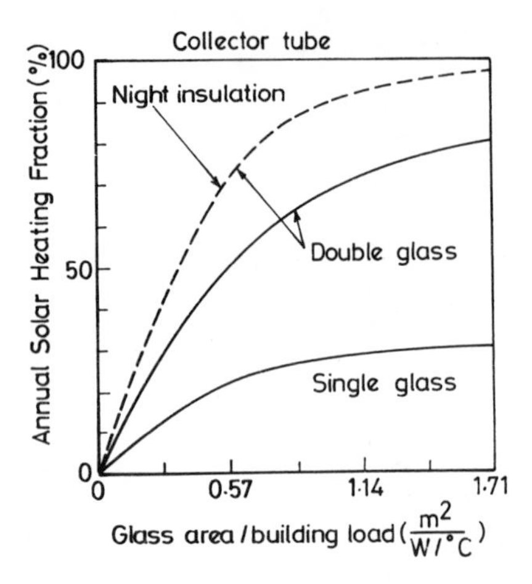

Fig. 7.8. Effect of glass area for a water wall case. (After Balcomb, 1981)

TABLE 7.4 Annual Results for Thermal Storage Walls (After Perry, 1977)

City	Annual Per cent Solar Heating				
	WW	SW	TW	TW(A)	TW(B)
Santa Maria	99.00	98.0	97.9	97.3	98.0
Dodge City	77.6	69.1	71.8	62.8	73.6
Bismarck	49.8	41.3	46.4	31.1	47.6
Boston	60.0	49.8	56.8	44.9	56.7
Albuquerque	90.8	84.4	84.1	81.8	87.5
Fresno	85.5	82.4	83.3	78.0	83.4
Madison	43.1	35.2	41.6	24.7	42.0
Nashville	68.2	60.7	65.2	54.1	65.4
Medford	59.0	53.3	56.1	42.2	56.8

WW: Water Wall (0.22 m thick)

SW: Solid Wall (no vents) (0.45 m thick)

TW: Trombe Wall (no reverse vent flow)

TW(A): Trombe Wall with vents open at all times

TW(B): Trombe Wall with thermostatic vent control

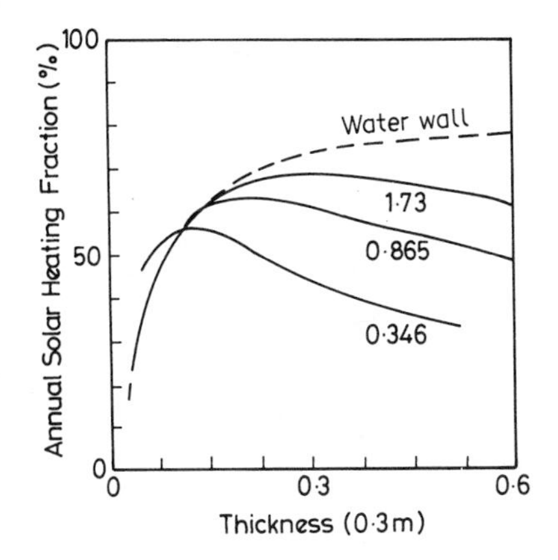

Fig. 7.9. Yearly performance of a thermal storage wall as a function of thickness for various thermal conductivities. The heat capacity was kept constant at a value of 189 kJ/m^2 ^{o}C. (After Balcomb, 1981)

with time the convection currents developed in the water columns reduced the effectiveness of the system. In order to rectify this problem the use of baffle plates was proposed in the water column. The experimental plate used in the test set-up was a semitransparent glass plate showing 80% absorptivity and 20% transmissivity for the solar radiation. Fuchs and McClelland (1979) compared the performance of transwall with the conventional direct/indirect gain concepts. The results for various configurations are presented in Table 7.5. Solar heating fraction (SHF) was calculated on the basis of design temperature at 21.1^{o}C, and the heat loss through walls/roof was assumed to be 3.95 MJ/day. The quenched transwall performs most satisfactorily, however, the transwall having high thermal conductivity shows equivalent performance to that of the concrete wall having thermosyphonic circulation. It is because the heat-losses through window using a transwall are significantly less. The concept is not yet used in actual building installations, however, the test results give a qualitative indication of the relative performance of the transwall configuration in comparison to the conventional wall configurations.

7.1.4 Thermic Diodes/Heat Pipes

The main drawback with most of the configurations discussed so far lies in the large heat losses from these systems to the surroundings. It occurs because the incident radiation absorbed at the outside surface raises its temperature which consequently increases heat losses. Thermic diodes which act as unidirectional heat-transfer devices can reduce these losses by transferring the collected energy at the outside surface to inside at a faster rate. The system which can be used as a thermic diode is shown in Fig. 7.10. The section of the tube positioned in the front plane are blackened and contain a liquid which can easily be evaporated by the application of heat energy. The vapours formed get condensed in the section of the tube which lies in the storage vessel and releases heat. The condensate then gets transferred to the evaporating section under the effect of gravity and the process gets repeated.

An experimental evaluation of the concept and its comparison with other convection suppressing mechanisms was made at Ispra Establishment of the European Communities (latitude 46°N) during the winter of 1979-80 by Baehr and Piwecki (1981). Figure 7.11 shows the temperature of the brick wall behind the collector/storage unit,

TABLE 7.5 Comparison of Heat-flows (MJ/day) and Solar Heating Fraction For Various Wall Configurations. (After Fuchs and McClelland, 1979)

	Solar input	Window heat loss	Wall heat loss	Suppl. heating	Supl. cooling (venting)	Solar heating fraction
Quenched Transwall*	9.10	6.28	3.00	0.47	0.27	88%
High cond. Transwall	9.10	6.85	2.96	0.94	0.23	76%
Trombe wall with concrete (air circ.)	9.10	7.13	2.83	0.86	0	78%
Solid concrete wall (no air circ.)	9.10	7.63	2.79	1.32	0	67%
Direct gain	9.10	7.61	3.07	1.71	0.13	57%

*quenched transwall means a transwall having baffle plates for suppressing convection inside it.

and the daily mean values of solar intensity and ambient temperature for various configurations of Fig. 7.4. It is seen that a building whose south facing wall is covered with a thermic diode system is the most effective one. It was reported that the wall-temperature dropped down below 20°C for 19 days with the present design, for 30 and 42 days for systems with honeycombs and inclined sheets respectively, and for 83 days with the vertical plates during the period from October 1, 1979 to April 30, 1980 for which experiments were carried out.

7.1.5 Conservatory

Conservatory design integrates the direct gain and indirect gain concepts. The sunspace consists of an enclosure having south facing wall and roof made up of transparent glass (Fig. 6.9d). This system cuts down the temperature swings of the habitant space significantly and provides an environment adequate for growing vegetation and taking a sunbath. In order to retain the heat gained during daytime, an insulation cover is used over the enclosure during night hours. Amongst others Sodha *et al.* (1982) have evaluated the thermal performance of a solarium using the periodic solution technique. The results obtained, corresponding to the climatic data of Boulder, Colorado and the set of parameters given in Table 7.2 are presented in Table 7.6. The temperature inside the conservatory increases when its area is kept fixed and the thickness of storage wall is increased. Use of movable insulation helps in cutting down the heat losses to the surroundings, and consequently results in the significant increase in the minimum temperature. It was reported that if the south wall area is increased along the breadth, thus proportionately increasing the ground area, increase in the temperature is smaller as compared to the case when area is increased along its height resulting in the increase of the area of east and west walls. The presence of thermal storage mass also showed a positive effect because it also absorbs direct solar radiation and retains it.

Performance of conservatories, water wall and Trombe wall concepts in terms of maxim minimum and average heat flux is given in Table 7.7. It is seen that conservatories

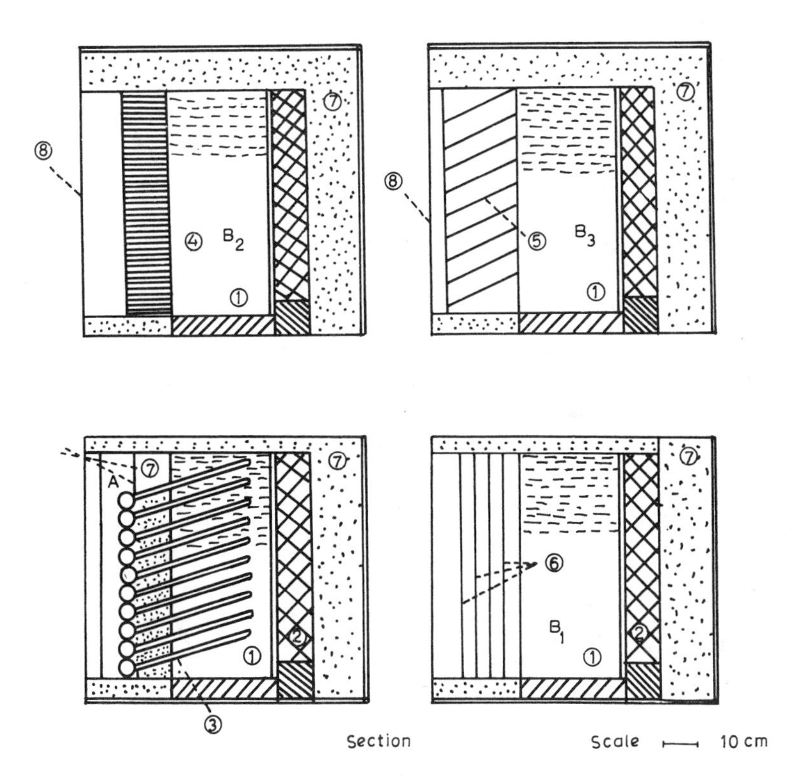

Fig. 7.10. Solar heated storage walls behaving as thermic diodes. (After Baehr and Piwecki, 1981)

performs best, when no night insulation is used. In the case of movable insulation, the maximum heat gain is obtained for a Trombe wall, however, the fluctuations in the heat flux coming into the room become large in this case.

7.2 PERFORMANCE OF PASSIVE COOLING CONCEPTS

In tropical and sub-tropical climates, the buildings are usually exposed to severe solar irradiation during the day for most part of the year. A number of concepts for passive cooling are in vogue, however, the crux of the problem lies

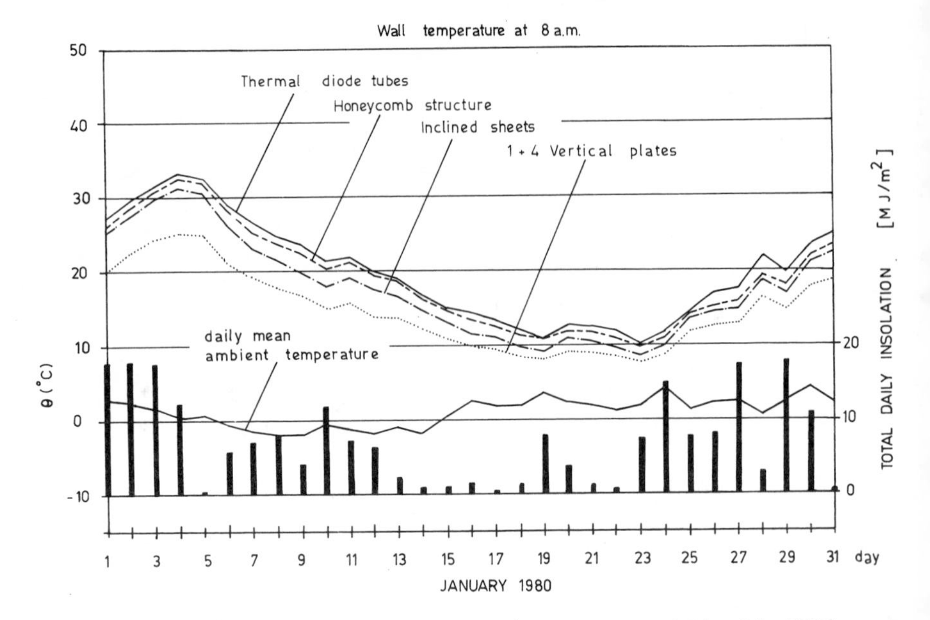

Fig. 7.11. Solar heated storage wall. (After Baehr and Piwecki, 1981)

in eliminating the undesired sun from entering the habitant space in summer. It should also be ensured that the proposed shade do not obstruct the winter sun to come in. The methods utilizing the evaporation of water and the earth's storage potential are seen to be the most effective cooling techniques. The former is practised by sprinkling/storing the water over the roof, while the latter can be implemented by constructing an underground tunnel for cooling the ambient air which is then drawn into the buildings. These two methods being of extreme importance are described separately in Chapter 8. In the present section, thermal performance of other passive cooling concepts is discussed.

7.2.1 Shading

An appropriate shading device can effectively reduce the vagaries of the outdoor weather so as to provide an acceptable indoor-climate. As discussed in Chapter 2, the shades on the windows may be used externally as well as internally. The roof, however, is shaded only externally.

7.2.1.1 Shading the windows. Ballantyne and Spencer (1961) have investigated the thermal performance of double glass windows when shades of different kind are used with it. The various designs considered were:

(i) *Unit C:* It consisted of two panes of clear polished glass with 6 mm air space between them. This unit was considered as the reference unit.

(ii) *Unit HA:* Its outside pane was of heat absorbing glass while the other specifications were similar as of Unit C.

(iii) *Unit HRI:* Inside glass pane was coated with a thin semi-reflecting metallic layer. Other specifications were similar as of Unit C.

(iv) *Unit HRO:* Inside surfaces of the outside glass pane were coated with the

TABLE 7.6 Comparison of the Temperatures reached in Solarium for Various Parameters in the Case of Non-insulated and Movable Insulation Solarium. (After Nayak *et al.*, 1982)

Isothermal mass (kg)	Wall area (m^2)			Thickness of mass wall (m)	Temperature of zone 1 (°C)					
	South	Floor	East/West		Solarium w/o insulation			Solarium with movable insulation		
					T_{max}	T_{min}	T_{av}	T_{max}	T_{min}	T_{av}
0.0	9	9	9	0.15	49.3	8.21	23.23	54.4	18.3	36.6
				0.30	50.4	9.86	24.58	56.6	21.2	39.8
				0.45	51.16	10.56	25.32	58.0	22.7	41.6
232.3	9	9	9	0.30	50.2	9.95	24.58	56.4	21.3	39.8
0.0	13.5	13.5	9	0.30	46.13	10.2	23.6	55.4	20.6	40.4
0.0	13.5	9	13.5	0.30	54.54	10.83	26.55	60.3	22.1	41.2

Average ambient temperature = -1°C

Average solar intensity = 311 W/m^2

semi-reflecting layer.

(v) *Unit LG:* It consisted of two glass sheets sandwiching stamped aluminium miniature louvers with a sun cut-off angle of 32°.

(vi) *Unit LB:* It consisted of two glass sheets sandwiching woven steel miniature louvers coated with black silicon enamel with a sun cut-off angle of 40.5°.

The experimental observations were carried out for March-April, 1960, and December, 1960 - January, 1961. Figure 7.12 depicts the amount of heat gain by convective and radiative exchanges at the inside surface, the heat gain by transmitted solar radiation, and the total heat gain which is the sum of two. By comparing the louver units with the heat reflecting units, it can be seen that the latter have the advantage of a lower heat-loss coefficient which ultimately leads

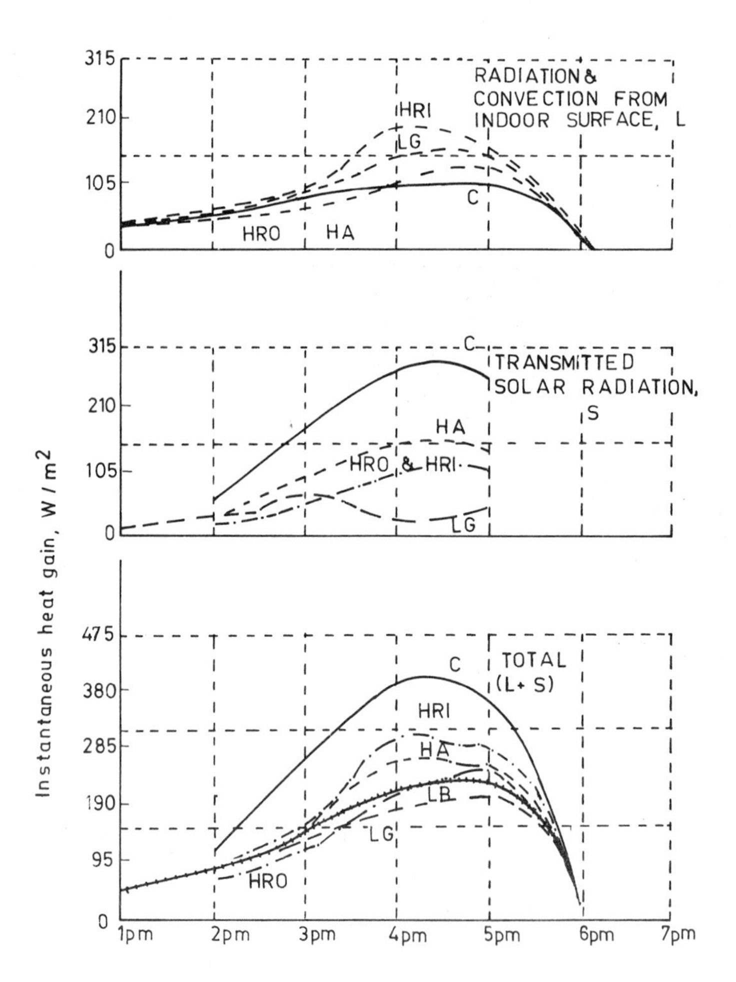

Fig. 7.12. Instantaneous heat gains for various shading devices on April 6, 1960. (After Ballantyne and Spencer, 1961)

TABLE 7.7 Maximum, Minimum and Average Heat Flux (in W/m²) into the Room for Various Passive Concepts. Average Solar Flux = 311.7 W/m². Average Ambient Temperature = -10°C. (After Nayak *et al.*, 1983)

Concrete thickness	Trombe Wall																		Water Wall						Solarium					
	Without vents						With vents																							
							Free Convection						Forced Convection																	
	I			II			I			II			I			II			I			II			I			II		
	Q_{max}	Q_{min}	Q_o	Q_{max}	Q_{min}	Q_o	Q_{max}	Q_{min}	Q_o	Q_{max}	Q_{min}	Q_o	Q_{max}	Q_{min}	Q_o	Q_{max}	Q_{min}	Q_o	Q_{max}	Q_{min}	Q_o	Q_{max}	Q_{min}	Q_o	Q_{max}	Q_{min}	Q_o	Q_{max}	Q_{min}	Q_o
0.15	276	-32	51.6	192	25.6	98.2	259	-79	55.1	278	-26	115	318	-702	-362	642	-289	111	80	19	50	121	79	101	109	-8.1	72	208	31	102
0.30	62.8	10.1	37	94	57.5	77	150	-27	45	181	14	111	310	-700	-362	607	-276	111	61	37	36	109	93	79	76	26.4	51	96	53.0	75
0.45	35	21.0	28	68	59	64	159	-21	39	175		100	277	-686	-362	603	-271	111	-	-	-	-	-	-	46	27.0	40	64	52.8	59

I Without Night Insulation

II With Night Insulation

to greater fuel savings. It was observed that the addition of internal blinds show a negligible effect on the amount of heat transmitted by the above-mentioned special glasses. It, however, protects against the radiation from hot surface and reduces the degree of discomfort and glare experienced. It may further be noted that the external shades, in general, are more effective than the internal shades. The energy absorbed by the internal shades gets reradiated to the internal environment and hence remain inside the house. However, a part of the incoming radiation gets reflected also from the internal shades which does not affect the internal environment. Therefore, the effectiveness of the internal shades is determined mainly by its reflectivity.

While reporting the results of full-scale laboratory experiments carried out in Israel, Givoni (1976) produced a curve (Fig. 7.13) showing the daily pattern of the indoor temperatures with different shading devices and made the following conclusions:

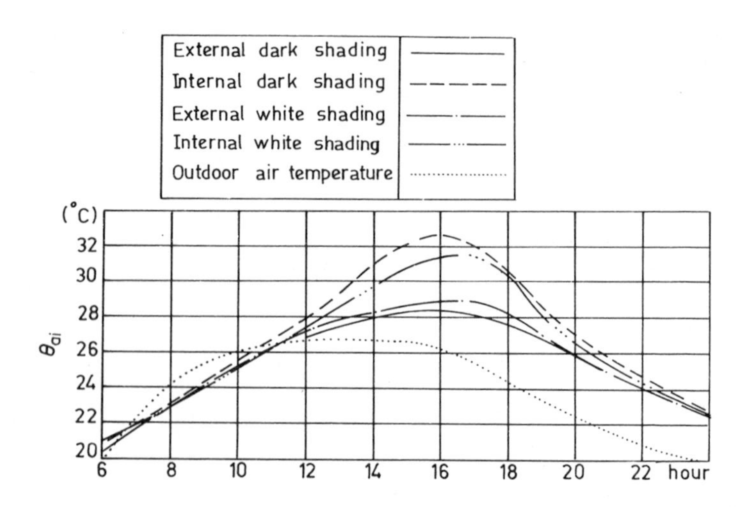

Fig. 7.13. Diurnal pattern of indoor temperatures in models with different shading devices. (After Givoni, 1976)

(i) External shading devices are much more efficient than the internal shading devices. The elevations in the maxima of indoor temperature above the outdoor temperature were 5.5 and 4.5°C respectively for green and white venetian blinds, while with external shading the elevations were 1.8 and 1.2°C respectively.

(ii) Colour of the shades has considerable effect on the performance. The difference in the performance using external and internal shades increases with the darkness of the colour. Light-coloured internal shades are seen to be more effective.

(iii) Efficient external shade can reduce the 90% heating effect of the solar radiation, while an inefficient internal shade would reduce only 20-25% of the solar heating effect.

7.2.1.2 Shading the roof. Usefulness of shading the roof has been studied experimentally by Ahmad (1980). The walls of the test room were 0.25 m thick made up of brick, and were plastered on both sides. The roof was consisting of a 0.10 m reinforced concrete slab, with a provision to install a movable reed panel to cover it. Reed panel is a locally produced material in Sudan, and is lightweight and light-coloured in nature. The doors of the room were kept closed during the whole day and opened from sunset to sunrise. Figure 7.14 illustrates the variation of ceiling temperature below the shaded and exposed part of the slab. Evidently, the maxima of ceiling temperature below the shaded slab is 5°C lower than that of the ceiling temperature below exposed portion. Without shading, the ceiling temperature is most of the time higher than the outside air temperature. For shaded roof, the ceiling temperature is below the outside air temperature from 9 a.m. to 5.30 p.m. Afterwards, the presence of the shading device obstructs the heat losses from the roof and hence the ceiling temperature is greater than the outside air temperature. Similar results were also reported by Nayak *et al.* (1982) who reported a reduction of 240 kWh in the cooling load of an air-conditioning plant when a movable canvas is used over the roof. The calculations were made for a typical summer day of Delhi, India.

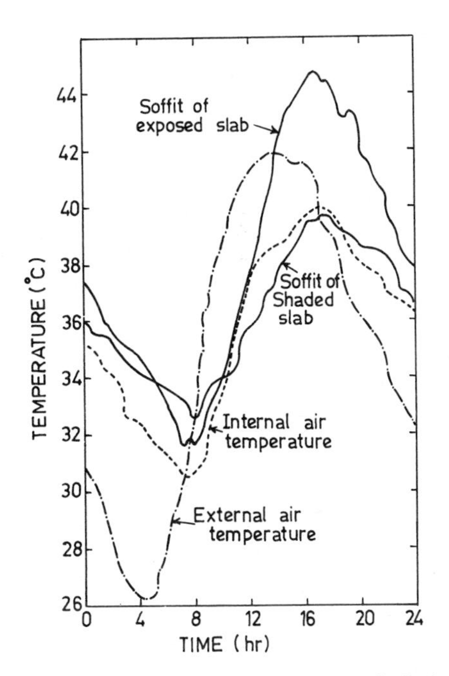

Fig. 7.14. Ceiling temperatures under exposed and shaded parts of the concrete slab. (After Ahmad, 1980)

7.2.1.3 Shading the walls. Like roof, walls can also be covered during off-sunshine hours by a movable canvas/insulation. Singh (1982) has studied the performance of the walls under shading by assuming these to be exposed to the outside environment on one side and the air-conditioned air (20°C) on the other side. The evaluation was based on the summer data of Delhi, India. Table 7.8 shows the results of the partial shading over the east, south, west and north walls. It is evident that shading can reduce an appreciable amount of daily cooling load.

7.2.2 Roof Treatment

Various options of reducing the heat gain through the roof have been investigated by Nayak *et al.* (1982) by assuming one side of the roof to be in contact with the inside air of an air-conditioned room. The roof configuration considered for the investigations was a typical one used in North India, the region where the ambient air temperatures of the order of 45°C occur during the summer. It was considered to be a three layer structure of concrete, mud-phuska* and the brick tiles. The methods studied include the use of vegetable pergola, inverted earthen pots, and roof garden over the roof. In rural India, the people usually grow some vegetable plants on some erected platforms of wooden mesh structures on which these are allowed to creep on. The height of the platform is normally kept about 2.5 m above the roof level. The vegetation covers whole of the roof; thicker the vegetation, greater is the shading achieved. Another method of cooling consists of the inverted earthen pots with white painted bottom kept over the roof. In addition to reflecting a part of the incident solar radiation, the presence of these pots provide air-insulation, and increased heat transfer by convection. Roof garden is a concept similar to water-film concept. The relative performance of these concepts is presented in Table 7.9. It is evident that in case of roof garden a phase shift of 18 hours is achieved because the heat capacity of the system is very high. The reduction in the amount of heat flux which enters through the roof is more in case of removable canvas than in the case of vegetable pergola. However, less swings are achieved when inverted earthen pots are used.

7.2.3 Reducing Conduction Heat-Transfer through Roof/Walls

The amount of heat flux which is being conducted in through the roof/walls may be reduced significantly by insulating the building components or by providing air cavities or both. The effect of each of them on the reduction of heat flux into the room is discussed below.

7.2.3.1 Thermal insulation. Insulating the building is a well-known and well-practised aspect used for energy conservation. The materials used for insulation can, in general, be divided into two classes viz. conductive and reflective. In case of the former the bulk properties, like the thermal conductivity and thermal diffusivity are the effective variables, whereas in case of the latter the surface properties are important. While investigating the effect of conductive insulation for an air-conditioned room, Sodha *et al.* (1979) reported that in order to achieve best thermal load-levelling**, equal thickness of the insulation should be used at both sides of wall/roof (Fig. 7.15). This result is an addition to the earlier established fact which states that the outside insulation gives more insulating effect as compared to the one used at the inside surface. Obviously, the difference between $\dot{Q}_{max}$ and $\dot{Q}_{min}$ is more for the case when thickness of the outside insulation is minimum i.e. zero, as compared to the case when there is no inside insulation. The variation of $\dot{Q}_{max}$ and $\dot{Q}_{min}$ can also be seen to be almost constant in the range $1/4 < (l_3/L) < 3/4$ where l_3 and L refer to the thickness of inside insulation and the total thickness of the roof respectively.

Considerable amount of work has been reported in literature from the point of view of minimizing thermal losses from a building in temperate and cold climates. Recommendations have been made in various countries to use insulated roofs and walls yielding maximum values of transmittances. These have been reported in Table 7.10. The choice of an insulation material should not merely be based on its thermal conductivity, but other properties like its density, specific heat, coefficient of thermal expansion, resistance to temperature extremes and vapour flow, durability, mechanical strength, affinity for moisture absorption, possibilities of vermin and insect infestation and influence of dust and air movement on its efficacy and appearance deserve to be considered proportionately

---

*mud-phuska is a locally used composition of mud mortar and hay.

**Best thermal load levelling is achieved when the difference between the amplitudes of maxima and minima of the heat flux entering the room is minimum.

TABLE 7.8 Effect of Shading on Infiltrating Heat Flux through a Massive Wall of 0.3 m on May 16, 1981 at Delhi, India. (After Singh, 1982)

%	East			South			West			North		
Shading in percentage	Q_o W/m^2	Q_{min} W/m^2	Q_{max} W/m^2	Q_o W/m^2	Q_{min} W/m^2	Q_{max} W/m^2	Q_o W/m^2	Q_{min} W/m^2	Q_{max} W/m^2	Q_o W/m^2	Q_{min} W/m^2	Q_{max} W/m^2
90	24.94	20.61 11 a.m.	29.06 11 p.m.	24.72	20.45 11 a.m.	28.85 11 p.m.	29.05	20.68 11 a.m.	29.41 11 p.m.	24.67	20.45 11 a.m.	28.78 11 p.m.
60	26.39	21.50 11 a.m.	30.80 10 p.m.	25.50	20.88 11 a.m.	29.95	26.81	21.77 12 noon	32.19 11 p.m.	25.30	20.87 11 a.m.	29.65 11 p.m.
30	27.84	22.38 10 a.m.	32.62 10 p.m.	26.29	21.32 11 a.m.	31.05 11 p.m.	28.57	22.83 12 noon	34.95 11 p.m.	25.93	21.29 11 a.m.	30.52 11 p.m.
0.0	29.29	23.12 10 a.m.	34.45 10 p.m.	27.07	21.75 11 a.m.	32.16 10 p.m.	30.33	23.90 12 noon	37.75 11 p.m.	26.57	21.71 11 a.m.	31.39 11 p.m.

TABLE 7.9 Relative Performance of Various Passive Cooling Concepts Applied to the Roof of an Air-conditioned Room at New Delhi, India. (After Nayak *et al.* 1982)

		Roof configuration	Q_{max} (W/m^2)	Q_{min} (W/m^2)	Phase shift (h)
1.	Normal roof	Concrete-mud-phuska-brick tiles (summer)	37.34	27.21	12
2.	Shaded roof	Vegetable pergola (summer)	37.34	27.21	12
		Shading (%)			
		30	32.44	24.21	12
		60	27.56	21.20	12
		90	22.68	18.17	12
3.	Removable canvas	Roof is covered during day (summer)	23.41	10.13	12
4.	Water film	Summer RH = 0.5	20.33	13.89	12
		RH = 0.3	14.83	9.31	12
		RH = 0.1	9.34	4.72	12
5.	Roof garden	Summer RH = 0.5	17.09	14.53	18
		RH = 0.3	12.42	10.18	18
		RH = 0.1	7.71	5.81	18
6.	Roof with inverted	Shading 90% h_{so} = 22.78 $W/m^2 {}^oC$	22.68	18.17	12
		h_{si} = 5.8 $W/m^2 {}^oC$			
		60% h_o = 22.78 $W/m^2 {}^oC$	24.22	20.00	12
		h_{si} = 5.7 $W/m^2 {}^oC$	27.39	20.03	12

RH = relative humidity

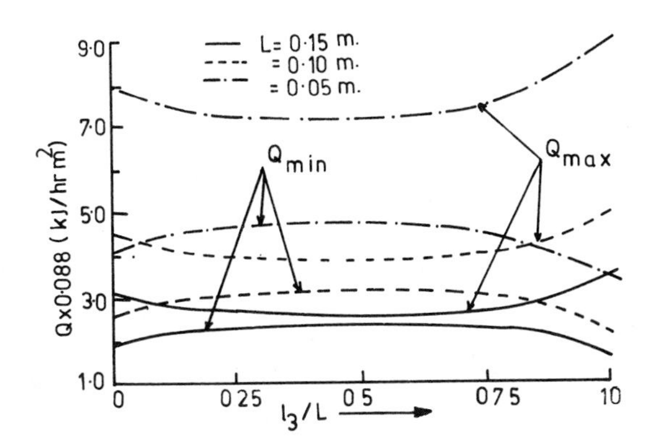

Fig. 7.15. The variation of $\dot{Q}_{max}$ and $\dot{Q}_{min}$ with l_3/L for a conditioned room corresponding to a typical data of Kuwait. (After Sodha *et al.*, 1979)

$\dot{Q}_{max}$: Maximum of the heat flux entering the room

$\dot{Q}_{min}$: Minimum of the heat flux entering the room

(Van Straatan, 1967). It is obvious that the better insulated a building is, the smaller will be the losses/or gains through the structural elements. From an economic point of view, however, there is a limit to the amount that should be spent on insulation.

For lightweight roofs, Lotz and Richards (1964) have studied the effect of various types of insulations on the ceiling temperature, indoor air temperature and the heat-flow through ceiling, corresponding to hot climates of South Africa. The houses over which the experiments were carried out were constructed using 280 mm cavity brick and galvanized corrugated iron roofs. The test house was not provided with any insulation. The effectiveness of insulation was investigated under both summer and winter conditions, and the results showing the mean daily values over the whole test period are given in Tables 7.11 and 7.12. The aluminium foil was fixed at the inside of the roof beams, and the bulk material was laid directly over the ceiling, care being taken to obtain a uniform thickness throughout. It is clear from the results that insulation has a significant effect on indoor air temperature, and the performance becomes better for higher values of insulation thicknesses. This is also noted that the insulation has more pronounced effect on ceiling temperatures than on the indoor air temperature.

For heavyweight roofs, the insulation should be applied on top of the slab rather than on the inner most surface and it should be kept dry in order to remain effective. As far as the optimum thickness of the insulation for heavyweight roof is concerned, the situation is much more complicated than for lightweight roofs, because of the damping effect of the relatively heavy concrete. For the same reason the concept of U value has little meaning for roofs of this type under conditions of large variations in outdoor air temperature and solar radiation.

It has already been discussed earlier that the response time of the building component gets considerably affected by the position of insulation i.e. either on top of the slab or on the innerside of the slab. For heating purposes, insulation on the inner side results in quicker response to the heating mechanism, while for reducing the heat gains placement of insulation on the exposed surface will increase the damping of the incident solar flux. An example is given below which shows the effect of insulation position on the time constant factor and resulting time lag.

TABLE 7.10 Suggested Averaged Maximum Values of Overall Thermal Transmittance for Domestic Buildings in Different Countries. (After Van Straatan, 1967)

Country	Thermal transmittance ($W/m^2\ ^oC$)			Remarks	Source of information
	External walls	Roofs	Floors		
England	0.85-1.14*	1.14-1.70	0.85	*Lower value for walls of warmed living-rooms	Egerton Comm. 1945
	1.70	1.42	1.42**	*Unglazed part of wall **Exposed to outdoor air	Local authorities By-laws in England and Wales
	2.38* 1.70**	1.14	1.14***	*Average over entire wall **Unglazed part of wall ***Exposed to outdoor air	Building standards (Scotland) Regulations
France	1.14-1.53* 1.42-1.76** 1.76-1.99***	0.91-1.02* 1.14-1.31** 1.42***	1.42-2.1+	*For elements with mass less than 1.26 kg/m^2 **For elements with mass from 1.26 to 2.94 kg/m^2 ***For elements with mass over 2.94 kg/m^2 +Exposed to outdoor air Lower values apply to areas with more extreme climates	R.F.E.F.58 Hydrothermique et Ventilation
Germany	1.19-1.53	-	-	Lower values for colder areas Values based entirely on prevention of surface condensation	DIN 4018, Thermal Insulation in Buildings
Holland	0.68-0.97* 0.68-1.14**	0.68-0.97+ 0.68-1.14++	0.68-0.97+ 0.68-1.14++	*For walls with mass less than 0.84 kg/m^2 **For walls with mass from 0.84 to 2.1 kg/m^2 ***For walls with mass over 2.1 kg/m^2 +Exposed to direct air and with mass less than 0.84 kg/m^2 ++Exposed to direct air and with mass over 0.84 kg/m^2	Norm NEN 1068 Thermal Properties of Dwellings

TABLE 7.10 (cont'd)

Country	Thermal transmittance ($W/m^2\ ^oC$)			Remarks	Source of information
	External walls				
Scandinavia	0.68-0.91* 0.45-0.57**	0.45-0.57	0.45-0.57*** 0.40-0.45****	*For walls with mass over 0.84 kg/m^2 **For walls with mass less than 0.84 kg/m^2 ***Facing unheated rooms ****Facing outdoor air Lower values apply to colder regions	Joint Scandinavian Building Code, 1964
Canada	0.57-0.85	0.45-0.68	0.57-0.85	For heating with gas or oil Lower values apply to colder areas and to areas where fuel cost is higher	Residential standards, NRC 8251, 1965, Supplement No. 5 to National Building Code
	0.45-0.51	0.45-0.51	0.28-0.40	For heating with electricity Lower values apply to colder regions	

TABLE 7.11 Effect of Ceiling Insulation on Various Performance Parameters Under Summer Conditions. (After Lotz and Richards, 1964)

Performance parameter		No insulation	Type of insulation			
			Draped Al. foil	50 mm Vermiculite	Mineral wool	
					50 mm	100 mm
Indoor air temperature (°C)	max	25.0	23.3	23.6	23.6	23.0
	min	20.6	20.6	20.6	20.6	20.6
Ceiling temperature (°C)	max	31.9	24.8	26.1	25.9	23.3
	min	20.6	20.0	20.3	20.4	20.6
Heat-flow through ceiling (W/m²)	max	51.46	13.58	22.10	18.94	11.05
	min	-15.78	-6.3	-5.37	-4.42	-0.95

TABLE 7.12 Effect of Ceiling Insulation on Various Performance Parameters Under Winter Conditions. (After Lotz and Richards, 1964)

Performance parameter		No insulation	Type of insulation			
			Draped Al. foil	50 mm Vermiculite	Mineral wool	
					50 mm	100 mm
Indoor air temperature (°C)	max	14.4	13.3	14.2	14.2	13.6
	min	10.0	10.6	10.8	10.8	10.6
Heat-flow through ceiling (W/m²)	max	26.82	11.04	12.62	6.31	-
	min	-23.66	-12.6	-9.47	-7.89	-

In contrast to the work performed for cold climates, the thermal design of roofs in most countries with warm climates and abundance of solar energy have often been given inadequate attention. Insulation over the roof in warm and hot climates can considerably cut down the heat flux due to solar radiation incident on the roof. Effect of insulated roofs over the indoor environment temperature depends on the condition whether the roofs are lightweight or heavyweight.

Example 7.1. Calculate the time constant for the configurations shown in Fig. 7.16 using the following thermophysical parameters.

$$R_{so} = 0.08 \text{ m}^2 \text{ }^\circ\text{C/W}$$

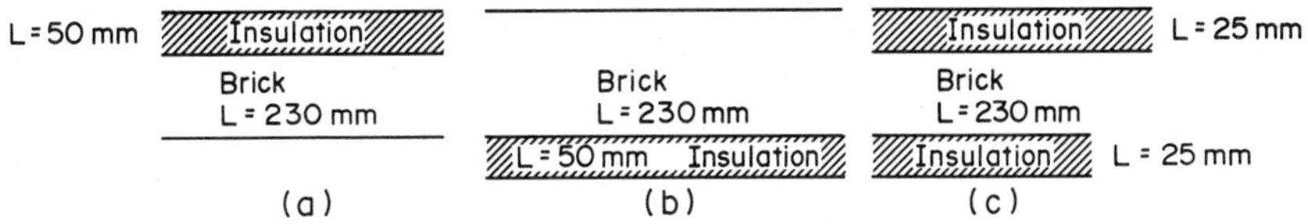

Fig. 7.16

Brick:

$K = 0.969\ W/m^oC$

$\rho = 1762\ kg/m^3$,

$C = 840\ J/kg^oC$,

Insulation (Vermiculite)

$K = 0.068\ W/m^oC$

$\rho = 122\ kg/m^3$

$C = 750\ J/kg^oC$

Solution: From equation (5.35), the time constant for various configuration may be calculated as follows:

Configuration 1 (Fig. 7.16a)

$$\frac{Q_1}{U_1} = \left(0.08 + \frac{0.05}{2 \times 0.068}\right) \times (0.05 \times 122 \times 750)$$

$$= 2048\ s$$

$$\frac{Q_2}{U_2} = \left(0.08 + \frac{0.05}{0.068} + \frac{0.23}{2 \times 0.969}\right) \times (0.23 \times 1762 \times 840)$$

$$= 317951\ s$$

$$\therefore \text{ Time constant} = \left(\frac{2048 + 317951}{3600}\right) \text{hr}$$

$$= 88.9\ \text{hr}$$

From Eq. (6.38), Time lag $= 1.18 + \frac{\pi}{12} \times 88.9 \simeq 24$ hr

Configuration 2 (Fig. 7.16b)

$$\frac{Q_1}{U_1} = \left(0.08 + \frac{0.23}{2 \times 0.969}\right) \times (0.23 \times 1762 \times 840)$$

$$= 67634\ s$$

$$\frac{Q_2}{U_2} = \left(0.08 + \frac{0.23}{0.969} + \frac{0.05}{2 \times 0.068}\right) \times (0.05 \times 122 \times 750)$$

$$= 3134\ s$$

$$\text{Time constant} = \left(\frac{67634 + 3134}{3600}\right) \text{hr} = 18.7\ \text{hr}$$

From Eq. (6.38)

$$\text{Time lag} = 1.18 + \frac{\pi}{12} \times 19.7$$

$$\simeq 6\tfrac{1}{2}\ \text{hr}$$

Configuration 3 (Fig. 7.16c)

$$\frac{Q_1}{U_1} = \left(0.08 + \frac{0.025}{2 \times 0.068}\right) \times (0.025 \times 122 \times 750)$$

$$= 603 \text{ s}$$

$$\frac{Q_2}{U_2} = \left(0.08 + \frac{0.025}{0.068} + \frac{0.23}{2 \times 0.969}\right) \times (0.23 \times 1762 \times 840)$$

$$= 192788 \text{ s}$$

$$\frac{Q_3}{U_3} = \left(0.08 + \frac{0.025}{0.068} + \frac{0.23}{0.969} + \frac{0.025}{2 \times 0.068}\right) \times (0.025 \times 122 \times 750)$$

$$= 1987 \text{ s}$$

$\therefore$ Time constant = 54.3 hr.

$$\text{Time lag} = 1.18 + \frac{\pi}{12} \times 54.3$$

$$\simeq 15 \text{ hr.}$$

7.2.3.2 Air cavities in the roof/walls. Cavities within walls or an attic space in the roof-ceiling combination inhibits the heat transfer, because the thermal conductivity of still air is very low. However, heat transfer by radiation constituting about 60 to 65% of the total heat transfer across the cavity, takes place from one boundary surface to the other. Some heat transfer also takes place due to convection currents within the air space itself. Heat transfer, by conduction, is negligible in air spaces thicker than 20 mm in comparison to other modes of heat transfer (Van Straatan, 1967).

The effect of air cavities in reducing the heat flux into the wall has been studied by Pratt (1981), Singh *et al.* (1981), Sodha *et al.* (1981a,b). The calculated values of Pratt (1981) for the amplitude ratio* and the time lag are given in Table 7.13. It is interesting to note that for the thickness specified, the respective properties of the two alternative arrangements of such dissimilar materials such as brick and timber differ by rather less than 20%.

The response of the inside temperature of the surface to step change in the outer surface temperature for a solid brick wall and a cavity wall is shown in Fig. 7.17. The reduction of the inside ceiling temperature by nearly 50% due to the presence of air cavity is noticeable from the figure.

It is possible to introduce more than one air gap in the wall/roof component. For a 250 mm layer of total concrete thickness, carrying different numbers of air gaps, the reduction in the heat flux coming into a room (maintained at balance point temperature) is shown in Fig. 7.18. This effect, in terms of the reduction of heat flux entering the room through roof has been investigated by Singh *et al.* (1981). The results reported in Fig. 7.18 show an appreciable reduction of heat flux with increasing air gaps. The effect initially is more (up to 3 air gaps) after which it is only marginal. Sodha *et al.* (1981c) have reported that the reduction in heat flux can further be enhanced by using reflecting sheets over the surfaces of the cavity (Fig. 7.19). For the typical data of Delhi, India, a

---

*The amplitude ratio measures the extent to which the periodic variation in the outside ambient temperature is attenuated in its passage through the wall to the inside surface.

TABLE 7.13 Calculated Values of Time Lag and Amplitude Ratio of Solid and Cavity Walls. (After Pratt, 1981)

Wall (outside to inside)	Time lag (hr)	Amplitude Ratio
11.5 cm brick (4½ in)	3.7	0.324
11.5 cm brick - cavity - 2.5 cm timber (1 in)	5.0	0.114
2.5 cm timber - cavity - 11.5 cm brick	5.7	0.093
11.5 cm brick - cavity 7.6 cm aireted concrete (3 in)	6.7	0.073

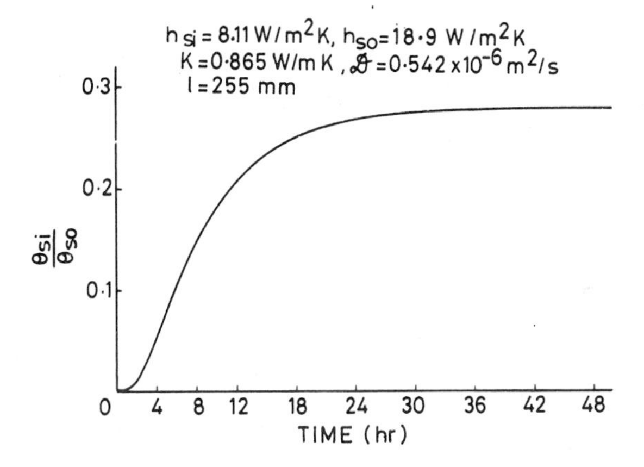

Fig. 7.17(a). Response of inside surface temperature of 220 mm solid brick wall to step-change in outside air temperature. (After Pratt, 1981)

reduction of 70% was seen to occur in Q_{max} by using a reflective insulation as against a reduction of 25% in cases of a single hollow roof. The effect of reflective surfaces of the cavities on the cooling load can be appreciated from Table 7.14. The table shows that the number of air cavities should not be more than two, and the reflective surfaces are very useful in cutting down the maximum heat flux entering the room which ultimately reduces the cooling load on the air-conditioning plant.

7.2.4 Use of Paints/Radiative Cooling

The colour of the outside surface influences the thermal performance of a building significantly because it governs the absorption and reflection of the incident

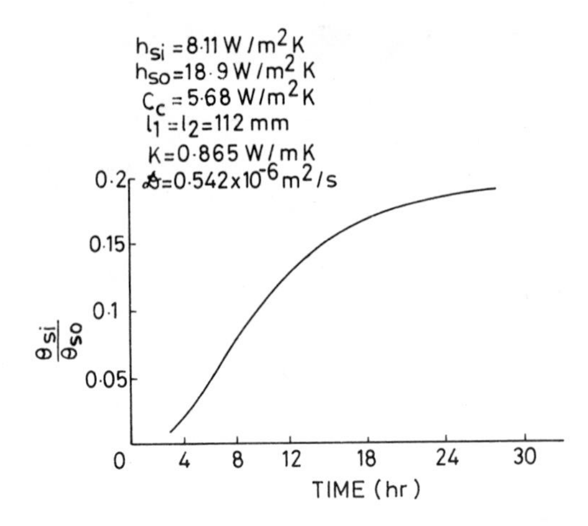

Fig. 7.17(b). Response of inside surface temperature of 280 mm cavity brick wall to step change in outside air temperature. (After Pratt, 1981)z

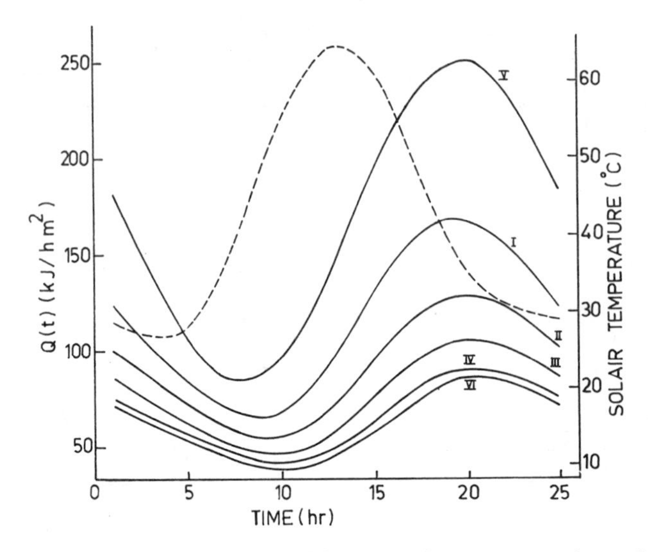

Fig. 7.18. The variation of $Q(t)$ for various roof configurations keeping the total air gap thickness to be constant. (After Singh *et al.*, 1981)

----- Solair temperature
——— $Q(t)$
I - Single hollow roof
II - Double hollow roof
III - Triple hollow roof
IV - Roof with four air cavities
V - Concrete roof
VI - Roof with five air gaps

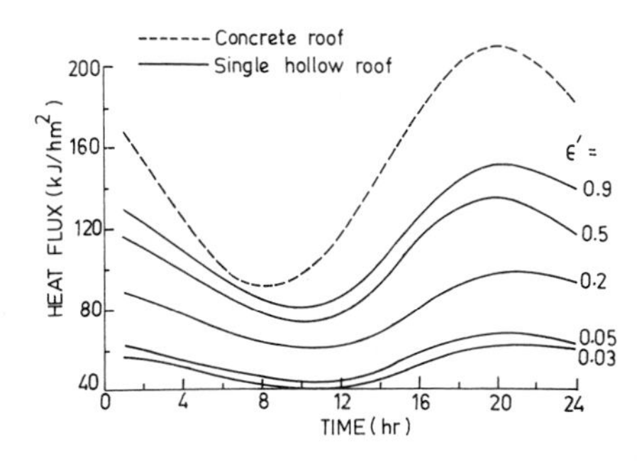

Fig. 7.19. Time variation of Q(t). (After Sodha *et al.*, 1981c)

solar radiation during daytime, and the emission of longwave radiation during night-time. Givoni (1976), however, has mentioned that the effect of external colour gets modified by changed thermal resistance and heat capacity of the structure (Fig. 7.20). The effect of colour, as expected, is more in case of the light structures which offer low resistance to heat-flow and has low thermal capacity. It can be verified from the figure that the differences between the ceiling temperatures of the grey and whitewashed roofs were much greater for the 70 mm thick roof than for a 200 mm thick one.

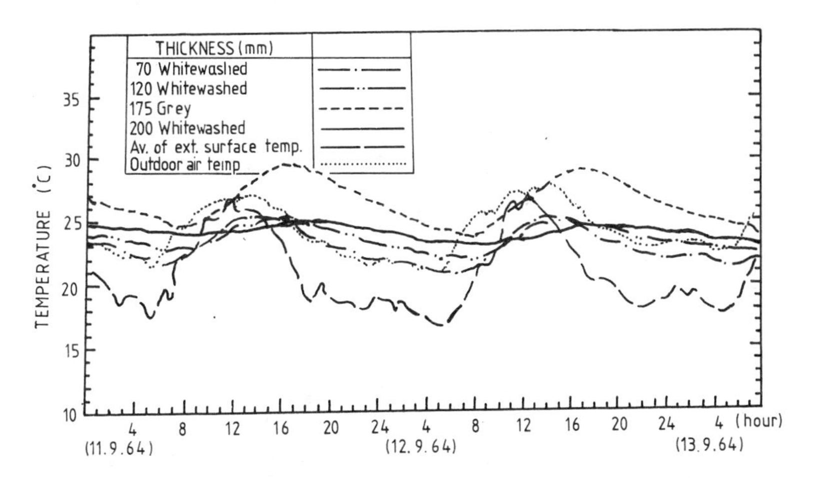

Fig. 7.20. Internal surface temperatures of horizontal panels of whitewashed and grey external colour. (After Givoni, 1976)

The internal air temperature also gets affected by the colour of external surface. Givoni and Hoffman (1973) have carried out some experiments with different colours in Israel and reported that a temperature difference of the order of 3°C and 1°C occurred when measured 0.1 m below the roof and 1.2 m above the floor respectively (Fig. 7.21) for grey and whitewashed roofs. The observations indicated that in case of a grey-coloured roof, the direction of heat flow is from roof to the room air, whereas the light-coloured roof shows a reverse flow of heat. This means that even the room heat is radiated to the atmosphere in the case of whitewashed roof.

By radiating energy to the clear night sky, the temperature of a body may fall

TABLE 7.14 Effect of Air Cavities having Reflective Surfaces on Heat Flux entering the Room (After Singh *et al.*, 1981)

Roof configuration		($kJ/h\ m^2$)	($kJ/h\ m^2$)	Phase shift* (h)	Daily inside heat flux (kJ/m^2)
Concrete roof		250	84	7	4008.00
Single hollow concrete roof	Air gap with bare surfaces	168.34	65.29		2755.44
	Air gap with reflecting surfaces				
Double hollow concrete roof	All air gaps with bare surfaces	128.22	54.05	8	2160.72
	One air gap with reflecting surfaces	51.79	25.42	9	924.00
	Both air gaps with reflecting surfaces	32.53	16.57	9	587.76
Triple hollow concrete roof	All air gaps with bare surfaces	104.31	46.61	9	1793.04
	One air gap with reflecting surfaces	56.74	27.72	9	1010.16
	Two air gaps with reflecting surfaces	38.98	19.77	9	703.20
	All air gaps with reflecting surfaces	29.64	15.34	9	539.28
Quadruple hollow concrete roof	All air gaps with bare surfaces	88.69	40.98	8	1546.93
	One air gap with reflecting surfaces	58.04	28.43	9	1034.16
	Two air gaps with reflecting surfaces	43.12	21.82	9	776.88
	Three air gaps with reflecting surfaces	34.23	17.70	9	622.08
	All air gaps with reflecting surfaces	28.37	14.88	10	518.64

*Relative to solair temperature

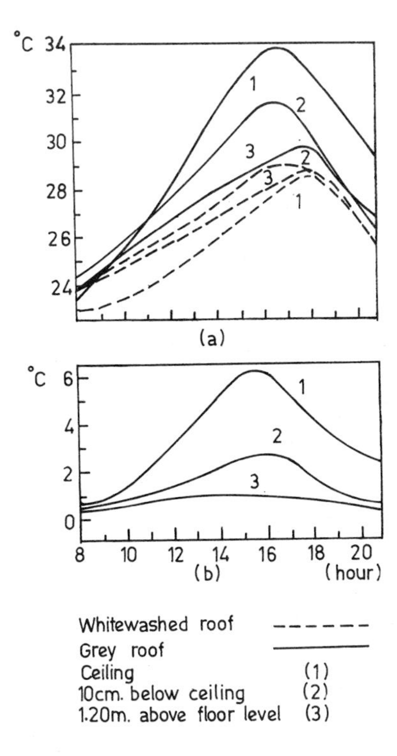

Fig. 7.21(a). Effect of flat roof colour on indoor temperatures
(b). Differences between grey and white roofs. (After Givoni and Hoffman, 1973).

considerably below the ambient temperature. It is reported in the literature that the use of selective surfaces which reflect all radiation outside the wavelength range 8-13 μm and absorb and emit within this limit as a black body, are very effective in this type of cooling. The lowest temperature available with a selective surface is believed to be considerably lower than those attainable with a black body radiator. Michell and Biggs (1979) have investigated experimentally the effect of radiative cooling by constructing two huts, the roofs of which had different configurations. One of the huts had a roof of galvanized steel decking painted white, while the other had a roof of aluminium decking to which aluminized "Tedlar" sheet was glued. The huts were constructed 2.4 m in height and covered a base area of 3 m^2. The paint used for making the surface white contained 17.5% of pigment (TiO_2) volume concentration in a binder of long soyabean modified alkyd corresponding to a concentration of 44.1% by weight on nonvolatiles. Selective properties were given to the roof of other huts by glueing 12 μm thick sheets of Tedlar aluminized on one side to the surface of decking. The cross-section of the experimental huts is shown in Fig. 7.22. The observations indicate that the roof-deck temperatures, 5.6°C below the outdoor ambient, can be obtained for short periods. Typically, the performance is shown in Fig. 7.23. This figure shows that the cooling of roof-decks starts at 5 p.m., and the temperature sharply falls 8°C below ambient at about 7 p.m. However, at this instant the fans start blowing due to which the temperature-fall reduces to 5°C only with respect to the ambient. However, to achieve effective cooling, the air can be circulated through the ceiling ducts. This cooled air provided a net cooling potential of 200 W at an air flow rate of 170 litres per second which

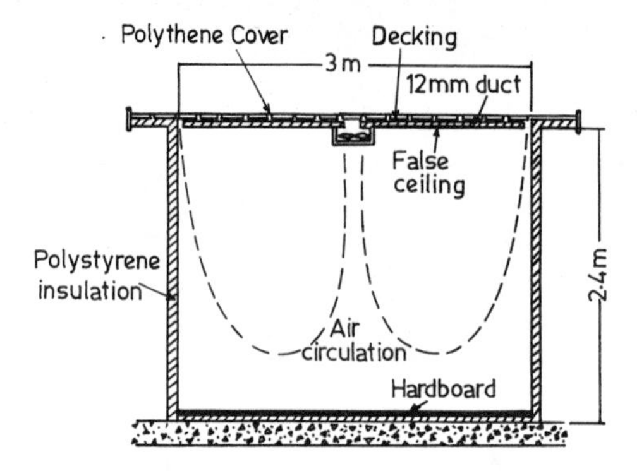

Fig. 7.22. Schematic cross section of experimental huts. (After Michell and Biggs, 1979)

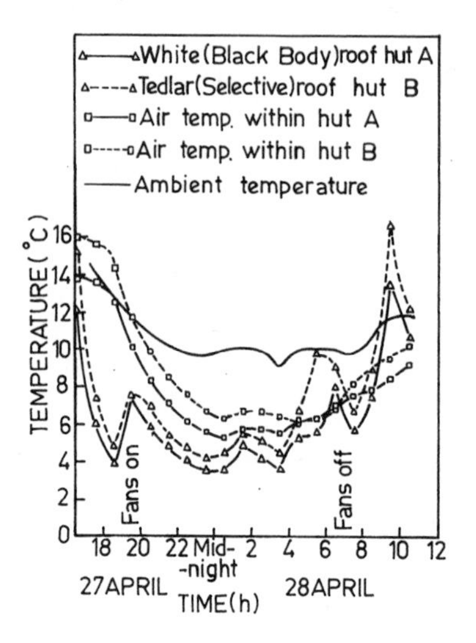

Fig. 7.23. Temperatures recorded during 27-28 April, 1978 at Melbourne (34°N), Australia. (After Michell and Biggs, 1979)

represents a cooling power of about 22 W/m^2. These observations, therefore, establish that a stream of air can be cooled within the room by thermal transfer between the air and a surface which is radiating to the clear sky.

7.2.5 Ventilation

In principle, ventilation introduces the pattern of outside air temperature into the inside environment. Air has a very low thermal capacity, therefore, in a building which is not ventilated, it assumes the temperatures nearly equal to that of the internal surfaces. When ventilation is allowed, air entering the indoor space has outdoor temperature, but in passing through the room it mixes with the indoor air and heat-transfer takes place. As seen earlier, the inside surface temperatures apart from being functions of the thermophysical properties of the structure also depend strongly on the colour of the outside surfaces.

Significant experimental studies showing the effect of ventilation on the internal environment are made in Israel by Givoni (1976). Three ventilation strategies were examined:

(i) No ventilation.

(ii) Ventilation during day and night (permanent ventilation).

(iii) No ventilation during sunshine hours, and ventilation during rest of the time (night ventilation).

The results are presented in Tables 7.15 and 7.16. It may be noted that when colour of the external surface is grey, permanent ventilation has a greater effect on the maxima of indoor temperature than night ventilation. In cases of white-coloured exterior surface, night ventilation is more effective than permanent ventilation, however, the overall effect is seen to be only marginal ($\simeq 1^{o}C$). The quantitative effect of ventilation also depended on the material and thermal capacity of the walls, especially when they were painted grey at the outside surface. Effect of permanent ventilation was more significant for lighter construction, however, night ventilation produced a better performance in the case of heavier construction.

It can be inferred from these studies that during the periods when outside temperature is below the indoor temperature, ventilation provides cooling. The extent of cooling will, however, depend upon the number of air changes per hour (ACH). Croome (1981, has produced a curve (Fig. 5.13) showing the effect of number of air changes on the indoor air temperature for various amount of glazing. It is evident that more than 3 air changes per hour show only a little change in the indoor air temperature. The exact results will in any case depend upon the type of construction and particular climate.

7.3 PERFORMANCE OF PASSIVE HEATING AND COOLING CONCEPTS

There are certain passive concepts which are useful for heating the building in winter and cooling it in summer as well. These are important, especially in the regions of mixed climates (hot summer and temperate cold winters) where the winter and summer seasons spread over a fair length of time period. These concepts have already been described in Chapter 4. An indication of their performance under actual and realistic conditions is given in the present section.

7.3.1 Roof-Pond/Sky Therm

The performance of this concept depends upon the radiation losses to the sky during night hours (for cooling), and storage of heat in water during daytime (for heating). The cooling performance can further be enhanced by sprinkling water over the plastic bags in summer. Hay and Yellot (1969) have carried out extensive experiments at Phoenix, Arizona, with the roof-pond system meant for providing thermal comfort all over the year. A detail of the experimental set-up, and the cooling performance of the system is given in Chapter 8. During the winter months (December and January), heating performance of the system is presented in Fig. 7.24. It can be seen that except for a time period of December 11-22, room temperature is always above 21°C. During this period, the outdoor temperature always remained below 10°C, and the amount of incident solar radiation was only about one-third of its value on other days. It was concluded that in winter the test building, whenever there was normal sunshine, could be maintained at indoor air temperature which has 17°C above the outdoor temperature. In summer,

TABLE 7.15 Effect of Ventilation on the Temperatures of the Indoor Air and Western Surface (Deviations from Maximum Outdoor Air Temperature, °C). (After Givoni, 1976)

External colour	Point	Ventilation conditions	Material and thickness						
			Concrete 12 cm	Concrete 22 cm	Hollow concrete blocks 20 cm	Ytong* 12 cm	Ytong 22 cm	Ordinary curtain walls 7 cm	Insulated curtain walls 16 cm
Grey	Indoor air	Without ventilation	8.0	3.3	3.5	4.6	2.2	5.8	5.1
		Night ventilation	7.2	1.9	1.7	3.4	0.2	4.8	3.7
		Permanent ventilation	1.0	-0.4	-0.3	0.5	-0.2	0.3	0.4
		Effect of night ventilation	-0.8	-1.4	-1.8	-1.2	-2.0	-1.8	-1.4
		Effect of permanent ventilation	-7.0	-3.7	-3.8	-.41	-2.4	-5.5	-4.7
	Western surface	Without ventilation	13.2	6.3	6.8	7.8	2.7	9.3	6.6
		Night ventilation	12.9	4.6	5.2	6.9	0.7	8.4	4.3
		Permanent ventilation	10.1	3.5	2.8	4.5	0.1	3.7	1.7
		Effect of night ventilation	-0.3	-1.7	-1.6	-0.9	2.0	0.9	-2.3
		Effect of permanent ventilation	-3.1	-2.8	-5.0	-3.3	2.6	-5.6	-4.9
White	Indoor air	Without ventilation	-1.0	-1.8	-2.3	1.8	3.2	1.1	-1.9
		Night ventilation	-2.7	-2.5	-2.8	2.7	4.8	0.5	-3.1
		Permanent ventilation	-1.1	-1.4	-1.4	0.8	2.0	0.0	-1.0
		Effect of night ventilation	-1.7	-0.7	-0.5	0.9	1.6	0.6	-1.2
		Effect of permanent ventilation	-0.1	0.4	0.9	1.0	1.2	1.1	0.9
	Western surface	Without ventilation	-1.1	-1.6	-2.2	1.7	3.1	1.2	-1.8
		Night ventilation	-2.1	-2.8	-2.8	2.2	4.5	0.3	-2.7
		Permanent ventilation	1.9	-3.1	-2.5	0.8	3.6	0.4	-0.9
		Effect of night ventilation	-1.0	1.2	0.6	0.5	1.4	0.9	0.9
		Effect of permanent ventilation	-0.8	-1.5	-0.3	0.9	-0.5	0.8	0.9

*Ytong refers to lightweight concrete.

TABLE 7.16 Effect of Ventilation on Minimum Temperatures of Indoor Air and Western Surfaces (Deviations from the Minimum Outdoor Air Temperature, °C). (After Givoni, 1976)

External colour	Point	Ventilation conditions	Material and thickness						
			Concrete 12 cm	Concrete 22 cm	Hollow concrete blocks 20 cm	Ytong 12 cm	Ytong 22 cm	Ordinary curtain walls 7 cm	Insulated curtain walls 16 cm
Grey	Indoor air	Without ventilation	2.6	4.4	4.4	2.3	4.6	2.0	3.4
		Night ventilation	1.0	2.2	1.7	1.0	1.4	0.8	0.9
		Permanent ventilation	0.6	1.6	1.2	0.7	0.9	0.7	0.7
		Effect of night ventilation	-1.6	-2.2	-2.7	-1.3	-3.2	-1.2	-2.5
		Effect of permanent ventilation	-2.0	-1.8	-3.2	-1.6	-3.7	-1.3	-2.7
	Western surface	Without ventilation	2.1	4.6	4.6	1.7	4.7	1.3	2.9
		Night ventilation	1.7	4.0	3.0	1.2	2.9	0.6	0.9
		Permanent ventilation	0.9	3.1	2.0	0.6	2.1	0.4	0.9
		Effect of night ventilation	-0.4	-0.6	-1.6	-0.5	-1.8	-0.7	-2.0
		Effect of permanent ventilation	-1.2	-1.5	-2.6	-1.1	-2.6	-0.9	-2.0
White	Indoor air	Without ventilation	2.2	3.7	3.4	2.5	3.4	2.2	3.5
		Night ventilation	0.8	1.7	1.9	1.1	1.0	1.7	0.9
		Permanent ventilation	0.2	0.8	1.1	0.3	0.7	1.0	0.3
		Effect of night ventilation	-1.4	-2.0	-1.5	-1.4	-2.4	-0.5	-2.6
		Effect of permanent ventilation	-2.0	-2.9	-2.3	-2.2	-2.7	-1.2	-3.2
	Western surface	Without ventilation	2.2	3.7	3.3	2.5	3.3	2.2	3.4
		Night ventilation	0.4	2.3	2.1	0.5	1.7	1.0	0.3
		Permanent ventilation	-0.2	1.3	1.1	-0.2	0.8	0.5	-0.1
		Effect of night ventilation	-1.8	-1.4	-1.2	-2.0	-1.6	-1.2	-3.1
		Effect of permanent ventilation	-2.4	-2.4	-2.2	-2.7	-2.7	-1.7	-3.5

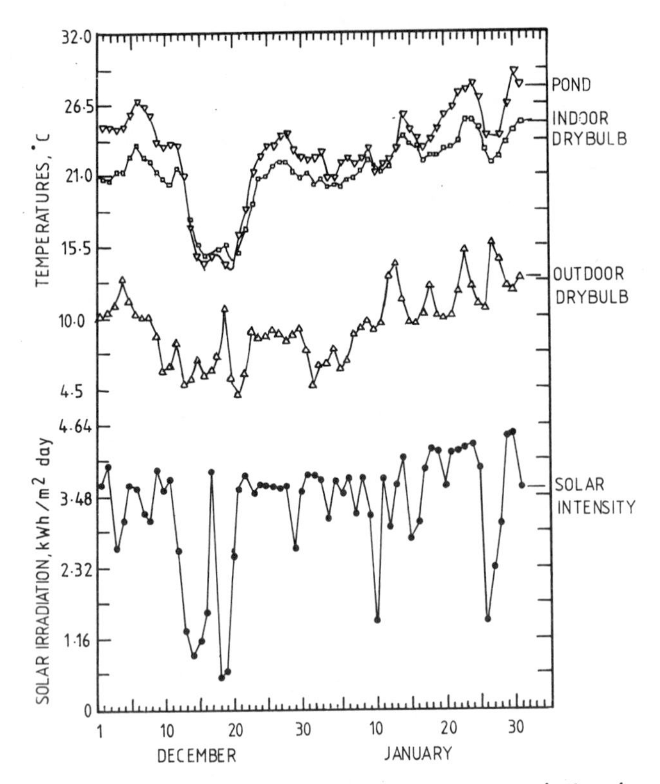

Fig. 7.24. Average temperatures of a roof-pond system in heating mode at Phoenix during December 1967 and January 1968. (After Hay and Yellot, 1969)

the building was kept within the comfort zone, or at worst in the slightly warm conditions by the cooling effect of the pond's water.

Skytherm concept developed by Hay in 1965, consists of a horizontal roof made up of steel plates and treated against rust. The operation of the concept is illustrated in Chapter 4 (Fig. 4.23). A full scale house conditioned by this concept was built in Atascadera, California, and was tested extensively. Haggard *et al.* (1975) has reported the thermal performance of this house. It was observed that on a typical winter day, when the outdoor temperature ranged from a minimum of 3°C to a maximum of 21°C, the indoor temperature varied only between 20°C and 24°C. On a typical summer day, the variation in outdoor temperature was noted to be from 13°C to 34°C, however, the indoor temperature ranged from 21°C to 23°C only.

Skytherm concept has its own limitations and probable difficulties in its implementation. Some of these are:

(i) the deterioration of the plastic with time reducing its transparency and hence the effectiveness of the concept.

(ii) corrosion of the metallic-roof, which may reduce the lifetime considerably.

(iii) susceptibility of the plastic bags against wind.

(iv) accumulation of the dust which is difficult to clean off.

Technically, however, skytherm concept is suitable for regions with relatively mild climates, especially in winter, and in regions of low wind velocities and dust.

Haggard (1977) has proposed several architectural applications of the roof-ponds and movable insulation systems Fig. 7.25. The classification is based on the climatic data of California, USA which represents a region where a wide range of variations are witnessed. Although the classification is very general and requires a great amount of design and analytical work, it illustrates the potential of the concepts for wider applicability.

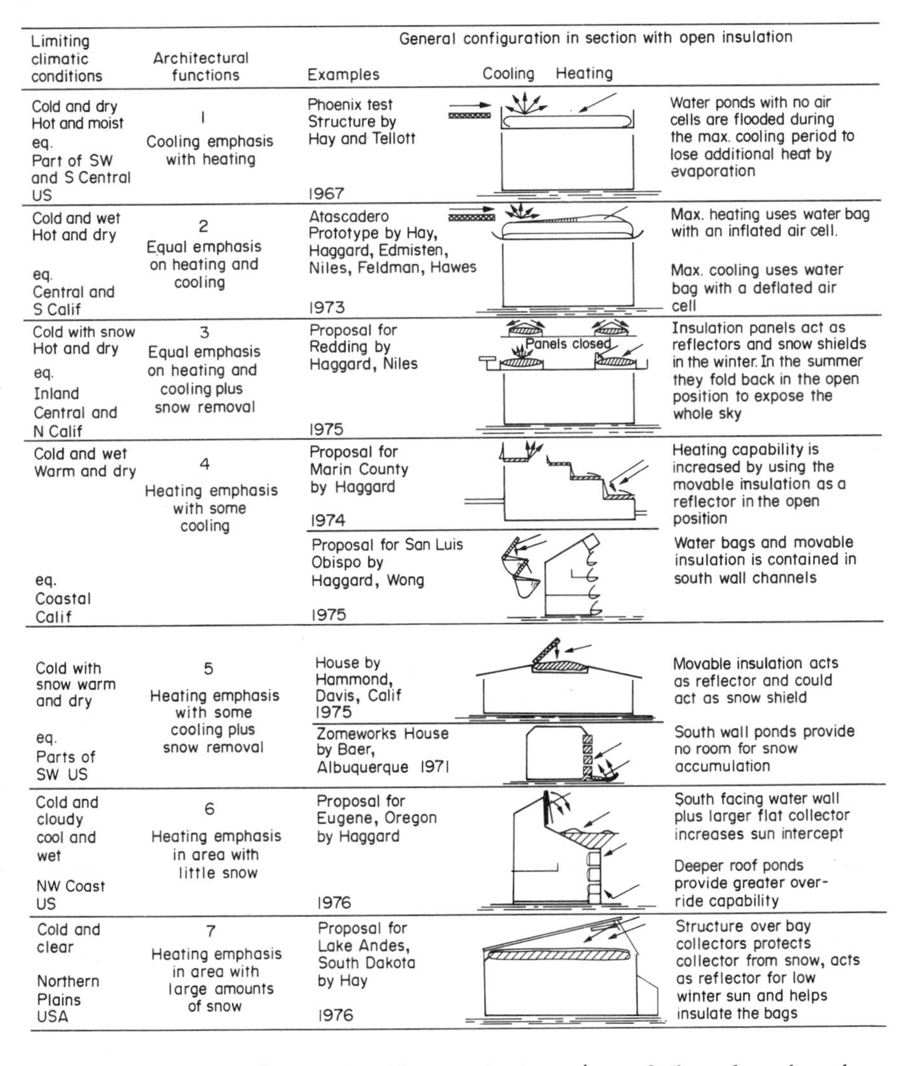

Limiting climatic conditions	Architectural functions	General configuration in section with open insulation: Examples	Cooling Heating	
Cold and dry Hot and moist eq. Part of SW and S Central US	1 Cooling emphasis with heating	Phoenix test Structure by Hay and Tellott 1967		Water ponds with no air cells are flooded during the max. cooling period to lose additional heat by evaporation
Cold and wet Hot and dry eq. Central and S Calif	2 Equal emphasis on heating and cooling	Atascadero Prototype by Hay, Haggard, Edmisten, Niles, Feldman, Hawes 1973		Max. heating uses water bag with an inflated air cell. Max. cooling uses water bag with a deflated air cell
Cold with snow Hot and dry eq. Inland Central and N Calif	3 Equal emphasis on heating and cooling plus snow removal	Proposal for Redding by Haggard, Niles 1975	Panels closed	Insulation panels act as reflectors and snow shields in the winter. In the summer they fold back in the open position to expose the whole sky
Cold and wet Warm and dry eq. Coastal Calif	4 Heating emphasis with some cooling	Proposal for Marin County by Haggard 1974		Heating capability is increased by using the movable insulation as a reflector in the open position
		Proposal for San Luis Obispo by Haggard, Wong 1975		Water bags and movable insulation is contained in south wall channels
Cold with snow warm and dry eq. Parts of SW US	5 Heating emphasis with some cooling plus snow removal	House by Hammond, Davis, Calif 1975		Movable insulation acts as reflector and could act as snow shield
		Zomeworks House by Baer, Albuquerque 1971		South wall ponds provide no room for snow accumulation
Cold and cloudy cool and wet NW Coast US	6 Heating emphasis in area with little snow	Proposal for Eugene, Oregon by Haggard 1976		South facing water wall plus larger flat collector increases sun intercept Deeper roof ponds provide greater over-ride capability
Cold and clear Northern Plains USA	7 Heating emphasis in area with large amounts of snow	Proposal for Lake Andes, South Dakota by Hay 1976		Structure over bay collectors protects collector from snow, acts as reflector for low winter sun and helps insulate the bags

Fig. 7.25. Comparisons of architectural adaptations of thermal ponds and movable insulation. (After Haggard, 1977)

7.3.2 Vary Therm Wall

The thermal resistivity of the walls can be changed by introducing an air cavity between the wall and outer layer, made from any light material like aluminium, concrete or wood and controlling the flow of air into the room or to the ambient by providing proper vents. Vary Therm Wall deriving its name from the variable resistance can be operated in three modes

1. No flow of air in the gap thus effectively reducing the system to an air gap within the wall.

2. Continuous flow of air into the room or to the atmosphere maintained by natural or forced convection.

3. No air flow during the day (or night) and creating an air flow by opening the vents during night (or day) time depending on the weather conditions as illustrated in Fig. 7.26.

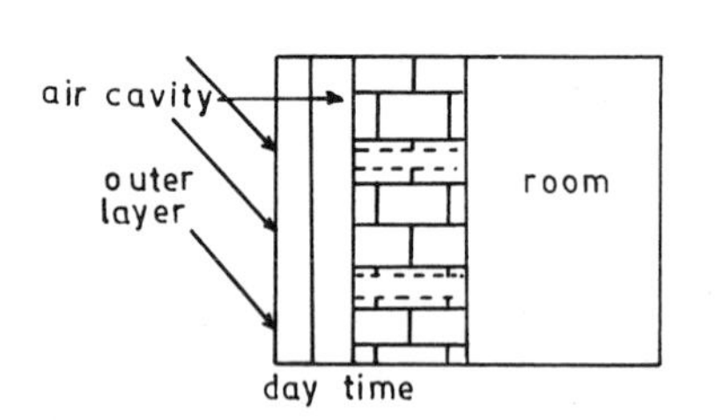

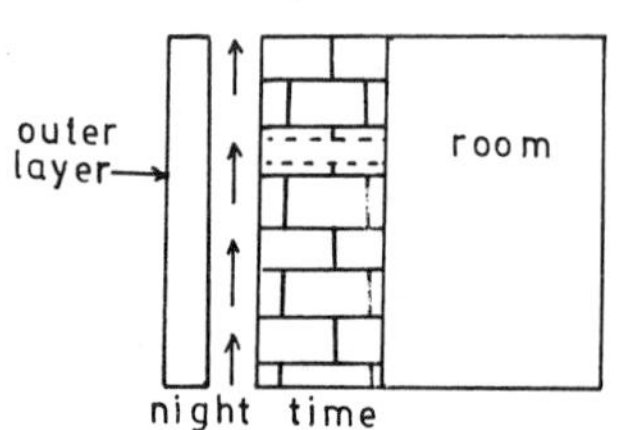

a. Summer period.

warm air

room

day time

outer layer

cavity

room

night time

b. Winter period.

Fig. 7.26. Summer and winter operations of vary therm wall

Experimental measurements on small test rooms, of volume (1.2 x 1.2 x 1.0) m^3, show that it brought an increase of 6.7°C in the minimum temperatures during winter, while in the summer the peak temperature dropped by a magnitude of 7 to 9°C. The performance for a typical day in February is shown in Fig. 7.27 while Fig. 7.28 shows the performance of this configuration under various modes for a typical summer day in June (maximum temperature 40°C) at Delhi (latitude 28.6°N).

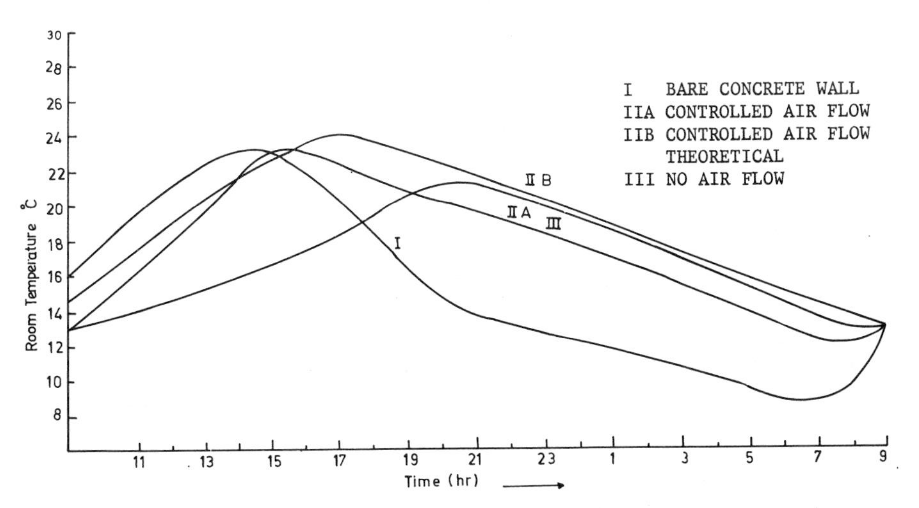

Fig. 7.27. Variation of room air temperature in the case of a vary therm wall, at Delhi, India on February 6, 1983

7.3.3 Roof Radiation Trap System

It is a passive/active system designed to provide winter heating and summer cooling to buildings with one (or two) storey having a flat roof. It is a composite system from passive point of view as it involves radiation cooling, direct-gain system and storage of heat simultaneously. The air, sucked by a fan under corrugations which are cooled by means of radiative losses, is blown either to the living space directly, or to the gravel store which acts as a storage of coolness; the air this way gets cooled even up to 4-6°C below the ambient temperature.

The radiation trap system can be designed as one unit over the whole roof or as a series of separate smaller units covering the whole or part of the roof area. Presently, no experimental data is available, however, some studies are in progress at the Building Research Station, Technion, Israel on small scale models (Givoni, 1977).

7.3.4 Earth-Air Tunnel System

Air tunnels constructed deep inside the ground exploit earth's storage potential, and can effectively be used for heating/cooling of the buildings. A description of these systems, along with their performance is given in Chapter 8 in detail.

7.4 PERFORMANCE OF BUILDING FORMS

Thermal performance of various building forms viz. Pavilion, Court and Street (as discussed in Chapter 3) has been studied by Gupta (1984) who developed an analytical model based on the following assumptions:

(i) The heat-flow through the various elements and the climatological parameters viz. the solar intensity and ambient air temperature are periodic functions of time.

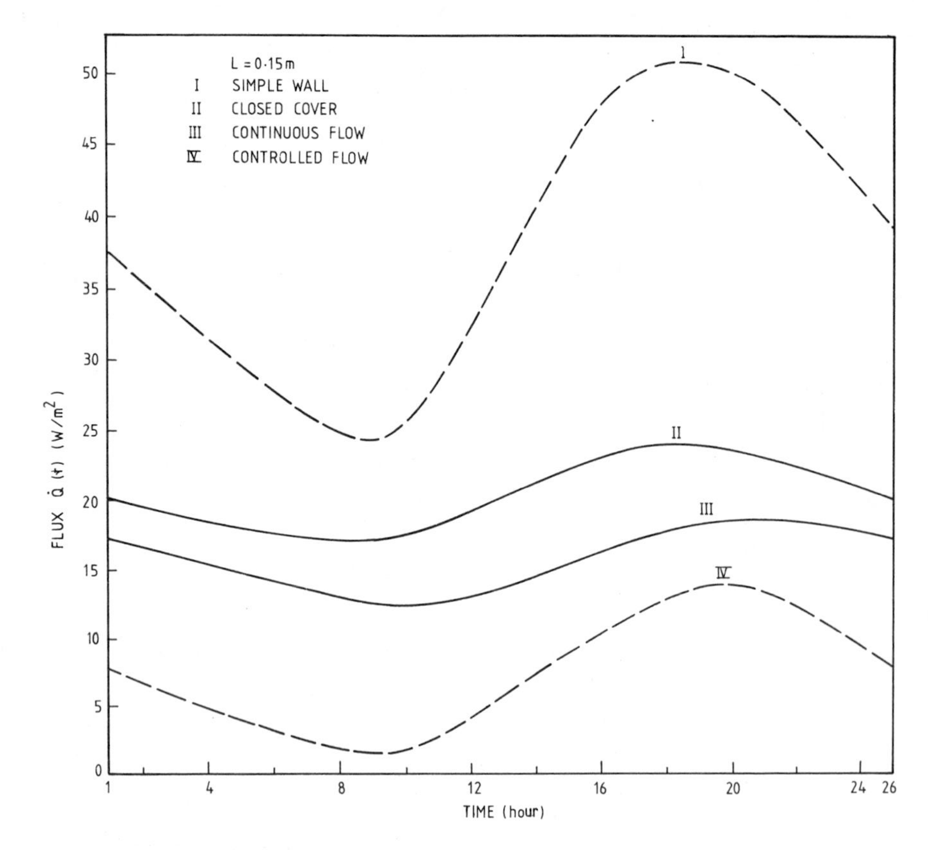

Fig. 7.28. Performance of vary therm wall in summer conditions. (Maximum air temperature = 40°C) at Delhi, India

(ii) Because of the difference in solar irradiance, the thermal behaviour of blocks in a cluster varies from the periphery (where there is no shading) to the centre (where there is shading). An average solar exposure equal to the total solar exposure of the cluster divided by the number of blocks has been assumed for each block.

(iii) The single building block has been assumed to be consisted of four parts viz. (i) roof, (ii) sunlit walls, (iii) shaded walls, and (iv) windows. Area of the sunlit part of the walls is assumed to be constant and the intensity of incident solar radiation is adjusted to give the calculated solar exposure for the partially sunlit surface.

(iv) The entire thermal mass of the building (floors and partitions, etc.) is assumed to be separated from the building envelope so that the heat transfer between the two is only convective and radiative. The effect of increased cellurity in deep plan buildings has been neglected, i.e. the building is considered as a single thermal zone.

(v) The space around the buildings, whether Street or Court, is assumed to be at a uniform temperature.

(vi) In contrast to the normal practice where it is assumed that the radiative heat loss from the walls is negligible because it "sees" other walls at similar temperatures, the actual area of sky "seen" by walls depending upon the portion of Streets or Courts is taken into account.

(vii) Floor of the building is considered to be a semi-infinite medium.

The physical and solar characteristics of the above-mentioned three building clusters, for which the numerical calculations were made, are given in Table 7.17. The results of the computations are presented in Figs. 7.29–7.31. It can be seen that for the Pavilion (Fig. 7.29) the fluctuation of air temperature is much greater than for the Court (Fig. 7.30) and Street (Fig. 7.31). This is probably due to the fact that the surface area of the Pavilion cluster is nearly twice that of either Court or Street. In all cases the effect of shading is to lower the internal temperatures while ventilation is accompanied by lower temperatures and an increased amplitude. Evidently, the thermal capacity of the structure does not have a significant effect on the performance, and therefore it would be advantageous in hot climates to have a larger surface area of buildings which did not have mechanical heating and cooling. As reported, the Pavilion form shows the best performance because its solar exposure per unit surface area is the lowest. The Court, which has least surface area as well as solar exposure, has the highest solar exposure per unit area and therefore performs poorly.

TABLE 7.17 Physical and Solar Characteristics of Three Typical Building Clusters. (After Gupta, 1984)

	PAVILION	COURT	STREET
1. PHYSICAL PROPERTIES:			
(a) Number of subdivisions	24	8	32
(b) Height (m)	12	12	12
(c) Width (m)	10.758	10	10
(d) Length (m)	10.758	36.042	208.33
(e) Roof area (sq. m.)	66668	66668	66668
(f) Wall area (external)	297435	160000	172916
(g) Total external surface area (m^2)	364103	226668	239584
(h) Surface area/volume	0.45	0.28	0.30
(i) Internal partition area (m^2)	697437	867202	867200
(j) Window area (m^2)	40000	40000	40000
2. SOLAR PROPERTIES:			
(a) Daily total summer solar exposure (m^2)	586709	493364	437488
(b) Efficiency E_s*	47.2	59.5	61.78
(c) Solar exposure/surface area	1.611	2.177	1.826
(d) Equivalent radiative loss area (with HW = 2.0) m^2	123180	100268	101251

*defined in Chapter 3.

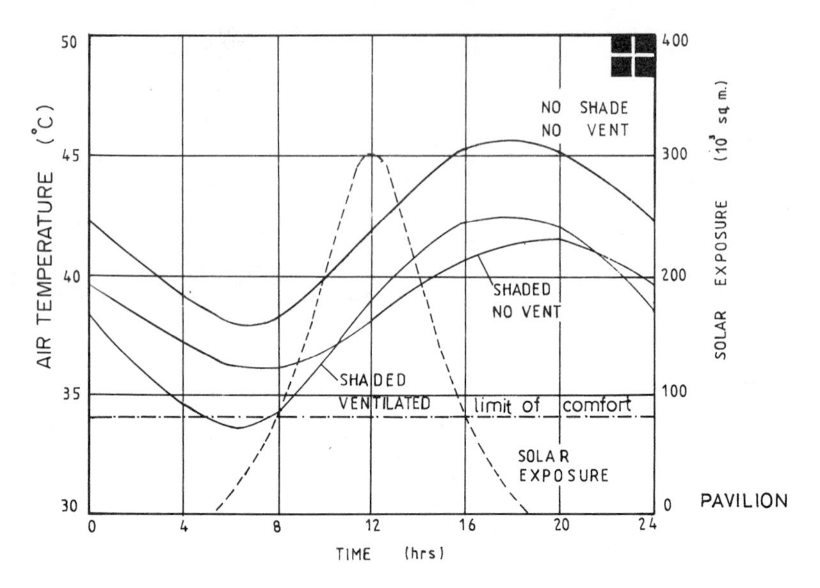

Fig. 7.29. Air temperatures and solar exposure for the Pavilion. (After Gupta, 1984)

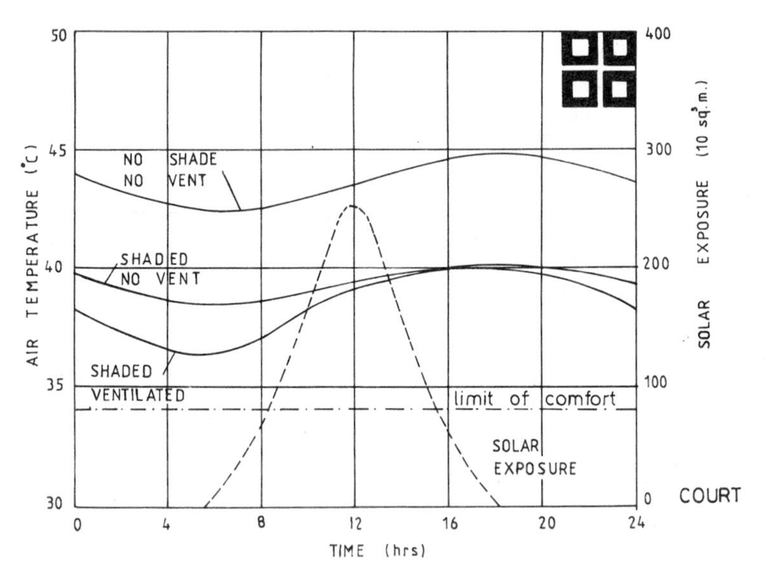

Fig. 7.30. Air temperatures and solar exposure for the Court. (After Gupta, 1984)

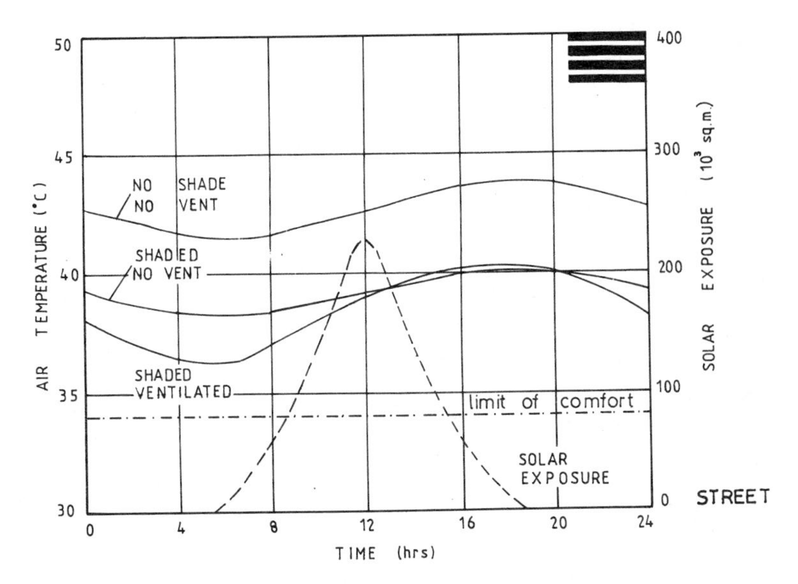

Fig. 7.31. Air temperatures and solar exposures for the Street. (After Gupta, 1984)

REFERENCES

Ahmad, A. M. (1980) The thermal performance of concrete roofs and reed shading panels under arid summer conditions, Building in Hot Climates - A selection of Overseas Building Notes, Prepared by the Overseas Division of the Building Research Establishment, UK, 401-407.

Ballantyne, E. R. and Spencer, J. W. (1961) Comparison of the thermal properties of some double glazed window units, Private Communication.

Baehr, A. and Piwecki, H. (1981) Passive solar heating with heat storage in the outside walls, Commission of the European Communities, Report No. EVR 7077 EN, Joint Research Centre, Ispra Establishment, Italy, page 24.

Balcomb, J. D., Hedstrom, J. C. and McFarland, R. D. (1977) Simulation analysis of passive solar heated buildings - Preliminary results, *Solar Energy, 19*, 277.

Balcomb, J. D. (1981) Passive solar energy systems for buildings, Chapter 16, In *Solar Energy Handbook*, Edited by Kreith and Kreider, McGraw Hill, NY.

Croome, D. J., Roberts, B. M. (1981) Airconditioning and Ventilation of Buildings, Pergamon Press, New York.

Fuchs, R. and McClelland, J. F. (1979) Passive solar heating of buildings using a transwall structure, *Solar Energy, 23*, 123.

Givoni, B. and Hoffman, M. E. (1973) Effect of roof design on indoor climate in hot arid zones, *Build. International, 6*, 525.

Givoni, B. (1976) *Man, climate and architecture*, Applied Science Publishers Ltd., London, Second Edition, page 483.

Givoni, B. (1977) Solar heating and night radiation cooling by a roof radiation trap, *Energy and Building, 1*, 141.

Gupta, V. K. (1984) A study of the natural cooling systems at Jaisalmer, Ph.D. Thesis, ITT, New Delhi, India.

Haggard, K. L. *et al.* (1975) Research evaluation of a system of natural air conditioning, Report by California Institute, San Luis Obispo, USA.

Haggard, K. L. (1977) The architecture of a passive system of diurnal radiation heating and cooling, *Solar Energy, 19*, 403.

Hay, H. R. and Yellot, I. J. (1969) Natural air conditioning with roof ponds and moveable insulation, *ASHRAE Trans. 75*, (Part I).

Lotz, F. J. and Richards, S. J. (1964) The influence of ceiling insulation on indoor thermal conditions in dwellings of heavy weight construction under South African conditions, CSIR Research Report No. 214, Pretoria, South Africa.

Mazaria, M., Baker, M. S. and Wessling, F. C. (1977) An analytical model for passive solar heated buildings, PAM 10.

Michell, D. and Biggs, K. L. (1979) Radiation cooling of buildings at night, *Applied Energy, 6*, 263.

Nayak, J. K., Srivastava, A., Singh, U. and Sodha, M. S. (1982) The relative performance of different approaches to the passive cooling of roofs, *Building and Environment, 17*(2), 145.

Nayak, J. K., Bansal, N. K. and Sodha, M. S. (1983) Analysis of passive heating concepts, *Solar Energy, 30*, 51.

Perry Jr., J. E. (1977) Mathematical modelling of the performance of passive solar heating system, Los Alamos Scientific Laboratory of the University of California, Los Alamos, New Mexico 875 45, Report No. LA-UR-77-2345, Page 37.

Pratt, A. W. (1981) *Heat Transmission in Buildings*, John Wiley and Sons, New York.

Singh, U., Kumar, A., Srivastava, A. and Tiwari, G. N. (1981) Passive cooling of buildings using multi-air-gaps, *Applied Energy, 9*(1), 33.

Singh, U. (1982) Analysis of some solar passive concepts, Ph.D. thesis, IIT Delhi, India, Page 157.

Sodha, M. S., Kaushik, S. C., Tiwari, G. N., Goyal, I. C., Malik, M. A. S. and Khatry, A. K. (1979) Optimum distribution of insulation inside and outside the roof, *Building and Environment, 14*, 47.

Sodha, M. S., S. P. Seth, N. K. Bansal, and A. K. Seth (1981a) Optimum distribution of concrete in a double hollow concrete slab, *Int. J. Energy Research, 5*, 289-295.

Sodha, M. S., S. P. Seth, N. K. Bansal and A. K. Seth (1981b). Comparison of the thermal performances of single and double hollow concrete slabs, *Applied Energy, 9*, 201-209.

Sodha, M. S., Singh, U., Kumar, A. and Tiwari, G. N. (1981c) Effect of using a reflecting sheet in an air-gap on the thermal performance of hollow walls and roofs, *Applied Energy, 8*, 67.

Sodha, M. S., Nayak, J. K., Bansal, N. K. and Goyal, I. C. (1982) Thermal performance of a solarium with removable insulation, *Building and Environment, 17*(1), 23-32.

van Straatan, J. F. (1967) *Thermal performance of buildings*, Elsevier Publishing Company, New York.

Wray, W. O. and Balcomb, D. (1979) Trombe wall vs. Direct gain: A comparative analysis of passive solar heating systems, Los Alamos Scientific Laboratory of the University of California, Los Alamos, New Mexico 87 545, USA, Report No. LA-UR-79-116, Page 7.

Wray, W. O. (1981) Design and analysis of direct gain solar heated buildings, Los-Alamos Scientific Laboratory, University of California, Los Alamos, New Mexico 97545, USA, Report No. LA-8885-MS, Page 113.

Chapter 8

EVAPORATIVE COOLING AND EARTH COOLING SYSTEMS

8.1 EVAPORATIVE COOLING

The state-of-the art for passive cooling of buildings lags behind that of heating. Providing cooling of the space inside the unconditioned building by passive means in the regions of hot and dry climates poses considerable problems to the building designers. Traditional methods to reduce the heat gains through the building viz. increasing the roof thickness, providing a false ceiling, roof shading and the use of reflecting paints help to reduce the cooling load to some extent, but these methods are ineffective in removing the heat which is present inside the building space due to various sources, like, metabolism, lighting, electrical appliances etc. In contrast, evaporative cooling has emerged as a technically viable and economically attractive method to maintain comfortable room temperatures in dry-hot zones. The basic principle behind evaporative cooling is the conversion of sensible heat into latent heat which causes evaporation of water. In nature the evaporative cooling occurs near water surfaces, like, waterfalls, flowing streams, lakes, oceans, and even upon wet skin. Most primitive humans observed it which resulted in the useful exploitation of the concept in many regions, especially, in the Near East, where hot, arid climate provided incentive as well as favourable conditions for its operation. The present chapter attempts to trace down the history of developments in evaporative cooling and discuss the feasibility, evaluation and performance of the concept in detail.

8.1.1 Historical Background

Evaporative cooling represents the oldest application of the air-conditioning principles. In its primitive age it flourished mainly in lightly populated, hot desert areas. It was known to the ancient Egyptians dating back about 2500 *B.C.* In India, the technique was even used to make ice. According to folklore, the king's throne in the Red Fort at Delhi was kept cool by keeping ice in a pool around it. Ice used to be made in the 19th century in Delhi in winter and stored till summer. The simple process involved the cooling of water by evaporation and by radiation in a shallow metal pan at night. The bottom of the pan was insulated from the ground by placing it upon a thick bed of blackish straw. The essential conditions were: a clear sky, still atmosphere and air temperature less than $6^{o}C$.

In Iran, the buildings are often cooled by this technique. These buildings contain pools of running water, the surface of which receives the outside air through a wind-catcher opening above it. The outside air is diverted across the water surface and into the subterranean rooms.

Another ancient practice, commonly used by American Indians, was to cover the tents with wet felt. However, in India the old *cocos-tatti* is still in great use. The *tatties* covered with dried *khuss-khuss* grass are hung in the doors/windows, and kept wet either by using recirculating pumps or manually.

The primitive efforts in the West were accomplished in covering a window, occasionally, with moistened burlap or drapes in doorways, or bed-sheets sprinkled with water for cooler sleep in the night. The arrangement's effectiveness depended, of course, upon the breeze.

The efforts made in modern times to achieve evaporative cooling, resulted in the mechanical evaporative coolers. Air washers and textile mill evaporative cooling, which did not have any notable antecedents in the ancient times, were developed for the first time in New England during 1900-1930. The attempts made around 1930 in USA, which resulted in the development of drip and slinger type evaporative coolers, were the derivatives of the earlier successfully tried concepts of cooling tower and the one practised by Indians and Mexicans, respectively. These air-washers, however, were replaced by an evaporative cooler designed by W. H. Carrier. These were spray chambers featuring multiple, pressure-type spray nozzles and eliminator plates which allowed air to pass through but kept the mist and spray confined. During the intermediate stages of the development, the most crucial function of designing these coolers was the humidity control. This function had special relevance in the context of textile mills where high indoor-humidity was required in summer.

Some other types of coolers also emerged out of the attempts put to avoid the addition of water vapour to the indoor environment. This resulted in the development of various kinds of heat exchangers viz. the plate and coil type heat exchangers of various buildings in Arizona in 1935. Such coolers were used on the rooftops of various buildings in Arizona in 1935, however, in the later course of development they were found to be vulnerable to hard water. The initial cost of this type of cooler was very high, and hence only the industries could use them.

Simultaneously, some investigations were also made to find the passive means of cooling, utilizing the same concept. Houghten *et al.* in 1940, and Hay and Yellot in the late sixties of 20th century made serious scientific efforts in this direction. Later on, the expanding economics of the conventional evaporative coolers, and their dependence on conventional fuels triggered off an additional interest in the methods of passive cooling. A considerable amount of work in this direction has been done at the Central Building Research Institute, Roorkee, India which will be discussed later.

8.1.2 Basic Principle and Classification

Evaporation is a process which is called to occur when liquid (in the present case, water) changes its state to a vapour. When water and unsaturated air are in free contact, heat and moisture transfer takes place between them. Some of the air's sensible heat is transferred to water and gets converted to latent heat of vaporization through the process of evaporation. This results in the lowering of dry-bulb temperature of the air and an increase in its humidity contents. The decrease in air temperature is directly proportional to the change in the amount of water vapour contained in it. It is also called adiabatic cooling because no external heat is involved and evaporation occurs purely by conversion of the air's existing sensible heat. This process continues until the temperatures and vapour pressure of the two (water and air) equalize.

On a psychrometric chart, the cooling process may be illustrated as shown in Fig.

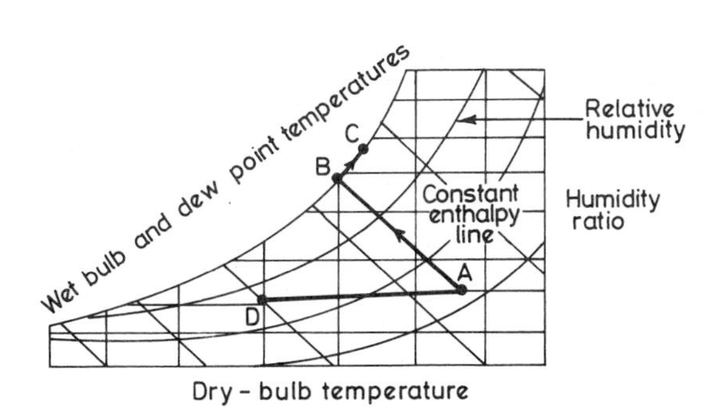

Fig. 8.1.

8.1. Point A represents the outdoor air entering a cooler. Point B represents the wet-bulb temperature of the outdoor air. When water evaporates into the air, the cooling occurs from point A to point B along a constant enthalpy line. This is true only if the water remains at the wet-bulb temperature of the outdoor air. However, in practical situations, water usually gains some heat because of the circulatory friction, heat transfer from the surroundings, and the incident solar radiation. The actual temperatures, usually reached, therefore, correspond to point C. The path in the figure shows sensible cooling which is along the constant humidity line.

Depending upon the choice of operating system, the humidity of the room air may or may not increase. Accordingly, the process may be classified in the following two categories:

(a) Direct evaporative cooling. When the room air is in direct contact with water surface, the evaporation of water into the air increases its humidity. This process is called the direct evaporative cooling. Several designs of direct-type evaporative coolers are in vogue, some of which are named below:

(i) Drip-type evaporative coolers,

(ii) Spray-type evaporative coolers,

(iii) Rotary pad evaporative coolers, and

(iv) Textile mill evaporative coolers.

(b) Indirect evaporative cooling. In this process, the evaporation occurs separately and air/water is cooled without humidity gain. The resultant cool air/water is then used to cool the room air. Following types of indirect evaporative cooling systems are existing:

(i) Simple dry surface,

(ii) Regenerative dry surface,

(iii) Plate-type heat exchanger,

(iv) Pennington heat-storage wheel

(v) Gafford "run around",

(vi) Indirect evaporative radiant, and

(vii) Two-stage indirect evaporative.

It may not be out of place here to define *saturation efficiency* which is usually used for rating the performance of various evaporative coolers. It normally depends upon the density of the pads used in the coolers, and may be calculated by the following expression:

$$\text{Saturation efficiency} = \frac{(\theta_1 - \theta_2)}{(\theta_1 - \theta_3)}$$

where θ_1, θ_2, and θ_3 represent the dry-bulb temperature of the entering and leaving air, and the wet-bulb temperature of the entering air, respectively. For denser pads and low air velocities, saturation efficiency is more, however, it has to be chosen in accordance with the requirements of indoor environment.

8.1.3 Climatic Conditions for Evaporative Cooling

In order to utilize the concept of evaporative cooling effectively, an analysis, based upon the comfort chart, should first be made to investigate whether the given climatic conditions permit achievement of thermal comfort. Earlier, the regions where wet-bulb temperature seldom exceeded 24°C and dry-bulb temperature was usually above 32°C, were considered the ideal ones for using evaporative coolers. However, later studies indicated that thermal comfort inside the buildings can be achieved in most of the climates, provided high indoor air velocities and relative humdities are acceptable. In the absence of any satisfactory definition for defining a climate which would lead to indoor thermal comfort conditions, Watt (1963) has proposed certain limitations which can be checked before recommending for an evaporative cooler. These are as follows:

(i) Direct evaporative coolers should have an average saturation efficiency of 70% or more, and the cooled air enters the indoor space without any heat gain.

(ii) The maximum indoor air-velocity induced by the cooled air is 1 m/s.

(iii) The cooled air should be able to raise its temperature at least by 3°C before its discharge.

(iv) The temperature of indoor space is about 2°C higher than the discharge air temperature, and humidity is 70% or below.

(v) The temperature of the cooled space is about 4°C below the outdoor dry-bulb temperature. This is necessary to counteract the entering radiant heat.

8.1.4 Direct-Type Evaporative Coolers

A number of designs are available for direct type coolers which range from drip-type to slinger and rotary types. All these coolers are useful for rooms, apartments, commercial buildings and even for larger buildings, however, sometimes the latter type has an edge over the former because it requires less attention and maintenance. All these coolers cool large volumes of outside air by evaporating water into it. This cooled air is then delivered to indoors where it absorbs heat from walls, ceilings, furnishings and the occupants, and finally it is discharged to outdoor. A brief discussion about the design of these coolers is given below:

8.1.4.1 Drip-type evaporative coolers. These coolers, also known as "desert coolers", contain a wetting pad, a water circulating pump and a fan (Fig. 8.2). The water pump lifts the sump water up to a distributing system from which it

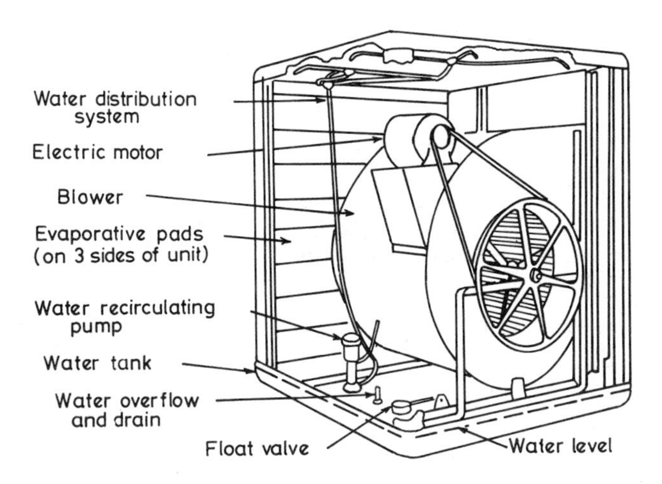

Fig. 8.2. Wetted pad-type evaporative cooler. (After Ashrae, 1972)

runs down through the pads and back into the sump. The wetting pad, usually made of aspen wood fibres, is stiffened on three sides by the co lers' walls in such a way that air enters through the pads only. Above the base of the cooler, the fan is mounted which may be a propeller type in the inexpensive units, and is a centrifugal blower in a good system. The fan discharge is either through the side of the cooler (the one in which a pad is not used) cabinet which is mounted on the outside wall of a building or through the sump bottom in case of room installations.

In these coolers, the choice of the evaporating pad is a critical factor in determining their performance. The coolers are usually designed for a face velocity of 1 m/s to 1.5 m/s with a pressure drop of about 30 N/m^2. In addition to providing cooling of the incoming air, the pads also act as air filters, and prevents the entry of the particles having a size greater than 10 μm. In order to increase the absorption of water and to prevent the growth of bacteria, fungi and other microorganisms, the pads are usually chemically treated. Sometimes when the cooler is used only for ventilating purposes, the supplementary fiber glass filters are also used. These coolers are usually made in sizes of 3400 m^3/hr to 3400 m^3/h of cooled air.

Water to the pads is supplied through the water troughs which are designed to ensure a uniform flow of water on all the pads. If water supply is good, all fresh water may be used with no recirculator. Also, the material used in the construction of pump, sump, water-distribution system and the casing of the system should necessarily be corrosion- resistive.

Some other designs of drip-type coolers have also been forwarded which include the "portable drip-type" cooler and the "exhaust-type window pad cooler". Former is used at the places where no fresh water supply is available. These types of coolers are the most inefficient ones. On the other hand, the exhaust-type window pad cooler is the most efficient evaporative unit, because it reduces the heat-gain of the building as well and removes the heat already present in the building. In this system, the outside of one or more windows is covered by a flat evaporating pad which may be stiffened to the louvered metal sheet, and the air is pulled inside through them using a large exhaust fan. This cool air sweeps the inhabited space and escapes through the ceiling shutters.

8.1.4.2 Spray-type evaporative coolers. These coolers are less popular than the drip-type coolers, and are usually made in the capacities of providing 5100-34000 m^3/hr of the cooled air. These contain a slinger pump which throws a spray of water into the air and on to the evaporative pad, and hence are also known as "slinger-type evaporative coolers". This action of slinger wets the evaporative pad and also washes the dirt and other foreign material on the pad into the tank. Slinger pump usually consists of a fractional horsepower motor and a water wheel that throws about 19 litres per minute. In big coolers, two slinger wheels are driven by a double shaft motor.

The outside air is drawn through the evaporative pad which is placed vertically to the direction of air-flow. This air picks up evaporated moisture from the pad and gets cooled. It is to be noted here that some of the water droplets are also carried away by this air which are usually eliminated by providing an additional pad, called eliminator pad (Fig. 8.3), at a distance of about 50-80 mm from evaporative pad. The evaporative and eliminator pads are usually the screen type made up of copper alloy. These have a thickness of the order of 30-50 mm and 25-40 mm respectively. The fan used to draw outside air is, in general, of centrifugal type, and is mounted in the cubical metal cabinets whose rear walls are either open or covered by screen, louvered panels, or dust filters. The pressure drop to air-flow in these types of coolers is approximately 61 N/m^2. These coolers are normally designed for a face air velocity of about 1.5 m/s. The air handling part of the cooler is similar to that of drip-type coolers.

In order to keep the concentration of minerals in the sump, an adjustable skimmer bleed-off may also be inducted in the system. During installation, it should be ensured that the cooler is properly levelled which is a necessary condition for the operation of the slinger.

It has been reported in the literature that capital and running costs of these coolers are more than that of drip-type coolers. Slinger-type evaporative coolers, however, require less maintenance and they last longer thus compensating for their higher initial cost. The other specific advantages of these coolers are as follows (Watt, 1963):

(i) The rate of pad clogging is very slow and, therefore, needs less human attention and labour.

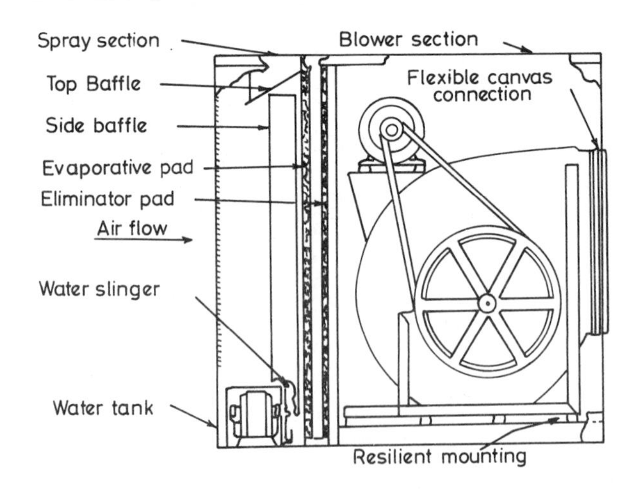

Fig. 8.3. Slinger-type evaporative cooler. (After Ashrae, 1972)

(ii) Water slingers are simpler, cheaper and more durable than equivalent pumps.

(iii) Since only the rear side requires exposure to fresh air, slinger coolers are easily built into buildings or located indoors.

(iv) Metal pads require no replacement when clogged, but can be cleaned and used again.

8.1.4.3 Rotary pad evaporative coolers. These coolers are more expensive than others, and are used by non-residential buildings only. These are usually available in two forms, one being the complete package unit including dust filters, revolving pad assembly and centrifugal fan, while the other is only the rotary pad assembly. In the latter case, one has to supply one's own enclosures, filters, fans and ducts which, in turn, offers a great deal of flexibility. Rotary-pad coolers are built in the capacities of providing 3400 m^3/h-20400 m^3/h of cooled air.

There are two types of rotary coolers viz. the rotary-disc cooler and the rotary-drum cooler. In the rotary-disc cooler (Fig. 8.4), the air-flow is at right angles to the face of a disc constructed of spirally wound layers of alternately flat and crimped bronze or copper screen wire. The disc usually has a thickness and diameter of 0.1-0.15 m and 0.9-1.5 m respectively, and revolves about a horizontal axis at a speed of about 2 rpm with its lower end immersed in the water. As a result of water lifted out from the tank in the form of a film on screen, the air coming into its contact gets cooled. The air-flow resistance of rotary pads is of the order of 125 N/m^2 which gets further increased by the addition of dust filters by about 25%.

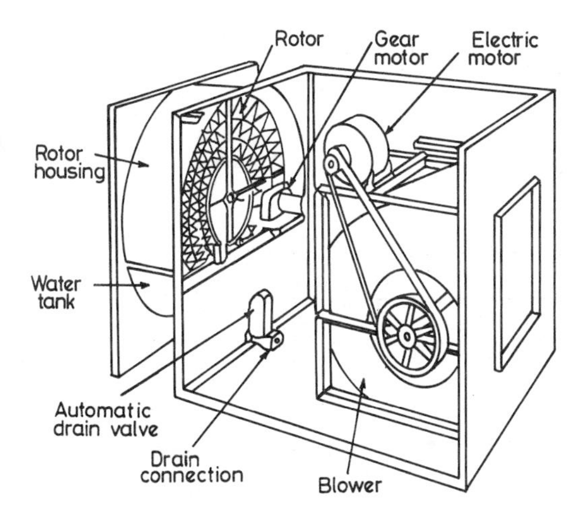

Fig. 8.4. Rotary-type evaporative cooler. (After Ashrae, 1972)

The other type of rotary-cooler uses a drum with the air entering the rotor through the sides of the drum. In this type of cooler, the saturation media is the pads formed of multiple wire mesh rolled into hollow cylinders about 0.9 m in diameter and 1.2 m long. The thickness of the wetting pad is about 50 mm while the air-velocity gets reduced to about half of its earlier value. This consequently reduces the power consumption producing requal cooling effectiveness, however, the overall load on the machine increases.

Both types of coolers are usually equipped with an automatic flush valve which keeps a check over the deposition of minerals in the water tank. For service and maintenance, a door is also provided in the fan compartment. It may also be noted here that the air-handling unit is the same in these coolers as that of drip-type coolers. These coolers have the following advantages (Watt, 1963):

(i) Little metal surface is exposed to water which reduces the corrosion to a minimum.

(ii) Metal pads are permanent and allow no sagging, decaying, odour formation, and are easily cleaned.

(iii) Scale formation is very slow, and therefore needs negligible attention.

(iv) Rotary-pad units can easily be installed upon former drip-type or sling-type coolers.

The rotary-pad type coolers have their most important disadvantage in being costly and requiring larger space. Compared to drip and slinger types, these have larger fan motors, but use less power. A typical comparison of the three types of coolers, based on the rated output and motor ratings, is given in Table 8.1. However, in view of practical considerations, the factors like the longer life and low maintenance cost place the rotary type coolers in competition with the other types.

TABLE 8.1 A Comparison of Drip-type, Slinger-type and Rotary-Pad Type Evaporative Coolers. (After Watt, 1963)

	Air delivered (m^3/h)/Total power (W)		
	Drip-type	Slinger-type	Rotary-type
Small sizes	24.46	9.63	9.59
Middle sizes	21.64	12.94	9.05
Large sizes	18.90	11.07	9.85
Average	21.67	11.21	9.50

8.1.4.4 Problems of direct-type coolers. Direct-type coolers are very common, however, their use over a period of time poses many operational problems. The single most important factor determining the effectiveness of the cooler is the evaporative pad. The requirements of an ideal evaporative pad and the possible ways of achieving them are discussed below briefly:

(i) The evaporative pad should present a maximum wet surface to the passing air, so that maximum cooling is achieved. For this, many materials have been in the use, however, the experience shows that redwood/aspenwood excelsior have proved satisfactory in drip-type coolers, glass fibre and metal screening in slinger coolers, while bronze and galvanized screening in rotary coolers is very efficient. The evaporative pad, among the three types of coolers, should have maximum porosity and capillarity in drip-type coolers, while minimum in rotary-type coolers because the water-supply in the former is limited while in the latter it is in abdundance.

(ii) The clogging and sagging of the evaporative pad should be as slow as possible. To avoid clogging, it is seen that wetting should be very satisfactory because of the reason that the flowing water takes away the dust and other foreign material trying to be deposited over the pad. With the type of material used in the drip-type coolers, it is oftenly recommended that the pad is replaced every year. On the other hand, the glass fibre and metal pads can be cleaned easily, and also resist sagging unless abused or corroded. In order to alleviate scale formation in the sump water either it has to be drained off increasing the water requirements, or some chemicals may be added which increase the solubility of the water. One commonly used chemical is sodium hexametaphosphate.

(iii) The evaporative pad should present minimum resistance to air-flow. The detrimental factors for this are the density and thickness of the pad. In general, the effect of these two parameters is complementary. The ideal pad should balance thickness against density to pass maximum air at desired per cent saturation.

(iv) The pad should filter all possible dust from the air. If, because of other considerations, the porosity and thickness of the pad does not allow it, additional filters are to be used in combination with the evaporative pads.

8.1.5 Indirect-Type Evaporative Coolers

If not properly designed direct-type evaporative coolers may pose the following problems:

(i) the cooled air may be excessively humid.

(ii) the high rate of air flow and large number of air changes, which are necessary for effective cooling, cause large variation in the air speed and the associated thermal sensation within the cooled space. This results in a waste of energy, which has been used to cool the discharged air.

Indirect-type evaporative coolers try to overcome these defects. Since the air, in these types of coolers, gets cooled without coming in direct contact with water, the problem of excessive humidity in the room air gets automatically solved. Simultaneously the required number of air changes also get reduced. In general, the indirect-type cooling methods fall in two categories viz. the active systems and passive systems. Some designs of the former class can be manufactured in the factories as the readymade units, while the latter method has to be employed upon the building itself. In the present section, only the active systems will be discussed, while the section 8.1.7 will cover the passive systems.

8.1.5.1 Simple dry-surface cooling system. The schematics of this type of cooling system are shown in Fig. 8.5. It consists of a large natural-draught cooling tower, a pump, a copper radiator and a fan circulating the air of the rooms to be cooled. The water is sprayed within the cooling tower, and the dry outside air cools it by evaporating some of this water into it. The cooled water is then pumped through the radiator, usually consisting of 4-8 rows of tubes, and transfers its coolness to the room air which is being circulated inside the room with the help of a fan. The outside air cooling the water in the cooling tower is rejected to the outside while the water heated in the radiator is ducted-back to the cooling tower. It is, however, seen that the water in the cooling tower is cooled within 2-3°C of the air's prevailing wet-bulb temperature, and the room air is cooled up to a temperature, generally 3-4°C above the outside wet-bulb temperature. This system, in the hot and dry climates, depending upon the quality and its size, can cool the indoor air temperatures by 10-16°C below the outdoor dry-bulb temperature. For best heat transfer, the radiator coils are

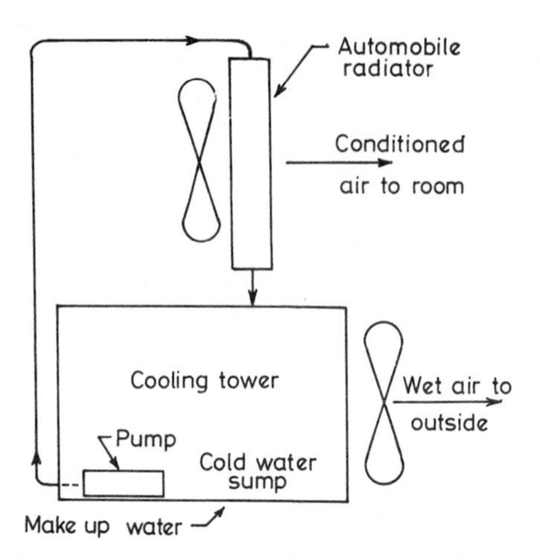

Fig. 8.5. Simple dry-surface cooling system. (After Watt, 1963)

usually arranged counterflow to the air and have eight rows of tubes. The velocities of water and air are maintained at about 1.25 m/s and 3.0 m/s respectively.

The most important disadvantage of these systems is the clogging and scaling of the tubes. Sometimes, algae growth also seems to be occurring in the cooling tower. These can be reduced by chemical treatment of the water and frequent washing of the tower sumps respectively.

8.1.5.2 Regenerative dry-surface cooling system. In the dry-surface cooling system, the temperature of the discharged air from room is about 6-8°C below the dry-bulb temperature of outdoor air. In order to make use of this wasted coolness, the discharged air is delivered to the cooling tower in place of outside air. This cools the water to a higher degree, and the system utilizing this principle is called as the "Regenerative dry-surface cooling system". A simple regenerative system is shown in Fig. 8.6 where its typical performance is also indicated. This system, however, has been tried only in few buildings, the reason being the bad connection between the exhaust room air and the cooling tower. Some modifications in the system were also tried out which permitted the use of the mixture of the exhaust room air and the outdoor air for cooling the water. It may also be noted here that unless the dew point is very low, the regenerative gain may not exceed the cost of achieving it.

8.1.5.3 Plate-type heat exchanger cooling systems. Several designs of this type of cooling system have been developed. In these, the exhaust room air is chilled in two-bank spray-type washers and then passed through the plate-type heat exchangers. Here, the outside air is cooled by giving its heat to the chilled water, and is distributed in the room. The wet air from air-washers is rejected to the atmosphere. In humid weather, the ice may be used to chill the water. The simplest kind of system is shown in Fig. 8.7. Since during cooling the water does not touch the surface of the heat exchanger, the problem of scaling is avoided.

8.1.5.4 Pennington heat-storage wheel system. This system has a resemblance with the Ljungstrom air-preheaters used in power plants, and Pennington's "Thermo-O-wheels" used at many places to preheat/precool the fresh outside air for ventilation (Watt, 1963). At any instant, the outside air (or the exhaust room air) flows through the wet evaporative pad which cools it. This cooled air

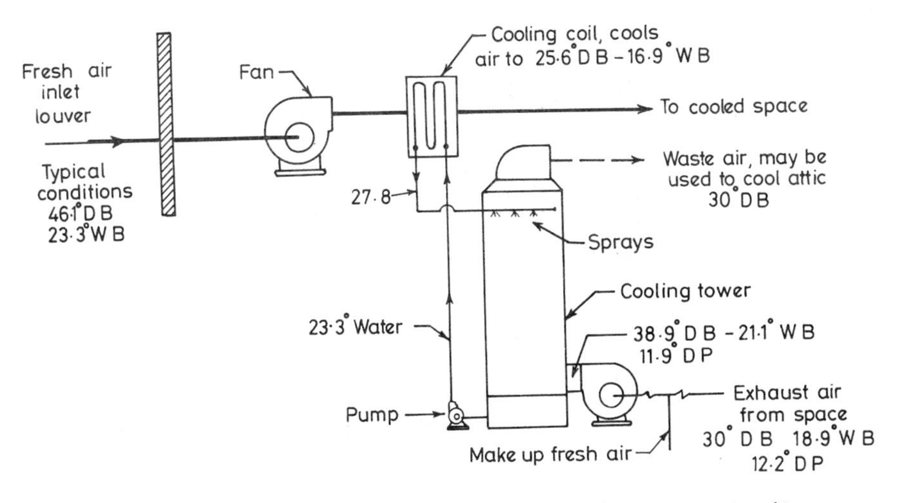

Fig. 8.6. Regenerative cooling system with cooling tower and coil. (After Watt, 1963)
DB - Dry bulb
WB - Wet bulb
DP - Dew point

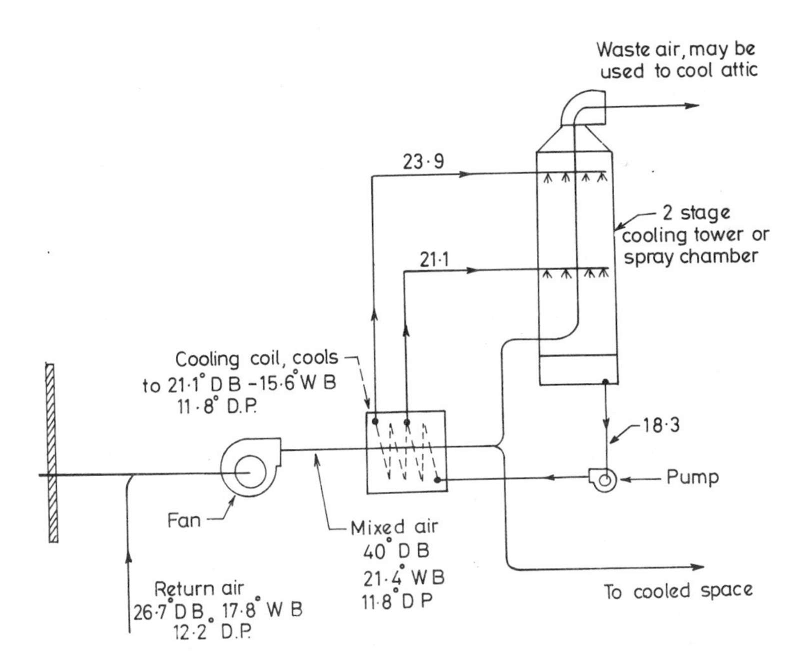

Fig. 8.7. Typical indirect-evaporative cooling system with air-to-air exchanger. (After Watt, 1963)

passes through half of the porous wheel (as shown in Fig. 8.8) and transfers its coolness to it. This wheel is usually made up of screening and filled with fine aluminium wool, and has a diameter and thickness of about 1.5 m and 0.3 m respectively. Fresh outside air, drawn through a filter, is blown over the cooled wheel and delivered to the room. The speed of the wheel is adjusted in such a manner that during half of its revolution the heat deposited upon a strand can flow to its own centre and back again, but not further. This is because each strand is exposed to cold air while being cooled, and to warm air while cooling it only long enough for heat to flow from its surface to its centre. In case the exhaust air of the room is used at the inlet, the indoor air temperatures even below the outside wet-bulb temperature may be obtained.

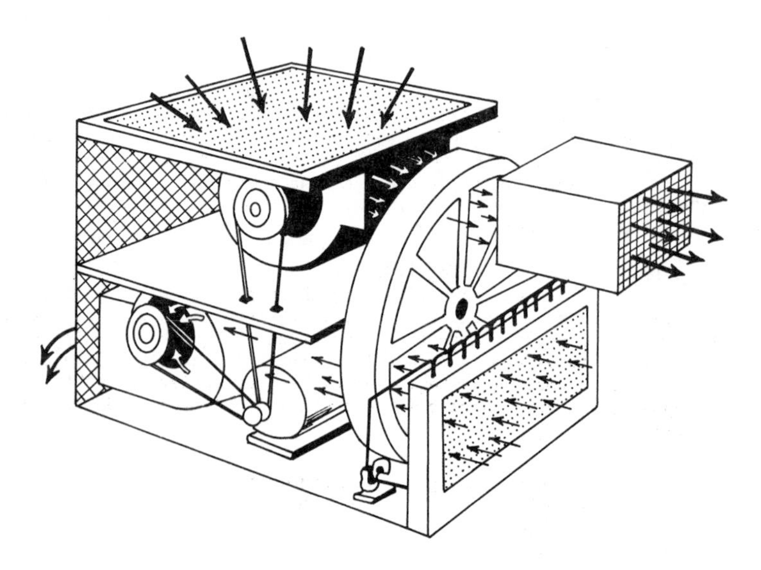

Fig. 8.8. Pennington heat-storage wheel. (After Watt, 1963)

8.1.5.5 Advantages and problems of indirect evaporative cooling. The important advantages of the indirect-type evaporative cooling are discussed below:

(i) Depending upon the performance of the system, the operating cost of the system gets reduced by 20%-60% below that of refrigerated air-conditioning.

(ii) Power consumption is less resulting in a sharp reduction of the running costs. Because of this reason, the indirect evaporative coolers can also be used where electricity is expensive or scarce.

(iii) It can be used as a precooler for refrigerated air-conditioning systems.

(iv) In this type of cooling, the exhaust room air can be delivered to the cooling tower as a result of which the lower water-temperature is obtained. This, in turn, produces more cooling.

In spite of above-mentioned advantages the indirect evaporative coolers suffers from two main problems during the operation stage. These are scaling and corrosion. To control the problem of scaling the following steps may be taken:

(i) A suitable filter should be used for the tower air which would stop the accumulation of dust particles.

(ii) Properly wet surface is less prone to scale formation.

(iii) Some chemicals may be used to increase the solubility of the water.

(iv) In case of water-spray over the tubes, the deposition can be brushed off.

(v) The size of the spray nozzle should be of 6 mm approximately. This will check clogging of the tubes.

The tubes are the most susceptible to corrosion, and hence should necessarily be of rust-resistant copper-bearing galvanized iron.

8.1.6 Two-Stage Evaporative Cooling Systems

Some of the designs consist of two stages of cooling which are particularly useful in humid climates. Two stages of cooling may comprise of direct-type and indirect-type coolers, or the indirect-type cooler coupled with a refrigerative stage. The important advantage of two-stage cooling is that the air's temperature is lowered sufficiently and hence less volume of air needs to be circulated through the rooms. This, in turn, implies that the sizes of filter, fan, duct, pump may be reduced.

A two-stage cooling system comprising indirect- and direct-type cooling systems as the first and second stages respectively is shown schematically in Fig. 8.9. This essentially consists of a number of plastic tubes covered with some wick-type material. The water is sprayed over the tubes as a consequence of which the air flowing through these tubes gets cooled. This dry cooled air is then again cooled by a direct-type cooling system. These types of coolers have been experimented upon in the hot desert climate of Phoenix, Arizona, USA (Sherman and Evans, 1981).

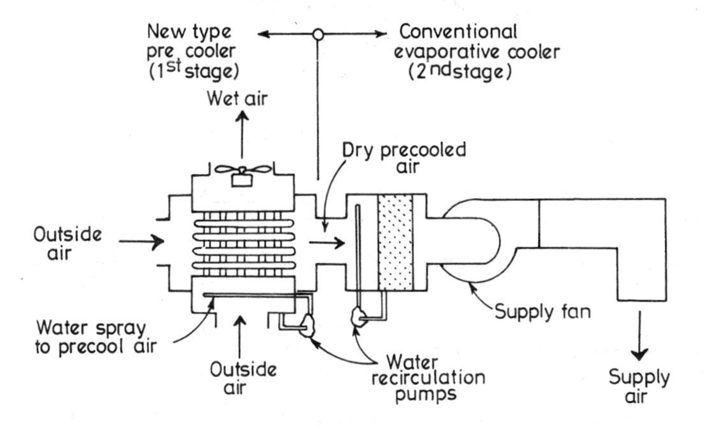

Fig. 8.9. Two-stage evaporative cooler with dry air evaporative first stage. (After Sherman and Evans, 1981)

In California, several two-stage coolers were installed which consisted of a refrigeration unit as the second stage. It was found in such installations that the indirect-stage (first stage) removed most of the air's sensible heat (about 70%) and the refrigerated coil removed the rest. Thus, another advantage of two-stage coolers was seen to be the fact that most of the sensible heat in room's exhaust air is removed at about 30% of the cost of refrigeration.

8.1.7 Indirect Evaporative Cooling: Passive Methods

Passive methods of indirect evaporative cooling are particularly useful in the regions where the cost of electricity is very high or it is scarce. These methods are much more inexpensive because these utilize only the incident solar radiation for the evaporation of water. Also, unlike the active systems the problems of scaling and clogging do not occur here, which in turn reduces the maintenance cost drastically. It has already been noted in Chapter 4 that the roof surface receives the maximum amount of solar radiation (about 45% of the total radiation) in summer and hence contributes the maximum to the cooling load of the building. Therefore, these techniques are usually employed over the roof which acts as a heat exchanger between the room air and outside environment.

Indirect evaporative cooling by passive methods can be achieved by adopting either of the following processes:

(i) By keeping the exterior surface of the roof wet. The sensible heat of the roof surface gets converted into latent heat of vaporization and the water evaporates. This results in the establishment of a temperature gradient from inside surface to the outside surface causing cooling of the indoor-environment. This concept is usually referred to as "Water-film over the roof". The critical factor determining its performance is the continuous wetness of the roof surface, and hence, needs special attention. Sometimes, the sprayers are used to achieve the constant wetness of the roof. In this case the technique is called the "roof-spray method".

(ii) By having a water pond over the roof surface. For the realization of this concept, the roof has to be made structurally sound. The incident solar radiation is absorbed by the roof surface which transfers heat to the water in its contact, and the water evaporation occurs at the water surface utilizing this heat. The effectiveness of the concept increases when the pond is covered during daytime by an insulation cover and is left exposed during the night hours. This helps in reducing the heat-gains during daytime and loses heat to the dry atmosphere during night-time. It may be noted here that covering the pond with a transparent material and reversing the strategy for the use of movable insulation i.e. by using it during night-time only, the same system can also be used for winter-heating.

(iii) By maintaining a moving water-film over the roof-surface. This concept utilizes the same principle for its operation; the evaporation process being helped by an increase in the relative speed between air and water surface, and is useful in very humid climates. The cooled water may be stored in the basement and then may be circulated within the room to cool it.

First two of these methods have been studied in detail, and are discussed in the subsequent subsections from experience and utility points of view. The third process has limited applications and therefore appears not to be investigated in detail.

8.1.7.1 Experience. Historically, the methods employing indirect evaporative cooling by passive techniques are the ancient ones, but the documentation of the performance data scientifically started probably in the beginning of present century only. The reduction of heat flux by using roof-pond was observed at the University of Texas, but the method suffered from structural problems. In 1940, Houghten *et al.* initiated a project to investigate the cooling produced by the roof-pond and roof-spray techniques over the roof. Both the methods were found to be equally effective in reducing the heat flux through all types of roofs. It was further observed that there was no significant change in the ceiling temperature (and hence the heat flux) when the depth of the pond was increased from 0.05 to 0.15 m. Based on the observations of Houghten (1940) and the analysis of Mackey and Wright (1944, 1946) of periodic heat flux through homogeneous and composite slabs, the ASHRAE guide (1958) prepared a table indicating the effectiveness of roof-pond and roof-spray techniques in the reduction of heat flux (Table 8.2).

It has been pointed out (Thappen, 1943; Holder, 1957; and Blount, 1958) that the cooling capacity of an air-conditioning plant can be reduced by 25% if a roof-spray cooling system is used. It was also seen that the spray system alone was fairly effective in keeping the building reasonably comfortable. In addition to the cooling of the roof, the air above the roof also gets cooled and being heavier than the hot air slides down the walls of the building. Much of these chilled air drifts in the building are due to infiltration and ventilation. It was, therefore, concluded that sprays were most efficient for buildings having large roof areas in proportion to the exposed area of the rest of the structure. Thus, more useful cooling occurs in the case of a large, single-storey furniture factory than that realized in a five-storey office building having the same floor area because only the roof was affected by this method. Sutton's (1950) observations that the surface-temperature of a roof which would reach 65.4°C if unsprayed, can be reduced to 42.2°C and 39.4°C by maintaining an open roof-pond of depth 0.05 and 0.15 m respectively, are in general agreement with Table 8.2.

Yellot (1969) carried out an experiment at Phoenix, Arizona, USA during the summer of 1965 to determine quantitatively the extent to which intermittent spraying could cool horizontal and tilted surfaces under conditions of intense insulation, high atmospheric temperature and low relative humidity. It was observed that under these conditions, intermittent sprays can effectively cool horizontal and tilted roofs. Using 1.5 kg of spray water $h^{-1}m^{-2}$, the roof temperature was reduced to the point where the day-long average differential between the roof and the ambient air become zero. Under the same conditions, the day-long average differential for a dry roof ranged from -1.1° to 4.4°C. The usefulness of roof-spray cooling was found to be most effective in buildings with lightly constructed, poor insulated roofs (as is evident from Table 8.2). The system used by Yellot (1969) is described below.

A 3.6 x 4.8 m test deck was constructed of 0.012 m plywood nailed to 0.6 x 1.8 m beams. Then a typical build-up roof with two layers of 6.81 kg felt, tar and an upper surface of 40.86 kg roofing was laid over the plywood. The roof was divided into a 3.6 m wide east section which was used as the wet portion and a 1.2 m wide west section which was kept dry. North half of the deck was insulated with 0.025 m polystyrene foam while the south half was left uninsulated. There was a provision to keep the roof in an inclined position. The east section was provided with two 0.012 m copper tubes with 15 spray perforations on 0.3 m centres in each tube. The tubes were connected to the two supply headers with rubber tubing so that the throw angle of the sprays could be adjusted easily. Temperatures at the deck surfaces were recorded on a strip-chart recorder with copper-constantan thermocouples.

Hay and Yellot (1969) also carried out the experiments with roof-ponds during the period, August, 1967 to July, 1968. A full scale test structure was erected to study the thermal effects produced by the roof-ponds with movable insulation and the building itself which is constructed using high heat capacity materials; thermal performance of the pond is shown in Fig. 8.10. It is obvious that the reduction in the peak of the outside dry-bulb temperature was as much as 14°C. It may be noted here that during most of the time, the room temperature was kept within 1-2°C of the pond temperature by circulating cool water from the ponds through a small fan coil unit which was suspended near the ceiling under the centre, adjacent to the inside north wall of the building. The fan was driven by a two-speed 90-watt motor, and a 2.5 gpm circulating pump, located outside the building, was driven by a 25 watt motor. The walls, as can be seen from Table 8.3, also played an important role in the cooling of the building. The heat lost by the walls made its way by convection and radiation to the pond water, because

TABLE 8.2 Total Equivalent Temperature Differentials* for 1 August (Clear Day) in 40°N Latitude with Maximum Air Temperature of 35°C, Minimum Air Temperature and Roof Air Temperature of 26.7°C. (After ASHRAE, 1958)

Sample No.	Roof	Sun time								
		a.m.			p.m.					
		8	10	12	2	4	6	8	10	12
1	Light construction	6.7	21.1	30.0	34.4	27.8	14.4	5.6	2.2	0
2	Light construction with 1" open roof-pond	0	2.2	8.9	12.2	10.0	7.8	5.6	1.1	0
3	Light construction with roof-spray	0	2.2	6.7	10.0	8.9	7.8	5.6	1.1	0
4	Heavy construction	2.2	3.3	13.3	21.1	25.6	24.4	17.8	10.0	6.7
5	Heavy construction with 1" open roof-pond	-1.1	-1.1	-2.2	5.6	7.8	8.9	7.8	5.6	3.3
6	Heavy construction with spray	-1.1	-1.1	1.1	4.4	6.7	7.8	6.7	5.6	3.3
7	Any roof with 6" open roof-pond	-1.1	0	0	3.3	5.6	5.6	4.4	2.2	0

*Total equivalent temperature differentials = Heat flux/overall heat transfer coefficient

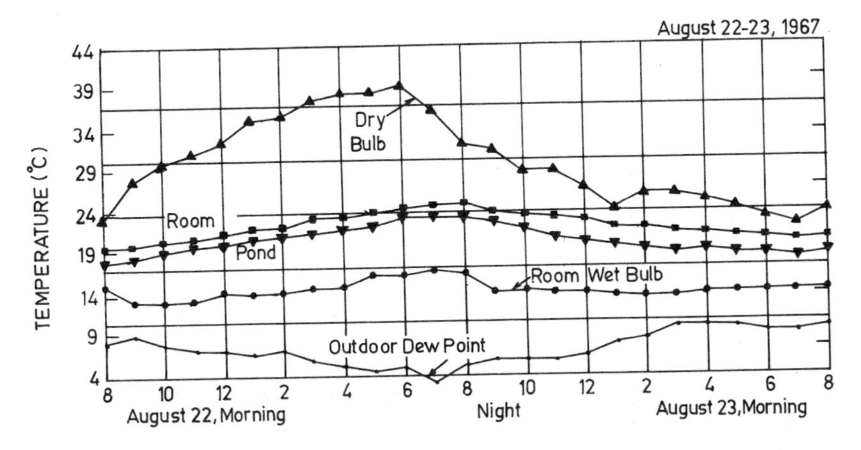

Fig. 8.10. Hourly variation of room and pond temperatures for August 22-23, 1967, at Phoenix, Arizona, USA. (After Hay and Yellot, 1969)

the outdoor air was generally warmer than the exterior wall surfaces, and these could not lose heat to the atmosphere. The north and south walls of the test building were made of ordinary concrete block with the voids filled with vermiculite. The east and west walls were, however, insulated externally by 0.04 m rigid urethane foam panels with the outer surfaces painted white. In contrast to the case of spray cooling, the high heat capacity of the walls and floor was reported to be a great asset in the case of roof-ponds.

Similar experiments with various evaporative cooling techniques have also been carried out at the Central Building Research Institute, Roorkee, India (latitude 28° 55' N). Jain and Rao (1974) and Jain (1977) have experimentally investigated in some detail the effect of roof-pond, roof-spray and wetted gunny bags on the temperature and heat flow through the reinforced concrete cement roof of air-conditioned and unconditioned rooms exposed to a hot-day sunny climate; experiments were carried out on four identical full sized (3.5 x 2.9 x 3.2 m) test rooms having the same dimensions. Roofs of the test rooms were 0.15 m whereas the walls were constructed using solid bricks and were of 0.23 m thickness; both sides of the walls and roofs were treated by 1.27 cm thick cement plaster. To affect cooling by roof-spray, an automatic intermittent spray system with a booster pump and time cycle arrangement which sprayed for $2\frac{1}{2}$ minutes at intervals of 27 minutes was designed and set-up. The time cycle arrangement consisted of a synchronous motor-driven clock having one revolution of 1 hour. The pump controlled by the time cycle arrangement and a solenoid valve were used to work the water spray. It was seen that, by roof-spray, the peak roof temperature decreased from 55° to 28°C as compared to reduction from 55° to 32°C in the case of roof-pond. This was obviously due to more effective evaporation of water at the roof surface. The ceiling surface temperature was observed to undergo a drop of the order of 15°C as compared to 13°C in the case of water pond. The indoor air temperature suffered a drop of the order of 3.5°C as compared to that of 3°C in the case of roof-pond. Jain and Rao (1974) and Jain (1977) have also investigated an alternate system of water evaporation viz. wetted gunny bags on the rooftop of a test room over the tar-felt and water was sprinkled three times a day to keep the surface wet round the clock. The results were compared to those in the case of roof-pond and roof-spray for selected clear days. First series of tests established the thermal equivalence of the rooms since the deviations in the air and ceiling surface temperatures were well within the accuracy of measurements. A comparison of drop of outside roof surface

TABLE 8.3 Heat Dissipation from Ponds during Night of August 22-23. (After Hay and Yellot, 1969)

Heat loss from pond water	11.8 kwh
Heat lost by walls	4.3 kwh
Infiltration air assuming 1 air change/hr	0.504 kwh
Conduction heat flow inward	1.41 kwh
Total heat dissipation	18.0 kwh
Average rate of heat dissipation	0.0088 kwh

temperature, inside ceiling-surface temperature and indoor air temperature corresponding to different methods of water evaporation is given in Table 8.4. Besides lowering the various temperatures, the roof treatments reduced the amount of heat-flux entering the room significantly. This, in turn, reduces the capacity of the air-conditioning plant and its cost. Jain and Rao (1974) have reported that a large reduction of the order of 85% in peak heat gains can be achieved by roof-spray and roof-pond systems. The total discomfort degree hours above a base temperature of 30°C are presented in Table 8.5. It is reported that the room temperature has always been less than 30°C throughout the daily cycle. While mentioning the results of the investigations on conditioned buildings, Jain (1977) has reported that for an air-conditioned building, the cooling load gets reduced significantly (Fig. 8.11).

TABLE 8.4 Temperature Drops on Account of Water Evaporation on Roof (After Jain and Rao, 1974)

Sample No.		Outside roof surface temperature (°C)	Inside ceiling surface temperature	Indoor air temperature (°C)
1	Open roof-pond	23	13	3
2	Spray roof	25	15	3.5
3	Wetted gunny bags	27	17	4

A wet gunny bag system has also been installed at the Bharat Heavy Electricals Limited factory at Haridwar, India, during the summer of 1979 at a capital cost of 5.50 Rupees (US$ 0.5) per m^2 and a running cost of 1.20 Rupees (US$ 0.12)/m^2/year. The building over which the cooling system was tried is a four-storeyed engineering building which has a large number of offices and rooms. The monitoring of the performance showed that a reduction of 17°C and 8°C was observed in the peak values of ceiling surface temperature and indoor air temperature respectively (Jain and Kumar, 1981). The results are shown in Table 8.6.

A Gunny-bag system was also installed on a roof of a single-storeyed building at Panjab University, Chandigarh, India. The building was located at the outer edges

TABLE 8.5 Degree Hours for Ceiling Temperature. (After Jain and Rao, 1974)

Test Room No.	Conditions	Total discomfort (Degree hours)	Duration (hr)	Peak	temp. hour 30°C	Average discomfort (Degree hours)
1	Untreated	153	19	15	1700	8
2	7.5 cm water pond on roof top (day and night)	4	4	1.4	1600	1
3	Roof-spray (daytime only)	2	3	1	1500	0.7
4	Gunny-bags wetted (during day only)	Nil	Nil	-2.2	1500	Nil (peak temp. is throughout less than 30°C)

TABLE 8.6 Performance of Evaporative Cooling System at B.H.E.L., Haridwar, India, in May 1979. (After Jain and Kumar, 1981)

Temperature of (°C)	Roof condition		
	Untreated	Treated	Drop
Ambient air	41.0	41.5	
Ceiling surface	45.0	28.0	17.0
Indoor air at living level	39.0	31.0	8.0

of a rectangular building. The temperature measurements taken before and after the installation of the gunny-bag system showed that a reduction of 10°C occurred in the indoor air temperatures.

8.1.7.2 Mathematical formulation.

Several workers have studied the performance of indirect cooling techniques experimentally, which is indicative of their effectiveness under typical climatic and operational conditions. So, there is a need for a mathematical model which would help in estimating the system performance under a given set of conditions prior to its actual installation. Considerable work has been done in this direction at IIT Delhi, India. Analytical models have been developed for thin water-film (Sodha *et al.*, 1978), roof-pond (Sodha *et al.*, 1980) and moving water-film (Sodha *et al.*, 1980). The model for roof-pond has also been validated by the experimental results (Sodha *et al.*, 1981a).

In this section, two analytical approaches will be discussed viz. (a) for "water-film" technique (b) a general model valid for all the three cooling techniques.

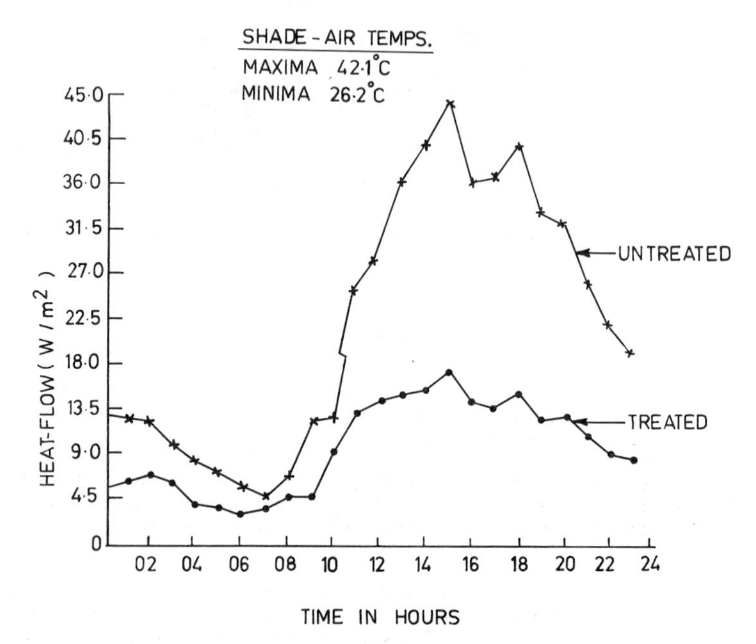

Fig. 8.11. Hourly heat flow in conditioned test rooms with and without roof-spray. (After Jain, 1977)

The former approach will also be useful in determining the ground temperature when it is wet. This will be discussed in section 8.2.

(a) Analytical model for water-film technique: Considering that the roof surface is adequately wet, the energy balance condition for it may be written as follows:

$$- k \left.\frac{\partial \theta_R}{\partial x}\right|_{x=o} = \tau_1 \; S(t) - Q_c - Q_r - Q_e \qquad (8.1)$$

The rates at which heat losses occur from roof surface to the surroundings by convection ($\dot{Q}_c$), radiation ($\dot{Q}_r$) and evaporation ($\dot{Q}_e$) are given by the following expressions:

$$\dot{Q}_c = h_c \; (\theta_R|_{x=o} - \theta_a), \qquad (8.2)$$

$$\dot{Q}_r = h_r \; (\theta_R|_{x=o} - \theta_a), \qquad (8.3)$$

$$\dot{Q}_e = 0.013 \; h_c \; (P(\theta_R|_{x=o}) - R_h P_{as}). \qquad (8.4)$$

The above-mentioned expressions for convective and radiative heat transfer are the standard relationships while the expression for Q_e is the same as used by Hay and Yellot (1970), Carrier (1968) and Dunkle (1961) for zero wind velocity.

From the observed temperature-dependence, the saturation vapour pressure can be assumed to be a linear function of temperature in the operating temperature range. Therefore,

$$P(\theta) = R_1\theta + R_2 \tag{8.5}$$

This assumption is valid for small variations of water and air temperatures. The constants R_1 and R_2 used in equation (8.5) can be obtained by least square curve fitting of the actual saturation vapour pressure data. For various temperature ranges of interest the values of R_1 and R_2 are given in Table 8.7. Substituting equation (8.5) in equation (8.4), one gets

$$Q_e = mL = 0.013\ h_c \left[R_2\ (1 - R_h) + R_1\ (\theta_R\big|_{x=o} - R_h\ \theta_a)\right] \tag{8.6}$$

Now, substituting of equations (8.2), (8.3) and (8.6) in equation (8.1) yields

$$-k \left.\frac{\partial\theta g}{\partial x}\right|_{x=o} = h_{eff}\ (\theta_{eff} - \theta_R\big|_{x=o}) \tag{8.7}$$

TABLE 8.7 Constants R_1 and R_2 obtained by Least Square Curve Fitting of Saturated Vapour Pressure Data in Certain Temperature Range

Temperature range (°C)	R_1 ($N/m^2\,°C$)	R_2 (N/m^2)	Correlation
10 - 30	148.7	- 478.8	0.9888
30 - 50	401.4	- 8316.0	0.992
50 - 70	948.4	-36160.0	0.993

where

$$h_{eff} = (h_c + h_r)\ (1 + 0.013\ R_1) \tag{8.7b}$$

$$\theta_{eff} = h_{eff}^{-1}\ [(h_c + h_r)(1 + 0.013\ R_h\ R_1)\theta_a + \tau_1 S(t)$$
$$- 0.013\ R_2\ (1 - R_h)(h_c + h_r)] \tag{8.7c}$$

The boundary condition for the inside surface of the roof, in contact with the room air, may be written as

$$-k \left.\frac{\partial\theta_R}{\partial x}\right|_{x=\ell_1} = h_{si}\ (\theta\big|_{x=\ell_1} - \theta_r) \tag{8.8}$$

On account of the periodic nature of ambient air temperature and solar intensity, the effective temperature (θ_{eff}) may also be expressed as Fourier series in time, viz.

$$\theta_{eff} = a_o^1 + \text{Re} \sum_{m=1}^{\infty} a_m^1 \exp(im\omega t) \tag{8.9}$$

Considering one dimensional heat-flow in the roof and assuming periodic solutions, the temperature distribution in the roof may be obtained with the method described in Chapter 6 by using the equations (8.7) and (8.8). θ_{eff} described by Eq. (8.7c) reduces to the earlier defined solair temperature θ_{sa} if the surface is not wet i.e. R_1 and R_2 are put as zero. An estimate of θ_{eff} directly helps to evaluate the effect of wet surface in reducing the heat flux into the room.

Example 8.1. A 300 m thick concrete slab with K = 0.72 W/m^2 °C is exposed to solar radiation of intensity 1000 W/m^2. Find theheat flux into the room maintained at 20°C for (i) dry surface (ii) wet surface if the ambient air temperature and relative humidity are 40°C and 40% respectively.

Solution: The heat flux into the room is given by the expression

$$\dot{Q} = \left(\frac{1}{h_{eff}} + \frac{\ell_1}{K} + \frac{1}{h_{si}}\right)^{-1} (\theta_{eff} - 20)$$

For dry surface

$$h_{eff} = h_c + h_2 \simeq h_{so} = 22 \text{ W/m}^2\text{°C}$$

$$h_{si} = 8 \text{ W/m}^2\text{°C}$$

$$\theta_{eff} = \theta_{sa} = \left(\frac{0.6 \times 1000}{22} + 40\right)$$

$$= 67.2\text{°C}$$

Hence

$$\dot{Q} = \left(\frac{1}{22} + \frac{0.30}{0.72} + \frac{1}{8}\right)^{-1} (67.2 - 20)$$

$$= 1.7 \times 47.2$$

$$= 80.4 \text{ W/m}^2$$

For wet surface

From Table 8.7, $R_1 = 401.4$

$R_2 = -8316.0$

$$\therefore \quad h_{eff} = 22\ (1 + .013 \times 401.4)$$

$$= 136.8 \text{ W/m}^2\text{°C}$$

It may be noted that the effective heat transfer from a wet surface gets considerably increased. For $R_h = 0.4$

$$\theta_{eff} = \frac{1}{136.8} \left[22\ (1 + 0.013 \times 0.4 \times 401.4)\ 20 + 0.6 \times 1000 + 0.013 \times 8316 \times 0.6 \times 22\right]$$

$$= 24.7^{o}C$$

In comparison to a solair temperature of 67.2°C, the effective temperature of a wet surface becomes 24.7°C. The net heat flux into the room becomes

$$\dot{Q} = \left[\frac{1}{136.8} + \frac{0.3}{0.72} + \frac{1}{8}\right]^{-1} (24.7 - 20)$$

$$= 8.56 \text{ W/m}^2$$

The heat flux gets reduced by a factor of 10 approximately.

(b) General model. In this section, a general mathematical model for passive cooling techniques is discussed. The model takes into account the time dependence of various parameters and is valid for all the three cooling concepts viz. the water-film, roof-pond and moving water-layer concepts.

Let us consider the roof of a building which is flooded with water and is exposed to the surroundings. Schematic configuration is shown in Fig. 8.12. The energy-balance for moving water-mass may be written as follows:

$$b\, d_{\omega}\, \rho_{\omega}\, C_{\omega} \frac{\partial \theta\omega}{\partial t} + \dot{m}_{\omega}\, c_{\omega} \frac{\partial \theta\omega}{\partial y}\, dy$$

$$= \left[\tau_2\, S(t) - Q_c - Q_r - Q_e + h_{sw}\, (\theta_R\big|_{x=o} - \theta_{\omega})\right] bdy \qquad (8.10)$$

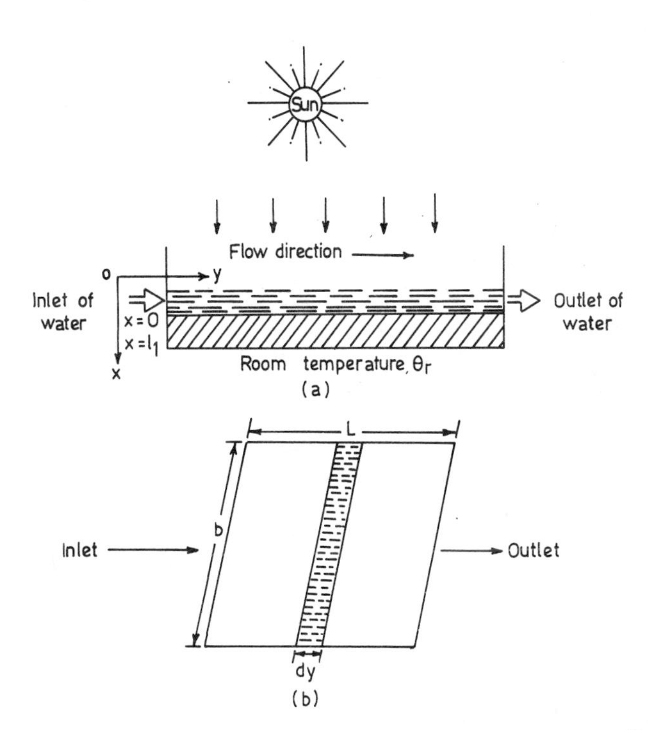

Fig. 8.12. (a) Schematic sketch of, "Flowing water over the roof" system. (b) Overside view of the flowing water system

Substituting equations (8.2), (8.3) and (8.6) and rearranging the terms, the equation (8.10) yields

$$M_{wo} \frac{\partial \theta w}{\partial t} + \dot{m}_w c_w \frac{\partial \theta w}{\partial y} = bH (\theta_c - \theta_w) \quad (8.11)$$

where

$$M_{wo} = b\, d_w\, \rho_w\, c_w,$$

$$H = h_{sw} + h_c + h_r + 0.013\, R_1\, h_c,$$

$$\theta_c = \frac{1}{H}\left[\tau_2\, S(t) + H_{sw}\, \theta_R\Big|_{x=o} + H_1\, \theta_a - 0.013\, R_2\, h_c\, (1 - R_h)\right],$$

$$H_1 = h_r + h_c + 0.013\, R_h\, R_1\, h_c$$

Equation (8.11) is a general equation for evaporative cooling systems which can be used for various systems by putting

(i) $\dot{m}_w = 0$ for open roof-pond, and

(ii) $M_{wo} = 0$, for water film

$\dot{m}_w = 0$ for water-film maintained by spray/gunny bags over the roof. It may also be noted here that the value of h_{sw} differs for the three cooling systems, and can be evaluated using appropriate relationships for forced convection mode or natural convection mode as the case may be. For gunny-bags, the same equation will be valid because the work on multiple-wick solar stills (Sodha *et al.*, 1981b) has shown that the heat and mass-transfer from an adequately wetted gunny-bag surface is identical to that from a free water surface.

The boundary condition for the outside roof-surface is

$$- K \frac{\partial \theta_R}{\partial x}\Big|_{x=o} = \tau_1\, S(t) - h_{sw}\, (\theta_R\Big|_{x=o} - \theta_w) \quad (8.12)$$

whereas for the inside roof-surface it is given by equation 8.8.

Following Chapter 6, the solar intensity and ambient-air temperature may again be assumed to be period. Using the method of periodic solutions and assuming one dimensional heat-flow in the roof, the water temperature, θ_w^*, can be evaluated as a function of flow-length and time. The initial condition used for the solution is

$$\theta_w\, (y,t) = \theta_j \text{ at } y = 0 \quad (8.13)$$

where θ_j is also expressed as a Fourier series in time, i.e.

$$\theta_j = \bar{\theta}_j + \sum_{n=1}^{\infty} \theta_{jn} \exp\{i(n\omega t - \sigma_n)\}$$

---

*Fourier series for water temperature is

$$\theta_w(y,t) = \bar{\theta}_w(y) + \text{Re} \sum_{n=1}^{6} \theta_{\omega n}(y).\ \exp(in\omega t)$$

The following expression is obtained for time-independent part of the water temperature:

$$\bar{\theta}_w(y) = \theta_o - (\theta_o - \bar{\theta}_j) \exp(-\beta y) \tag{8.14}$$

where

$$\theta_o = \left[\frac{\tau_2 a_o + M_3(h_c + h_r) + H_1 b_o - 0.013\, h_c\, R_2(1 - R_h)}{H_2 - M_4(h_c + h_r)}\right],$$

$$\beta = \left[\frac{b(H_2 - M_4\, h_c - M_4\, h_r)}{\dot{m}_w\, c_w}\right],$$

$$M_1 = h_{si}\, [(h_c + h_r)\theta_r - \tau_1\, a_o]/D,$$

$$M_2 = -\, [(h_c + h_r)h_{si}/D],$$

$$M_3 = [K\, h_{si}\, \theta_r + \tau_1\, a_o\, (h_{si}\, \ell_1 + K)]/D,$$

$$M_4 = [h_o\, (h_{si}\, \ell_1 + K)/D],$$

$$D = [(h_c + h_r)(h_{si}\, \ell_1 + K) + h_{si}\, K],$$

$$H_1 = [h_r + h_c + 0.013\, R_h\, R_1\, h_c],$$

$$H_2 = [h_o + h_c + h_r + 0.013\, R_1\, h_c],$$

where a_o and b_o are the time-independent parts of the Fourier series for solar intensity and ambient air temperature respectively.

For time-dependent part also, the expressions may be obtained in a similar way.

The rate of heat flux entering the room through the roof can be written as

$$\dot{Q}(y,t) = h_{si}\left[\theta\Big|_{x=\ell_1} - \theta_r\right] \tag{8.15}$$

Assuming the roof surface temperature to be uniform along y-direction, the mean heat flux (averaged over y) entering the room can be obtained as

$$\dot{Q}(t) = \frac{1}{L}\int_o^L \dot{Q}(y,t)\, dy \tag{8.16}$$

The temperature of the water at the end of the roof, and the amount of heat flux retrieved by the water may also be evaluated as follows

$$\theta_{wf}(t) = \theta_w(y,t)\Big|_{y=L}$$

and

$$\dot{Q}_w(t) = \frac{\dot{m}_w\, C_w}{A}(\theta_{wf} - \theta_j)$$

8.1.7.3 Experimental validation of the model. The analytical model developed in the previous section was validated experimentally by Sodha *et al.* (1981a). A test enclosure of the inside dimensions 2.2 x 2.2 x 0.4 m was constructed whose roof was made up of concrete. The experiments were carried out during April-May, 1980 at Delhi, India. In order to establish the steady-state periodic conditions, the pond was left exposed to the atmospheric conditions continuously for eight days. Various perofrmance parameters viz. the amount of heat flux entering the enclosure through roof, ceiling temperature and water temperature were measured. These are depicted by circles in Figs. 8.13-8.15.

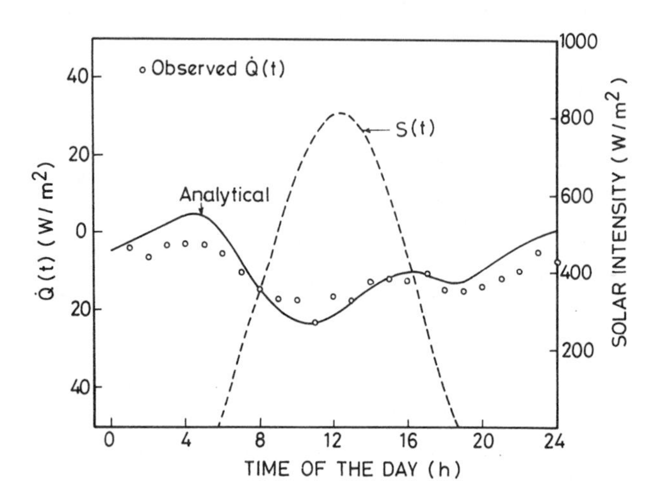

Fig. 8.13. Time variation of heat flux entering the room. (After Sodha *et al.*, 1981a)

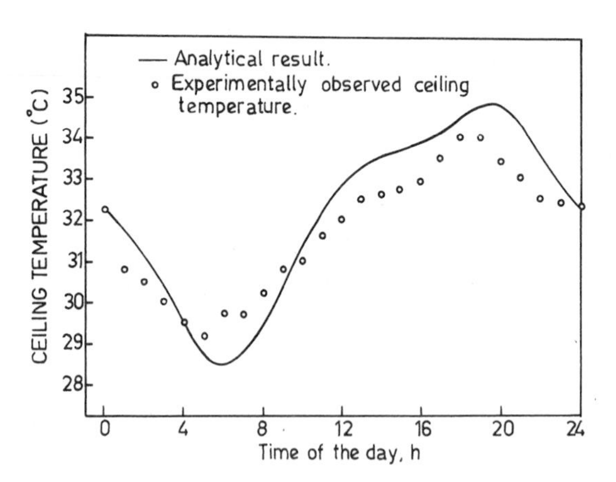

Fig. 8.14. Variation of ceiling temperature with time. (After Sodha *et al.*, 1981a)

Numerical calculations were also carried out for a typical sunny day (2 May, 1980) at Delhi, India corresponding to an actual set of parameters during the experiment. These parameters are given in Table 8.8.

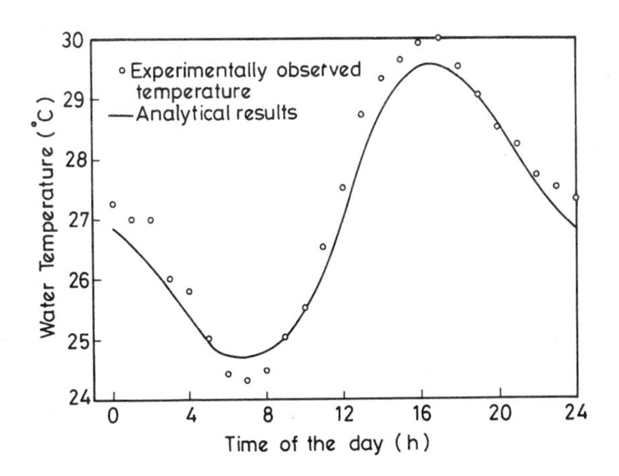

Fig. 8.15. Variation of water temperature with time. (After Sodha *rt al.*, 1981a)

TABLE 8.8 Parameters Related to the Experiment at IIT, Delhi, India

Roof material (concrete)	Water mass
$K = 0.192$ W/m°C	$d_w = 0.21$ m
$\rho = 1900$ kg/m^3	$\tau_2 = 0.05$
$c = 840$ J/kg°C	$\varepsilon_2 = 0.9$
$\ell_1 = 0.09$ m (roof thickness)	
General heat transfer coefficients	
$(h_r + h_c) = -$ 9.997 W/m^2°C (still air)	
$h_{sw} =$ 135.04 W/m^2°C	
$h_{si} =$ 9.235 W/m^2°C	
$R_1 =$ 325.17 N/m^2°C	
$R_2 = -5154.9$ N/m^2	
$R_h =$ 0.1	

The negative sign for the values of $(h_r + h_c)$ shows that although the water temperature is lower than the ambient temperature, it still loses heat by radiation because the sky temperature is lower than the water temperature. For these parameters, the heat gained by the water surface by convection is smaller in magnitude than that lost by radiation. The corresponding Fourier coefficients are given in Table 8.9.

TABLE 8.9 Fourier Coefficients of Daily Variation of Solar Intensity, Ambient Air Temperature and Enclosure Air Temperature on a Typical Day (May 2, 1980) at Delhi, India

Number of harmonics	Solar intensity		Ambient air temperature		Enclosure air temperature	
	Amp (W/m^2)	Phase*	Amp (^oC)	Phase*	Amp (^oC)	Phase*
0	258.32	-	39.96	-	33.0	-
1	399.50	3.216	7.29	3.790	3.32	4.380
2	162.38	0.157	2.26	0.450	0.99	5.653
3	2.32	4.301	0.63	1.911	0.63	2.015
4	26.36	0.262	0.69	4.039	0.24	1.930
5	2.40	0.961	0.54	5.941	0.20	2.931
6	6.26	6.123	0.26	1.390	0.55	3.480

*Phase factor is in radians

The results of the numerical calculations are plotted as a function of time in Figs. 8.13-8.15. Obviously, the theoretical results, shown by continuous curves, are in good agreement with the experiments.

8.1.7.4 Comparison of evaporative cooling techniques: water-film, roof-pond and moving water-film. The relative performance of the three concepts of evaporative cooling systems was investigated theoretically by Tiwari *et al.* (1982). The system parameters chosen were

$M_{wo} = 93646\ J/m^{2o}C$ (moving water mass)

$= 2809394\ J/m^{2o}C$ (roof-pond)

$h_r = 21.3\ W/m^{2o}C$, $h_{si} = 6.8\ W/m^{2o}C$

$K = 0.72\ W/m^oC$ $\theta_r = 20^oC$

$\tau_1 = 0.54$ $\tau_2 = 0$

$\ell_i = 0.23\ m$ $b = 4.5\ m$

Area of roof $= 20\ m^2$

$\dot{m}_w = \rho_w\ b\ d_w\ v$, $h_c = 5.7 + 3.8\ v$.

The numerical calculations were made for a typical summer day (19 June, 1980). The corresponding Fourier coefficients are given in Table 8.10.

The effectiveness of the roof-pond/water-film and flowing water layer system is presented in Table 8.11 and 8.12 respectively. It is clear that the performance is a strong function of the relative humidity of the outside air. Moving water layer is less effective because, as expected, less evaporation occurs from the water surface. Identical behaviour is observed in the case of roof-pond and

TABLE 8.10 Fourier Coefficients of Solar Intensity and Ambient Air Temperature for 19 June, 1980 at Delhi

n	Solar Intensity		Ambient air temperature	
	Amplitude (W/m)	Phase factor (radian)	Amplitude (°C)	Phase factor (radian)
0	316.79	-	36.69	-
1	462.24	3.326	6.67	4.016
2	144.37	0.383	1.10	6.132
3	21.64	0.543	1.04	1.360
4	14.90	4.300	0.30	2.720
5	15.08	3.536	0.30	3.233
6	2.94	0.256	0.16	0.440

water-film systems because the heat transfer coefficients are assumed to be identical for the two cases. An average reduction of 66% in $\dot{Q}_o$ was predicted for Delhi-like climates by using a roof-pond/water-film system while the moving water-layer was anticipated to reduce $\dot{Q}_o$ by 46% only. A reflective bottom was also seen to be effective in reducing the heat-flux.

However, the roof-pond system, as mentioned earlier, is seen to be less effective, because the solar radiation passes through the water to be absorbed at the underlying roof surface, and the evaporation takes place at some distance of maximum heating. Apart from this, the roof-pond system also suffers from some operational problems e.g. the structure has to be excessively strong, algae growth etc. Therefore, the water-film concept appears to be the most efficient one, further discussion is restricted to water-film system only.

8.1.7.5 Realization of the water-film concept. Water-film concept can be realized by spraying the roof intermittently or continuously depending upon the water sprinkling capacity of the sprayer. Before the installation of the system it has to be ensured that the roofs are treated with water-proofing material adequately. This material should, however, possess the property of absorbing and retaining sufficient quantities of water. Jain (1976) has carried out a detailed investigation over a wide range of materials for the purpose. It was reported that the use of materials, e.g. the gunny-bags, woven-coconut or coir matting, coconut husk, brick-ballast, sintered-fly-ash, is quite effective and economical. On account of their porosity, these materials when wet, behave like a free-water surface for evaporation. The durability of such materials is rather good, but these will have to be treated for fire safety. A comparative idea of the cost is presented in Table 8.13; material requirement is calculated for treating a roof of (3.6 x 3.0) m^2 area. The life of the gunny-bag system was reported to be 2-3 years whereas the life of other materials is fairly long. Final choice of the material, however, depends upon the availability of the material locally, convenience of the user and the investment that one would like to make. The choice of the material would also affect the frequency and time period of the spray.

The sprayer should be adequately designed and should possess the following characteristics (Jain, 1977):

TABLE 8.11 Reduction of Daily Heat Flux by Water-Film/Open Pond over Roof for June 19, 1979 in 28.3°N, New Delhi, India

Sample Number	v (m/s)	R_h	ℓ_1 (m)	τ_2	$\dot{Q}_o$ (kJ/m²/day)	$\dot{m}_d$ (kg/m²/day)	$-\Delta\dot{Q}_o/\dot{m}_d$ (kJ/kg)	$\Delta\dot{Q}_o/\mathring{Q}_o$
1	0.0	0.4	0.23	0.54	273.6(5306)+	-	-	0.95
2	2.5	0.4	0.23	0.54	1174.0(2967)	4.32	648	0.70
3	5.0	0.4	0.23	0.54	1242.0(3640)	8.96	267	0.66
4	7.5	0.4	0.23	0.54	1267.0(3492)	13.50	164	0.64
5	10.0	0.4	0.23	0.54	1282.0(34.06)	14.36	118	0.62
6	4.47	0.1	0.23	0.54	158.0(3578)	9.72	352	0.96
7	4.47	0.2	0.23	0.54	518.0(3578)	9.14	335	0.86
8	4.47	0.4	0.23	0.54	1235.0(3578)	7.99	292	0.66
9	4.47	0.6	0.23	0.54	1951.0(3578)	6.84	237	0.46
10	4.47	0.8	0.23	0.54	2671.0(3578)	5.72	159	0.25
11	4.47	0.4	0.05	0.54	2585.0(7326)	7.45	635	0.65
12	4.47	0.4	0.10	0.54	1983.0(5749)	7.70	489	0.66
13	4.47	0.4	0.15	0.54	1609.0(4730)	7.85	398	0.66
14	4.47	0.4	0.23	0.54	1354.0(4176)	7.96	335	0.66
15	4.47	0.4	0.25	0.54	1166.0(3492)	8.03	390	0.67
16	4.47	0.4	0.30	0.54	1026.0(3089)	8.10	255	0.67
17	4.47	0.4	0.23	0.1	936.0(3578)	3.31	797	0.74
18	4.47	0.4	0.23	0.3	1073.0(3578)	5.44	461	0.70
19	4.47	0.4	0.23	0.5	1206.0(3578)	7.56	314	0.66
20	4.47	0.4	0.23	0.7	1343.0(3578)	9.72	230	0.62
21	4.47	0.4	0.23	0.9	1476.0(3578)	11.84	178	0.59

*The bracketed values are for single concrete slab of same thickness.

(i) uniform and constant wetting because full cooling effect can be experienced only after three days.

(ii) convenience of operating the system from lower floors,

(iii) the capability of the system to work at low pressures (7000-30000 N/m²), and

TABLE 8.12 Reduction of Daily Heat Flux by Water Moving over the Roof for 19 June, 1979 in 28.3°N, New Delhi, India. (After Tiwari *et al.*, 1982)

Sample Number	v (m/s)	u (m/s)	R_h	ℓ_1 (m)	τ_i	$\dot{Q}_o$ (KJ/m²/day)	$\dot{m}_d$ (kg/m²/day)	$\Delta Q_o/\dot{m}_d$ (KJ/kg)	$\Delta\dot{Q}_o/\dot{Q}_o$
1	0	0.015	0.4	0.23	0.54	2592(5306)*	8.06	337	0.51
2	2.5	0.015	0.4	0.23	0.54	2156(3967)	19.01	95	0.46
3	5.0	0.015	0.4	0.23	0.54	1904(3640)	25.74	67	0.48
4	7.5	0.015	0.4	0.23	0.54	1753(3492)	30.92	56	0.50
5	10.0	0.015	0.4	0.23	0.54	1656(3406)	35.50	49	0.51
6	4.47	0.015	0.1	0.23	0.54	1375(3578)	37.73	58	0.62
7	4.47	0.015	0.2	0.23	0.54	1566(3578)	33.34	60	0.56
8	4.47	0.015	0.4	0.23	0.54	1948(3578)	24.52	67	0.46
9	4.47	0.015	0.6	0.23	0.54	2329(3578)	15.70	80	0.35
10	4.47	0.015	0.8	0.23	0.54	2711(3578)	6.91	126	0.24
11	4.47	0.15	0.4	0.23	0.54	1948(3578)	24.52	67	0.46
12	4.47	0.050	0.4	0.23	0.54	2416(3578)	35.10	33	0.32
13	4.47	0.500	0.4	0.23	0.54	2700(3578)	38.66	23	0.25
14	4.47	1.00	0.4	0.23	0.54	2714(3578)	35.28	24	0.24
15	4.7	2.00	0.4	0.23	0.54	2716(3578)	27.83	37	0.24
16	4.47	0.015	0.4	0.05	0.54	1422(7326)	24.12	133	0.44
17	4.47	0.015	0.4	0.10	0.54	3146(5749)	24.23	107	0.45
18	4.47	0.015	0.4	0.15	0.54	2545(4730)	24.30	90	0.46
19	4.47	0.015	0.4	0.20	0.54	2135(4018)	24.44	77	0.47
20	4.47	0.015	0.4	0.25	0.54	1840(3492)	24.48	68	0.47
21	4.47	0.015	0.4	0.30	0.54	1616(3089)	24.52	60	0.48
22	4.47	0.015	0.4	0.30	0.10	932(3578)	3.28	808	0.74
23	4.47	0.015	0.4	0.30	0.20	997(3578)	4.32	598	0.72
24	4.47	0.015	0.4	0.30	0.40	1127(3578)	6.48	378	0.68
25	4.47	0.15	0.4	0.30	0.60	1260(3578)	8.64	268	0.65
26	4.47	0.015	0.4	0.30	0.80	1390(3578)	10.76	203	0.61
27	4.47	0.015	0.4	0.30	0.00	1454(3578)	11.84	179	0.59

*The bracketed values are for single concrete slab of same thickness.

TABLE 8.13 Cost of the Various Water-Retentive Materials. (After Jain, 1976)

S. No.		Thickness (mm)	Quantity of material required	Total cost of the material (Rs.)*
1	Gunny bags	50 (double layered)	35 cement bags (empty)	26.25
2	Coir matting	50 (double layered)	10.8 m^2	108.00
3	Brick-ballast	25	0.28 m^3	10.00
4	Coconut-husk (loose)	25	20 kg	10.00
5	Coconut rubberized matting	25	10.8 m^2	140.00
6	Sintered fly-ash	25	0.25 m^3	16.00

*12 Indian Rupees ≃ 1US$

(iv) the capability of covering the maximum possible roof area at these low pressures.

To meet the above-mentioned requirements, a water actuated automatic switch was developed at Central Building Research Institute, Roorkee, India which operates according to the wet or dry conditions of the material used over the roof. An idea of the roof area covered by the sprayer operating at various pressures and mounted at different angles is given in Table 8.14.

TABLE 8.14 Diameter of the Area Covered at Different Pressures and Angle of Spray. (After Jain, 1977)

Sprayer angle (degree)	Pressure (N/m x 10^{-})	Diameter covered (m)
0	6.86	1.675
	13.73	2.690
	20.59	4.150
	27.46	4.575
15	6.86	1.980
	13.73	3.300
	20.59	4.165
	27.46	5.490
30	6.86	1.980
	13.73	3.505
	20.59	4.775
	27.46	6.405
45	6.86	2.03
	13.73	3.76
	20.59	5.185
	27.46	6.910

8.2 EARTH COOLING

The earth behaves as a large reservoir of solar energy. Its thermal capacity is such that the diurnal variations of the surface temperature do not penetrate much deeper than 0.5 m, and seasonal variations not much deeper than 4.0 m. Beyond this depth, the earth's temperature, therefore, remains constant. The value of this temperature is usually seen to be equivalent to the all-year mean of solair temperature of its surface. From an architectural point of view, the earth's potential may be exploited either by direct coupling of the building envelopes with the ground, or through indirect coupling of these buildings with the earth by means of earth-air heat exchanging devices. Out of these, the former technique results in the underground and earth-bermed structures, while the example of latter technique are earth-air tunnels and cooling pipes inside the ground.

The design and construction of the structures exploiting the storage potential of the earth must accomplish the following points:

(i) the walls and roofs must be strong enough structurally to retain the pressure of the earth,

(ii) the waterproofing layer should be used; this serves as the vapour barrier between the earth and the structure,

(iii) the earth contact surface system must be conductive in nature, because only then the benefit of earth's potential will be achieved.

The experience of the workers in the field indicated that the construction and maintenance of the underground buildings on a large scale is usually a difficult task. Labs (1981) and Labs and Watson (1981) have mentioned that the underground placement of the buildings can more accurately be thought of as one of the several mechanisms for reducing heat-gain from the exterior, rather than as a positive cooling source. The problems, like, water seepage into the ground, condensation into the structures, and the inhabitants' dislike of being away from the nature, are also the factors limiting the scope of the underground dwellings. On the other hand, the systems, like, earth-air tunnels spreading over a large portion of the ground can effectively retrieve the coolness of the earth, and be used as an effective way of cooling. Since the performance of underground structures is controlled by the ground temperature distribution the forthcoming sections, deal with the evaluation of ground temperature distribution. Performance studies on the earth-air tunnel systems follow in the subsequent sections.

8.2.1 Undisturbed Temperature Distribution of Ground

A knowledge of the undisturbed temperature distribution of the ground gives an indication of the initial cooling value of the ground. Since the data for daily and annual variation of the surface and inside temperature of the ground is scarce, it is worthwhile to obtain an expression for the ground temperature making use of the actual data of solar radiation and ambient air temperature. Amongst others, Khatri *et al.* (1978) obtained such an expression which was based on assumed periodic variation of the ground temperature. The analysis, however, could not appreciate the effect of varying surface conditions on the ground temperature distribution. By treating the earth's surface suitably, one can bring down the earth's temperature to an acceptable value. In general, the surface temperature of the earth is determined by the balance of several heat transfer modes, viz.

(i) absorption of incident solar radiation during daytime

(ii) heat loss upwards by the long wave radiation to the sky

(iii) heat exchange by convection with the ambient air

(iv) heat loss by water evaporation from the surface.

Cooling of a given area of the earth surface below its natural level is usually achieved by several means and their combinations, such as shading the surface to eliminate its heating by the sun, controlled irrigation to increase evaporative cooling, and others. In order to appreciate this phenomenon, Kusuda (1980) carrie out some experiments in Washington, DC. He observed that the average surface temperature of the earth covered by long grass was lowered by 4°C as compared to the uncovered earth. The total effect of shading and evaporation was observed to be 11°C decrease in the average surface temperature. To incorporate this effect, Bharadwaj and Bansal (1981) developed an analytical model and calculated the daily and annual variations of the ground temperature at Delhi, India. The model is described below.

8.2.1.1 Analytical model. Assuming that the ground surface is wet, the energy-balance equation for this may be written as equation (8.7) with the only differenc that $\theta\ (x,t)$ now represents the temperature distribution of the ground, instead of roof. This equation is a general equation which, however, can be applied to various surface conditions by substituting the following in the expression for θ_{eff} (Eq. 8.7).

(i) $R_1 = R_2 = 0$ for dry and sunlit surface

(ii) $R_1 = R_2 = 0$, $S(t) = 0$, for dry and shaded surface

(iii) $S(t) = 0$, for wet and shaded surface

(iv) the Eq. 8.7 is valid for wet and sunlit surface.

Now, assuming ground to be a semi-infinite media, which means that

$$\theta_G(x,t) = \text{finite as } x \to \infty \tag{8.17}$$

and using the periodic method of solutions, the expression for ground temperature distribution gets reduced to the following form:

$$\theta_G(x,t) = A_o + \text{Re} \sum_{n=1}^{\infty} A_n \exp\left[i(n\omega t - \alpha_n x)\right] \tag{8.18}$$

Substituting for θ_{eff} and θ_G from equations (8.9) and (8.18) respectively, into equation (8.7), one obtains

$$\theta_G = a_o^1 + \sum_{n=o}^{\infty} B \exp(-n^{\frac{1}{2}} \alpha_o x) \operatorname{Cos}(n\omega t - n^{\frac{1}{2}} \alpha_o x - \psi_n - \beta_n), \tag{8.19}$$

where

$$B_n = a_n \left[(1 + n^{\frac{1}{2}} \mu)^2 + n\mu^2\right]^{-\frac{1}{2}},$$

$$\mu = \left(\frac{K\alpha_o}{h_{eff}}\right),$$

$$\beta_n = \tan^{-1}\left[\frac{n^{\frac{1}{2}}\mu}{1 + n^{\frac{1}{2}}\mu}\right]$$

$$\alpha_o = \sqrt{\frac{\omega\rho c}{2K}}$$

Using the above expression and substituting the value of relevant parameters, the temperature distribution inside the ground may be evaluated as a function of depth and time.

8.2.1.2 Discussion of the model. Bharadwaj and Bansal (1981) carried out numerical calculations to evaluate the annual variation of ground temperature as a function of depth. The atmospheric data used in the calculations was of Delhi, India. Properties of the ground were taken to be as given in Table 8.15.

TABLE 8.15 Properties of the Ground. (After Bharadwaj and Bansal, 1981)

Condition of the surface	K (W/m °C)	ρ(kg/m^3)	C (kJ/kg °C)
Dry	0.510	2050	1.8423
Wet	1.456	2460	2.2353
(assuming 50% moisture content)			

It was found that the earth's temperature at a depth of 4 m shows an insignificant variation in its magnitude with time ($\approx$ months). For the four cases viz. dry and sunlit, wet and sunlit, dry shaded and wet shaded, the ground temperature was reported to be 29.5, 18.7, 22.0 and 17.3° respectively at a depth of 4 m with a variation of 1.5°C approximately. Expression (8.19) allows the determination of the inside temperature of the earth as a function of depth at any location and for various surface conditions. It may be noted that if the ground surface is maintained wet throughout the year, sufficiently low temperatures can be achieved.

8.3 EARTH-AIR TUNNEL SYSTEM

The use of earth as a heat source/sink with buried pipes or underground tunnels as a direct heat exchanger is a concept that has existed in Islamic and Persian Architecture for a number of centuries. The systems based on this concept have been repeatedly used in architectural design for natural conditioning of the air and maintaining internal comfort. However, the non-availability of the performance data of the actual systems restricted the extensive use of the concept. In recent years, the concept has again started getting due attention as a result of which some studies indicating the performance of such systems are seen.

The earth-air heat exchanger system utilizes the storage capacity of the earth, and conditions the air. Damp earth being a good conductor of earth, can effectively exchange heat with the passing air in its contact. The amount of heat exchanged between the air and the surrounding soil, however, is a function of following parameters.

(i) Surface area of the tunnel walls,

(ii) Length of the tunnel,

(iii) Inlet air temperature,

(iv) Water contents of the inlet air,

(v) Temperature of earth

(vi) Water contents of the earth,

(vii) Air-velocity

(viii) Surface conditions of the tunnel walls,

(ix) Material of the wall,

(x) Depth of the tunnel from ground surface.

To achieve air movement through the earth-air tunnel, additional systems e.g. thermal chimney (Figs. 8.16 and 8.17) and convective air drive are usually integrated with the tunnel. In Persian architecture, the thermal chimney appears to be very common for this purpose, a typical example is shown in Fig. 8.17. In conventional approach, the mechanical blowers are also used for creating air movement inside the tunnels (Hourmanesh *et al.*, 1980).

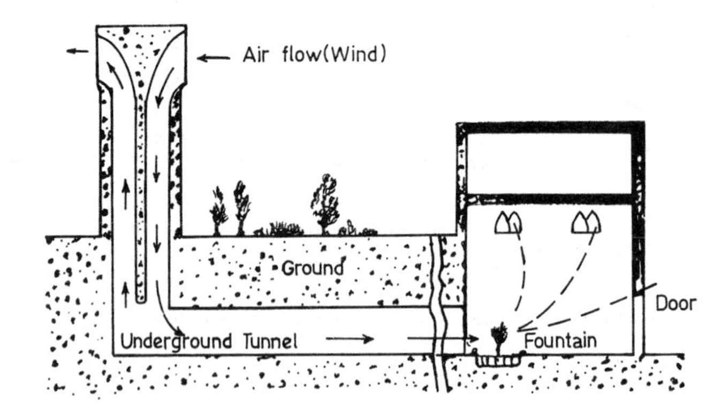

Fig. 8.16. Earth-air exchanger systems with thermal chimney system

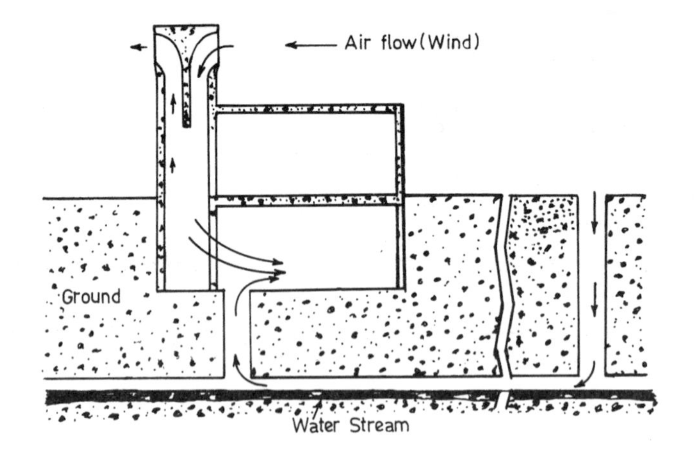

Fig. 8.17. Passive design in Persian architecture. (After Hourmanesh *et al.*, 1980)

8.3.1 Experience

The concept has not been used very extensively, however, it has come out of the natural curiosity phenomenon to the scientific realization in the past few decades. One of the first scientific attempts for constructing the actual earth-air tunnel system was made at Forman Christian College, Lahore, Pakistan where outside air was drawn into the windows of a cellar and then down an air-well to an underground tunnel. This tunnel made a rectangular circuit having a running-length of 35 m along its central line; this line was about 4 m below the ground surface. The air was then drawn up through a central duct by means of a single fan and distributed by other ducts to adjustable grills, one of which opened to each of the ten rooms of the house. The air paths from tunnel to occupants were made as short as the design would feasily allow and in order to reduce losses on the way, the ducts were made of wood. The fan was of low power and the system was not expected to be used for more than two rooms at a time. The earth tempered air was shut off from unoccupied rooms by damper values: one located suitably behind each grill. Exploratory studies and experiments indicated that, by and large, the earth cooled air stream provided reasonable comfort in summer weather. The benefits in winter were, however, found to be less marked, though the warm air stream certainly helped to the extent of removing the chill from a room on some winter days. At Clara Swaine Hospital, Bareilly in India, the operating rooms were successfully cooled by an air-conditioner whose effectiveness was substantially increased because the intake-air was drawn from an underground tunnel (Thoburn, 1983).

The later attempts, however, modified the approach of realizing the concept. Instead of digging the tunnel inside the ground, the pipes, made up of the materials like, plastic, concrete, clay or corrugated metal, were laid down into the earth. The pipes may have circular or square cross-sectional shape. Among others (Hendrick, 1980, Abrahms and Benton, 1980, Sinha *et al.*, 1981), Strayer (1979) studied the theoretical and experimental performance of such a system and presented the performance data (Table 8.16). The system had the cross-sectional area of 0.15 m x 0.5 m, and it was reported that the actual performance of the system was close to the calculated values.

More rigorous attempts for studying the performance of "Earth-cooling tubes" in USA and "earth-air tunnel system" in India were, however, made by Francis (1984) and Sodha *et al.* (1984) respectively. These are described below:

8.3.1.1 In United States of America. Francis (1984) has reported recently that a tube cooling system was installed in Illinois for the annual space-conditioning of a farrowing house. The system consisted of two plastic field-tile tubes, having a diameter of 0.2 m. The length of the tubes was 120 m and these were buried 3 m deep into the ground. The tubes sloped downhill, away from the farrowing house to a drainage creep, so that the condensate is automatically removed. A fan was used to draw the air through these tubes to a concrete manifold tube and was then ducted to an air-distribution system. The whole arrangement is shown in Fig. 8.18. The capacity of the fan was 746 watts. Performance of the system is presented in Table 8.17. It is obvious that most of the cooling occurred up to 60 m of its length. The conditions at the time of test were as follows:

Inlet air temperature = 35°C (dry-bulb)
= 31.7°C (wet bulb)

Air velocity = 3.425 m/s

Air volume = 388.8 m^3/hr

Condensate from the tube = 8.27 kg/hr

Cooling output = 9.46 kW per tube

The most important drawback of the system was found to be the small diameter and large length of the tubes. This forced the owner of the system to use a fairly high-powered fan. This was also essential to obtain a suitable amount of cooling capacity.

TABLE 8.16 Performance Data of Earth-air Tunnel System in Terms of Temperature Difference Between the Inlet and Outlet Temperatures of the Air. (After Strayer, 1979)

Tunnel length (m)	Ambient air temperature (°C)					
	-17.8	-6.7	4.4	26.7	32.2	37.8
0.3	1.1	0.7	0.3	0.5	0.7	0.9
0.6	2.2	1.4	0.6	1.0	1.4	1.8
0.9	3.3	2.1	1.0	1.4	2.0	2.6
1.2	4.3	2.8	1.3	1.9	2.6	3.4
1.5	5.3	3.4	1.5	2.3	3.2	4.2
1.8	6.2	4.0	1.8	2.7	3.8	4.9
2.1	7.2	4.6	2.1	3.1	4.4	5.6
2.4	8.0	5.2	2.3	3.5	4.9	6.3
2.7	8.9	5.7	2.6	3.8	5.4	7.0
3.0	9.7	6.3	2.8	4.2	5.9	7.6
3.3	10.5	6.8	3.0	4.5	6.4	8.3
3.6	11.3	7.3	3.2	4.8	6.9	8.9
3.9	12.0	7.7	3.4	5.1	7.3	9.4
4.2	12.7	8.2	3.6	5.4	7.7	10.0
4.5	13.4	8.6	3.8	5.7	8.1	10.5
4.8	14.0	9.0	4.0	6.0	8.5	11.0
5.1	14.7	9.4	4.2	6.3	8.9	11.5
5.4	15.3	9.8	4.4	6.5	9.3	12.0
5.7	15.8	10.5	4.5	6.8	9.6	12.5
6.0	+16.4	10.6	+4.7	-7.0	-10.0	-13.0
Heat output $(W/m^2 {}^oC) \times 10^{-3}$	79.7	51.3	22.8	34.2	48.4	62.6

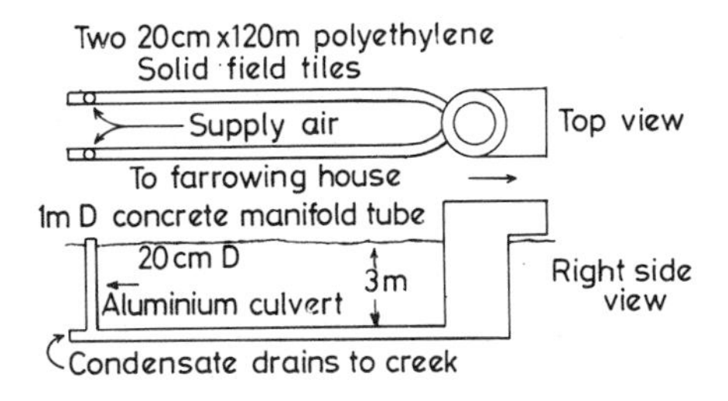

Fig. 8.18. Case I, cool-tube array. (After Francis, 1984)

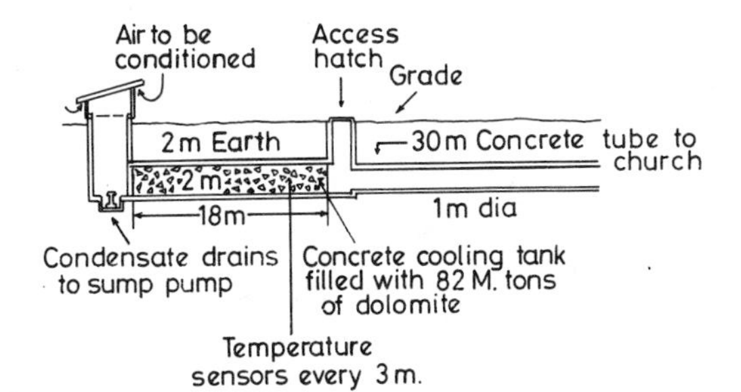

Fig. 8.19. Mennonite Church cooling system, normal Illinois. (After Francis, 1984)

TABLE 8.17 Performance of a Cooling-Tube System for a Farrowing House. (After Francis, 1984)

Distance into tube (m)	Dry-bulb temp. (°C)	Wet bulb temp. (°C)
3.0	30.0	-
6.0	26.1	-
9.0	25.0	-
12.0	23.9	-
30.0	22.8	-
60.0	20.6	-
90.0	18.3	-
120.0	18.3	13.9

The other systems installed were of smaller capacities. One system was made of a smooth surface PVC tube having a diameter of 0.15 m and 16.5 m in length. It was also sloped down away from the house to ensure automatic removal of the condensate. Inside the building, a special air-handling system, as shown in Fig. 8.19, controlled the operation. In typical hot weather conditions it could cool the ambient air from a temperature of 26.4°C to 20.6°C. The corresponding wet-bulb temperatures were 24.4°C and 18.9°C respectively. This system provided useful cooling.

8.3.1.2 In India. Sodha *et al.* (1984) have carried out the rigorous experimental studies with a large earth-air tunnel system situated at Mathura, India. The system consisted of a main tunnel and several subsidiary tunnels as shown in Figs. 8.20 and 8.21. This system was basically erected to cool the entire hospital complex below which the tunnel is situated. Vertical ducts were also provided in each of the rooms with a provision to draw the tunnel air into the rooms. The cross-sectional area of the tunnel varies from (3.6 x 4.5) m^2 to (0.9 x 0.9) m^2. Ambient air could be drawn inside the tunnel through a wind-tower located at the beginning of the tunnel. The tunnels provide convenient passageways for cables and water mains which are thus easily serviced and protected from corrosion in the alkaline soil.

Experimental results are reported only for a portion (situated between skylights B and C in Fig. 8.21) of the main tunnel whose length is 80 m. During the experiment, the air was drawn inside the tunnel by using mechanical fans. The wet-bulb and dry-bulb temperatures of the air were measured at inlet, outlet and at every 10 m length of the tunnel. The temperatures of the tunnel surface were also recorded at every 20 m length of the tunnel. The results of their experiments are summarized in Table 8.18 for the typical summer days. On an average, the temperature of the ambient air passing through the tunnel dropped by about 15°C providing the required thermal-comfort conditions (see Table 8.18).

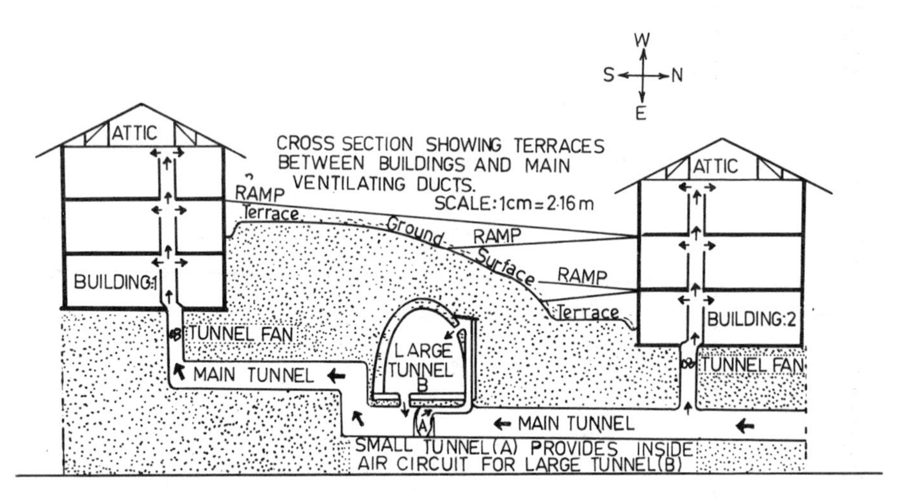

Fig. 8.20. Earth-air tunnel system at St. Methodist Hospital, Mathura, U.P. (India)

8.3.2 Thermal Modelling of Earth-Air Tunnel System

Various models have been developed which describe the thermal performance of earth-air tunnel system under certain specific conditions. These are discussed below:

8.3.2.1 Simplified model: (Bansal *et al.*, 1983).

(i) The Analytical Model. This model is based on the fact that the earth's

TABLE 8.18 Performance of Earth-Air Tunnel (Mathura, India). (After Sodha *et al.*, 1984)

Date	Air temperature (^{o}C)												Surface temperature (^{o}C)		
	Inlet						Outlet								
	Dry-bulb temperature			Wet-bulb temperature			Dry-bulb temperature			Wet-bulb temperature					
	Max.	Min.	Ave.	Max.	Min.	Ave.	Max.	Min.	Ave.	Max.	Min.	Ave.	Max.	Min.	Ave.
15.6.83	39.7	26.2	32.8	24.3	20.0	21.5	27.2	23.6	26.0	24.0	21.0	23.0	23.0	18.8	21.3
16.6.83	34.0	24.1	29.1	23.5	20.3	22.7	25.7	23.2	24.3	22.3	20.8	21.7	24.3	21.0	22.7
17.6.83	37.1	26.8	29.8	22.5	16.5	20.5	26.8	23.6	26.6	24.0	21.3	22.3	23.3	19.8	21.7
18.6.83	38.0	23.4	30.9	24.0	20.0	21.7	27.2	23.1	24.8	24.5	22.3	23.1	24.3	20.0	21.6
19.6.83	39.2	25.7	33.0	25.5	21.0	22.5	27.8	23.6	25.7	23.8	21.5	23.0	25.5	13.5	21.5
20.6.83	41.7	28.2	34.4	25.3	21.3	23.7	28.2	24.7	26.6	25.5	23.0	24.3	24.3	23.3	23.7
21.6.83	41.7	28.2	35.0	25.5	20.8	22.7	27.8	24.7	26.2	24.8	22.0	23.3	24.0	22.5	23.7
22.6.83	42.5	28.8	35.0	25.5	21.5	22.9	27.8	24.7	26.2	25.5	20.3	23.0	23.3	21.8	22.9
23.6.83	43.2	30.3	36.1	27.5	21.5	24.7	28.2	25.2	26.6	26.5	22.0	23.7	27.5	22.3	23.7
24.6.83	36.5	30.3	31.8	25.3	20.5	22.8	27.2	25.2	25.7	25.0	20.3	23.9	24.3	22.8	23.7

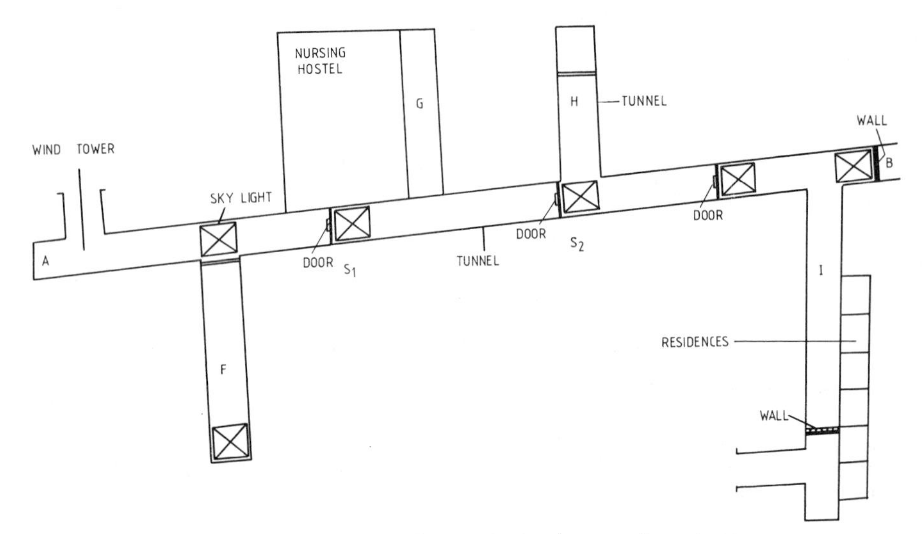

Fig. 8.21. Schematic of earth-air tunnel system

temperature at larger ($\approx$ 4 m) depths remains constant, and is equal to the all-year average value of solair temperature of the effective temperature defined by Eq. 8.7c.

Let us consider an infinitesimal element of tunnel in the direction of air-flow (Fig. 8.22). The energy balance over this element of width may be written as

$$\dot{m}_a \, c_a \frac{\partial \theta_f}{\partial y} \cdot dy = \dot{Q} \; b \; dy \tag{8.20}$$

where $\dot{Q}$ is the amount of heat transferred per unit area of the tunnel surface.

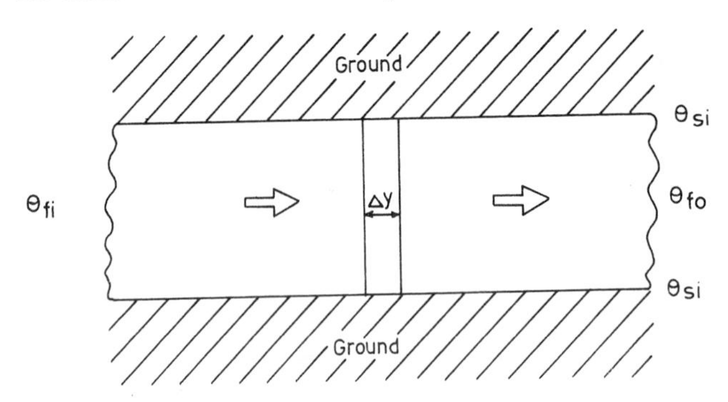

Fig. 8.22. Elemental cross-section of an underground tunnel

Assuming that the heat extraction does not disturb the temperature distribution of the surrounding earth and the earth is dry the heat will mainly be transferred by convection i.e.

$$\dot{Q} = h_c\ (\theta_{si} - \theta_f) \tag{8.21}$$

where θ_{si}, the surface temperature of the tunnel, is assumed to be greater than the inlet air temperature.

On solving the equations (8.30) and (8.31) in combination with the initial condition, that

$$\theta_f = \theta_{fj} \text{ at } y = 0,$$

one gets

$$\theta_f(y) = \theta_{si} - (\theta_{si} - \theta_{fj})\ \exp\ (-\ h_c\ by/\dot{m}_a C_a) \tag{8.22}$$

The rise in tunnel air temperature may then be calculated by the expression

$$\Delta\theta = (\theta_{si} - \theta_{fj})\ [1 - \exp\ (-\ h_c bL/\dot{m}_a C_a)] \tag{8.23}$$

where L is the length of the tunnel. It is obvious with this expression that if inlet air temperature is higher than the surface temperature of the tunnel, a negative $\Delta\theta$ will be obtained which shows that the air is getting cooled.

In case the tunnel surface is wet, the vapour transfer also takes place, which may be evaluated as follows

$$\dot{m}_w = h_D\ (W_s - W)$$

where W_s is the humidity ratio of the air in saturation state, and h_D is the mass transfer coefficient which depends upon convective heat-transfer coefficient as given below.

$$\frac{h_c}{h_D} = \frac{k}{\mathcal{D}_m\ \rho_w} \left(\frac{\mathcal{D}m}{\mathcal{D}}\right)^c$$

Following Threlkeld (1974), one has

$$W_s - W = \frac{Ca}{L}\ (\theta_f - \theta_{si})$$

Therefore

$$\dot{m}_w = \frac{h_D Ca}{L}\ (\theta_f - \theta_{si}) \tag{8.24}$$

and the amount of heat associated with the water vapours is

$$\dot{Q}_e = h_D Ca\ (\theta_f - \theta_{si}) \tag{8.25}$$

This adds to the cooling effect of the tunnel air, and hence

$$\dot{Q} = -\ (h_c + h_D\ C_a)\ (\theta_f - \theta_{si}) \tag{8.26}$$

where, the value of C_a is the sum of specific heat of dry air and that of the water vapour associated with it. Thus,

$$C_a = (1.005 + 1.884\ W) \times 10^3$$

(ii) Experimental Validation. Sodha *et al.* (1984) have evaluated the performance of the earth-air tunnel system using the above-described model, and compared with the actual performance of the system. The system under experimentation was the same which has already been described in section 8.3.1.2. The surface temperature of the tunnel was measured at various points along the length of tunnel; an average of these observations was used in the calculations. Other relevant parameters used were:

Velocity of air = 4.89 m/s

Density of air = 1.2 kg/m^3

h_c = 9.83 W/m^2 ^{o}C

h_D = 10.978/C_a kg/s.m^2

The results are presented in Figs. 8.23 and 8.24. It can be seen that the predicted values are in close agreement with that of the experimental ones.

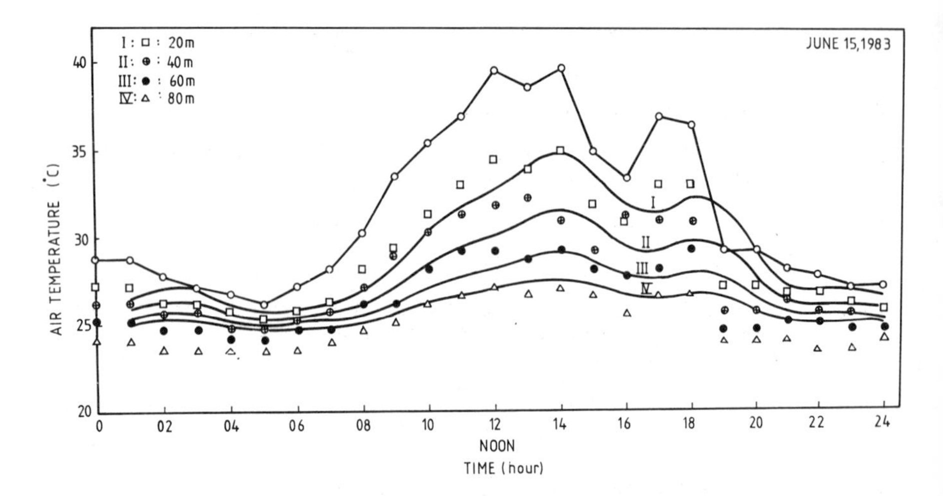

Fig. 8.23. Experimental performance of earth-air tunnel system. (After Sodha *et al.*, 1984)

The results of the theoretical investigations (Bansal *et al.*, 1983) carried out with respect to a change of various operational parameters viz. tunnel length, air-velocity, and inlet air temperature are presented in Tables 8.19 and 8.20 for sunlit ground surface (for winter heating) and shaded ground surface (for summer cooling) respectively. These tables may be used to assess the effectiveness of the concept as a function of various parameters.

8.3.2.2 Transient model (Pratt and Daws, 1958). Various mathematical models were developed by Pratt and Daws (1958) which were of special interest to the cases of heat transfer calculations in underground structures. Here, only the model useful for ventilated underground tunnels is discussed.

(i) The Analytical Model. Following assumptions were made in developing the analytical model:

(a) The tunnel is of uniform circular cross-section

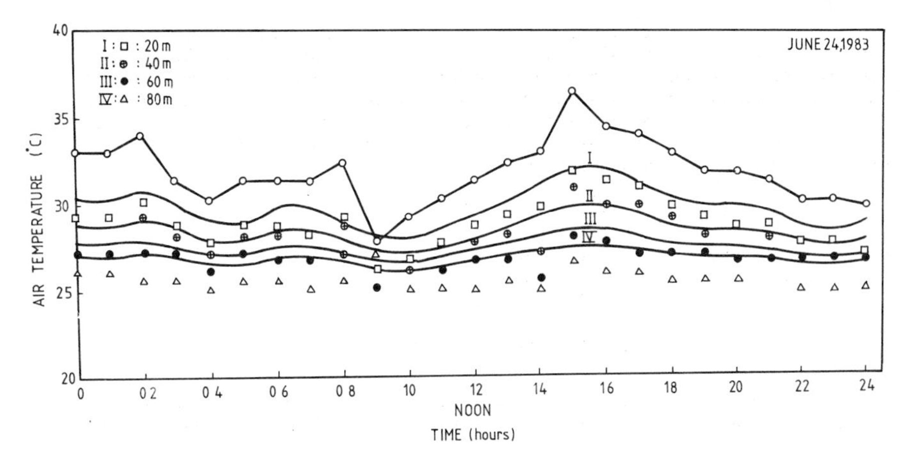

Fig. 8.24. Experimental performance of earth-air tunnel system. (After Sodha *et al.*, 1984)

(b) Longitudinal heat-flow is neglected.

(c) The surface temperature of the surrounding medium changes uniformly when heat is extracted from the tunnel.

(d) The material of the surrounding medium is homogeneous.

(e) Initially, the temperature distribution of the earth is uniform.

In view of the above assumptions, the temperature distribution of the surrounding earth may be assumed to be expressed by Fourier equation viz.

$$\frac{\partial^2\theta_G(r_d,t)}{\partial r_d^2} + \frac{1}{r_d} \cdot \frac{\partial\theta_G(r_d,t)}{\partial r_d} = \mathcal{D}\,\frac{\partial\theta_G(r_d,t)}{\partial t} \tag{8.24}$$

where $r_d > r_t$, r_t being the radius of the tunnel cross-section.

Now, we assume that the heat is generated by the machinery at a constant rate ($\dot{Q}$), and the heat given off by the occupants is proportional to the difference of body temperature (assumed to be constant) and the tunnel air temperature; it is also assumed that the tunnel is ventilated at a constant rate. The boundary conditions may, therefore, be written as follows:

$$Q + h_b\,[\theta_{c\ell} - \theta_f] = C^1\,[\theta_f - \theta_{fj}] - K\,\frac{\partial\theta_G(r_t,t)}{\partial r_d} \tag{8.25}$$

$$-K\,\frac{\partial\theta_G(r_t,t)}{\partial r_d} = h_c\,[\theta_f - \theta_G\,(r_t,t)] \tag{8.26}$$

where C' is the rate at which the heat is removed by the outside air, and h_b is heat-transfer coefficient from the body to the tunnel air.

On solving the equation (8.24) with the help of boundary conditions (equations 8.25 and 8.26), the solutions may be obtained in dimensionless form as follows:

TABLE 8.19 Temperature Gain by the Air for Glazed Sunlit Ground Surface (After Bansal *et al.*, 1983)

$r_t = 0.3$ m
$\theta_{sa} = 53.5^{o}C$

L (m)	Air Flow Velocity (m/s)															
	1.0				3.0				5.0				7.0			
	Increase in temperature for various inlet temp.				Increase in temperature for various inlet temp.				Increase in temperature for various inlet temp.				Increase in temperature for various inlet temp.			
	$4^{o}C$	$8^{o}C$	$15^{o}C$	$18^{o}C$	$4^{o}C$	$8^{o}C$	$15^{o}C$	$18^{o}C$	$4^{o}C$	$8^{o}C$	$15^{o}C$	$18^{o}C$	$4^{o}C$	$8^{o}C$	$15^{o}C$	$18^{o}C$
0.0	0.0	0.0	0.0	0.0	0.0	0.0	0.0	0.0	0.0	0.0	0.0	0.0	0.0	0.0	0.0	0.0
0.5	1.3	1.2	1.0	0.9	0.8	0.8	0.7	0.6	0.7	0.7	0.6	0.5	0.7	0.6	0.5	0.5
1.0	2.6	2.3	2.0	1.7	1.7	1.5	1.3	1.2	1.5	1.4	1.2	1.1	1.4	1.3	1.1	1.0
1.5	3.8	3.5	3.0	2.7	2.5	2.3	1.9	1.8	2.2	2.0	1.7	1.6	2.1	1.9	1.6	1.5
2.0	5.0	4.6	4.0	3.6	3.3	3.0	2.5	2.4	2.9	2.7	2.3	2.1	2.8	2.5	2.2	2.0
2.5	6.2	5.7	4.8	4.4	4.1	3.7	3.2	2.9	3.6	3.3	2.8	2.6	3.4	3.2	2.7	2.5
3.0	7.3	6.8	5.7	5.3	4.8	4.4	3.8	3.4	4.3	4.0	3.4	3.1	4.1	3.8	3.2	2.9
3.5	8.5	7.8	6.4	6.1	5.6	5.1	4.3	4.0	5.0	1.6	3.9	3.6	4.7	4.4	3.7	3.4
4.0	9.5	8.8	7.4	6.8	6.3	5.8	4.9	4.5	5.7	5.2	4.4	4.1	5.4	4.9	4.2	3.8
4.5	10.6	9.8	8.2	7.6	7.1	6.5	5.5	5.1	6.3	5.8	4.9	4.5	6.0	5.5	4.7	4.3
5.0	11.6	10.7	9.0	8.3	7.8	7.3	6.1	5.6	7.0	6.4	5.4	5.0	6.6	6.1	5.2	4.8
5.5	12.6	11.6	9.8	9.1	8.5	7.8	6.6	6.1	7.6	7.0	5.9	5.5	7.2	6.7	5.6	5.2
6.0	13.6	12.5	10.6	9.8	9.2	8.5	7.2	6.6	8.3	7.6	6.4	5.9	7.8	7.2	6.1	5.6
6.5	14.5	13.4	11.3	10.4	9.9	9.1	7.7	7.1	8.9	8.2	6.9	6.4	8.4	7.8	6.6	6.0
7.0	15.5	14.2	12.0	11.1	10.6	9.7	8.2	7.6	9.5	8.7	7.4	6.8	9.0	8.3	7.0	6.4

TABLE 8.19 (cont'd)

L (m)	Air Flow Velocity (m/s)															
	1.0				3.0				5.0				7.0			
	Increase in temperature for various inlet temp.				Increase in temperature for various inlet temp.				Increase in temperature for various inlet temp.				Increase in temperature for various inlet temp.			
	4°C	8°C	15°C	18°C	4°C	8°C	15°C	18°C	4°C	8°C	15°C	18°C	4°C	8°C	15°C	18°C
7.5	16.4	15.0	12.7	11.7	11.2	10.3	8.7	8.0	10.1	9.3	7.8	7.2	9.6	8.8	7.5	6.9
7.6	17.2	15.9	13.4	12.4	11.8	10.9	9.2	8.5	10.7	9.8	8.3	7.7	10.2	9.3	7.9	7.3
8.5	18.1	16.6	14.1	13.0	12.5	11.5	9.7	9.0	11.3	10.4	8.8	8.1	10.7	9.9	8.3	7.7
9.0	18.9	17.4	14.7	13.6	13.1	12.1	10.2	9.4	11.8	10.9	9.2	8.5	11.3	10.4	8.8	8.1
9.5	19.7	18.1	15.3	14.2	13.7	12.6	10.7	9.9	12.4	11.4	9.7	8.9	11.8	10.9	9.2	8.5
10.0	20.5	18.9	16.0	14.7	14.4	13.2	11.2	10.3	13.0	11.9	10.1	9.3	12.4	11.4	9.6	

TABLE 8.20 Temperature Loss by the Air for Wet-Shaded Ground Surface (After Bansal *et al.*, 1983)

r_t = 0.3 m
θ_{sa} = 17.3°C

L (m)	Air Flow Velocity (m/s): 1.0 — Decrease in temperature for various inlet temp.				3.0 — Decrease in temperature for various inlet temp.				5.0 — Decrease in temperature for various inlet temp.				7.0 — Decrease in temperature for various inlet temp.			
	30°C	35°C	40°C	45°C	30°C	35°C	40°C	45°C	30°C	35°C	40°C	45°C	30°C	35°C	40°C	45°C
0.0	0.0	0.0	0.0	0.0	0.0	0.0	0.0	0.0	0.0	0.0	0.0	0.0	0.0	0.0	0.0	0.0
0.5	1.2	1.7	2.1	2.6	0.8	1.1	1.4	1.7	0.7	1.0	1.2	1.5	0.7	0.9	1.2	1.4
1.0	2.3	3.1	4.0	4.9	1.5	2.1	2.7	3.3	1.3	1.9	2.4	2.9	1.3	1.8	2.3	2.8
1.5	3.2	4.5	5.8	7.0	2.2	3.0	3.9	4.7	1.9	2.7	3.5	4.2	1.9	2.6	3.3	4.0
2.0	4.1	5.7	7.3	9.0	2.8	3.9	5.0	6.1	2.5	3.5	4.5	5.5	2.4	3.3	4.3	5.2
2.5	4.9	6.8	8.8	10.7	3.4	4.7	6.1	7.4	3.1	4.3	5.5	6.7	2.9	4.1	5.2	6.4
3.0	5.6	7.8	10.1	12.3	4.0	5.5	7.1	8.7	3.6	5.0	6.4	7.8	3.4	4.8	6.1	7.5
3.5	6.3	8.8	11.2	13.7	4.5	6.3	8.0	9.8	4.1	5.7	7.3	8.9	3.9	5.4	7.0	8.5
4.0	6.9	9.6	12.3	15.1	5.0	7.3	8.9	10.9	4.5	6.3	8.1	9.9	4.3	6.5	7.8	9.5
4.5	7.4	10.3	13.3	16.2	5.5	7.6	9.8	11.9	5.0	7.0	8.9	10.9	4.8	6.7	8.5	10.5
5.0	7.9	11.0	14.1	17.3	5.9	8.2	10.5	12.9	5.4	7.5	9.7	11.8	5.2	7.2	9.3	11.3
5.5	8.4	11.7	14.9	18.2	6.3	8.8	11.3	13.8	5.8	8.1	10.4	12.6	5.6	7.8	9.9	12.1
6.0	8.8	12.2	15.7	19.1	6.7	9.3	12.0	14.6	6.1	8.6	11.0	13.5	6.0	8.3	10.6	12.9
6.5	9.1	12.7	16.3	19.9	7.1	9.8	12.5	15.4	6.7	9.1	11.7	14.2	6.3	8.7	11.2	13.7
7.0	9.5	13.2	16.9	12.6	7.4	10.3	13.2	16.1	7.2	9.6	12.2	14.9	6.6	9.2	11.8	14.4

TABLE 8.20 (Cont'd)

L (m)	Air Flow Velocity (m/s)															
	1.0				3.0				5.0				7.0			
	Decrease in temperature for various inlet temp.				Decrease in temperature for various inlet temp.				Decrease in temperature for various inlet temp.				Decrease in temperature for various inlet temp.			
	30°C	35°C	40°C	45°C	30°C	35°C	40°C	45°C	30°C	35°C	40°C	45°C	30°C	35°C	40°C	45°C
7.5	9.8	13.6	17.4	21.3	7.7	10.8	13.8	16.8	7.5	10.0	12.8	15.6	6.9	9.6	12.4	15.1
8.0	10.0	14.0	17.9	21.9	8.0	11.2	14.3	17.5	7.8	10.4	13.3	16.3	7.2	10.0	12.9	15.7
8.5	10.3	14.3	18.4	22.4	8.3	11.6	14.8	18.1	8.0	10.8	13.8	16.9	7.5	10.4	13.4	16.3
9.0	10.5	14.6	18.8	22.9	8.6	11.9	15.3	18.7	8.3	11.2	14.3	17.5	7.8	10.8	13.9	16.9
9.5	10.7	14.9	19.1	23.4	8.8	12.9	15.8	19.2	8.5	11.5	14.7	18.0	8.0	11.2	14.3	17.5
10.0	10.8	15.3	19.5	23.8	9.1	12.6	16.2	19.8	8.7	11.9	15.1	18.6	8.2	11.5	14.7	18.0

For moderately small periods of time:

$$\phi(r_t,t) \simeq N\left[\frac{2T'^{\frac{1}{2}}}{\pi^{\frac{1}{2}}} - (\tfrac{1}{2} + N)\,T' + (\tfrac{3}{8} + N + N^2)\,\frac{4T'^{3/2}}{3\pi^{\frac{1}{2}}}\right.$$

$$- (\tfrac{3}{8} + N + \frac{3N^2}{2} + N^3)\,T'^2 + (\frac{126}{256} + \frac{261}{256}N$$

$$\left. + \frac{15N^2}{8} + \frac{7}{4}N^3 + N^4)\,\frac{8T'^{5/2}}{15\pi^{\frac{1}{2}}} + ----\right] \tag{8.27}$$

For large values of time:

$$\phi(r_t,t) \simeq 1 - \frac{2}{Ny}\left[1 + \frac{\nu}{y} - \frac{(\pi^2/6 - \nu^2)}{y^2} + \frac{2.404 + (3\nu\pi^2/2) + \nu^3}{y^3}\right.$$

$$+ \frac{1}{T'}\left[\frac{N+2}{N^2y^2} - \frac{2\{(1+\nu)N(N+2)+2\}}{N^3\,y^3} + ---\right] + --- \tag{8.28}$$

where,

$$\phi_a(t) = (1 - \eta) + \eta\,\phi\,(r_t,t),$$

and

$$N = [r_t\,h_c\,(c' + h_b)/K\,(c' + h_c + h_b)],$$

$$\eta = [h_c/(c' + h_c + h_b)],$$

$$\phi_a(t) = \theta_G(r_t,t)/\theta_{final},$$

$$\phi(t) = \theta_f(t)/\theta_{final},$$

$$\theta_{final} = (Q + h_b\,\theta_{c\ell} + c'\,\theta_{fc})/(c' + h_b),$$

$$T' = \mathcal{D}t/r_t^2,$$

$$y = \ln\Delta\theta_G - 2\nu + 2/N,$$

ν = Euler's constant ($\simeq 0.5772$)

θ_{final} = final temperature.

It may be noted from the equations (8.27) and (8.28) that $\phi(r_t,t)$ depends mainly on two dimensionless variables T' and N, and hence has been plotted as a function of these variables (Fig. 8.25). For various values of η, the function $T'(t)$ is also represented in Fig. 8.26. These curves are very useful, because for a given set of system parameters, $\phi(r_t,t)$ can be evaluated from Fig. 8.25 corresponding to the relevant values of T' and N, which consequently can be used to determine the value of air temperature with the help of Fig. 8.26.

(ii) Experimental Validation of the Model. Experimental investigations were carried out with an underground tunnel system which was basically meant for providing the suitable ventilation to the underground structures. The tunnel

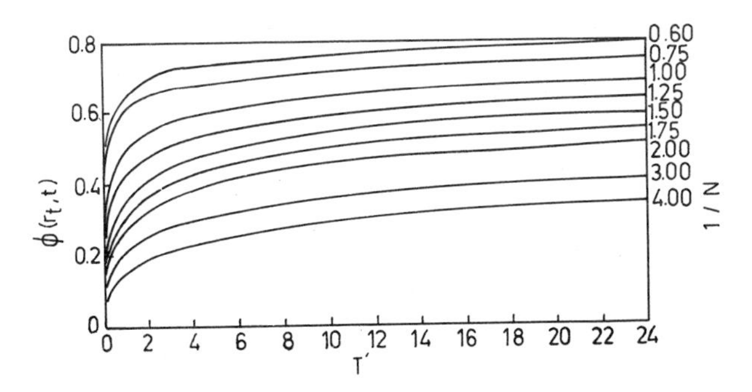

Fig. 8.25. Theoretical solution giving tunnel air and surface temperatures with continuous heating and ventilating conditions. (After Pratt and Daws, 1958)

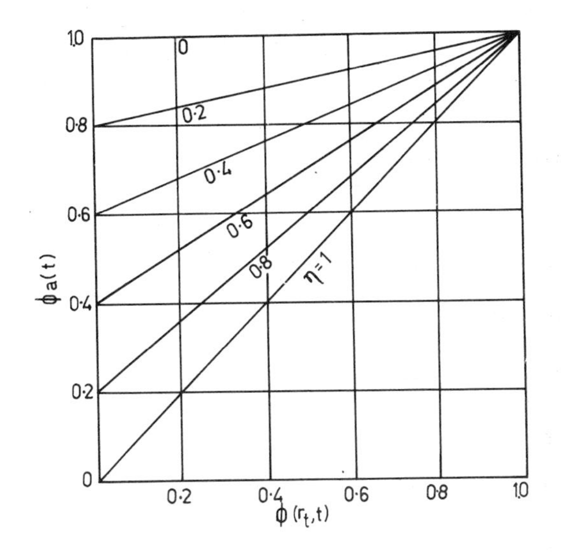

Fig. 8.26. Theoretical solution giving tunnel air and surface temperatures with continuous heating and ventilating conditions. (After Pratt and Saws, 1958)

system was approximately 18 m long, 3.3 m wide and 3 m high. In all its directions, but one, the tunnel was surrounded by the damp chalk. The floor was of dense concrete, and the various surfaces were of hewn chalk. Heat load was simulated using forty heaters of 240 W in the test portion of the tunnel; the heaters were hung 0.15 m above the floor. The positions at which the measurements were made are shown by X in Fig. 8.27. Outlet air temperature, as predicted by the theory and observed experimentally is shown in Fig. 8.28. It can be noted that the theoretical curve explains the experimental observations quite satisfactorily.

8.3.2.3 Rigorous model (Sodha *et al.*, 1984). In contrast to the earlier models, this model takes into account the temperature distribution of the surrounding medium along the flow-direction also. In the light of this addition to the model, it seems to predict the behaviour of earth-air tunnel systems more satisfactorily, however, the theoretical results have not been tested against the experimentally

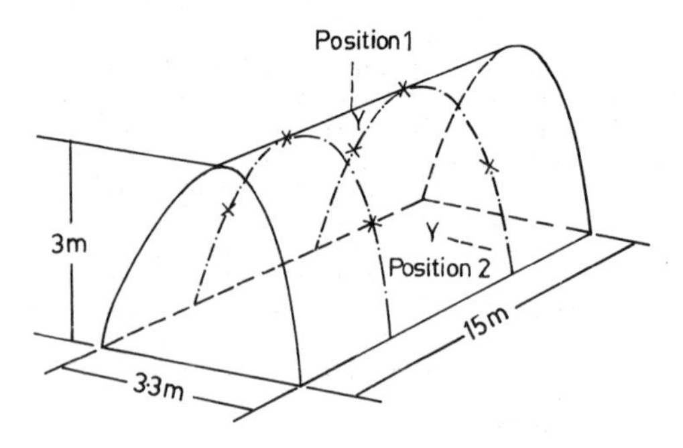

Fig. 8.27. Diagram showing positions at which temperatures were measured. (After Pratt and Daws, 1958)

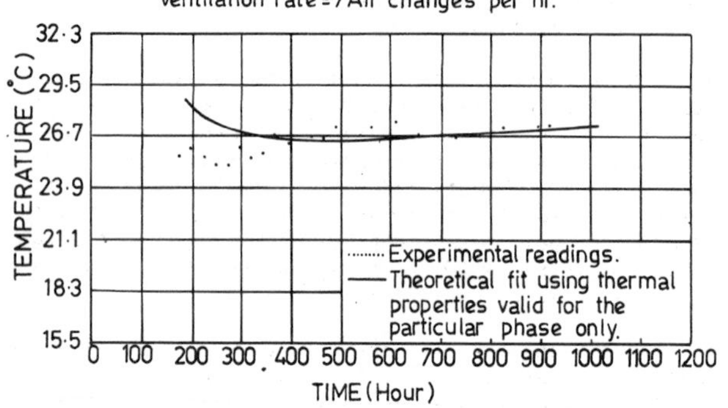

Fig. 8.28. Outlet air temperature. (After Pratt and Daws, 1958)

observed results.

(i) The Analytical Model. In developing the model, the following assumptions were made:

(a) The tunnel is of cylindrical geometry (Fig. 8.29).

(b) The surrounding medium is an infinite medium.

(c) The temperature of the surrounding medium at infinite is constant, and is equal to the annual average of the solair temperature.

(d) The ends of the air tunnel system across its length are perfectly insulated.

(e) The time-dependence of various performance parameters is insignificant.

In view of above assumptions, the steady state temperature distribution of the surrounding medium may be assumed to be represented by

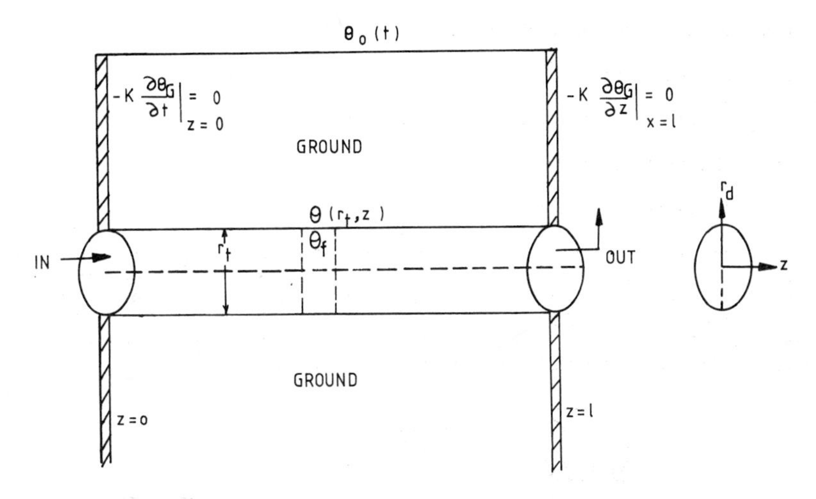

Fig. 8.29. Tunnel in cylindrical geometry. (After Sodha *et al.*, 1984)

$$\frac{1}{\xi}\ \frac{\xi}{\partial\xi}\ (\xi\ \frac{\partial\phi}{\partial\xi}) + \frac{\partial^2\phi}{\partial\eta^2} = 0 \tag{8.29}$$

where the dimensionless parameters ξ, η and ϕ are defined below

$$\xi = r_d/L,\ \eta = z/L,\ \phi = \theta_G/\theta_o$$

where θ_o is the value of earth's temperature at $\xi \to \infty$, i.e.

$$\phi\ (\xi,\eta)\big|_{\xi\to\infty} = 1 \tag{8.30}$$

Other boundary conditions may be written as

$$\left(\frac{\partial\phi}{\partial\eta}\right)_{\eta=o} = \left(\frac{\partial\phi}{\partial\eta}\right)_{\eta=1} = 0 \tag{8.31}$$

$$\left(\frac{\partial\phi}{\partial\xi}\right)_{\xi=o} = \frac{h_c}{k}\left[\phi\ \left(\frac{r_t}{L}\right),\ \eta\ - \frac{\theta_f}{\theta_o}\right] \tag{8.32}$$

where h_c is the heat-transfer coefficient between the surrounding medium and the earth.

The solution of equation (8.29) subject to the boundary conditions given by equations (8.31) and (8.32), may be written as

$$\phi(\xi,\eta) = 1 + F'\ (\xi,\eta) \tag{8.33}$$

where

$$F'\ (\xi,\eta) = \sum_{p=1}^{\infty} A_p\ K_o\ (p\pi\xi).\ \mathrm{Cos}\ (p\pi\eta)$$

and K_o represents the modified Bessel's function.

The temperature distribution of air inside the tunnel is governed by the equation

$$\frac{\partial \phi a}{\partial \eta} = \beta \left[\phi(\xi_o,\eta) - \phi_a(\eta)\right] \tag{8.34}$$

where

$$\phi_a = \frac{\theta_f}{\theta_o}, \quad \xi_o = \frac{r_t}{L}, \quad \beta = \frac{2\pi r_t L h_c}{\dot{m}_a C_a}$$

On solving equation (8.34) in combination with the initial condition that,

$$\text{at } \eta = 0, \ \phi_a = \phi_j = \frac{\theta_{fj}}{\theta_o},$$

one obtains

$$\phi_a(\eta) = (\phi_j - 1)\exp(-\beta\eta) + 1 + F(\eta,\beta), \tag{8.35}$$

where

$$F(\eta,\beta) = \sum_{p=1}^{\infty} A_p \frac{\beta K_o(p\pi\xi_o)}{(\beta^2 + \pi^2 p^2)} \left[\beta.\text{Cos}(p\pi\eta) + p\pi \,\text{Sin}(p\pi\eta) - \beta \exp(-\beta\eta)\right]$$

Now, substituting equations (8.33) and (8.34) in equation (8.32) one gets

$$\sum_{p=1}^{\infty} p\pi A_p K'_o(p\pi\xi_o)\,\text{Cos}(p\pi\eta)$$

$$= \frac{h_c L}{k}\left[(1 - \theta_{fj})\exp(-\beta\eta) + \sum_{p=1}^{\infty} A_p K_o(p\pi\xi_o)\,\text{Cos}(p\pi\eta) - \sum_{p=1}^{\infty} A_p \frac{\beta K_o(\pi p\xi_o)}{(\beta^2 + \pi^2 p^2)}\left\{\beta\,\text{Cos}(p\pi\eta) + p\pi\,\text{Sin}(p\pi\eta) - \beta\exp(-\beta y)\right\}\right] \tag{8.36}$$

In order to evaluate the unknown constants A_p, the above equation is multiplied by cos $(m\pi\eta)$ and integrated w.r.t. η from $\eta = 0$ to $\eta = 1$. Hence, one gets

$$\left[m\pi(\beta^2 + m^2\pi^2)K_o(p\pi\xi_o) - \frac{h_c L m^2\pi^2}{k} K_o(m\pi\xi_o)\right] Am$$

$$- \sum_{p=1}^{\infty} \frac{h_c L p}{2k}\left(\frac{\beta^2 + m^2\pi^2}{\beta^2 + \pi^2 p^2}\right)\left\{\frac{\text{Cos}\{(m+p)\pi\} - 1}{m + p} + \frac{\text{Cos}\{(p-m)\pi\} - 1}{(p - m)}\right\} K_o(p\pi\xi_o) A_p$$

$$= \frac{h_c L\beta\ (1 - \theta_{fi})}{k}\ \{1 - \exp\ (-\ \beta)\ \mathrm{Cos}\ m\pi\} \tag{8.37}$$

(ii) Numerical appreciation of the model. Equation (8.35) expresses the temperature distribution of the tunnel air as a function of β and η. For a typical set of values, the parameter β varies from 0.5 to 1.3. The variation of the function $F(\eta,\beta)$ with these typical values is shown in Fig. 8.30. The amount of cooling/heating produced by the tunnel may be evaluated by reading the value of $F(\eta,\beta)$, and adding it to the quantity plotted in RHS scale of the Fig. 8.30.

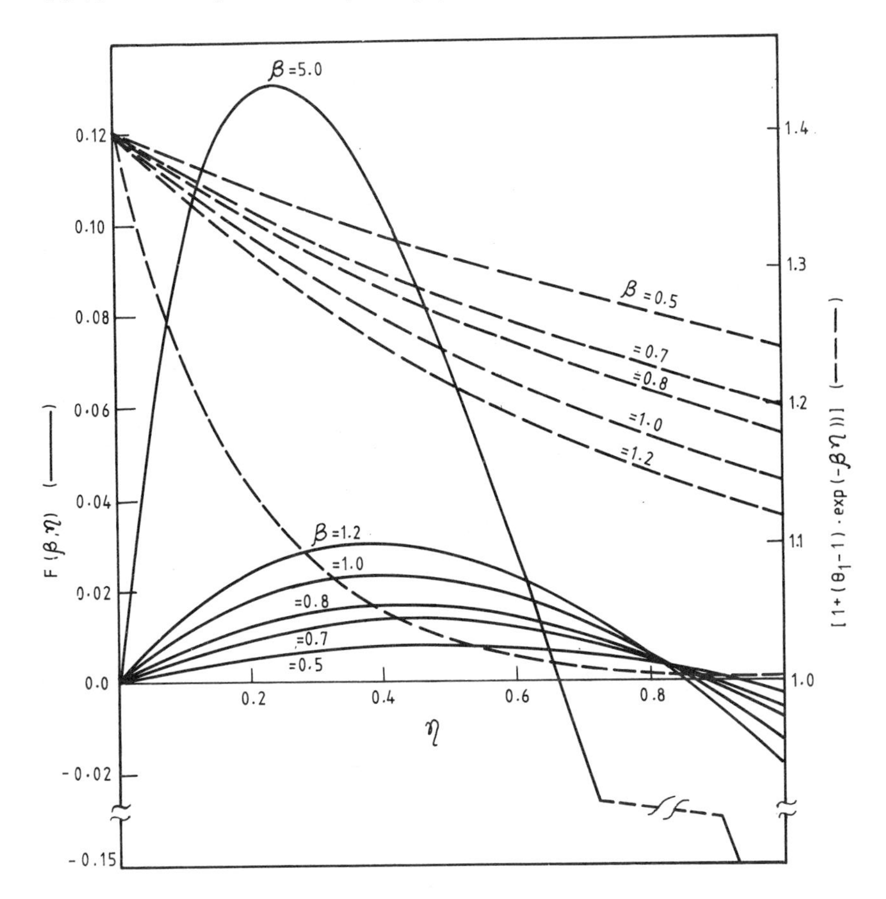

Fig. 8.30. Variation of $F(\eta,\beta)$ and the function $[1+ (\theta_1 - 1)\ \exp\ (-\ \beta\eta))]$ for various values of η and β. (After Sodha *et al.*, 1984)

If one decreases the radius of the tunnel by a factor of $1/n$, the cross-section area gets changed by a factor of $1/n^2$. Such a change ultimately changes the value of the parameter β by a factor of n and this affects the performance of the tunnel. For example, if we take $\beta = 0.5$ (say for $r_t = 1$ m), air temperature at the outlet comes out to be 30.94, whereas $\beta = 1.0$ ($r_t = 0.5$ m) it becomes only 27°C and for $\beta = 5.0$ ($r_t = 0.2$ m) the air gets cooled down to the maximum possible value of 25°C. This result essentially brings out the fact that if instead of a large size tunnel, one can use a number of smaller size tunnels, the tunnel can be cooled or heated more effectively, a conclusion which is drawn out by using a simpler model also (Bansal *et al.*, 1983).

The variation of the function $F'(\xi,\eta)$, directly yielding the variations in the ground temperature distribution with η and ξ is shown in Fig. 8.31. At $\xi = 0.25$, it is seen that the function has a very small value which essentially means that the temperature of the medium becomes constant at a distance of about 2.5 m from the surface of the tunnel. This gives us an idea about the distance at which the tunnel should be located below the surface of the ground to get a constant temperature.

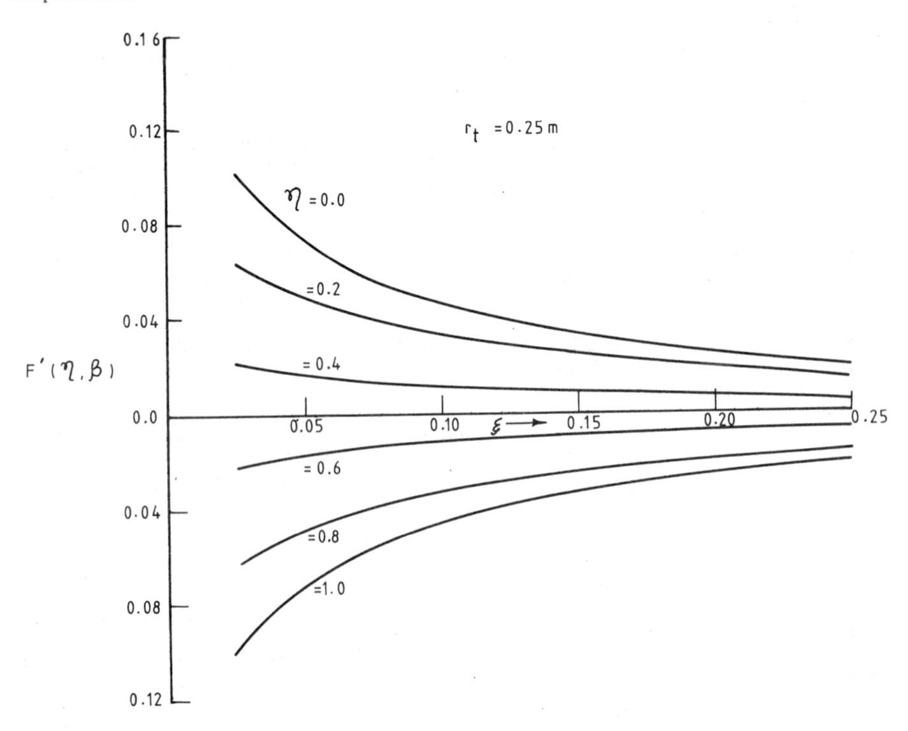

Fig. 8.31. Variation of the function $F'(\eta,\xi)$ with and .
(After Sodha *et al.*, 1984)

8.3.3 Conclusions

The experience in the field indicates that the concept of using the earth-air tunnels has much potential to be used for conditioning the outside air throughout the year. The concept is seen to be quite effective in summer, however, in winter, the heating may not be sufficient. One important reason for this is that the cold air in winter tends to sink down into the tunnel through any available openings, thus reducing the warmth of the tunnel air. The air tunnels also inherit a big advantage of removing the unpleasant odours from the living space.

For a successful implementation of the earth-air tunnel concept, Thoburn (1983) have listed the following recommendations:

(i) It should be ensured that the water-table level is well below 6-7 metres of the ground surface. This depth is chosen because for the maximum efficiency of the system, the tunnel should be located at about 4 m depth of the ground.

(ii) It should be ensured that the rain-water can easily flow away from the tunnel entrances, and from the area of the ground under which the air tunnel is situated. This means, no standing water should be allowed above the tunnel.

(iii) Mechanical fans should be used to ensure the proper functioning of the system.

(iv) One-way valves should be used at the grills, which will prevent the entry of the room air into the tunnel. Also, the ducts leading the tunnel air to the rooms should be properly insulated.

(v) The tunnels should be adequately lighted for servicing and occasionally sweeping. It should also be kept clear of obstructions. If the tunnel has to be used for water pipes and electric service lines, these must be supported preferably near the roof of the tunnel.

(vi) The tunnels should be fitted with suitably placed doors to restrict the entry of unwanted elements viz. the pest animals, and the unauthorized persons.

Nevertheless, the serious problems have also been encountered in the existing earth-air tunnel systems. These include:

(a) Since both the systems at Lahore in Pakistan, and Mathura in India, are situated at the banks of rivers, floods caused serious maintenance problems. It has almost become impractical to keep the flood water out of the tunnel, which also carried the silt along with it into the tunnel; these undesirable elements hamper the performance of the tunnel drastically.

(b) The neglect of regular maintenance was observed which caused some operating problems, bursting of the water-pipe line in the tunnel, and the entrance of pests was reported to be the cause for these difficulties.

On the other hand, the earth cooling tubes promise less maintenance problems. For a successful erection of such a system, Francis (1984) has made the following recommendations:

(i) On the practical side of balancing cost and performance, a 1.5 m - 2.5 m depth should be deep enough for most of the applications.

(ii) The tube length may be around 45 m - 75 m. It has been seen that the longer tubes do not help much in improving the performance.

(iii) It is recommended that air-velocity should be about 2 - 2.5 m/s. The tube diameter may then be calculated based upon the total volume of air required for the load.

(iv) The most popular earth-tube materials are plastic drainage tile, both perforated and solid, steel and aluminium culverts, concrete culverts, and clay tile. When cost is the consideration factor, clay tiles are preferred, however, from installation points of view plastic pipes are preferred.

REFERENCES

ASHAE (1958) American Society of Heating, Air-conditioning Engineers Guide, Cobling Road, Chapter 13, p. 304.

ASHRAE Handbook of Fundamentals (1972) American Society of Heating, Refrigerating and Air-conditioning Engineers, Chapter 4, Air-washers and Evaporative Air-cooling Equipment, p. 29.

Abrahms, D. S. and Benton, C. C. (1980) Simulated and measured performance of earth cooling tubes, Proc. of the 5th National Passive Solar Conference of the American Section of the International Solar Energy Society, Amherst, Massachusetts October, 737-741.

Bansal, N. K., Sodha, M. S. and Bharadwaj, S. S. (1983) Performance of Earth-Air Tunnel System, *Int. J. Energy Research, 7*(4), 333.

Bharadwaj, S. S. and Bansal, N. K. (1981) Temperature Distribution Inside Ground for Various Surface Conditions, *Building and Environment, 16*(3), 183-192.

Blount, S. M. (1958) A Report on Sprayed Roof Cooling System, *Ind. Exp. Prog. Facts for Industry Ser.*, Bull. No. 9, North Carolina State College, North Carolina, USA.

Carrier, W. H. (1968) The Temperature of Evaporation, *Trans. ASHVE, 24*, 24.

Dunkle, R. V. (1961) Solar Distillation: The Roof-type Still and Multiple Diffusion Still, Int. Heat Transfer Conference, Part V, Int. Developments in Heat Transfer, University of Colorado, Colorado.

Francis, C. E. (1984) Earth cooling tubes: case studies of three Midwest installations, Proc. of International Hybrid and Passive Cooling Conference of American Section of International Solar Energy Society, Miami Beach, 171.

Hay, H. R. and Yellot, I. J. (1969) Natural air-conditioning with roof ponds and moveable insulation, *ASHRAE Trans., 75* (Part I).

Hay, H. R. and Yellot, I. J. (1970) A naturally air-conditioned building, *Mech. Engng., 92*(1), 1902.

Hendrick, P. L. (1980) Performance evaluation of a terrestial heat exchanger, Proc. of the 5th National Passive Solar Conference of the American Section of the International Solar Energy Society, Amherst, Massachusetts, October, 732-736.

Holder, L. H. (1957) Automatic roof cooling, Aril Showers Company, Washington, DC, 2.

Houghten, F. C., Olsen, H. T. and Gutberlet, C. (1940) Summer Cooling Load as Affected by Heat Gain Through Dry, Sprinkled and Water Covered Roof, *ASHVE Trans., 46*, 231.

Hourmanesh, Mo, Hourmanesh, Ray and Elmer, D. B. (1979) Earth-air Heat Exchanger System, Proc. of the 4th National Passive Solar Conference, Kansas City, Missouri, pp. 146-149.

Jain, S. P. and Rao, K. R. (1974) Experimental Studies on the Effect of Roof-System Cooling on Unconditioned Building, *Building Science, 9*, 9.

Jain, S. P. (1976) Cooling of Buildings by Roof-Surface Evaporation: Choice of Materials for Retention of Water, *Building Digest*, No. 117, C.B.R.I., Roorkee (U.P.), India.

Jain, S. P. (1977) Simple and Effective Roof-Spraying System of Cooling Buildings in Hot Dry Climates, *Building Digest No. 124*, C.B.R.I., Roorkee (U.P.), India.

Jain, S. P. and Kumar, V. (1981) Energy Saving in Cooling and Air-conditioning of Buildings Utilizing Natural Energies by Implementing Roof Surface Evaporation, National Seminar on Energy Conservation, Vigyan Bhavan, New Delhi, EC III 14, p. 113, March.

Labs, K. (1981) Direct-coupled ground cooling: Issues and opportunities, Proc. of International Hybrid and Passive Cooling Conference of the American Section of International Solar-Energy Society, Miami Beach, 131.

Labs, K. and Watson, D. (1981) Regional suitability of earth tempering, Proc. of the Technical Conference on Earth Shelter Performance and Evaluation, Edited by L. L. Boyer, Oklahoma State University, Stillwater.

Khatry, A. K., Sodha, M. S. and Malik, M. A. S. (1978) Periodic Variation of Ground Temperature with Depth, *Solar Energy, 20*, 425.

Kusuda, T. (1980) Earth temperatures beneath five different surfaces, NBS Report 10373, Washington, DC, 197.

Mackey, C. O. and Wright (Jr), L. T. (1944) Periodic Heat-Flow: Homogeneous Walls/Roof, *ASHVE Trans.*, *50*, 293.
Mackey, C. O. and Wright (Jr), L. T. (1946) Periodic heat-flow: Composite walls/roof, *ASHVE Trans.*, *52*, 263.
Pratt, A. W. and Daws, L. F. (1958) Heat Transfer in Deep Underground Tunnels, Research Paper No. 26, National Building Studies, Department of Scientific and Industrial Research, London, p. 37.
Sherman, C. E. and Evans, D. L. (1981) Systems simulation of six conventional and hybrid evaporative cooling alternatives for residential applications in a hot desert climate, Proc. of Internation Hybrid and Passive cooling conference of the American Section of International Solar Energy Society, Miami Beach, 231.
Sinha, R. R., Goswami, D. Y. and Kleit, D. E. (1981) Theoretical and experimental analysis of cooling technique using underground air-pipe, Solar World Forum, Brighton.
Sodha, M. S., Khatry, A. K. and Malik, M. A. S. (1978) Reduction of Heat-Flux Through a Roof by Water-Film, *Solar Energy*, *20*, 189.
Sodha, M. S., Kumar, A., Singh, U. and Tiwari, G. N. (1980) Reduction of Heat-Flux by a Flowing Water-Layer over an Insulated Roof, *Building and Environment*, *15*, 133.
Sodha, M. S., Singh, U., Srivastava, A. and Tiwari, G. N. (1981a) Experimental Validation of Thermal Model of Open Roof Pond, *Building and Environment*, *16*, 93.
Sodha, M. S., Kumar, A., Tiwari, G. N. and Tyagi, R. C. (1981b) Simple Multi-wick Solar Still: Analysis and Performance, *Solar Energy*, *26*, 127.
Sodha, M. S., Goyal, I. C., Bansal, N. K. and Kumar, A. (1984) Temperature Distribution in an Earth-Air Tunnel System (communicated).
Sodha, M. S., Sharma, A. K., Singh, S. P., Bansal., Bansal, N. K. and Kumar, A. (1985) Evaluation of Earth-Air Tunnel System at St. Methodist Hospital, Mathura, India, Building and Environment, *20* (2), 115-122.
Strayer, R. D. (1979) Cost Effective Solar Heated Earth Structures and Earth-Air Tunnels, Proc. 4th National Passive Solar Conference, Oct. 3-5, 1979, Vol. *4*, Ed. G. Grants, Published by American Section of ISES, Inc. McDowell Hall, University of Delaware, Delaware, NY.
Sutton, G. E. (1950) Roof Spray for Reduction in Transmitted Solar Radiation, ASHVE, Trans., *131*, September.
Thappen, A. B. (1943) Excessive Temperature in Flat-Top Building, *Refrigerating Engineering*, 163.
Thoburn, W. C. (1983) Performance of Earth-Air Tunnel System, (Private communication).
Threlkeld, J. L. (1970) *Thermal Environmental Engineering*, Prentice-Hall, Inc., Englewood Cliffs, New Jersey.
Tiwari, G. N., Kumar, A. and Sodha, M. S. (1982) A Review - Cooling by Water Evaporation over Roof, *Energy Conversion and Management*, *22*, 143.
Watt, J. R. (1963) *Evaporative Air-Conditioning*, The Industrial Press, New York 13, NY.
Yellot, J. I. (1969) Roof Cooling with Intermittant Water Sprays, ASHRAE 73rd Annual Meeting in Toronto, Ontario, Canada, June 27-29.

Chapter 9

RULES OF THUMB: DESIGN PATTERNS*

The design, development and construction of all the buildings, large or small, are based on certain thumb rules evaluated from various calculations.

This section lists twenty-nine patterns for the application of passive solar energy systems to building design. The patterns are ordered in a rough sequence, from large-scale concerns to smaller ones. Each pattern is connected to other patterns which relate to it. All the patterns, together, form a coherent picture of a step-by-step process for the design of a passive solar heated building. The patterns can also be used to analyse or critique existing buildings or proposed designs. One may select the patterns most useful to the project, more or less in the sequence presented here.

All the patterns can be identified with various stages of the design. They are basically organized into categories (1) site planning, (2) envelope design, (3) interior design and (4) mechanical equipment. In Table 9.1, the energy saving potential of each design pattern is rated either as major, moderate or minor.

The various passive systems described above need further consideration as follows:

DIRECT GAIN SYSTEMS

1. Solar Windows
2. Clerestories and Skylights
3. Masonry Heat Storage
4. Interior Water Wall

THERMAL STORAGE WALL SYSTEMS

1. Sizing the Wall
2. Wall Details

ATTACHED GREENHOUSE SYSTEMS

1. Sizing the Greenhouse
2. Greenhouse Connection

---

*This chapter derives its information mainly from Mazaria (1979)

TABLE 9.1 Design Guidelines Matrix

Site planning	Envelope design	Interior design	
1. Building location	1. Building shape	1. Choosing the design	1. Mechanical equipment efficiency
2. Orienta-tion	2. North side	2. Appropriate materials	2. Mechanical equipment control
3. Cluster-ing	3. Protected entrance	3. Solar passive system	3. Ventilation
	4. Window location	a. Direct gain systems	4. Domestic hot water
	5. Shading	b. Thermal storage wall	
	6. Insulation	c. Attached greenhouse	
	7. Exterior surface colour	d. Roof-pond	
		e. Greenhouse	
	8. Infiltration reduction	4. Combining systems	
		5. Cloudy day storage	
		6. Movable Insulation	
		7. Thermal mass	
		8. Reflectors	
		9. Shading devices	
		10. Summer cooling	
		a. Evaporative cooling	
		b. Earth-air tunnel system	
Major	Moderate	Minor	

ROOF-POND SYSTEMS

1. Sizing the Roof-Pond
2. Roof-Pond Details

GREENHOUSE

1. South-Facing Greenhouse
2. Greenhouse Details.

A Word of caution should be exercised in using these patterns. The patterns always evolve and change with time as the new information becomes available. One should also change the word "south" with "north" for locations in the southern hemisphere.

9.1 BUILDING LOCATION AND ORIENTATION

For cold climates, where heating is required for most part of the year, one should locate the building where it receives most sun during the houses of maximum solar radiation - 9 a.m. to 3 p.m. (sun time). Placement of the building in the northern portion of the sunny area ensures that (i) outdoor areas and gardens placed to the south will have adequate winter sun and (ii) the possibility of shading the building by future off-site developments will be minimized.

Solar sun charts help to visualize the obstruction of the sun by the site construction, trees etc. Figure 9.1 illustrates several of the other major planning considerations. Buildings like D below do not all have to face due south nor be elongated on the east-west axis. Solar collection apertures can vary within 30° of true south with minor effect on building energy performance. Buildings should be situated so that one building does not shade the solar apertures of another during the primary solar collection hours. Also, the building itself should be designed to minimize unwanted self shadowing from wing-walls and other building appendages (like buildings R and B in Fig. 9.1). L shapes and other building shapes are by no means precluded if one is careful about window placement and shading from steep roofs. East or west rotation of irregular shaped buildings can sometimes serve to minimize or prevent self-shading.

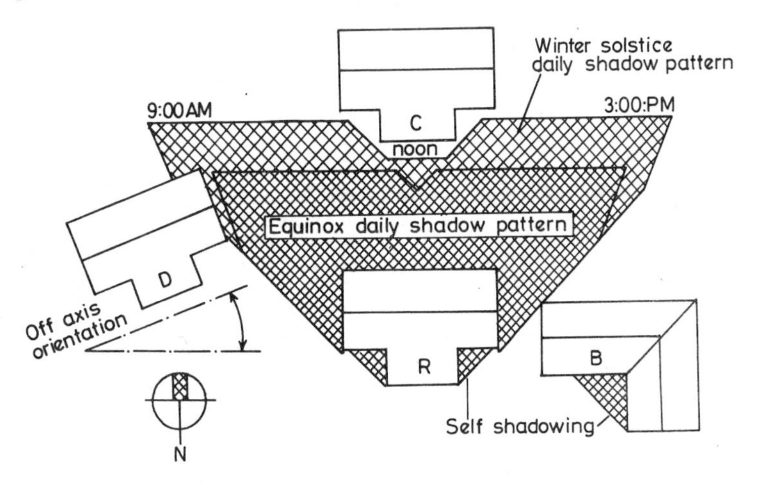

Fig. 9.1. Site planning considerations

9.2 CLUSTERING

Building clusters as opposed to single isolated buildings have the advantage of shading each other and thus help in minimizing the incident solar energy on the building envelope. The mutual shading can be used to maximum advantage when the streets between the buildings are narrow and when buildings are built around courtyards. An even heating for the courtyard and street causes thermal air currents to be set up, which ventilate the house. At night, the radiative cooling of roof-surface causes the air immediately in contact with the roof to cool down. This cooler and heavier air flows down and cools the house. During the day, the outdoor air is hot and therefore lighter. It cannot displace the cooler and denser air in the building.

From the point of view of avoiding shadowing (O'Cathain, 1982) of the sun for effective solar passive heating, if 40% of the site is devoted to house plots a density of below 30 hours per $10^4 m^2$ (1 hectare) plot area of three-storeyed buildings is recommended. This allows no more than 10% of solar energy to be lost. If the proportion of the site to the houses is increases, one can increase the density to 40 houses. For two-storeys houses and for 40% of site devoted to 40 houses. For two-storey houses and for 40% of site devoted to houses, more than 30 houses (between 30 and 40) are permissible per $10^4 m^2$ of site area.

9.3 ENVELOPE DESIGN

The design of the building envelope directly influences auxiliary heating requirements and the potential for using passive solar energy. The design strategies discussed in this section corresponding to building shape, insulation, infiltration reduction, exterior surfaces colour, glazing selection and location, and shading. Through the appropriate selection, sizing and installation of building materials and products, the building envelope becomes an energy efficient filter between the outside and inside environments.

9.3.1 Building Shape and Orientation

It is common belief that minimization of heating requirements in winter (or cold climates) and cooling in summer, requires building elongation along the east-west axis. This orientation helps in a large exposed area towards south, helping in collection of solar energy in winter, while in summer it receives only marginal radiation.

Design for energy efficiency, however, does not limit the flexibility in building shape to a great extent if one considers the surface area to volume ratio aspect. Figure 9.2 illustrates that the surface area of an L-shaped plan increases by only 6% over that of a square plan. Therefore, the designer is free to conceive of floor plan arrangements appropriate to interior spatial needs. For direct gain concepts, however, the depth of the space should not exceed 2½ times the height of the window from the floor area for an effective solar entrance to the living space. This rule also provides adequate day lighting of the interior surfaces. For buildings with thermal storage wall, the depth should be of the order of 4.5 m to 6 m, the maximum distance for effective heating from a radiant wall.

Spaces, which require larger depth should not have large south-facing windows. South-facing clerestory windows or skylights are preferred in this case.

Multistoreys result in significant surface area savings in comparison to single-storey arrangements. For the planaria shown in Fig. 9.2, a two-storey building would have 20% less surface area. As long as adequate solar exposure and interior thermal zoning are provided, Row houses are multifamily building

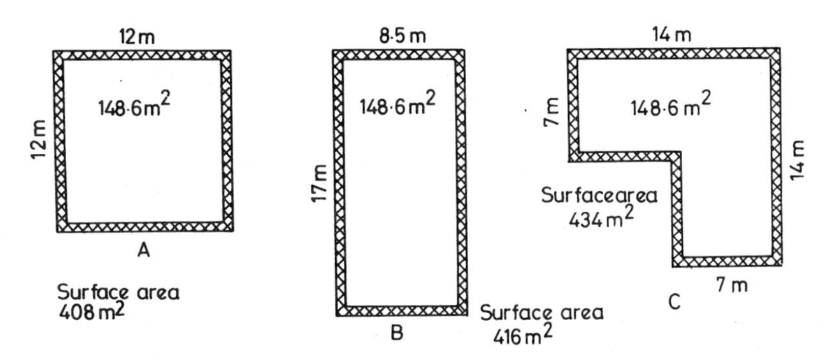

Fig. 9.2. Plan shape alternatives

configurations have the same flexibility as single family housing as long as adequate solar exposure and interior thermal zoning are provided.

9.3.2 North Side

The north side of the building does not receive any radiation during winter. It is then best to shape the building in such a way that its north side slopes towards the ground. It is desirable, whenever possible, to slope the building from south to north and/or berm earth against the north face of a building to minimize the portion of the exposed area of the north wall. The nearby structures or walls to the north of the building if made light-coloured, help to reflect sunlight into north-facing rooms and outdoor spaces. A shorter north wall, also reduces the shadow casted by the building.

9.3.3 Protected Entrance

One of the major sources of heat loss/heat gain into the building is the infiltrated outdoor air through cracks around the entrance door and frame as well each time the door is opened. The main entrance should therefore be made a small enclose space (vestibule, foyer), which provides an air lock between the building and the exterior. Such an air lock will prevent a large quantity of warm (or cool) air from leaving the building each time a door is opened, because only the air within the enclosed space can escape. Due to the still air space between the interior and exterior, the infiltration is virtually eliminated. The entrance, preferably, should be oriented away from the prevailing winds or a windbreak is provided to reduce the wind's velocity against the entrance. The entry space can be utilized for storing items that do not require heating e.g. winter clothing, or for activities that require little space heating.

9.3.4 Window Location

Window location is one of the most important considerations in the building design from an energy conservation point of view. Improperly placed windows are a source of large heat losses into the building. To admit winter sun the major window openings should be located south-east, south and southwards as per requirements of the inside space. On the east, west and especially north side, the window area should be kept small and double glass should be used to minimize heat losses. If possible, windows can be recessed to reduce heat losses.

9.3.5 Shading

Shading devices on the windows can add a degree of thermal and solar control, not possible with fixed building components. The choice of these patterns is determined by the cost effectiveness, durability, reliability and ease of maintenance.

9.3.6 Insulation

Adding insulation to the building elements (walls, floor, ceiling and foundation), is one of the most effective ways of energy savings and it is one of the most cost-effective energy design strategies. At many locations, enough sunshine is not available during the heating season. Higher insulation levels are therefore a must for the buildings in such locations. Table 9.2 lists insulation values for three levels of thermal integrity. For the North European climates with more diffuse radiation, level 2 is generally recommended.

For achieving envelope thermal integrity, construction detailing is very important. Major building element connections such as the wall-floor connection shown in Fig. 9.3, should be detailed to avoid thermal bridges in insulation or vapour barrier.

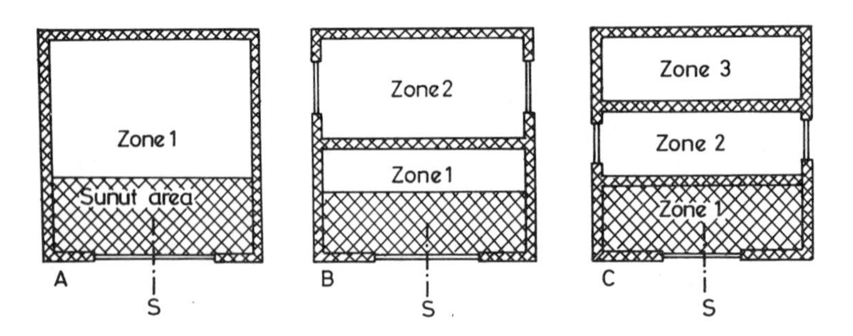

Fig. 9.3. North-south zoning effects

TABLE 9.2 Insulation Level Selection

Thermal integrity	R values m^2 °C/W				
	Walls	Ceiling	Floor		Basement
			Crawl space	Slab	
Level 1	1.94	3.35	1.94	0.88	0.88
Level 2	3.35	5.28	1.94	1.76	1.76
Level 3	4.23	6.69	3.35	1.76	1.76

9.3.7 Exterior Surface Colour

The choice of an exterior surface colour will have a minor to negligible influence on both heating and cooling energy performance. The selection of exterior surface treatment therefore is determined by appearance, durability and ease of maintenance rather than for energy performance.

9.3.8 Infiltration Reduction

Reduction of air infiltration helps in saving energy and improves comfort level. Minimizing infiltration requires careful attention to design, construction detailing and product selection. Table 9.3 defines a range of infiltration rates, specified in air changes per hour (ACH). An infiltration rate of 1 ACH is typical of current construction practice. A rate of 0.5-0.7 ACH is commonly found in energy efficient homes. An infiltrate rate of less than 0.5 ACH is not recommended or allowed presently in Air Force housing due to concern for indoor air quality. It is generally recommended that mechanical ventilation, usually with an air-to-air heat exchanger, be used when infiltration rates are below 0.5 ACH.

TABLE 9.3 Infiltration Level Section

	Air changes per hour	Design and construction guidelines
Level 1	0.8-1.0	Standard design and construction practice All doors and windows caulked and weather stripped.
Level 2	0.6-0.7	LEVEL 1 practices plus • Low infiltration windows • All ceiling, wall and floor penetrations sealed • Continuous vapour barrier on walls, floor and ceiling. • Outside air to combustion type heating system • Rough opening seals. • Storm doors or airlock entry • Minimum ceiling, wall and floor penetrations • Continuous air barrier on exterior wall sheathing
Level 3	0.5	

Higher levels of envelope insulation are typically accompanied by lower rates of air infiltration. Since achieving a lower infiltration rate has construction cost implications, Table 9.3 gives the most common techniques used to reduce infiltration.

9.4 LOCATION OF INDOOR SPACES

Interior space planning and design is the most overlooked element in energy efficient home design. Effective interior design will result in better utilization of passive solar energy to create greater levels of thermal comfort.

To take maximum advantage of the sun, the indoor spaces should be placed along south-face of the building. Rooms should be placed south-east, south, south-west as per requirements of sunlight. Spaces such as corridors, closets, laundry rooms and garages, requiring minimal heating and lighting should be placed along the north-face of the building. These spaces then act as a buffer between the heated space and the colder north-face.

Figure 9.3 illustrates the typical interior zoning that occurs when rooms are organized in north-south fashion. Floor plan B is the most common condition. The partition wall and door openings generally provide adequate coupling for moving passive solar heat from the south to the north zone. Unless lighted by a south-facing clerestory, zone 3 in floor plan C should not be used for the primary living spaces. Properly sized east and west windows in the north zones are important to energy performance and comfort of these zones. South-facing clerestory windows in north zones can enhance or replace east and west windows.

Figure 9.4 illustrates typical east-west interior zoning. The partition wall between zones, is of massive material, can become the primary thermal storage element in the design. This approach is particularly good for row house design. For most designs, one has to have both north-south and east-west zoning conditions. The problem then arises of thermally coupling these zones. Figure 9.5 illustrates two design strategies. In plan arrangement A, open architecture is used to thermally couple a south zone with a north zone. In plan arrangement B, a fan is used. Conduction through lightweight or heavyweight partition walls and interior window is also an effective means for thermally coupling interior zones, especially for a sunspace which typically experiences larger temperature variations than other building spaces.

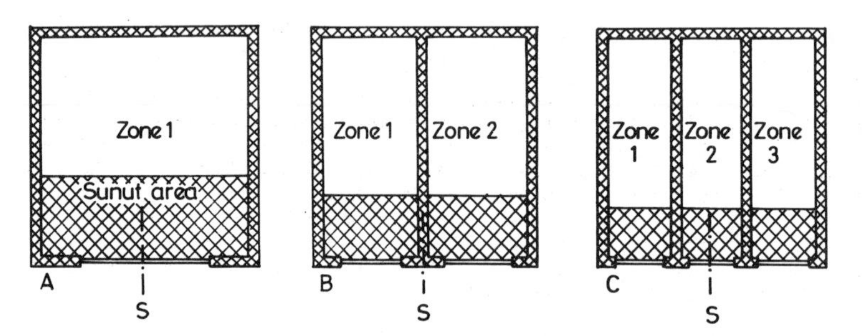

Fig. 9.4. East-west zoning effects

9.4.1 Systems Choice

The choice of a particular passive heating (or cooling) system for a building design is one of the most difficult choices. Each building requires a particular system that fulfils its thermal needs. The passive heating (or cooling) system should be so chosen as to satisfy most of the design requirements generated for each space. Different systems can be used for different spaces or a combination of systems can be used for one space.

For passive heating three configurations are mainly employed namely (1) direct gain windows, (2) thermal storage walls and (3) sunspaces.

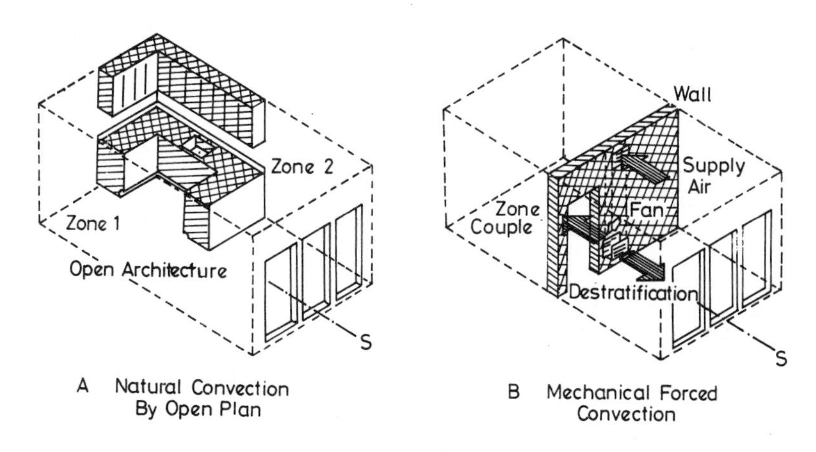

Fig. 9.5. Zone thermal coupling

Any home with windows facing within 30° of south utilizes direct gain passive solar energy. The sizing and location of direct gain has to be, however, carefully done. Direct gain systems should have, in general, efficiencies of the order of 30 to 75%.

The direct gain system employing glazed area towards south have efficiencies of the order of 30 to 75%. Direct gain systems are however characterized by large daily indoor temperature fluctuations. Only when a conventional forced air heating system is added to a space, uniform interior temperatures can be maintained.

For systems employing thermal storage wall, the depth of the space is limited to approximately 5-6 metres since this is considered the maximum distance for effective radiant heating from the solar wall. This limitation of the space depth, requires a linear arrangement of spaces along the south wall of the building. The storage material is either water or masonry and double-glazing in front of the wall is necessary unless night insulation is employed. For the same wall area and storage capacity, a water wall is more efficient than a masonry wall. The obtained efficiencies are of the order of 30-35%. For cloudy winter climates like the one prevailing in North Europe, thermal storage wall systems cannot be recommended.

An attached greenhouse towards southside (solarium) approximately requires 1½ times greenhouse area than the one required in thermal wall storage systems. Overall efficiency of the greenhouse area is the same as that of the direct gain concept viz. 60 to 75%, while the percentage of heat supplied to the adjoining spaces is roughly 10 to 30% of the incident energy on the collector face.

In climates with more diffuse radiation, solarium performs approximately 5 to 9% more effectively than a direct gain system alone. For such climates, usually the following recommendations are made.

1. Double-glazing of sunspace windows is recommended as the minimum level; triple-glazing is preferred.

2. The sunspaces of solarium should always be used in combination with windows in the wall separating sunspace and the living space. The glazing area of the sunspace should always be at least equal to or more than the window area of the living space. Usually the following relationship is followed:

$$\frac{\text{Window area of living space}}{\text{Glazing area of sunspace}} = 0.35 \text{ to } 1.0$$

3. Thermal mass is required in the sunspace. Typically, the thermal mass in the form of the concrete floor slab is adequate. The sizing relationship between sunspace glazing area and thermal mass floor area is as follows

$$\frac{\text{Area of floor mass in sunspace}}{\text{Area of glazing in sunspace}} = 1$$

The sizing relationship between the thermal mass in sunspace and the living space is as follows:

$$\frac{\text{Thermal mass area of living space}}{\text{Thermal mass area of sunspace}} = 0 \text{ to } 1$$

4. The partition wall between the sunspace and the living space can be either an insulated frame wall or a mass wall. The mass wall should be solid and approximately 10-15 cm in thickness.

   A window is recommended in the common wall for view and light. It should be single-glazed and no more than 25-45% of the total common wall area.

5. Total glazing are on the east and west walls of the sunspace should be less than 10% of the total sunspace floor area.

6. Sloped glazing in the sunspace should not be used. Sloped glazing creates a difficult solar control problem (summer and winter overheating) and it does not improve sunspace performance.

7. The sunspace floor is the primary thermal mass location and the following design guidelines should be followed:

   (a) insulate the perimeter of the sunspace foundation wall; insulate underneath floor slab if warranted by high water table or soil conditions.

   (b) start sunspace glazing within 15 cm of floor to adequately illuminate/charge thermal mass floor.

   (c) Surface colour of floor slab should have absorptivity of at least 0.65 (value of uncoloured concrete),

   (d) minimum shading of floor slab (no carpets or excessive funiture). However, 15-25% of the floor area can be covered with area rugs or plants without reducing sunspace performance.

   (e) Sunspace may be provided with an auxiliary heating system. Its set point should be between 7-13°C. Some means to vent the sunspace during both heating and cooling season must be provided. An automatic venting, set at 25-28°C, is provided during the heating season and between 15-21° during the cooling season.

8. Four coupling between the sunspace and the living space is required to augment the passive heat distribution. The recommended flow rate of air is 90-100 l/s.

Roof-pond systems usually have 15 to 30 cm deep water. The building structure therefore must support about 150 to 300 kg/m^2 load of the roof-pond. A structural metal deck, which also acts as finished cooling and radiating surface is the most commonly used support. Roof-pond heating and cooling is characterized by small temperature fluctuations. Masonry wall buildings have temperature fluctuations between 2.5 and 4.4°C, while a lightweight structure building has fluctuation of the order of 5-8°C. The efficiency of roof-pond system is of the order of 30-45%.

9.4.2 Appropriate Materials

Low energy consuming and biodegradable materials should be used for the construction of buildings. Locally available materials should be preferred to keep the transportation energy consumption to a minimum value. For thermal mass and bulk materials, usual materials are adobe, soil cement, brick, stone and concrete or water in containers. For finished materials, wood, plywood, particle board and gypsum board are used.

9.4.3 Sizing Solar Windows

Solar windows employed in the direct gain concept should have an area of 0.19 to 0.38 m^2 of south-facing glass for each square metre of the space floor area for cold climates (average winter temperature -6°C to 1°C). In temperate climates (winter temperatures - 1.5°C to 7°C), these should have 0.11 to 0.25 m^2 area of south-facing glass for each m^2 of floor area. This amount of glazing should be sufficient to keep the space at an average temperature of 18° to 20°C during winter. A solar window oriented 25° east or west of the true south, still intercepts over 90% of the solar radiation incident on a south-facing surface.

Table 9.4 gives the ratio of glass area for different climates for well-insulated houses.

TABLE 9.4 Solar Window Sizes for Different Climatic Conditions*

Average winter outdoor temperature (°C) (degree day/month)	Glass window area for 1 m^2 of floor area**
cold climates	
-9.5°C (1500)	0.27 - 0.42 (w/o night insulation)
-6.5°C (1350)	0.24 - 0.38 (w/o night insulation)
-3.8°C (1200)	0.21 - 0.33
-1.1°C (1050)	0.19 - 0.29
Temperature climates	
1.6°C (900)	0.16 - 0.25
4.5°C (750)	0.13 - 0.21
7°C (600)	0.11 - 0.17

*Calculations for space heat loss of 33 - 42 W/m^2 °C

**For lower latitudes, lower values may be used, while for higher latitudes higher range of values should be used.

9.4.4 Clerestories and Skylights

In many cases, the admittance of solar radiation through south-facing windows is not feasible. In such cases clerestories (facing south) or skylights are used for admitting sunlight. The ceiling of clerestory is made of light colour and appropriate shading devices are used to both clerestories and skylights for summer control.

9.4.5 Heat Storage Masonry

To minimize temperature fluctuations inside the living space in a direct gain concept, the interior walls and floors are made of at least 10 cm thick masonry. The masonry floor is made dark-coloured. The walls of masonry can be of any colour. Direct sun radiation should be avoided on dark-coloured masonry surfaces for long times. Wall-to-wall carpeting is not used on the masonry floors.

To avoid overheating during daytime and cooling (below comfort range) during night, heat should be stored; the percentage of stored heat depends upon the location, size and distribution of thermal mass. Three cases are possible:

1. A dark-coloured concrete mass is placed against the rear wall or in the floor of space in direct sunlight. The surface area of concrete exposed to sunlight is $1\frac{1}{2}$ times the area of the glazing. An increase in masonry thickness beyond 20 cm results in little improvement. Increasing the thickness from 10 cm to 20 cm, maximum air temperatures are relatively unchanged while minimum air temperatures are changed slightly; 20 cm masonry wall increases the minimum room air temperatures by 2.5°C. Increasing the thickness to 40 cm has little impact on air temperatures. The fluctuations in space temperatures in this configuration are of the order of 20°C.

2. In other configurations a dark-coloured concrete mass is either placed in the rear wall or in the floor; the surface area of the storage exposed to direct sunlight is 3 times the glazing area.

   Corresponding to maximum ambient temperature of 50°C, the maximum room temperature is seen to be 30°C and the minimum room temperature to be 12.5°C (in comparison to a minimum outdoor temperature of -6.5°C). With increasing thickness of the mass, no change in maximum temperature was found, while an increase of wall thickness from 10 to 20 cm increases the minimum temperature of the room by 1.6°C.

3. If the surface area of the storage mass exposed to direct sunlight is 9 times the glazing area, the maximum and minimum room temperatures remain almost constant at a value of 23.8°C and 15.5°C respectively, provided the thickness of the mass wall is more than 10 cm.

   In conclusion, every m^2 of direct sunlight should be diffused to about 9 m^2 of the storage surface to maintain comfortable temperatures.

9.4.6 Water Wall

Water wall is used for controlling the indoor temperature fluctuations; its size and the colour of the absorbing surface determine its effectiveness. When used for heat storage, it should be located such that it receives direct solar radiation, and should contain 1 m^3 of water for each m^2 area of solar window. Colour of the water container should be dark. For a clear winter day of New York city, the indoor temperature fluctuated by 10°C and 6.5°C for a space having 1 and 3 m^3 of water volume for 1 m^2 of glass area respectively. Table 9.5 shows that for greater absorption of solar radiation, the fluctuations are smaller in

TABLE 9.5 Indoor Temperature Fluctuations (°C) as a Function of Absorptivity and Thickness of Water Wall

Fraction of the incident energy absorbed	Volume of water wall (m^3)			
	0.0283	0.0424	0.0566	0.0849
0.75 (dark colour)	9.4	8.3	7.2	6.7
0.90 (black)	8.3	6.7	5.6	5.0

9.4.6.1 Sizing the wall. Proper sizing of the thermal storage wall helps in keeping the indoor temperature within comfortable range. The required area of the wall depends upon the climate and latitude of the place; for the places with cold climates and larger latitudes bigger walls are required. Table 9.6 presents the recommended ratios of the double-glazed south-facing thermal storage wall to the floor area for different climates. The criterion used for calculating these ratios was that the energy transmitted through the wall is sufficient to maintain an average indoor temperature of 18-24°C over the 24-hour period. For the regions with higher latitudes and poorly insulated buildings, higher ratios should always be chosen. For thermal walls with night insulation, 85% of the above mentioned values are recommended. For thermal walls with a horizontal specular reflector (equal to the height of the wall in length) alone, and a night insulation also, these values are reduced to 67% and 57% respectively. In order to take into account the unpredicted cloudy weather, the walls should be slightly oversized.

Table 9.6 Sizing a Thermal Storage Wall for Different Climatic Conditions

Average winter outdoor temperature (°C) (degree-days/month)	Square metre of wall needed for each square metre of floor area	
	Masonry wall	Water wall
Cold climates		
- 9.4 (1,5000)	0.72 - 1.0	0.55 - 1.0
- 6.7 (1350)	0.60 - 1.0	0.45 - 0.85
- 3.9 (1200)	0.51 - 0.93	0.38 - 0.70
- 1.1 (1050)	0.43 - 0.78	0.31 - 0.55
Temperate climates		
1.7 (900)	0.35 - 0.60	0.25 - 0.43
4.4 (750)	0.28 - 0.46	0.20 - 0.34
7.2 (600)	0.22 - 0.35	0.16 - 0.25

9.4.6.2 Wall details. The efficiency of a thermal storage wall is determined by its thermal capacity, and the colour of the outside surface. The recommended thicknesses of a wall made up of different building materials are given in Table 9.7. The colour of the outside surface of the wall is made dark. Thermocirculation vents, of roughly equal size, at the top and bottom of a masonry wall are provided in cold climates. Total area of each row of vents is

TABLE 9.7 Recommended Thicknesses of a Thermal Storage Wall for Various Building Materials

Material	Thickness (m)
Adobe	0.20 - 0.30
Brick (common)	0.25 - 0.35
Concrete (dense)	0.30 - 0.45
Water	0.15 or more

kept equal to 1% of the wall area. The vents are closed during the night hours. In mild climates, however, the vents are not required since winter daytime temperatures are comfortable and the heating is not required at that time. For the equal size wall and heat storage capacity, the water wall performs slightly better than a masonry wall.

9.4.7 Sizing the Greenhouse

A greenhouse is usually built to produce food in an efficient and economic way, and supplies heat to a building when attached to the south side of the building. However, to design it accurately is a difficult task since it involves the complicated heat-transfer processes between the building and the attached greenhouse system. The recommended glass areas of the greenhouse required to heat up a certain area of the building are given in Table 9.8. It is assumed here that the greenhouse glass is on the south-facing side, and is doubly-glazed. Higher values are chosen for the buildings which are poorly insulated and/or located in the regions of higher latitudes. For greenhouse alone, the efficiet shape for solar collection is the one which is elongated along the east-west axis. In the houses which are retrofitted by attaching a greenhouse on the south side are mostly useful during daytime.

TABLE 9.8 Sizing the Attached Greenhouse for Different Climatic Conditions

Average winter outdoor temperature (^{o}C) (degree-days/month)	Greenhouse glass area needed for 1 m^2 of floor area (m^2)	
	Masonry wall	Water wall
Cold climates		
- 6.7 (1350)	0.90 - 1.50	0.68 - 1.27
- 3.9 (1200)	0.78 - 1.30	0.57 - 1.05
- 1.1 (1050)	0.65 - 1.17	0.47 - 0.82
Temperate climates		
1.7 (900)	0.53 - 0.90	0.38 - 0.65
4.4 (750)	0.42 - 0.69	0.30 - 0.51
7.2 (600)	0.33 - 0.53	0.24 - 0.38

9.4.7.1 Greenhouse connection. Thermal performance of a building connected with a greenhouse depends mainly upon the thermal behaviour of the joining section of the two. In a passive system, a thermal storage wall is usually used as the medium for the heat-transfer from greenhouse to the building. In this case, the recommended thicknesses of the wall are given in Table 9.9.

TABLE 9.9 Recommended Thicknesses of the Thermal Wall used for Connecting the Greenhouse with the Building

	Recommended thickness (m)
Adobe	0.20 - 0.30
Brick (common)	0.25 - 0.35
Concrete (dense)	0.30 - 0.45
Water	0.20 or more

Colour of the wall on the exposed side is chosen to be dark. It should be ensured that the direct solar radiation is not blocked from reaching the wall. In cold climates, the vents are also provided which allow the transfer of heat (by means of thermocirculation) to the interior of the building during the daytime also. When masonry wall is the only element for heat transfer and storage, the indoor temperature of the building fluctuates by 10-12°C, which can be controlled by providing additional storage in the greenhouse. However, in case of water wall no additional mass is needed in the greenhouse if 0.67 m^3 of water is used for each 1 m^2 of south-facing glazing. In temperate climates, the excess heat collected in greenhouse is transferred to the interior of the building by using some active system where it is stored and used during night hours.

9.4.8 Sizing the Roof-pond

Roof-pond can be used for both heating and cooling of the buildings. For heating the recommended ratios of roof-pond collector area for one m^2 of space floor area are given in Table 9.10; all the configurations mentioned in the table use night insulation. At higher latitudes (> 36 °N) reflectors are used for augmenting the solar energy collection. This is usually accomplished with a movable insulation folding in half and becoming a reflector in the open position. The optimum angle of the reflector to the pond is about 80°-90° in winter.

TABLE 9.10 Recommended Area of Roof-pond Corresponding to 1 m^2 of Floor Area

Average winter outdoor temperature (°C)	(-9-(-4) °C	(-4)-(2) °C	(2)-(7) °C
Double-glazed pond	-	0.85 - 1.0	0.60 - 0.90
Single-glazed pond with reflector	-	-	0.33 - 0.60
Double-glazed pond with reflector	-	0.50 - 1.0	0.25 - 0.45
South-sloping collector cover	0.6 - 1.0	0.40 - 0.60	0.20 - 0.40

Recommended ratio of the roof-pond area to the space floor area for cooling in hot-dry climate is 0.75 which reduces to 0.33 - 0.50 if the cooling is augmented by evaporation of open water. For hot-humid climate the required pond area is equal to the space floor area.

9.4.8.1 Roof-pond details. The water-pond is usually supported on a water proofed metal roof or concrete deck. Water is contained either in transparent plastic bags or in fibre-glass tanis that form the roof. Top of the container is kept transparent while the inner surfaces are made blackened. Depth of the water varied between 0.15 - 0.30 m. For a south-sloping collector system, the inclination angle of south-glazing is kept equal to the latitude of the place plus 15°. Most commonly used insulating panels are formed by 0.05 m thick polyurethane foam reinforced with fibre-glass strands and sandwiched between aluminium skins. It should be ensured that the panels are sealed tightly over the ponds when closed.

9.4.9 South-Facing Greenhouse

The large glass area of the conventional greenhouse contributes significantly to the overall heat losses. In cold northern and temperate climates, the greenhouse is erected along the east-west axis. North wall of the greenhouse is made up of opaque materials while the south-facing wall and the roof are made completely transparent. To prevent one-sided plant growth, the ceiling and/or the upper part of the north wall is painted with light-coloured paint which reflects the radiation back on to the plant canopy. To reduce further heat losses the north wall is insulated by 0.05-0.07 m thick insulation.

9.4.9.1 Greenhouse details

Excess solar heat collected by a conventional greenhouse is vented out of it. This excess heat may be stored for night use by providing suitable thermal storage inside the greenhouse. For the purpose, the opaque walls and floor are constructed of solid masonry having a thickness of 0.25 m or more. Masonry, however, is not sufficient storage, so additional thermal mass (e.g. water in containers) is provided. The quantity of additional thermal mass is determined by the acceptable range of temperature fluctuations. External surface of the water containers is blackened. An interior water wall is also used to dampen the temperature fluctuations inside the greenhouse. This is achieved by integrating the water wall into the north wall using roughly 0.5-1 m^3 of water for 1 m^2 of south-facing glass. The expected range of fluctuations in the interior temperature corresponding to the quantity of water used are given in Table 9.11. Here, it is assumed that 75% of the sunlight entering the space is absorbed by the water wall.

TABLE 9.11 Temperature Fluctuations in a Greenhouse using a Water Storage Wall System

Volume of water for 1 m^2 of south-facing glass (m^3)	Temperature fluctuations (°C)
0.33	18 - 23
0.50	16 - 20
0.67	15 - 18
1.00	13 - 16
1.33	11 - 16

Active system can also be used for thermal storage in which warm air is ducted from a high place in the greenhouse, and is passed through a rock bed. The rock bed system is usually located beneath the concrete floor or in the north wall which during night-time functions as a heat radiating panel. Whenever, the greenhouse temperature is $5^{o}C$ more than the rock, the warm air is passed through rock bed. The recommended volume of rock is 1.5-3.0 m^3 and 1 m^2 of south-facing glass area. The additional advantages of the ventilation are the control of humidity and the carbon dioxide level in the greenhouse.

9.4.10 Combining Systems

In a practical situation, more than one passive concept may be used for heating the space. Sizing details, however, are available only for individual concepts. It is recommended that for the same amount of heating the areas required for direct gain glazing system, thermal storage wall and the attached greenhouse system are in the ratio 1:2:3 i.e. each 1 m^2 of direct glazing equal 2 m^2 of thermal storage wall and 3 m^2 of solarium common wall area.

9.4.11 Cloudy Day Storage

(a) *Direct gain system*

To provide heat storage for one of two cloudy days, the following rules are applicable:

(i) the south-glazing area is increased by 10 to 20%,

(ii) interior walls and floors are made of 20 cm thick masonry,

(iii) for each m^2 of south glazing, an interior water wall of 2 to 3 m^3 capacity is used.

(b) *Indirect gain system*

(i) collector area is increased by 10 to 20%,

(ii) a thick masonry wall of large thermal conductivity is preferred,

(iii) for each one m^2 of collector area 1 m^3 of more of water wall is used,

(iv) roof-ponds having a depth of 0.24 to 0.30 m are used.

9.4.12 Movable Insulation

Movable insulation should preferably be used over entire glass area during off-sunshine hours. It, however, becomes necessary if only single-glazing is used. The percentage of heat losses in double-glazed and single-glazed windows with and without night insulation is presented in Table 9.12 given below. It may be noted here that a single-glazed window with night insulation is more effective than a double-glazed window. The insulation is more effective if it makes a tight and well sealed cover for the glazed opening.

The application of movable insulation can basically be divided into three categories viz.

1. hand-operated
2. thermal sensitive, and
3. motor-driven

TABLE 9.12 Percentage of Heat-Losses through Various Types of Windows (Resistance = 10.55 kJ)

No. of glazings	Heat losses in percentage	
	without night insulation	with night insulation
Single	100	40
Double	57	20

Hand-operated devices include sliding panels, hinged shutters and drapes. The initial cost of these is low. Typical examples of thermally sensitive devices are skylid* (a Freon activiated movable louver system), heat motors and large bimetallic strips. All these devices convert heat into a mechanical movement. They are more expensive than the hand-operated devices. The motor-driven movable insulations can either be manually-activated or controlled by automatic timers, thermostats or light-sensitive devices. Beadwall** is a typical example of motor-driven applications, the main advantage of which is a possible automatic operation. The main drawback of motor-driven movable insulation is somewhat more complicated equipment and higher initial maintenance costs.

9.4.13 Reflectors

When a large collector area is not feasible it is advisable to use, for vertical glazings, a reflector equal to the width of glazing but 1 to 2 times higher than the glazing opening in length. For south-sloping windows, the reflector is placed above the window with an angle of 100° between the plane of glazing and the plane of reflector with its length and width equal to that of glazing. By using reflectors, the average winter solar radiation incident on vertical glazing can be increased by roughly 30 to 40% during the winter months. The reflectors could also be adjusted for summer months to act as the shading devices.

9.4.14 Shading Devices

For a south-facing glass window a horizontal overhang is the most appropriate one. The projection of the overhang can be approximately calculated by the formula

$$\text{Projection} = \frac{\text{Window height}}{F}$$

where the factor F is calculated from the following table

North latitude	F
28°	5.6 - 11.1
32°	4.0 - 6.3
36°	3.0 - 4.5
40°	2.5 - 3.4

*Patented device of Stephen Baer, Zomeworks Corp., Albuquerque, N.M.
**Patented device of David Harrison, Zomeworks Corp., Albuquerque, N.M.

44°	2.0 - 2.7
48°	1.7 - 2.2
52°	1.5 - 1.8
56°	1.3 - 1.5

For different window areas, the percentage of day-light factor is given in Fig. 9.6 which corresponds different percentages of external shading.

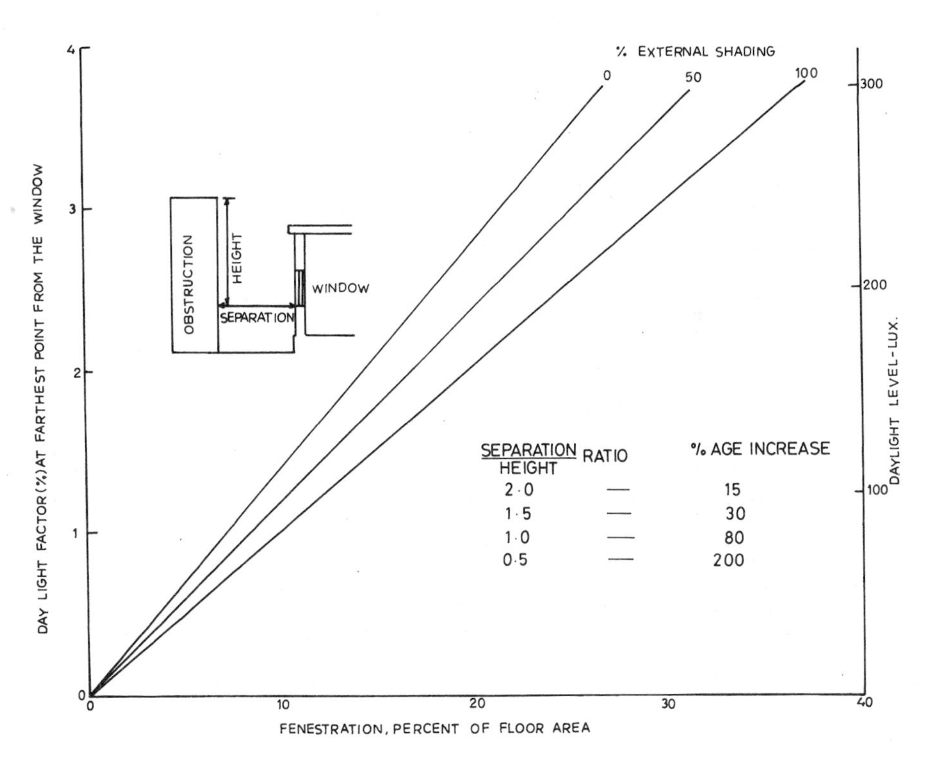

Fig. 9.6. Daylight availability for different window sizes

NOTE: This graph gives the window areas for isolated buildings. In situations where adjoining buildings etc. are quite high and obstruct the light these values should be increased by percentage given above.

9.4.15 Daylighting

Daylighting is often overlooked in buildings. Daylighting in residential buildings can create a stimulating living and working environment. Shared lighting can be created through the use of interior windows - from sunspace into a living space. This borrowed light can enliven an otherwise dark room as well as improve its thermal performance at little or no cost. Figure 9.7 illustrates some daylighting strategies.

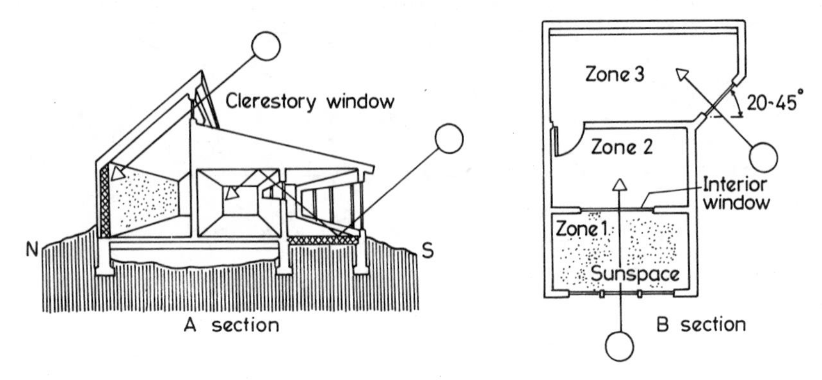

Fig. 9.7. Daylighting strategies

9.4.16 Summer Cooling

In climates with hot dry summer

(i) the roof should be of light colour, preferably white,

(ii) the building is opened at night by means of operable windows or vents,

(iii) the building is closed during daytime,

(iv) air cavities or insulation over the roof are provided.

In hot and humid climates

(i) the building is opened up during the day and evening for letting in the summer breezes,

(ii) inlets and outlets are arranged at the bottom and top of the walls, the area of the outlet being slightly larger than the inlet area.

9.4.17 Evaporative Cooling

Evaporation of the open water from the roof surface is an effective way of reducing the cooling load in summer conditions. The system for evaporative cooling is to be installed over the roof. For its realization, it is desired that

1. the roof is treated with water-proof material adequately,

2. the roof is covered with a water absorptive and retentive material. Such materials are gunny-bags, brick-ballast, sintered fly-ash, coconut husk and coir matting.

3. the required water quantity is $\sim$ 10 kg/day per m^2 of roof area during peak summer conditions,

4. a water sprayer which would keep the roof surface wet throughout the day. The sprayer can be manually-operated or controlled by an automatic device sensing the moisture of the water-retentive material.

The sprayer usually works at low water pressure which can be achieved either by a water head of the storage tank at the roof, or by a small water pump.

9.4.18 Earth-air tunnel system

This concept utilizes the storage potential of the earth. For an effective earth-air tunnel system, following considerations should be met.

1. The tunnel should be constructed at a depth of 4 m from the surface inside the ground.

2. The water-table level should be below 6-7 m of the ground surface.

3. A one-way valve should be used at the points from where the tunnel air enters into the room.

4. A mechanical fan should be used for blowing the air.

9.5 MECHANICAL EQUIPMENT

The proper selection and design integration of mechanical equipment will make or break all the energy saving features which have been included in the design. If an oversized or poorly-designed mechanical system is installed, energy consumption could be very high. The care which one has exercised in the design of a building of mechanical systems require the consideration of the following.

9.5.1 Mechanical Efficiency

The available fuel choices for auxiliary heating (or cooling) should be identified. Next, the mechanical heating (or cooling) systems that provide least life cycle cost should be selected. The rate structure of the supplying utility, e.g. the existence of a demand rate or time of day charge, should be carefully examined. If electrical resistance heating is being considered, the current or near term rate structure will greatly influence the system design and control strategy. For liquid fuels, a furnace or a boiler with an Annual Fuel Utilization Efficiency value of 80-95 should be selected. For air-conditioning, equipment with a Seasonal Energy Efficiency Ratio (SEER) of at least 8.0 should be selected.

Once the fuel type and equipment type is selected, it must be fully integrated with architectural design. Duct and pipe runs should be as short as possible and through conditioned space.

9.5.2 Mechanical Equipment Controls

The thermostat is the key device for controlling the amount of heating (and cooling) supplied by mechanical equipment to the home. It is controlled by the occupant. Energy savings of approximately 10% can be realized at no cost by simply lowering the night-time temperature setting during the heating season and raising it during the cooling season.

9.5.3 Ventilation

Ventilation, both natural and fan forced, is adequate to handle all cooling heads in temperate climates. A whole house fan is generally better. The fan could be centrally located to minimize the air path from any room in the house. Use of ceiling paddle fans to augment the whole house fan is an effective strategy, since they directly cool the occupants. If natural ventilation is used, the inlets and outlets for the air must be properly sized and located to ensure adequate flow rate (Fig. 9.8).

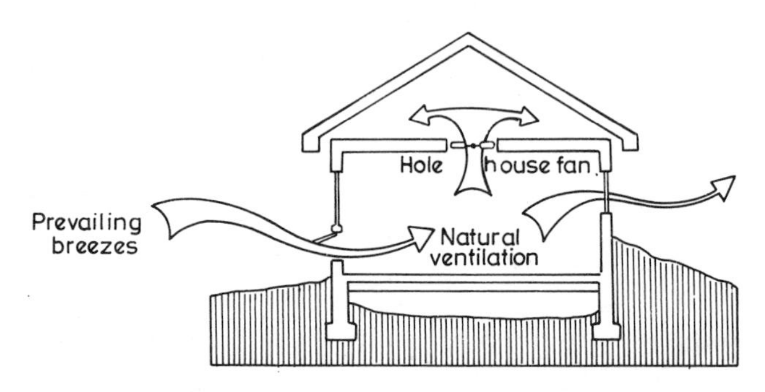

Fig. 9.8. Ventilation Strategies

9.5.4 Domestic Hot Water (DHW)

Use of hot water contributes nearly 20 to 30% of a home's total energy consumption. Energy savings can be achieved through four options (i) flow reduction, (ii) tank insulation, (iii) lowered set point temperature, (iv) solar domestic hot water systems. A flow rate of 5.5 to 11.5 l/min will result in significant hot water savings at little inconvenience. Insulation on the storage tank reduces heat losses and a solar DHW can provide nearly 70-75% of the hot water system. One should however exercise caution in choosing the system. The manufacturer's warranty and reputation should be evaluated.

REFERENCES

1. Mazaria, E. (1979) *The Passive Solar Energy Book*, Rodale Press, Emmans. Pa., USA.
2. Cathain, C. S. O' (1982) Exploration with a model of passive solar passive housing, *Energy and Buildings*, *4*, 181-183.

Chapter 10

TYPICAL DESIGNS OF SOLAR PASSIVE BUILDINGS

Design, performance and other characteristic features of some typical buildings, which illustrate different passive approaches, are described below. These buildings fall into four categories depending upon the climatic zones namely cold, tropical, arid and hot and humid.

10.1 COLD CLIMATE

10.1.1 The Hodges Residence

The Hodges residence (Hodge, 1983) is the first earth-sheltered passive solar home located in Ames, Iowa (latitude 42.0°N, longitude 93.8°W) which is scientifically designed to incorporate simultaneously the devices for the solar energy collection, storage and its distribution incorporating direct gain concept. It meets nearly 85% of the total heating requirements of the house by solar energy only. The house occupies a total area of 204 m^2 subdivided into two equal levels (102 m^2), each extending 13.4 m long in the east-west direction and 7.6 m deep in the north-south direction. The south side of the house is double-glazed over 37 m^2; 1.5 m of glass on the lower level has direct exposure on the ground. The lower level contains bedrooms only for coolness and tornado protection whereas the upper level is relatively quite open and contains living-room, dining-room, kitchen, family room, pantry and bathroom. The floor separating the two levels, consists of 200 m thick concrete cored slab topped by 50 mm of exposed aggregate and acts as a thermal storage unit for the house.

A family of four persons (two adults, two minors) has occupied the Hodges residence since September 1979 and has monitored the observations with the help of three gas meters, two electric meters, two hygrometers and several other thermometers.

Performance. The meteorological data for the 1979-80 heating season are listed in Table 10.1; the heating degree days (described in Chapter 6) are computed for a base of 18.3°C using the maxima and minima of daily-recorded temperatures.

The direct gain system provided adequate comfort to both the upper and lower levels despite cloudiness in extreme winter when the cold waves outside were drifting the snow whereas the inside air temperature was at 21°C. For instance, on January 31, 1980, the ambient temperature varied from (-21°C to -11°C) and

TABLE 10.1 Meteorological Data at Ames (Iowa) for the 1979-80 Heating Season (After Hodges, 1983)

Month	Heating Degree-days (°C)		Average Daily Horizontal Insolation (kWh/m²)	
	Actual	Normal	Actual	Normal
October	211	213	3.04	3.21
November	459	472	1.83	2.06
December	586	687	1.36	1.59
January	727	775	1.40	2.02
February	711	662*	2.43	2.91
March	531	531	3.70	3.94
April	244	265	4.80	4.47
May	82	121	5.09	5.46
Total	3551	3726	23.65	25.66

*Corrected for 29-day month.

total insolation on horizontal surface was 3.5 kWh/m^2, however, average inside air temperature was maintained at 24.4°C. The lower level was found to be slightly cooler (by 2-3°C) than the upper level. This difference in temperature did not cause any problems to the family as they used to enjoy the sun during the day in winter months in the upper level only. Moreover, it was very pleasant during summertime. Natural ventilation was sufficient to keep the house comfortable even during hot sunny periods. The actual data estimates of 1979-80 have indicated that the heating requirement of the house was met up to 60% by solar energy. The auxiliary heat requirement may be further lowered if Klegecell* for night insulation would have been used on glass panes of south side.

10.1.2 Warehouse: Direct Gain and Water Wall Storage

The Benedictine Monastery Warehouse (Paul, 1979) at New Mexico (latitude 36°N) where the ambient temperature drops below -7°C, essentially consists of two parts (Fig. 10.1), the warehouse and the office space. The warehouse is heated directly by solar gain whereas the offices are heated by direct gain as well as a water filled drumwall which acts as a collector/storage system.

The house occupies a total space of 929 m^2, the south and north portions of the building being 260 m^2 and 511 m^2 respectigely. The insulated glass windows occupying an area of 144 m^2 are located on the south side of the house for direct

*Klegecell is a polyvinyl chloride foam which acts as a vapour barrier. It can be held against the glass along the edge and no condensation appears on the glass even after a very cold night. It is fabricated in ten different densities by American Klegecell Corporation, 204 North Dooley St., Grapevine, Texas 76051

gain. The outer walls of the building are made of concrete masonry blocks; all surfaces being well insulated with styrofoam/fibreglass. The warehouse is partitioned from the office space by a concrete block (filled with sand) providing an additional thermal storage unit.

The important feature of the house is the water filled drumwall which extends along the entire length of the building on the south side and consists of 138 standard 0.25 m^3 black-painted barrels. The wall is provided glazing from the lower row of windows (40.9 m^2). Additionally, aluminium reflecting doors are also provided on the outer side of the south wall for reducing the heat losses. The warehouse is heated by direct gain of solar energy as well as the heated air venting in from the office space which is warmed by direct gain through windows and the heat released by the drumwall. The south-east portion of the building houses a double-glazed flat plate solar collector for water heating.

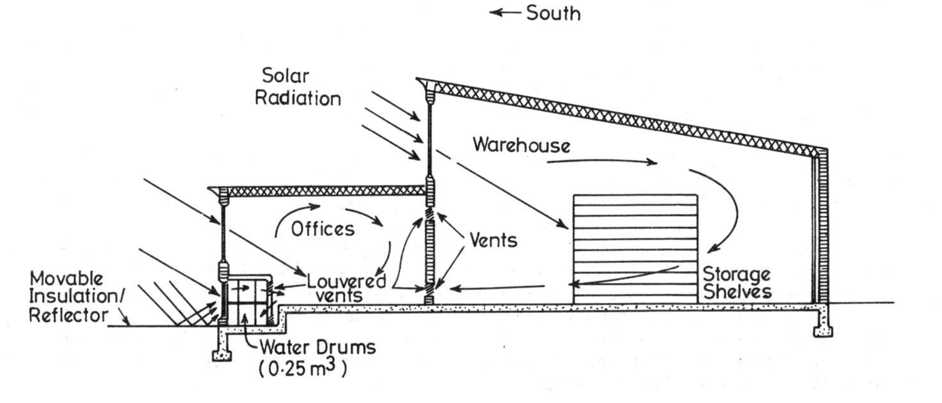

Fig. 10.1. Solar heating system of Benedictine Monastery Warehouse

Performance. Although the warehouse was designed to maintain a minimum temperature of 7.2°C, the average minimum temperature attained in the house was 8.9 to 10°C in contrast to the temperature of the office which ranged between 16.7 to 25.6°C when the doors were closed (Fig. 10.2). The lowest temperature seen in the offices was 12.8°C which was observed only during nonworking hours. The performance of the building was observed to be beyond expectation particularly in the situation where 5 to 7 people have worked comfortably under adverse winter climatic conditions with ambient temperature varying from -6.7°C in the day to -17.8°C during the night.

The economic data of the building is also very encouraging. The total construction cost of the building is evaluated to be around \$140/m^2 out of which approximately \$8,500 was spent for solar units.

The back-up heating system, consisting of nine 1440 W radiant electric heating panels, was found to be rarely used, keeping the fuel bill to a minimal value. One full-time occupant, who resided in the building during the entire winter period, did not find it necessary to use the electrical back-up heating system at all. Only during several cloudy days in March-April, the back-up system was used during working hours.

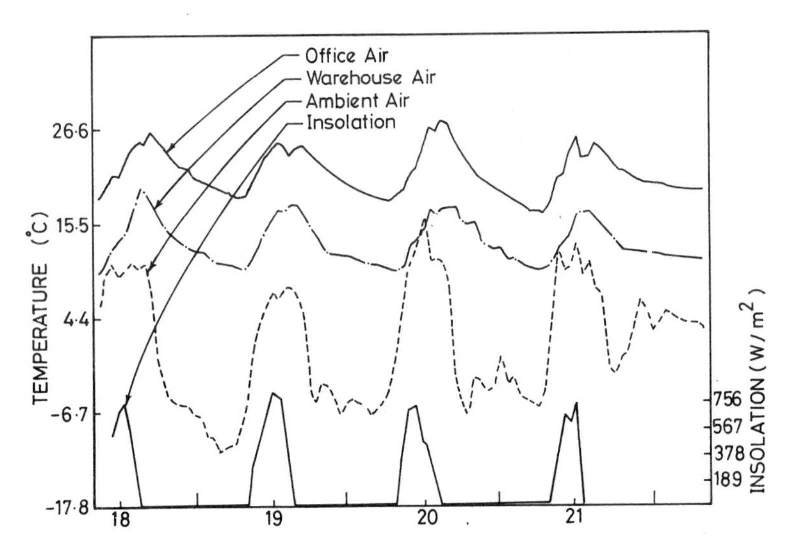

Fig. 10.2. Variation of various temperatures and insolation collected during the period February 18-22, 1977 in warehouse

10.1.3 Unit One, First Village: Solarium and Trombe Wall (Haskins and Stromberg, (1979)

The Unit 1, First Village is located in Santa Fé and is designed for cold climate of New Mexico (latitude: 36°N, altitude: 2347 m, annual heating degree (for 18.3°C base temperature) days: 6000, Ambient temperature = -23.9°C). The unit is divided into two floors, the plan of which is described in Fig, 10.3.

The glazing on the south side serves as greenhouse for direct solar gain with total area of 38.3 m^2. The greenhouse is separated from the house by an abobe mass wall which acts as a thermal trombe wall and distributes the heat by radiation and convection into the interior. Another storage is provided by pumping the hot air of the greenhouse to two storage bins underneath the floor and exhausting the cold air again to the greenhouse. An automatic arrangement has been provided to switch on the fan in situations when the air temperature exceeds the rock temperature by 8.5°C (see thermal flow diagram in Fig. 10.4). During the day, the solar heat is absorbed by the greenhouse and trombe wall; the heat of the trombe wall is conducted away and is transferred to interior by convection and radiation. Heat absorbed by 0.36 m thick wall takes nearly 7 to 8 hours to conduct inside. However, the heat absorbed by the floor of the greenhouse keeps the space warmer than the ambient due to infrared thermal radiations emitted by the floor. During the summer, excess heat is vented out through two of the glass panels existing on south wall which allows outside air to enter the greenhouse. Also, a window is provided on the top of the staircase which acts as a chimney for venting the excess heat of the greenhouse to the outside. A flat plate collector is also installed for hot water supply to the house.

Another feature of this house is the seasonal temperature control provided by projecting the greenhouse roof and the balcony floor (in horizontal planes) on the south side of the adobe wall. During the summer, when the sun's midday altitude angle is high, the projecting planes shade the adobe wall from direct incidence of sun's rays. During the winter, the low altitude angle of the sun allows the sun's ray to undershoot these projecting planes and fall directly on to the adobe wall, which acts as an efficient solar collector-storage system.

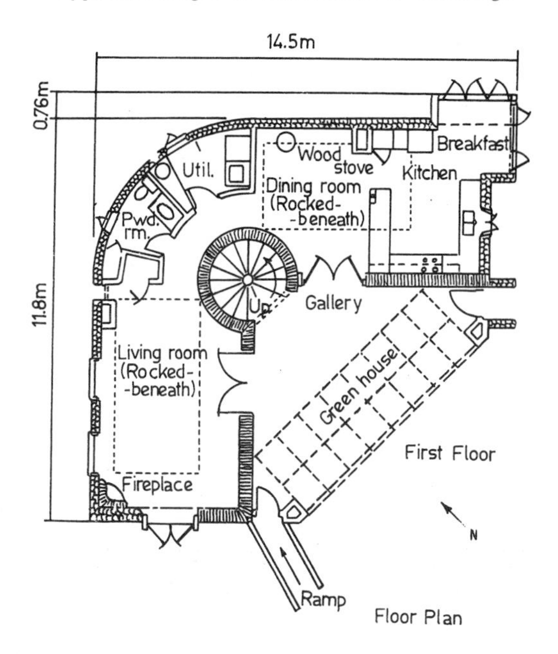

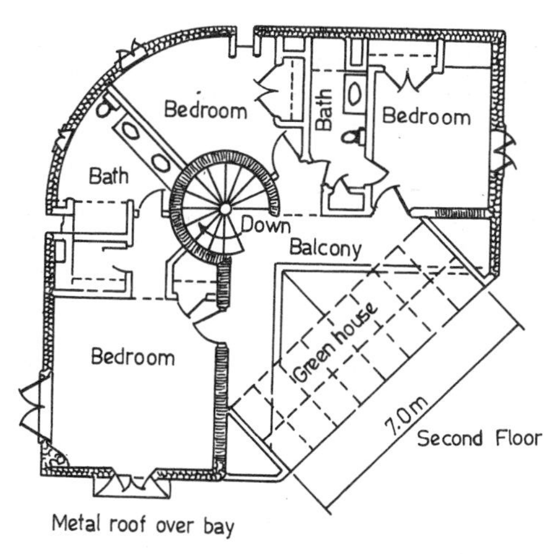

Fig. 10.3. Floor plan of unit 1, First Village (After Haskins and Stromberg, 1979)

The overall effect of the system is that the unneeded summer solar energy is largely prevented from entering the greenhouse or living spaces, while winter solar energy is efficiently collected, stored and kept in the living spaces at comfortable temperatures.

The building is provided with auxiliary heat through base-board electric heaters, a fireplace and wood burning stove. A differential thermostat controls the fan

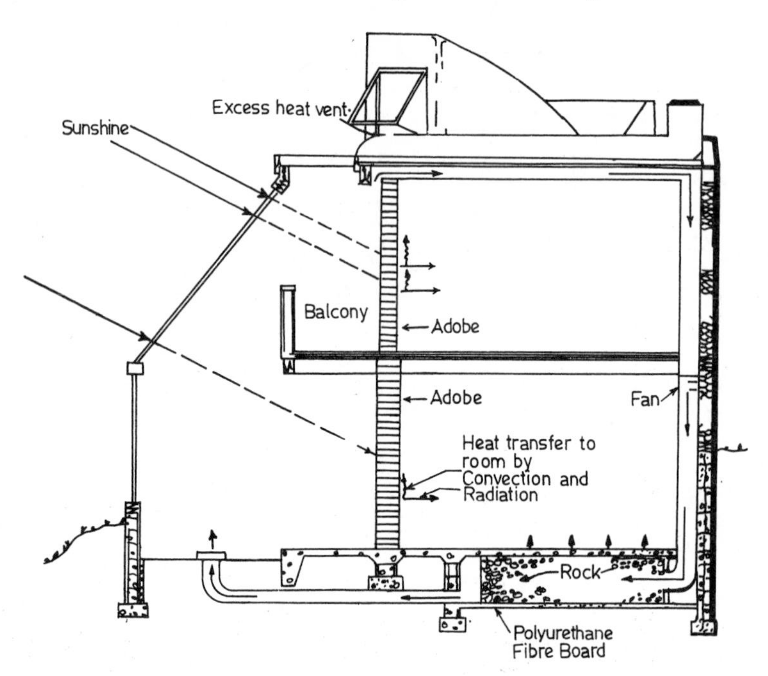

Fig. 10.4. Thermal flow diagram

forcing air flow into the rockbed storage. The baseboard electric heaters had individual thermostats for controls.

Performance. The performance of the house during cold winter season of 1979-80 has been very satisfactory. The experimental observations of the performance are illustrated in Fig. 10.5. It is interesting to note that the temperature of the dining-room air is maintained at 21°C against the ambient air temperature varying between -18°C to 4.4°C with negligible electric input for heating.

10.1.4 The Dr. Hollinger Building

The Dr. Hollinger building (Arnoth, 1983) was built in 1978 in Albuquerque, New Mexico and is commonly known as Comanche Place. The building comprises of examination rooms, reception, lounge, cast area in the middle and offices along the north side of the building (see Figs. 10.6 and 10.7).

The direct solar gain/heating systems of the building include the south-facing verticle windows, clerestories and a greenhouse space having a glazing on south side at an inclination of 30° from the horizontal. Two trees are planted to shade the greenhouse glazing during summertime. The other details of the house are illustrated in Table 10.2.

Performance. The illustration of the monthly energy used in the Dr. Hollinger building is shown in Fig. 10.8. The predicted values are based on the Los Alamos Scientific Laboratory (Arnoth, 1983) computer calculations whereas ACTUAL represents the experimental values.

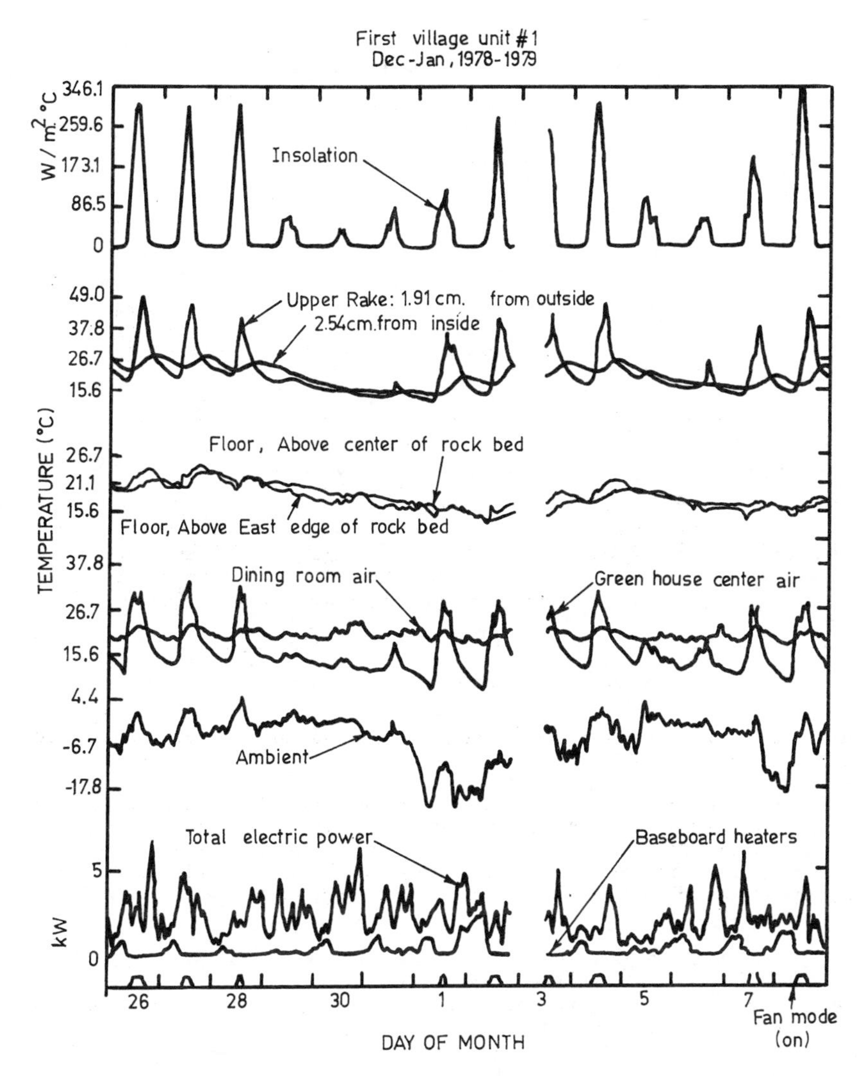

Fig. 10.5. Recorded data for the Balcomb House

10.1.5 Srinagar Dispensary

Srinagar, the Capital of the State of Jammu and Kashmir of India, lies in the Kashmir Valley, in the lap of the North-Western Himalayas. The City (latitude 34° 05'N, longitude 40° 50') is an example of cool humid or cold climate. The yearly variations are from a cold and humid period in winter to a temperate and humid late summer. During the spring and early summer periods, the climate is dry temperate and most comfortable and these are the seasons when tourists come to spend a comfortable vacation in the town, to escape from the dry heat of the North Indian plains.

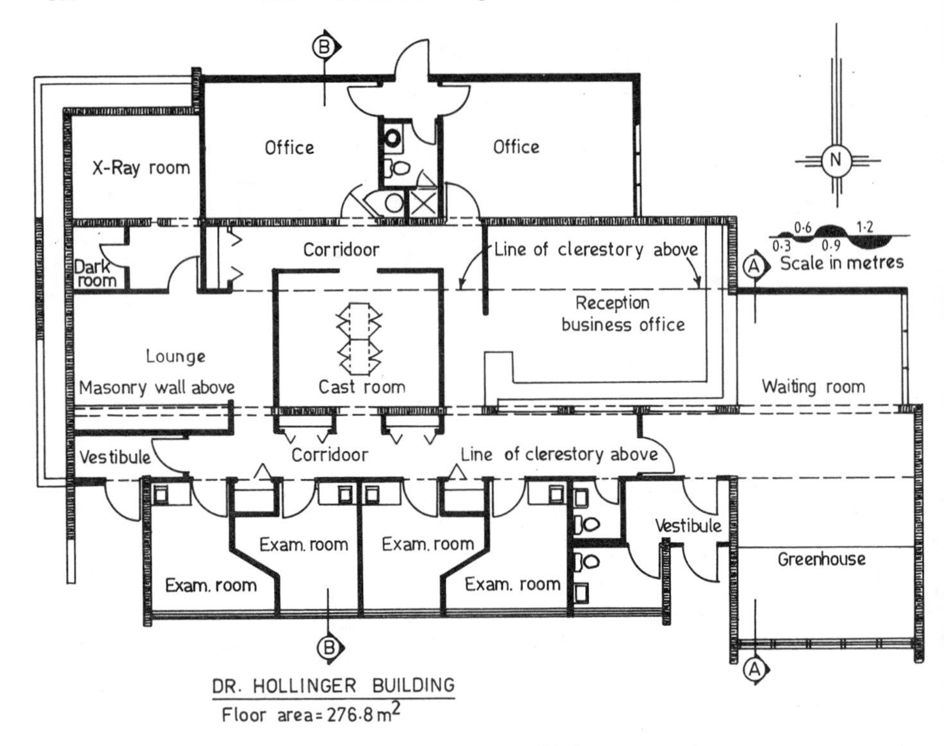

Fig. 10.6. Dr. Hollinger building floor plan

Table 10.3 of the climatic data of Srinagar shows that the diurnal variation of the temperature during summer and winter is small due to relatively high humidities during this month. Except during early winter, rainfall is evenly distributed with snowfall during the month of January and February. Wind speed is moderate at Srinagar $\sim$ 4 kmph with the most predominant direction as south-east and north-west.

The radiation pattern is typical of the latitude. Direct radiation during winter months especially gets moderated by the cloud cover. The mean cloud cover even during the comparatively clear October month is as much as 3.5 hours in a day. The south wall receives maximum radiation during the winter months and the minimum during the summer period.

The climatic pattern of Srinagar indicates the basic requirement of providing heating during the winter months. The obvious heating method in this case therefore is to use south collection apertures. Unwanted summer radiation can be easily excluded by appropriate horizontal overhangs on each designs.

Cloud cover appreciably reduces the effectiveness of this radiation collection method. Therefore, insulation and heat conservation will play an important role in the overall energy balance of the buildings. Insulation is specially important for the north wall because this wall does not receive any radiation and its outer surface temperature will be close to the ambient. Roof insulation is also important because roof does not receive very high radiation during winter and in summer, radiation is to be effectively prevented from entering into the living space. Another component, which should receive proper attention, is various windows which if not properly oriented and protected would be the major

TABLE 10.2 Details of Dr. Hollinger House

Typical number of occupants:	8
Typical hours of operation:	9:00 to 5:00 Monday through Thursday
	9:00 to 12:00 Friday
Thermostat setting:	21.1°C day
	15.6°C night
Internal heat rate:	443551.5 kJ/day
Floor area:	276.8 m^2
Type of auxiliary heating:	Heat pump with electric resistive back-up
Building load coefficient:	29168.6 kJ/DD
Passive solar system type:	Net glazing area:
	Clerestory: 20.5 m^2
	Greenhouse (30 degree tilt): 12.5 m^2
	Direct Gain (exam. rooms): 9.7 m^2
Total net area of collector:	42.6 m^2
Load collector ratio:	684.7 kJ/DDm^2
Collector area to floor area ratio:	16%

source of heat losses.

The proposed passive building structure at Srinagar is an extension of the existing Regional Engineering College Dispensary located on the bank of famous Dal Lake. The final design of the building is sketched in Fig. 10.9. The components of the extended passive structure include rooms for guard, ward, doctor, isolation room, dark room, store room, x-ray room and toilets. The roof is made of steeply sloping galvanized iron sheet (along with a false ceiling) to take care of snow during winter months. The solar heat gain system is essentially a Trombe wall, made of masonry painted with radiation absorbing colour and glazed. Trombe wall is provided with slit windows to admit some daytime radiation directly, thus helping to admit more solar radiation, provide natural lighting and also provide a better view to the outside.

Area of Trombe wall has been kept 40 m^2 including 5.5 m^2 of direct gain window area. The floor area of the ward immediately connected to the wall is 30.4 m^2 and its volume 120 m^3. In addition, a "greenhouse" or solarium had been added to the isolation room west of the main ward and also faces south. This is the application of a typical isolated gain concept.

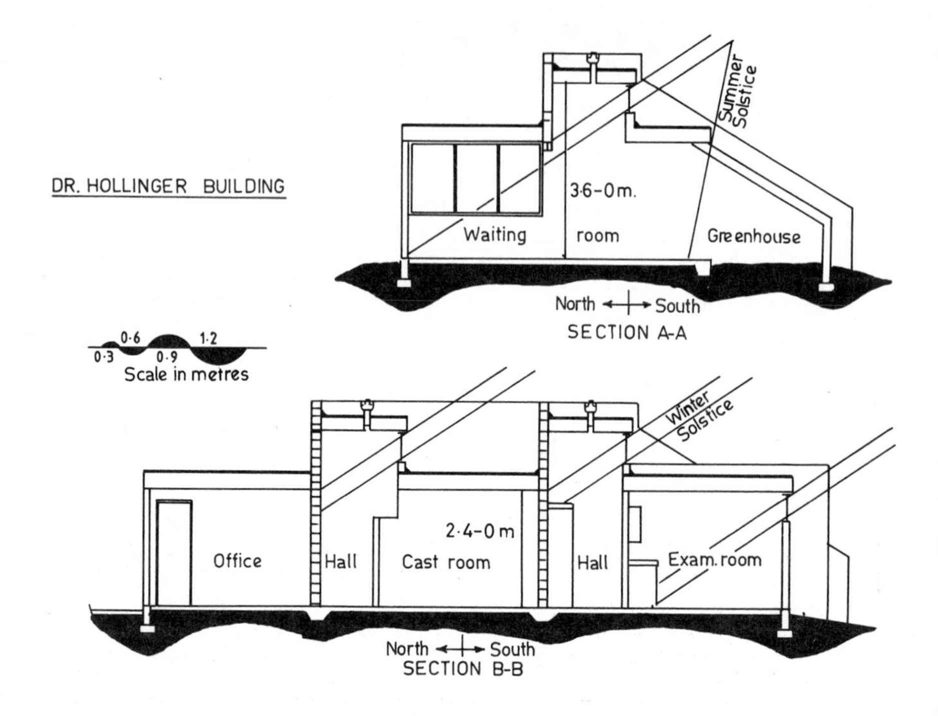

Fig. 10.7. Dr. Hollinger Building Sections

Unlike fully-glazed solariums, it is proposed to have glass on the walls only. The roof will be made of the same materials as the ward. This will save vastly on cost and maintenance. Solarium can be made to serve the function of waiting in pleasant surroundings. Movable insulation coverings would be desirable for this area.

The various measurements for the room-solarium set is as below:

Floor area of isolation room	= 12.3 m^2
Floor area of solarium	= 11.2 m^2
Glass area, solarium, facing south	= 5.3 m^2
Glass area, solarium, facing west	= 6.3 m^2
Glass area, solarium, facing south-east	= 4.5 m^2
Window area between room and solarium	= 1.2 m^2
External window on room (faces west)	= 1.3 m^2

There is a 45° chamber given to the solarium nearer the end, close to the ward side. This is to prevent shading of the Trombe Wall in the winter afternoons. The solarium also could not be extended westwards because the sewage pipes run below the ground there and the new sewage pipes will also run alongside these. The septic tank (12 in Fig. 10.9) has also just been avoided.

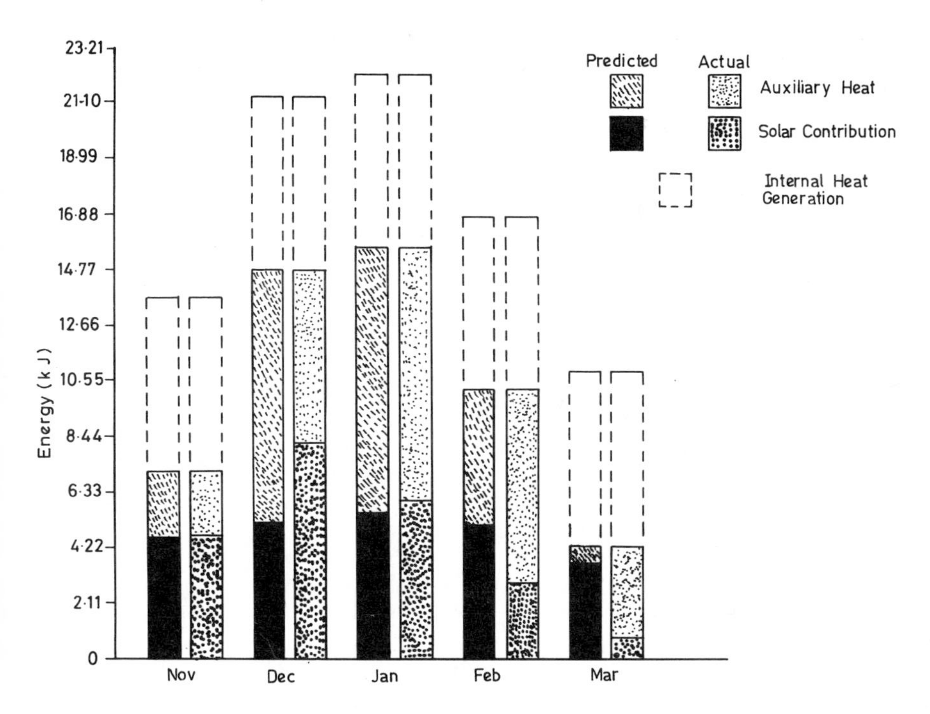

Fig. 10.8. Dr. Hollinger Building monthly energy bar graph-predicted and actual

With the location of the solarium on the south and its use as a waiting room, an entrance from the east and south of the building will be useful. The atmosphere of spaces at this entrance would also be pleasing, and will probably not look like a typically dark or smelly entrance one associates with hospitals and dispensaries.

To aid in direct gain in winter and to provide required ventilation during humid summer months, the ward and isolation room areprovided with windows (1.3 m^2) on east and west faces respectively. The north wall of the ward (the only north wall being built) has been kept windowless. This will help in reducing heat losses and also fits in well with the internal layouts of beds, which abut against this wall.

Insulation: Walls and Windows. As the temperature difference between the outside and the inside should desirably be high (15°C in winter), any simple thermal insulation would pay dividends in preventing heat losses. Wherever new external walls are being built, they are proposed to be cavity walls. The size of the cavity, determined by the minimum it is possible to construct, will be 50 mm. The two leaves will be combinations of 230 mm + 12 mm plaster and 115 mm + 12 mm plaster.

The windows to be provided in the new structure will be one layer of glass and an openable shutter of wood. The shutters can be opened during the time the desirable radiation comes in through the windows; at other times the wood shutters should be closed to avoid undesirable heat losses.

Performance. The actual building is presently under construction stage. The

TABLE 10.3 Climatic Data of Srinagar: 34°N, 75°, 1586 m above MSL (after Seshadri *et al.*, 1979)

	January	May	August	November
Mean Diurnal Temperature range (°C)	9.3	14.5	12.9	17.5
Mean Maximum Temperature (°C)	5.0	25.0	30.3	16.9
Mean Minimum Temperature (°C)	-4.3	10.5	17.4	-0.6
Sky cover (%)	78.4	39.0	45.5	28.8
Morning Mean RH (%)	89	81	86	87
Evening Mean RH (%)	75	39	48	49
Mean Rainfall (mm)	73.7	60.5	61.5	11.2
Mean Monthly Wind Speed (Kmph)	3.54	3.51	3.70	2.90
Mean radiation on horizontal surface (W/m^2)	115 115	294 294	277 277	131 131
Mean Sunshine hours	2.5	7.2	7.7	6.7
Mean Cloudy hours	7.6	6.3	5.8	3.6

performance of the building, based on simulation results is however shown in Figs. 10.10 and 10.11 for typical winter and summer days respectively. The temperatures of air in various rooms are seen to be fairly stable against vast fluctuations in the ambient temperatures. The computations for the winter day correspond to the worst day (average solar intensity only 54 W/m^2), even then the temperature of the ward, isolation room and the toilet are always much higher than the ambient temperature. On sunny days room temperatures are expected to acquire a very comfortable range. In summer period (Fig. 10.11) ambient air temperature varies between 16°C and 28°C; the general ward and the isolation ward remain however around 20°C.

10.2 TROPICAL CLIMATE

10.2.1 Atascadero House: Skytherm System

The skytherm system of solar heating and natural cooling (Fig. 10.12) was invented and fabricated by Hay (1976) at Atascadero, California. The characteristic feature of the system is its ability to keep the house warm in winter and cool in summer in a climate where ambient temperature extremes have varied from -12.2°C in the winter to 43.3°C in summer, without any use of electricity other than that used for food preparation, hot water heating, and lighting etc.

The floor plan of the house is illustrated in Fig. 10.13. The floor occupies an

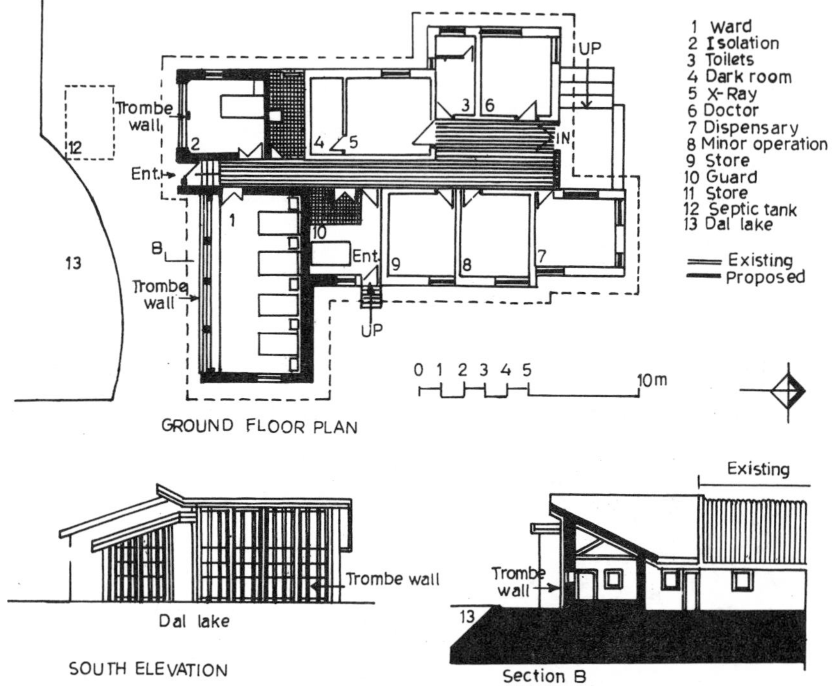

Fig. 10.9. Solar heated additions to dispensary at Srinagar

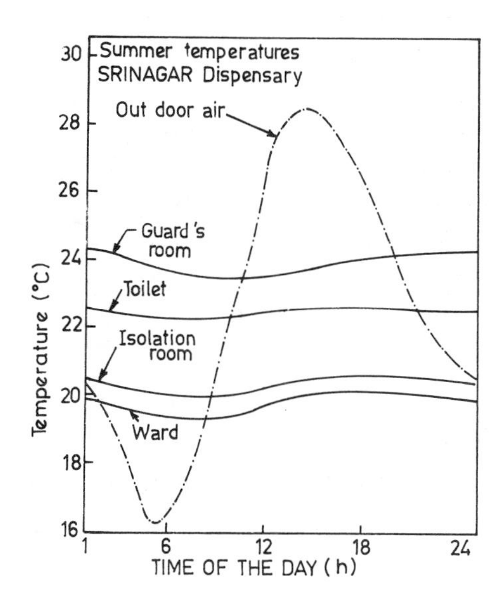

Fig. 10.10. Hourly variation of various room temperatures in Summer in Srinagar Dispensary

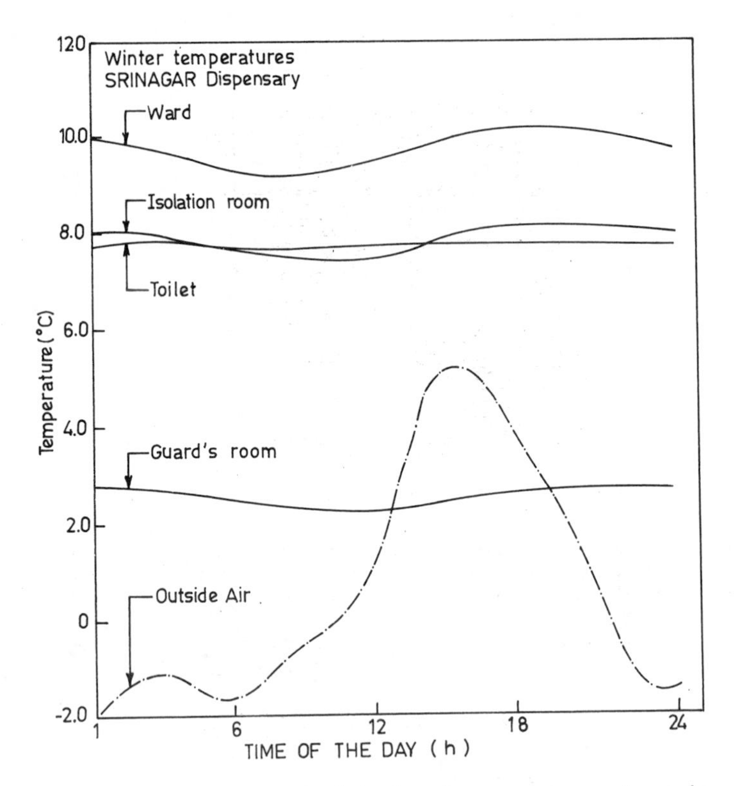

Fig. 10.11. Hourly variation of various room temperatures in Winter at Srinagar Dispensary

Fig. 10.12. Skytherm residence built at Atascadero, California

area of 111.5 m^2 and is made of reinforced concrete slabs. The exterior perimeter of the footing walls and the slab edges are insulated with polystyrene. The south wall is not provided with any window (except in the bathroom) but all windows on other sides (comprising 21% of the total wall area) are double-glazed. The ceiling is insulated to avoid condensation during summertime.

The basic principles of the skytherm system in the summer and winter months were illustrated in Chapter 4. Essentially, it consists of a metal roof which serves as the ceiling of the room below. PVC bags (20 mm thick) filled with water are placed directly on the deck to store sun energy. An additional layer of PVC is

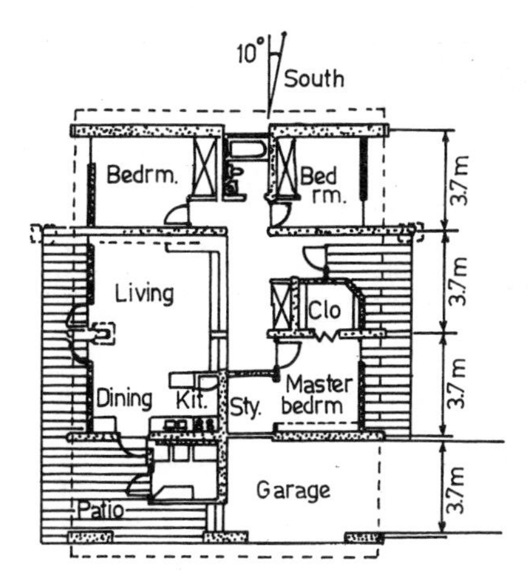

Fig. 10.13. Floor plan of the Atascadero House

spread over bags with space between the layers being inflated to provide greenhouse effect. The 51 mm thick urethane panels covered with aluminium foil are used as movable insulation; the panels may be rolled along the roof beams horizontally to expose or cover the thermoponds (see Fig. 10.14). The total collector area is 102.2 m^2 and the average water depth is 216 mm (approximate capacity 27.3 m^3).

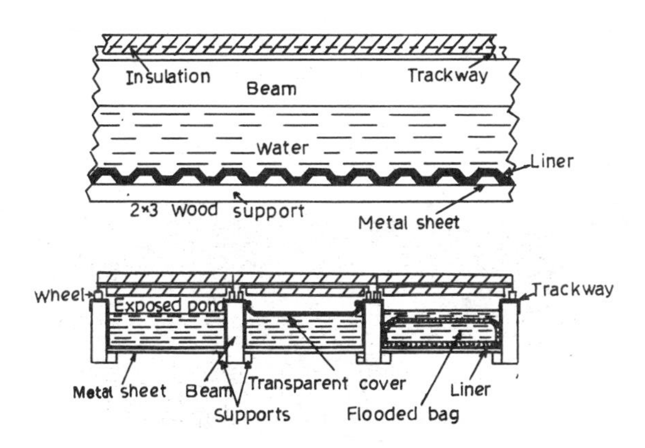

Fig. 10.14. Cross-section of ceiling ponds

The thermopond involves three combinations in itself: (a) heat collection, (b) heat storage and (c) heat transfer to the structure. During the winter months, the thermoponds are exposed to solar radiations in the day in order to collect solar energy and are covered with movable insulation in the night to retain the stored heat. Heat is transferred from the ceiling to the floor and walls primarily by radiation.

Conversely, in summer months, the greenhouse cover is removed and the panel positions are reversed in order to reject solar heat which is generated in the living space. The thermoponds are exposed to ambient in the night to lose their heat to the cool night sky and are covered by movable insulation in the day to avoid the heating. Generally, the water bags are flooded with excess water to produce cooling effect of evaporation.

Performance. The skytherm system has proved its ability to maintain thermal comfort conditions under wisely fluctuating temperature conditions and has provided 100% of heating and cooling requirements of the building. During the test period from May 1, 1974 through September 23, 1974, the skytherm noctural system was very effective in maintaining the indoor temperature at 22.2°C against the maximum outdoor average temperature of 31°C (Fig. 10.15). Simultaneously, the system was able to impede the temperature fluctuations in the living space to less than 2.5°C in winter and 0.5°C in summer, thereby, maintaining the indoor temperature between the extremes of 18.9°C and 23.3°C.

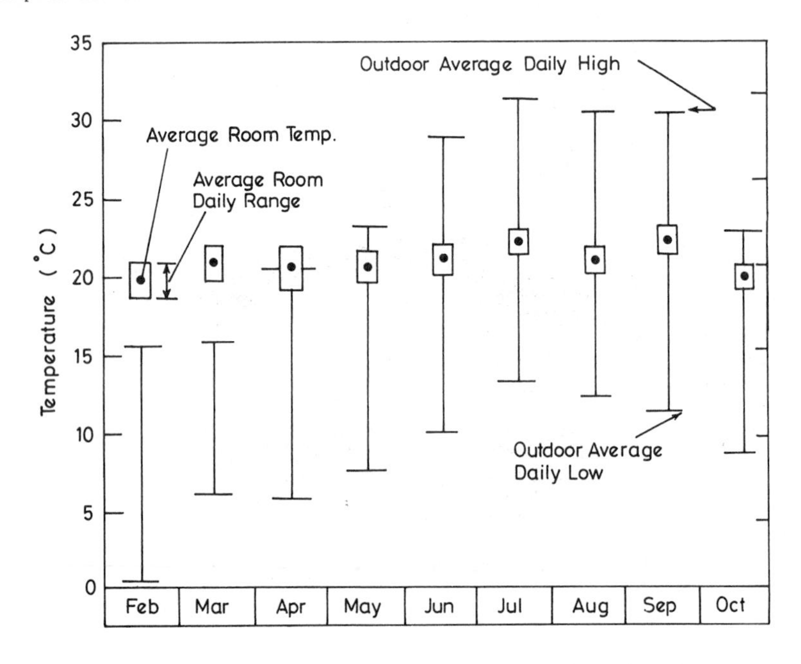

Fig. 10.15. Average 9-month performance of skytherm system

10.2.2 Hostel for Research Scholars: IIT Delhi

Delhi (29°N, 77°E, 218 m above MSL), the capital of India, is a representative of a composite tropical climate, experiencing three distinct seasons: the hot-dry period (April to June), the warm-humid period (June to September) and the cold period (November to February). Delhi's climate is in fact characterized by extremes as the maximum temperature touches 45°C during summer and minimum 4°C during winter; every year the minimum record being sub-zero. The relative humidity is normally high and the wind speeds are fairly low flowing in the direction SSE/NNW (see Fig. 10.16). Therefore, the comfort requirements of the hostel building are based on the following three aspects in three different seasons:

(i) Cooling/humidification in hot-dry season.

(ii) Ventilation/dehumidification in the warm-humid seasons and

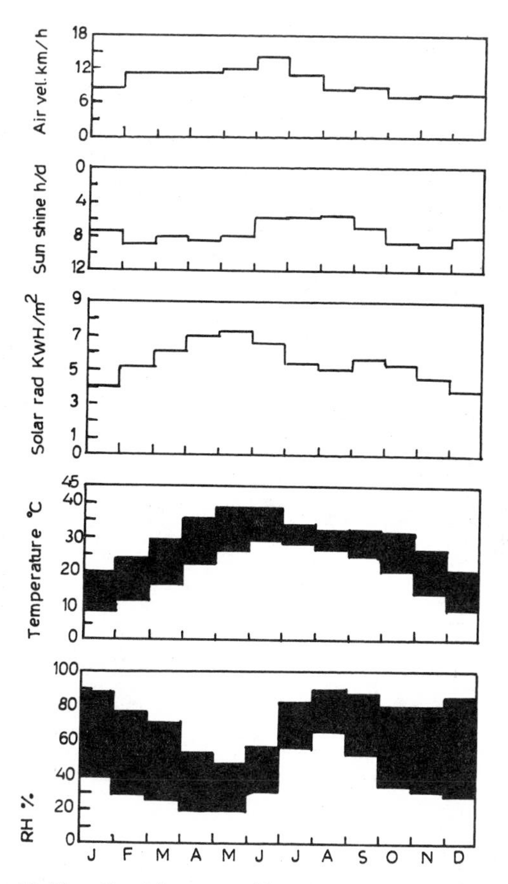

Fig. 10.16. Monthly mean climate parameters for Delhi

(iii) Heating in winter season

Incorporating appropriate passive concepts, a hostel building is designed for married research scholars at IIT Delhi which is under fabrication stage and is likely to be completed by the end of 1984 (Fig. 10.17). The building in all has 12 residential units; each unit consisting of two rooms, one kitchenette, one toilet and a verandah. The dimensions of each component are:

Room 1	13 m^2
Room 2	13 m^2
Kitchenette	3.3 m^2
Toilet	3.3 m^2
Verandah	5.5 m^2

The building is basically a two-storey south-facing structure having an entrance on the north side. The structure of the building is partially sunk into the

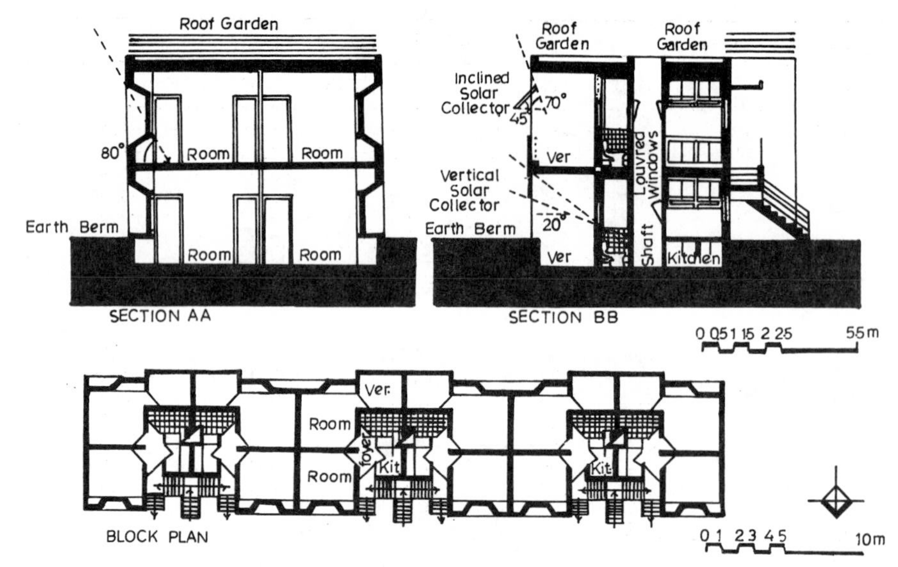

Fig. 10.17. Hostel building at Delhi

ground (about 1 m), with water spray and vegetable pergola on the roof for shading to achieve coolness during the summer months. Each unit is provided with south-facing windows shaded with a horizontal projection which may provide good winter heating. In order to reduce summer daytime heat gain and winter night-time heat losses, adequate arrangements of movable insulation or shade are being made. Cross ventilation also takes place through each room in the bank through grilled window which assures a constant air movement at the human level when both the north and the south windows remain open. Additionally, separate units of solar flat plate collectors based on the principle of thermosyphon, will be installed to provide hot water to the residents during all the three seasons.

Performance. Numerical computations have been made to predict the thermal performance of the hostel building on two typical days, January 26 and May 16, 1981 of New Delhi climate. The corresponding variation of room air temperature is shown in Figs. 10.18 and 10.19 respectively.

It is evident from the figure that when the outside air temperature is as low as 9 to 10°C during the winter night, the room air temperatures are as high as 17-18°C. This is due to the fact that most of the heat losses through windows to the cold night sky during off-sunshine hours have been checked by removable shutters during that period.

Conversely, in summer seasons when the ambient air temperature touches 45°C and remains always more than 27°C, the room air may be maintained at an average temperature of 33°C (upper floor rooms) and 32°C (ground floor rooms) with the use of passive cooling approaches. These results are to be validated with the actual experimental data of the building.

10.3 ARID CLIMATE

Hostel for Married Scholars: University of Jodhpur. Jodhpur (26°N, 73°E, 224 m above Mean Sea Level) lies at the edge of the Great Indian Desert (Thar). It is a representative of arid or dry climate with a severe summer and a rather mild

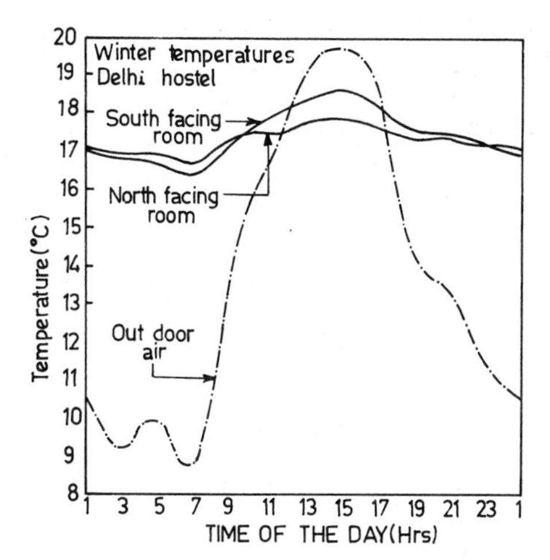

Fig. 10.18. Variation of room air temperature (on January 26, 1981) of hostel building, New Delhi

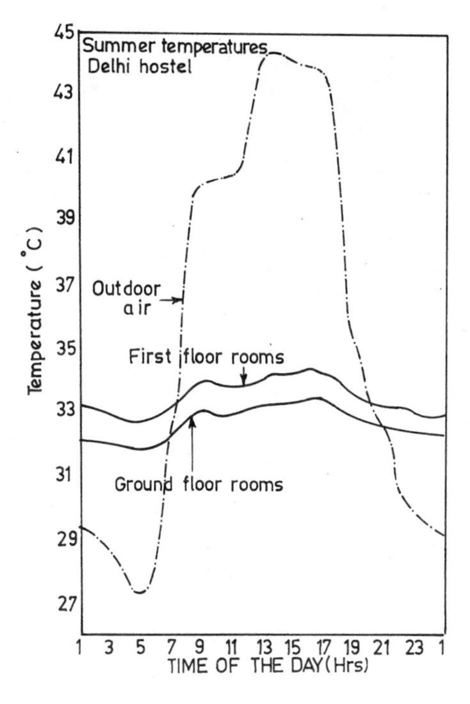

Fig. 10.19. Variation of room air temperature (on 16 May, 1981) of hostel building, New Delhi

winter. It experiences a short hot humid season (August monsoon) as well. The diurnal range of temperatures is large (nearly 15°C), especially during the hot-dry period, and the wind velocity is high throughout the year (see Fig. 10.20). The flow direction of wind is south-west except in winter when it is just the reverse, i.e. north-east.

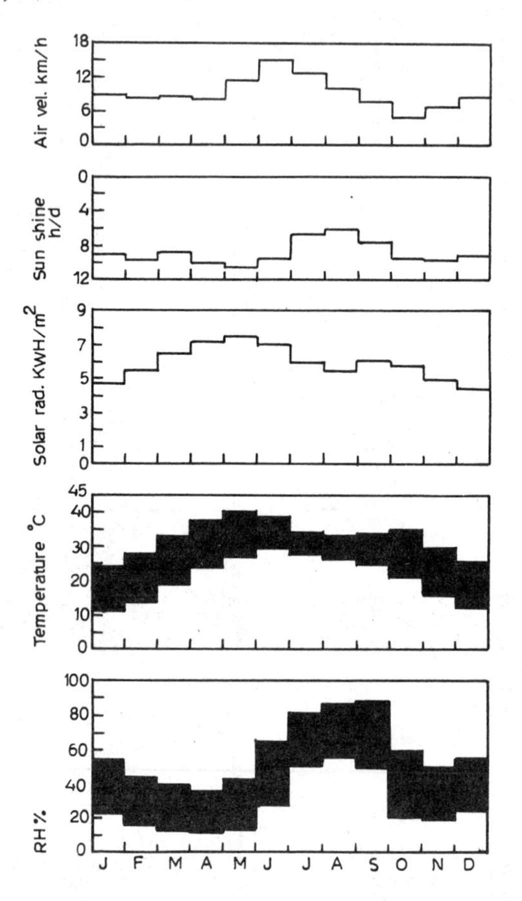

Fig. 10.20. Monthly mean climatic parameters for Jodhpur

The overall climate parameters indicate that the main comfort requirements would indicate a design thrust in the direction of sensible cooling during the hot-dry early summer and dehumidification and/or ventilation during the monsoon. Due to acute scarcity of water, water spray or other related systems cannot be adopted for cooling purpose but instead, cooling may be achieved by available high wind velocity either by ventilation (direct or through wind catchers) or pretreatment of ventilation air (by passing through underground ducts or evaporative cooling).

The design incorporates a massive building, partially sunk into the ground (see Fig. 10.21). About 20-single-room suits have to be provided as University of Jodhpur, Rajasthan (India). The area for each room may be 15-18 m^2 as they are all double rooms for married research scholars. With each room, there is an attached kitchenette (about 4 m^2) and a toilet (about 4 m^2); one lobby and a courtyard is provided with a set of seven rooms which are to be constructed in the very first phase. The kitchenette could be incorporated with the main room or pantry.

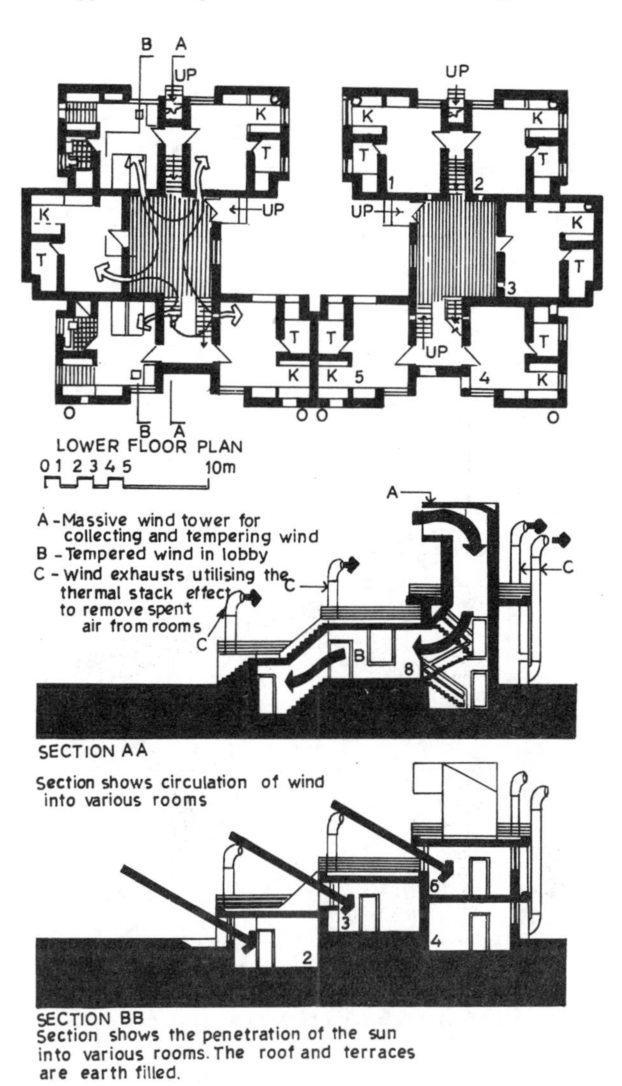

Fig. 10.21. Hostel building at Jodhpur

A wind tower with built-in-evaporative cooling has been proposed for sucking air into the rooms. A wind tower is adopted to reduce the sand coming into the space in a large quantity and is built at a height (over the lobby) unobstructed by the structure to catch the winds. This tower has a large mass to aid in stabilizing the temperature of the incoming air and acts in conjunction with a number of wind exhausts above each room facing the toward direction to cause suctions and, thereby, to improve air flows into the building. The system thus acts as a passive central evaporative cooling device for a set of seven rooms which are all clustered around the lobby.

South apertures may be exposed to sunlight for moderate heating in winter months.

However, shading of walls and roofs becomes important for the summer. Generally, light colouring is preferred on the outer surfaces with adequate arrangement of over-hangs or movable insulation.

Performance. The temperatures expected in various rooms of the building have been plotted in Fig. 10.22 for a typical summer day when the ambient temperature fluctuates between 28 and 40°C. Room temperatures, in contrast, are seen to be fairly stable; the maximum temperature being only 28.5°C in Room No. 1. Room No. 5 is seen to be always at a temperature around 24°C. All the room temperatures are within the acceptable range of thermal comfort.

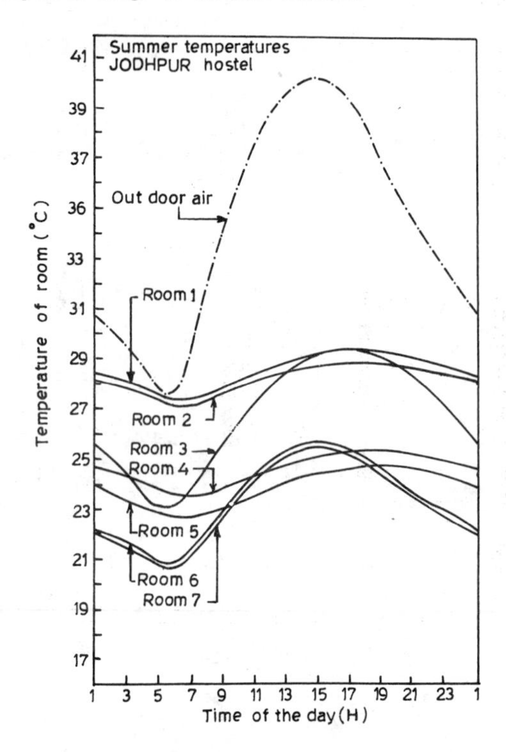

Fig. 10.22. Hourly variation of hostel room air temperatures in summer climate of Jodhpur

10.4 HOT AND HUMID CLIMATE

Solar Passive House: Baroda. A solar house (Yardi and Jain, 1983) is constructed in the hot and humid climate of Baroda to provide thermal comfort conditions throughout the year. The attractive features of the house are the roof lawns, optimized overhangs, cavity walls, verandahs, solar cooler, solar hot water heater and solar photovoltaic system.

Baroda (22°N, 73°E and 35 m above MSL) is a modern city located in the Gujrat State of India. The climate is usually comfortable in winter months when the ambient air temperatures lie between 10 to 30°C where it is most uncomfortable during summer months when the outside air temperature normally touches 40 to 44°C with relative humidity ranging between 60 to 70%. The wind flows in south-west or west direction at a speed of 8 to 10 m/s in summer season. Conversely, the wind speed is usually low (3 to 4 m/s) in winter months flowing in the direction from north and north-east. Therefore, the house has been designed primarily on passive cooling aspects.

Solar heat load calculations have revealed that the roof, east, south, west and north wall of a building in Baroda receive respectively 50%, 20%, 2%, 20% and 4% of solar heat. In view of this analysis, the orientation of the building has been kept east-west so that the building may receive minimum solar heat. This may be further reduced to zero by having extended verandahs of east/west walls of the ground floor. The verandah on the east side is used for relaxing whereas on the west side, it has a laundry and a greenhouse which are used for washing the clothes and the drying of papads, chillies and other products (see Fig. 10.23). Overhangs are provided on north and south walls to effectively reduce the solar heat gain. However, the roof is protected from solar heat gain by having roof lawns (see Fig. 10.24); the roof is first covered with a thin plastic sheet and then a soil layer is placed over it to avoid leakage of water. In addition to this, the house has other characteristic features such as trees, ponds and lawns for proper shading as well as fresh and cool air.

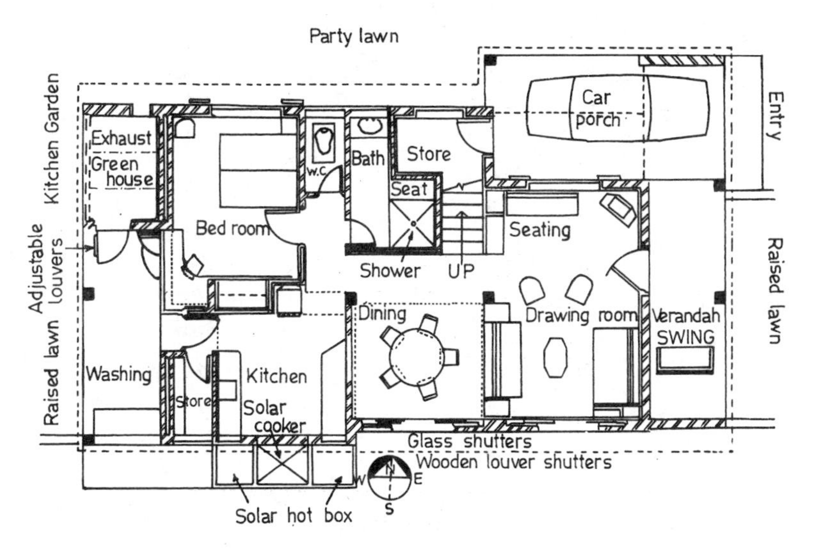

Fig. 10.23. Ground floor plan of a "solar house"

Windward direction windows and doors are provided for proper ventilation. A water pond has also been built in front of the kitchen so that air entering the kitchen gets cooled and saturated. The hot air of the kitchen is exhausted through the chimney (provided at the top of the kitchen) for better ventilation. The other interesting feature of the kitchen is the solar cooker which has been incorporated in the south wall with solar hot boxes on each side for keeping the food warm (see Fig. 10.25). The cooker is operated from inside rather than going to outside.

The living-room has large south-facing wall and windows. The windows are kept open to allow the solar radiation to enter in winter months but they are shut during summer months with movable wooden louvers which allow wind to pass through. A separate unit of thermosyphon solar water heater is also installed on the ceiling of the house which has a capacity of 350 litres and supplies hot water at 60°C.

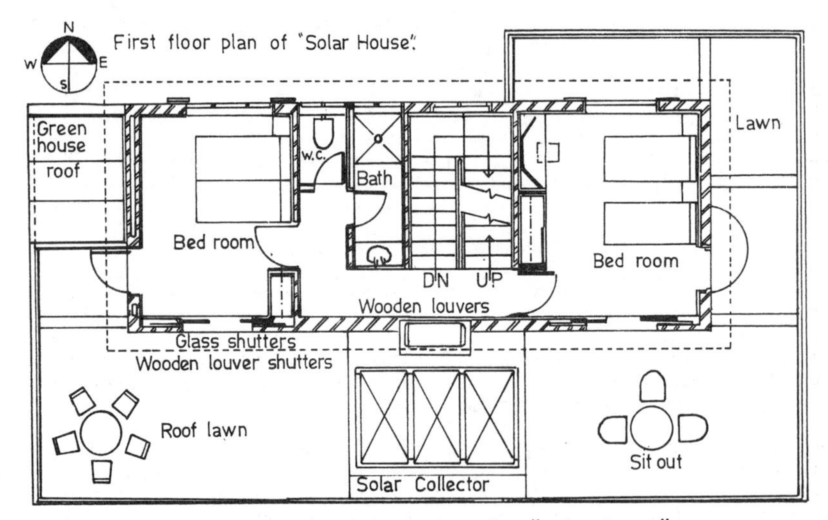

Fig. 10.24. First floor plan of a "solar house"

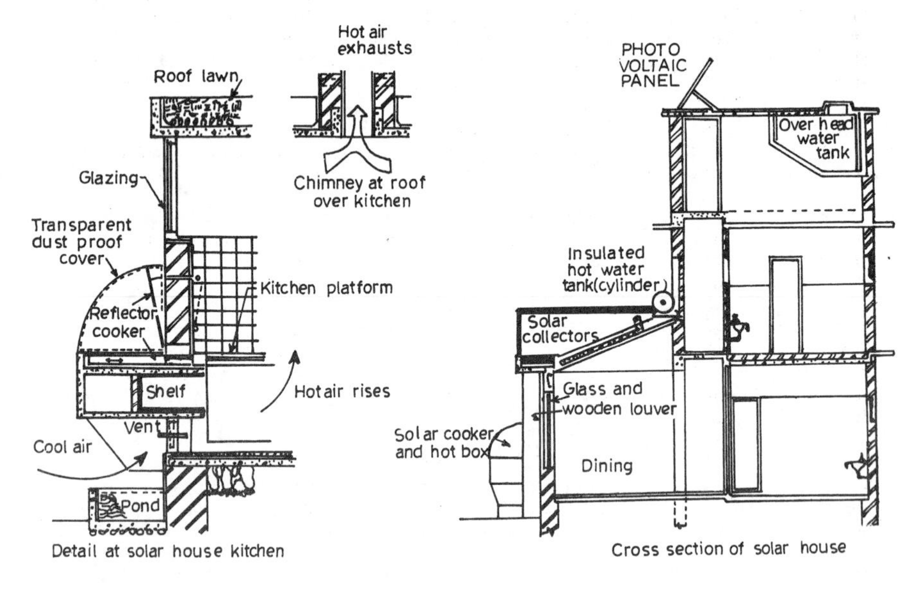

Fig. 10.25. Cross-sectional details of a "solar house"

The tube lights, fans and other equipment of entertainment are run by a small solar photovoltaic panel, connected with 24 V batteries through electronic regulators. This photovoltaic system can provide about 1500 W hours of electrical energy per day.

Performance. On the basis of 1 year's performance of the solar house, the owner's experience has been quite satisfactory. In summer months, room air temperatures of solar house are found to be 5 to 6°C lower than the temperatures in other conventional houses. The roof lawns are aesthetically attractive and have contributed significantly to the reduction of solar heat gain. Also, the use of louvered doors on all windows has made the inside environment quite comfortable and the owner did not make use of fans during most of the summer nights. Additional units involving hot water heating system, solar cooker, greenhouse and solar photovoltaic have also performed satisfactorily. The passive systems have lowered the overall energy requirement from about 200 kWh - a month to about 140 kWh - a month.

10.5 TYPICAL EUROPEAN HOUSE

St. George's School. St. George's School at Wallasey near Liverpool, UK, (ϕ = 53° 25'N) was constructed in 1962. The daily average environmental temperature varies between -4.5°C and 22.5°C. The important feature of the building has been a large fully glazed south-facing solar wall and its ability to maintain comfortable temperature inside the dwelling in cold weather without the use of any conventional central heating.

Structural Details. The school is a complex of massive well-insulated buildings of two-storeys with an approximately east-west alignment; its annexe consists of an assembly hall at the west and together with few utility rooms, a series of classrooms downstairs and upstairs, a solar wall (measuring 7) m x 8 m) on the south and a boy's gymnasium at the extreme east end (also provided with a solar wall). A corridor runs the entire length of the building along the north wall on the ground floor. The solar wall on the south is double-glazed with 61 cm distance between the two leaves. Each classroom is provided with two or three openable single-glazed windows. The ground floor, the intermediate floor and the roof slab of the school are built of massive concrete of thicknesses 25 cm, 23 cm and 18 cm respectively. To provide better insulation and large heat flow, the entire roof slab and the brick portions of the north wall are pasted with 13 cm thick expanded polystyrene.

Observations. A section of the annexe consisting of a typical classroom downstairs together with the art room above, was selected for observations. Thermocouples were used to measure the temperatures of the internal surfaces of both the rooms together with those of the adjacent outer and inner leaves of the solar wall. Measurements were also made for external variables viz. solar radiation received on vertical walls and in the art room, dry- and wet-bulb temperatures of the ambient air and wind speed etc. Temperature of the inside air was measured in Room 10 at ceiling level, off the east wall and near the floor. The instantaneous mean of these three values was used as the average indoor temperature.

Performance. In order to predict the thermal comfort conditions (Davies, 1970), the variation of ceiling temperature, average air-temperature (mean of 8 locations), globe temperature (at a height of about 2.9 m) serving as a weighted mean of air temperature and mean radiant temperature, the difference between ceiling and floor surface temperatures and the ambient air temperature are exhibited in Figs. 10.26 and 10.27 for a sunny winter day (February 13, 1969) and sunny summer day (June 13, 1969) respectively. On the sunny winter day, the maximum cyclic variation of ceiling temperature was noted to be 1.6°C and subsequently, if the windows were shut, air temperature was increased by nearly 5°C. Conversely, on a sunny summer day, windows were kept open to allow light breezes (around 4 m/s) to enter in and restrain the variation in air temperature.

Surveys conducted from teachers and students for the thermal comforts of the conventionally heated main school and the partly heated annexe of St. George's School (Wallasey), had revealed that the annexe attained temperatures of 16°C and above and was warmer than the main school in winter but some overheating was experienced in September-October, 1971 due to exceptionally warm and sunny

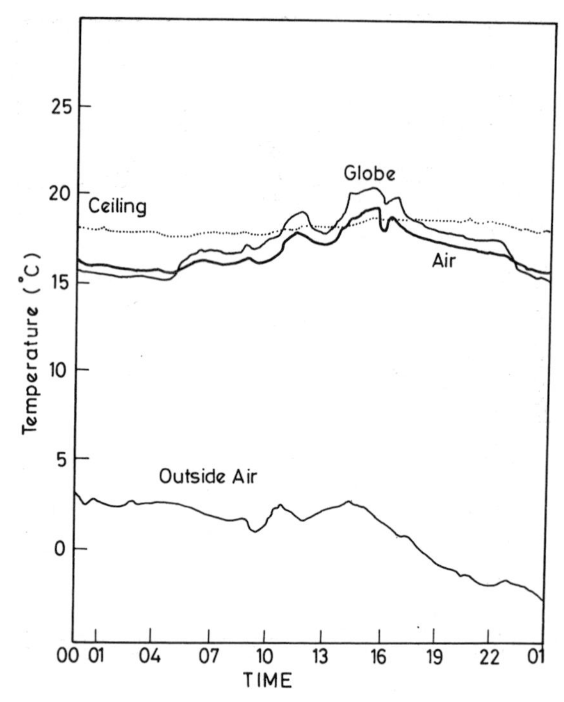

Fig. 10.6. Hourly variation of various temperatures on a sunny winter day (12 February, 1969) at St. George's School

weather outside when daily mean air temperatures up to 24.5°C were observed with higher peak values (Davies, 1976). However, the highest mean indoor temperature (which was 27.4°C) could not be reduced by ventilation during the occupied period in July, as the ambient air temperature was 29°C.

10.6 CONCLUSIONS

It is thus seen that typical buildings, designed and constructed by incorporating passive concepts are able to maintain acceptable room temperatures even in the extreme climatic conditions. Various workers in this field however should appreciate that the application of solar passive concepts to agriculture and rural buildings should be based on an appreciation of the traditional techniques, with special effort to recognize passive features which have stood the test of time.

Thus an important aspect of future research should be a serious and scientific study of traditional architecture, identification of the features which are desirable; compilation of thermal properties of noncommercial building materials is absolutely necessary in this context. After such a study, the means for increasing the effectiveness of traditional features and incorporating other desirable features of the architecture should be investigated. The feasibility of additional passive means for getting the desired thermal environment, within the existing constraints of resources and different levels of technology associated with agriculture, should be explored. This study should include as far as possible utilization of material available on the side. One should also seriously consider cross-fertilization of desirable features of traditional

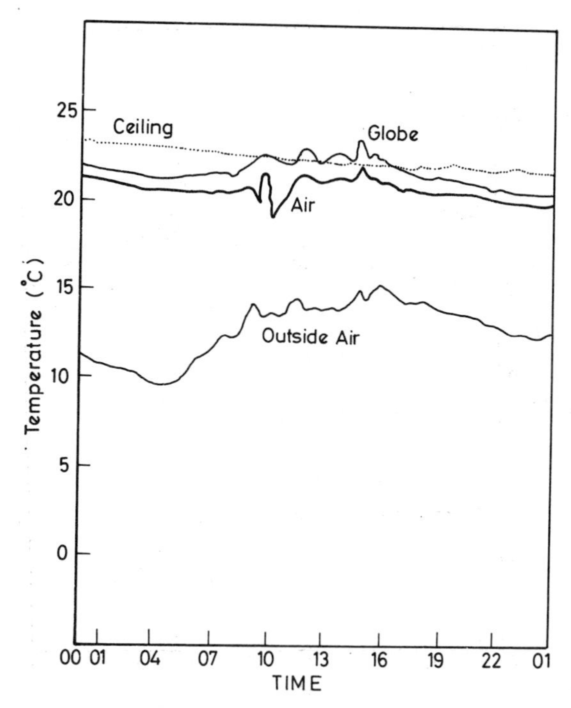

Fig. 10.27. Hourly variation of various temperatures on a sunny summer day (June 17, 1969) at St. George's School

architecture in countries with similar climatic characteristics. The ultimate attempt should be to evolve different sets of design patterns (rules of thumb) applicable to different countries' climates with minimum utilization of conventional building materials and technology.

An effective program on "solar passive" should start with small scale model experiments, validating the passive concepts, applicable to different situations. The experimental results should be checked against the theoretical models. After this stage typical design of buildings in different regions of the country be undertaken. The designs may be started on the basis of rules of thumb and then refined by increasing order of sophistication in numerical simulation.

There is a need to investigate some passive concepts of cooling a building (e.g. earth ducts, evaporative cooling etc.) in detail, both experimentally and analytically. These concepts should also be incorporated in the numerical simulation programs.

A great deal of work has been done on optimization to secure a given air temperature in the building. Studies should be extended to get a desirable comfort level, determined by air temperature, air flow, humidity, floor temperature and temperature of walls. If one takes all these factors into account in optimization studies to maximize thermal comfort for a given investment or minimize investment to give thermal comfort, new challenges, options and opportunities are encountered.

REFERENCES

Arnoth, D. E. (1983) An Evaluation of the Los Alamos Scientific Laboratory's Residential Passive Solar Methods Applied to Small Comercial Buildings, Private Communication.

Davis, M. G. (1970) Model Studies of St. George's School, Wallasey, Summer Conference, *JIHVE*, Vol. *39*, pp. 77-87.

Davis, M. G. (1976) The Contribution of Solar Gain to Space Heating, *Solar Energy, 18*, pp. 361-367.

Haskins, D. and Stromberg, P. (1979) Passive Solar Buildings, Report No. SAND 79-0284.

Hay, H. (1976) Atascadero Residence, Passive Solar Heating and Cooling, Conference and Workshop Proceedings, May 18-19, Albuquerque, New Mexico.

Hodges, L. (1983) The Hodges Residence: Performance of a Direct Gain Passive Solar Home in Iowa, Private Communication.

Paul, J. K. (1979) *Passive Solar Energy Design and Materials*, Noyes Data Corporation, Park Ridge, NJ, USA.

Seshadri, T. N. *et al.* (1979) Climatological and Solar Data for India, Central Building Research Institute Roorkee, India.

Yardi, N. R. and B. C. Jain (1983) Solar House for Hot and Humid Climate, Proc. IInd Int. PLEA Conf., held at Crete (Greece); edited by Simons Yannas, Pergamon Press, London, June 28-July 1.

APPENDIX 1

Predicted Mean Vote. (After Fanger, 1970)

Activity Level 58 W/m^2

Clothing clo	Ambient Temp. °C	Relative Velocity (m/s)								
		0.10	0.10	0.15	0.20	0.30	0.40	0.50	1.00	1.50
0	26	-1.62	-1.62	-1.96	-2.34					
	27	-1.00	-1.00	-1.36	-1.69					
	28	-0.39	-0.42	-0.76	-1.05					
	29	0.21	0.13	0.15	0.39					
	30	0.80	0.68	0.45	0.26					
	31	1.39	1.25	1.08	0.94					
	32	1.96	1.83	1.71	1.61					
	33	2.50	2.41	2.34	2.29					
0.25	24	-1.52	-1.52	-1.80	-2.06	-2.47				
	25	-1.05	-1.05	-1.33	-1.57	-1.94	-2.24	-2.48		
	26	-0.58	-0.61	-0.87	-1.08	-1.41	-1.67	-1.89	-2.66	
	27	-0.12	-0.17	-0.40	-0.58	-0.87	-1.10	-1.29	-1.97	-2.41
	28	0.34	0.27	0.07	-0.09	-0.34	-0.53	-0.70	-1.28	-1.66
	29	0.80	0.71	0.54	0.41	0.20	0.04	-0.10	-0.58	-0.90
	30	1.25	1.15	1.02	0.91	0.74	0.61	0.50	0.11	-0.14
	31	1.71	1.61	1.51	1.43	1.30	1.20	1.12	0.83	0.63
0.50	23	-1.10	-1.10	-1.33	-1.51	-1.78	-1.99	-2.16		
	24	-0.72	-0.74	-0.95	-1.11	-1.36	-1.55	-1.70	-2.22	
	25	-0.34	-0.38	-0.56	-0.71	-0.94	-1.11	-1.25	-1.71	-1.99
	26	0.04	-0.01	-0.18	-0.31	-0.51	-0.66	-0.79	-1.19	-1.44
	27	0.42	0.35	0.20	0.09	-0.08	-0.22	-0.33	-0.68	-0.90
	28	0.80	0.72	0.59	0.49	0.34	0.23	0.14	-0.17	-0.36
	29	1.17	1.08	0.98	0.90	0.77	0.68	0.60	0.34	0.19
	30	1.54	1.45	1.37	1.30	1.20	1.13	1.06	0.86	0.73

Clothing clo	Ambient Temp. °C	Relative Velocity (m/s) 0.10	0.10	0.15	0.20	0.30	0.40	0.50	1.00	1.50
0.75	21	-1.11	-1.11	-1.30	-1.44	-1.66	-1.82	-1.95	-2.36	-2.60
	22	-0.79	-0.81	-0.98	-1.11	-1.31	-1.46	-1.58	-1.95	-2.17
	23	-0.47	-0.50	-0.66	-0.78	-0.96	-1.09	-1.20	-1.55	-1.75
	24	-0.15	-0.19	-0.33	-0.44	-0.61	-0.73	-0.83	-1.14	-1.33
	25	0.15	0.12	-0.01	-0.11	-0.26	-0.37	-0.46	-0.74	-0.90
	26	0.49	0.43	0.31	0.23	0.09	0.00	-0.08	-0.33	-0.48
	27	0.81	0.74	0.64	0.56	0.45	0.36	0.29	0.08	-0.05
	28	1.12	1.05	0.96	0.90	0.80	0.73	0.67	0.48	0.37
1.00	20	-0.85	-0.87	-1.02	-1.13	-1.29	-1.41	-1.51	-1.81	-1.98
	21	-0.57	-0.60	-0.74	-0.84	-0.99	-1.11	-1.19	-1.47	-1.63
	22	-0.30	-0.33	-0.46	-0.55	-0.69	-0.80	-0.88	-1.13	-1.28
	23	-0.02	-0.07	-0.18	-0.27	-0.39	-0.49	-0.56	-0.79	-0.93
	24	0.26	0.20	0.10	0.02	-0.09	-0.18	-0.25	-0.46	-0.58
	25	0.53	0.48	0.38	0.31	0.21	0.13	0.07	-0.12	-0.23
	26	0.81	0.75	0.66	0.60	0.51	0.44	0.39	0.22	0.13
	27	1.08	1.02	0.95	0.89	0.81	0.75	0.71	0.56	0.48
1.25	16	-1.37	-1.37	-1.51	-1.62	-1.78	-1.89	-1.98	-2.26	-2.41
	18	-0.89	-0.91	-1.04	-1.14	-1.28	-1.38	-1.46	-1.70	-1.84
	20	-0.42	-0.46	-0.57	-0.65	-0.77	-0.86	-0.93	-1.14	-1.26
	22	0.07	0.02	-0.07	-0.14	-0.25	-0.32	-0.38	-0.56	-0.66
	24	0.56	0.50	0.43	0.37	0.28	0.22	0.17	0.02	-0.06
	26	1.04	0.99	0.93	0.88	0.81	0.76	0.72	0.61	0.54
	28	1.53	1.48	1.43	1.40	1.34	1.31	1.28	1.19	1.14
	30	2.01	1.97	1.93	1.91	1.88	1.85	1.83	1.77	1.74
1.50	14	-1.36	-1.36	-1.49	-1.58	-1.72	-1.82	-1.89	-2.12	-2.25
	16	-0.94	-0.95	-1.07	-1.15	-1.27	-1.36	-1.43	-1.63	-1.75
	18	-0.52	-0.54	-0.64	-0.72	-0.82	-0.90	-0.96	-1.14	-1.24
	20	-0.09	-0.13	-0.22	-0.28	-0.37	-0.44	-0.49	-0.65	-0.74
	22	0.35	0.30	0.23	0.18	0.10	0.04	0.00	-0.14	-0.21
	24	0.79	0.74	0.68	0.63	0.57	0.52	0.49	0.37	0.31
	26	1.23	1.18	1.13	1.09	1.04	1.01	0.98	0.89	0.84
	28	1.67	1.62	1.58	1.56	1.52	1.49	1.47	1.40	1.37
Activity Level 70 W/m²										
0	25	-1.33	-1.33	-1.59	-1.92					
	26	-0.83	-0.83	-1.11	-1.40					
	27	-0.33	-0.33	-0.63	-0.88					
	28	0.15	0.12	-0.14	-0.36					
	29	0.63	0.56	0.35	0.17					
	30	1.10	1.01	0.84	0.69					
	31	1.57	1.47	1.34	1.24					
	32	2.03	1.93	1.85	1.78					
0.25	23	-1.18	-1.18	-1.39	-1.61	-1.97	-2.25			
	24	-0.79	-0.79	-1.02	-1.22	-1.54	-1.80	-2.01		
	25	-0.42	-0.42	-0.64	-0.83	-1.11	-1.34	-1.54	-2.21	
	26	-0.04	-0.07	-0.27	-0.43	-0.68	-0.89	-1.06	-1.65	-2.04
	27	0.33	0.29	0.11	-0.03	-0.25	-0.43	-0.58	-1.09	-1.43
	28	0.71	0.64	0.49	0.37	0.18	0.03	-0.10	-0.54	-0.82
	29	1.07	0.99	0.87	0.77	0.61	0.49	0.39	0.02	-0.22
	30	1.43	1.35	1.25	1.17	1.05	0.95	0.87	0.58	0.39

Clothing clo	Ambient Temp. °C	Relative Velocity (m/s)								
		0.10	0.10	0.15	0.20	0.30	0.40	0.50	1.00	1.50
0.50	18	-2.01	-2.01	-2.17	-2.38	-2.70				
	20	-1.41	-1.41	-1.58	-1.76	-2.04	-2.25	-2.42		
	22	-0.79	-0.79	-0.97	-1.13	-1.36	-1.54	-1.69	-2.17	-2.46
	24	-0.17	-0.20	-0.36	-0.48	-0.68	-0.83	-0.95	-1.35	-1.59
	26	0.44	0.39	0.26	0.16	0.01	-0.11	-0.21	-0.52	-0.71
	28	1.05	0.98	0.88	0.81	0.70	0.61	0.54	0.31	0.16
	30	1.64	1.57	1.51	1.46	1.39	1.33	1.29	1.14	1.04
	32	2.25	2.20	2.17	2.15	2.11	2.01	2.07	1.99	1.95
0.75	16	-1.77	-1.77	-1.91	-2.07	-2.31	-2.49			
	18	-1.27	-1.27	-1.42	-1.56	-1.77	-1.93	-2.05	-2.45	
	20	-0.77	-0.77	-0.92	-1.04	-1.23	-1.36	-1.47	-1.82	-2.02
	22	-0.25	-0.27	-0.40	-0.51	-0.66	-0.78	-0.87	-1.17	-1.34
	24	0.27	0.23	0.12	0.03	-0.10	-0.19	-0.27	-0.51	-0.65
	26	0.78	0.73	0.64	0.57	0.47	0.40	0.34	0.14	0.03
	28	1.29	1.23	1.17	1.12	1.04	0.99	0.94	0.80	0.72
	30	1.80	1.74	1.70	1.67	1.62	1.58	1.55	1.46	1.41
1.00	16	-1.18	-1.18	-1.31	-1.43	-1.59	-1.72	-1.82	-2.12	-2.29
	18	-0.75	-0.75	-0.88	-0.98	-1.13	-1.24	-1.33	-1.59	-1.75
	20	-0.32	-0.33	-0.45	-0.54	-0.67	-0.76	-0.83	-1.07	-1.20
	22	0.13	0.10	0.00	-0.07	-0.18	-0.26	-0.32	-0.52	-0.64
	24	0.58	0.54	0.46	0.40	0.31	0.24	0.19	0.02	-0.07
	26	1.03	0.98	0.91	0.86	0.79	L.74	0.70	0.57	0.50
	28	1.47	1.42	1.37	1.34	1.28	1.24	1.21	1.12	1.06
	30	1.91	1.86	1.83	1.81	1.78	1.75	1.73	1.67	1.63
1.25	14	-1.12	-1.12	-1.24	-1.34	-1.48	-1.58	-1.66	-1.90	-2.04
	16	-0.74	-0.75	-0.86	-0.95	-1.07	-1.16	-1.23	-1.45	-1.57
	18	-0.36	-0.38	-0.48	-0.55	-0.66	-0.74	-0.81	-1.00	-1.11
	20	0.02	-0.01	-0.10	-0.16	-0.26	-0.33	-0.38	-0.55	-0.64
	22	0.42	0.38	0.31	0.25	0.17	0.11	0.07	-0.08	-0.16
	24	0.81	0.77	0.71	0.66	0.60	0.55	0.51	0.39	0.33
	26	1.21	1.16	1.11	1.08	1.03	0.99	0.96	0.87	0.82
	28	1.60	1.56	1.52	1.50	1.46	1.43	1.41	1.34	1.30
1.50	12	-1.09	-1.09	-1.19	-1.27	-1.39	-1.48	-1.55	-1.75	-1.86
	14	-0.75	-0.75	-0.85	-0.93	-1.03	-1.11	-1.17	-1.35	-1.45
	16	-0.41	-0.42	-0.51	-0.58	-0.67	-0.74	-0.79	-0.96	-1.05
	18	-0.06	-0.09	-0.17	-0.22	-0.31	-0.37	-0.42	-0.56	-0.64
	20	0.28	0.25	0.18	0.13	0.05	0.00	-0.04	-0.16	-0.23
	22	0.63	0.60	0.54	0.50	0.44	0.39	0.36	0.25	0.19
	24	0.99	0.95	0.91	0.87	0.82	0.78	0.76	0.67	0.62
	26	1.35	1.31	1.27	1.24	1.20	1.18	1.15	1.08	1.05

Activity Level 80 W/m^2

Clothing clo	Ambient Temp. °C	0.10	0.10	0.15	0.20
0	24	-1.14	-1.14	-1.35	-1.65
	25	-0.72	-0.72	-0.95	-1.21
	26	-0.30	-0.30	-0.54	-0.78
	27	0.11	0.11	-0.14	-0.34
	28	0.52	0.48	0.27	0.10
	29	0.92	0.85	0.69	0.54
	30	1.31	1.23	1.10	0.99
	31	1.71	1.62	1.52	1.45

Clothing clo	Ambient Temp. °C	Relative Velocity (m/s)								
		0.10	0.10	0.15	0.20	0.30	0.40	0.50	1.00	1.50
0.25	22	-0.95	-0.95	-1.12	-1.33	-1.64	-1.90	-2.11		
	23	-0.63	-0.63	-0.81	-0.99	-1.28	-1.51	-1.71	-2.38	
	24	-0.31	-0.31	-0.50	-0.66	-0.92	-1.13	-1.31	-1.91	-2.31
	25	0.01	0.00	-0.18	-0.33	-0.56	-0.75	-0.90	-1.45	-1.80
	26	0.33	0.30	0.14	0.01	-0.20	-0.36	-0.50	-0.98	-1.29
	27	0.64	0.59	0.45	0.34	0.16	0.02	-0.10	-0.51	-0.78
	28	0.95	0.89	0.77	0.68	0.53	0.41	0.31	-0.04	-0.27
	29	1.26	1.19	1.09	1.02	0.89	0.80	0.72	0.43	0.24
0.50	18	-1.36	-1.36	-1.49	-1.66	-1.93	-2.12	-2.29		
	20	-0.85	-0.85	-1.00	-1.14	-1.37	-1.54	-1.68	-2.15	-2.43
	22	-0.33	-0.33	-0.48	-0.61	-0.80	-0.95	-1.06	-1.46	-1.70
	24	0.19	0.17	0.04	-0.07	-0.22	-0.34	-0.44	-0.76	-0.96
	26	0.71	0.66	0.56	0.48	0.35	0.26	0.18	-0.07	-0.23
	28	1.22	1.16	1.09	1.03	0.94	0.87	0.81	0.63	0.51
	30	1.72	1.66	1.62	1.58	1.52	1.48	1.44	1.33	1.25
	32	2.23	2.19	2.17	2.16	2.13	2.11	2.10	2.05	2.02
0.75	16	-1.17	-1.17	-1.29	-1.42	-1.62	-1.77	-1.98	-2.26	-2.48
	18	-0.75	-0.75	-0.87	-0.99	-1.16	-1.29	-1.39	-1.72	-1.92
	20	-0.33	-0.33	-0.45	-0.55	-0.70	-0.82	-0.91	-1.19	-1.36
	22	0.11	0.09	-0.02	-0.10	-0.23	-0.32	-0.40	-0.64	-0.78
	24	0.55	0.51	0.42	0.35	0.25	0.17	0.11	-0.09	-0.20
	26	0.98	0.94	0.87	0.81	0.73	0.67	0.62	0.47	0.37
	28	1.41	1.36	1.31	1.27	1.21	1.17	1.13	1.02	0.95
	30	1.84	1.79	1.76	1.73	1.70	1.67	1.65	1.58	1.53
1.00	14	-1.05	-1.05	-1.16	-1.26	-1.42	-1.53	-1.62	-1.91	-2.07
	16	-0.69	-0.69	-0.80	-0.89	-1.03	-1.13	-1.21	-1.46	-1.61
	18	-0.32	-0.32	-0.43	-0.52	-0.64	-0.73	-0.80	-1.02	-1.15
	20	0.04	0.03	-0.07	-0.14	-0.25	-0.32	-0.38	-0.58	-0.69
	22	0.42	0.39	0.31	0.25	0.16	0.10	0.05	-0.12	-0.21
	24	0.80	0.76	0.70	0.65	0.57	0.52	0.48	0.35	0.27
	26	1.18	1.13	1.08	1.04	0.99	0.95	0.91	0.81	0.75
	28	1.55	1.51	1.47	1.44	1.40	1.37	1.35	1.27	1.23
1.25	12	-0.97	-0.97	-1.06	-1.15	-1.28	-1.37	-1.45	-1.67	-1.80
	14	-0.65	-0.65	-0.75	-0.82	-0.94	-1.02	-1.09	-1.29	-1.40
	16	-0.33	-0.33	-0.43	-0.50	-0.60	-0.67	-0.73	-0.91	-1.01
	18	-0.01	-0.02	-0.10	-0.17	-0.26	-0.32	-0.37	-0.53	-0.62
	20	0.32	0.29	0.22	0.17	0.09	0.03	-0.01	-0.15	-0.22
	22	0.65	0.62	0.56	0.52	0.45	0.40	0.36	0.25	0.18
	24	0.99	0.95	0.90	0.87	0.81	0.77	0.74	0.65	0.59
	26	1.32	1.28	1.25	1.22	1.18	1.14	1.12	1.05	1.00
1.50	10	-0.91	-0.91	-1.00	-1.08	-1.18	-1.26	-1.32	-1.51	-1.61
	12	-0.63	-0.63	-0.71	-0.78	-0.88	-0.95	-1.01	-1.17	-1.27
	14	-0.34	-0.34	-0.43	-0.49	-0.58	-0.64	-0.69	-0.84	-0.92
	16	-0.05	-0.06	-0.14	-0.19	-0.27	-0.33	-0.37	-0.50	-0.58
	18	0.24	0.22	0.15	0.11	0.04	-0.01	-0.05	-0.17	-0.23
	20	0.53	0.50	0.45	0.40	0.34	0.30	0.27	0.17	0.11
	22	0.83	0.80	0.75	0.72	0.67	0.63	0.60	0.52	0.47
	24	1.13	1.10	1.06	1.03	0.99	0.96	0.94	0.87	0.83

Clothing clo	Ambient Temp. °C	Relative Velocity (m/s)								
		0.10	0.10	0.15	0.20	0.30	0.40	0.50	1.00	1.50
Activity Level 90 W/m²										
0	23	-1.12	-1.12	-1.29	-1.57					
	24	-0.74	-0.74	-0.93	-1.18					
	25	-0.36	-0.36	-0.57	-0.79					
	26	0.01	0.01	-0.20	-0.40					
	27	0.38	0.37	0.17	0.00					
	28	0.75	0.70	0.53	0.39					
	29	1.11	1.04	0.90	0.79					
	30	1.46	1.38	1.27	1.19					
0.25	16	-2.29	-2.29	-2.36	-2.62					
	18	-1.72	-1.72	-1.83	-2.06	-2.42				
	20	-1.15	-1.15	-1.29	-1.49	-1.80	-2.05	-2.26		
	22	-0.58	-0.58	-0.73	-0.90	-1.17	-1.38	-1.55	-2.17	-2.58
	24	-0.01	-0.01	-0.17	-0.31	-0.53	-0.70	-0.84	-1.35	-1.68
	26	0.56	0.53	0.39	0.29	0.12	-0.02	-0.13	-0.52	-0.78
	28	1.12	1.06	0.96	0.89	0.77	0.67	0.59	0.31	0.12
	30	1.66	1.60	1.54	1.49	1.42	1.36	1.31	1.14	1.02
0.50	14	-1.85	-1.85	-1.94	-2.12	-2.40				
	16	-1.40	-1.40	-1.50	-1.67	-1.92	-2.11	-2.26		
	18	-0.95	-0.95	-1.07	-1.21	-1.43	-1.59	-1.73	-2.18	-2.46
	20	-0.49	-0.49	-0.62	-0.75	-0.94	-1.08	-1.20	-1.59	-1.82
	22	-0.03	-0.03	-0.16	-0.27	-0.43	-0.55	-0.65	-0.98	-1.18
	24	0.43	0.41	0.30	0.21	0.08	-0.02	-0.10	-0.37	-0.53
	26	0.89	0.85	0.76	0.70	0.60	0.52	0.46	0.25	0.12
	28	1.34	1.29	1.23	1.18	1.11	1.06	1.01	0.86	0.77
0.75	14	-1.16	-1.16	-1.26	-1.38	-1.57	-1.71	-1.82	-2.17	-2.38
	16	-0.79	-0.79	-0.89	-1.00	-1.17	-1.29	-1.39	-1.70	-1.88
	18	-0.41	-0.41	-0.52	-0.62	-0.76	-0.87	-0.96	-1.23	-1.39
	20	-0.04	-0.04	-0.15	-0.23	-0.36	-0.45	-0.52	-0.76	-0.90
	22	0.35	0.33	0.24	0.17	0.07	-0.01	-0.07	-0.27	-0.39
	24	0.74	0.71	0.63	0.58	0.49	0.43	0.38	0.21	0.12
	26	1.12	1.08	1.03	0.98	0.92	0.87	0.83	0.70	0.62
	28	1.51	1.46	1.42	1.39	1.34	1.31	1.28	1.19	1.14
1.00	12	-1.01	-1.01	-1.10	-1.19	-1.34	-1.45	-1.53	-1.79	-1.94
	14	-0.68	-0.68	-0.78	-0.87	-1.00	-1.09	-1.17	-1.40	-1.54
	16	-0.36	-0.36	-0.46	-0.53	-0.65	-0.74	-0.80	-1.01	-1.13
	18	-0.04	-0.04	-0.13	-0.20	-0.30	-0.38	-0.44	-0.62	-0.73
	20	0.28	0.27	0.18	0.13	0.04	-0.02	-0.07	-0.23	-0.32
	22	0.62	0.59	0.53	0.48	0.41	0.35	0.31	0.17	0.10
	24	0.96	0.02	0.87	0.83	0.77	0.73	0.69	0.58	0.52
	26	1.29	1.25	1.21	1.18	1.14	1.10	1.07	0.99	0.94
1.25	10	-0.90	-0.90	-0.98	-1.06	-1.18	-1.27	-1.33	-1.54	-1.66
	12	-0.62	-0.62	-0.70	-0.77	-0.88	-0.96	-1.02	-1.21	-1.31
	14	-0.33	-0.33	-0.42	-0.48	-0.58	-0.65	-0.70	-0.87	-0.97
	16	-0.05	-0.05	-0.13	-0.19	-0.28	-0.34	-0.39	-0.54	-0.62
	18	0.24	0.22	0.15	0.10	0.03	-0.03	-0.07	-0.20	-0.28
	20	0.52	0.50	0.4	0.40	0.33	0.29	0.25	0.14	0.07
	22	0.82	0.79	0.74	0.71	0.65	0.61	0.58	0.49	0.43
	24	1.12	1.09	1.05	1.02	0.97	0.94	0.92	0.84	0.79

Clothing clo	Ambient Temp. °C	Relative Velocity (m/s)								
		0.10	0.10	0.15	0.20	0.30	0.40	0.50	1.00	1.50
1.50	8	-0.82	-0.82	-0.89	-0.96	-1.06	-1.13	-1.19	-1.36	-1.45
	10	-0.57	-0.57	-0.65	-0.71	-0.80	-0.86	-0.92	-1.07	-1.16
	12	-0.32	-0.32	-0.39	-0.45	-0.53	-0.59	-0.64	-0.78	-0.85
	14	-0.06	-0.07	-0.14	-0.19	-0.26	-0.31	-0.36	-0.48	-0.55
	16	0.19	0.18	0.12	0.07	0.01	-0.04	-0.07	-0.19	-0.25
	18	0.45	0.43	0.38	0.34	0.28	0.24	0.21	0.11	0.05
	20	0.71	0.68	0.64	0.60	0.55	0.52	0.49	0.41	0.36
	22	0.97	0.95	0.91	0.88	0.84	0.81	0.79	0.72	0.68
Activity Level 100 W/m^2										
0	22	-1.05	-1.05	-1.19	-1.46					
	23	-0.70	-0.70	-0.86	-1.11					
	24	-0.36	-0.36	-0.53	-0.75					
	25	-0.01	-0.01	-0.20	-0.40					
	26	0.32	0.32	0.13	-0.04					
	27	0.66	0.63	0.46	0.32					
	28	0.99	0.94	0.80	0.68					
	29	1.31	1.25	1.13	1.04					
0.25	16	-1.79	-1.79	-1.86	-2.09	-2.46				
	18	-1.28	-1.28	-1.38	-1.58	-1.90	-2.16	-2.37		
	20	-0.76	-0.76	-0.89	-1.05	-1.34	-1.56	-1.75	-2.39	-2.82
	22	-0.24	-0.24	-0.38	-0.53	-0.76	-0.95	-1.10	-1.65	-2.01
	24	0.28	0.28	0.13	0.01	-0.18	-0.33	-0.46	-0.90	-1.19
	26	0.79	0.76	0.64	0.55	0.40	0.29	0.19	-0.15	-0.38
	28	1.29	1.24	1.16	1.10	0.99	0.91	0.84	0.60	0.44
	30	1.79	1.73	1.68	1.65	1.59	1.54	1.50	1.36	1.27
0.50	14	-1.42	-1.42	-1.50	-1.66	-1.91	-2.10	-2.25		
	16	-1.01	-1.01	-1.10	-1.25	-1.47	-1.64	-1.77	-2.23	-2.51
	18	-0.59	-0.59	-0.70	-0.83	-1.02	-1.17	-1.29	-1.69	-1.94
	20	-0.18	-0.18	-0.30	-0.41	-0.58	-0.71	-0.81	-1.15	-1.36
	22	0.24	0.23	0.12	0.02	-0.12	-0.22	-0.31	-0.60	-0.78
	24	0.66	0.63	0.54	0.46	0.35	0.26	0.19	-0.04	-0.19
	26	1.07	1.03	0.96	0.90	0.82	0.75	0.69	0.51	0.40
	28	1.48	1.44	1.39	1.35	1.29	1.24	1.20	1.07	1.00
0.75	12	-1.15	-1.15	-1.23	-1.35	-1.53	-1.67	-1.78	-2.13	-2.33
	14	-0.81	-0.81	-0.89	-1.00	-1.17	-1.29	-1.39	-1.70	-1.89
	16	-0.46	-0.46	-0.56	-0.66	-0.80	-0.91	-1.00	-1.28	-1.44
	18	-0.12	-0.12	-0.22	-0.31	-0.43	-0.53	-0.61	-0.85	-0.99
	20	0.22	0.21	0.12	0.04	-0.07	-0.15	-0.21	-0.42	-0.55
	22	0.57	0.55	0.47	0.41	0.32	0.25	0.20	0.02	-0.09
	24	0.92	0.89	0.83	0.78	0.71	0.65	L.60	0.46	0.38
	26	1.28	1.24	1.19	1.15	1.09	1.05	1.02	0.91	0.84
1.00	10	-0.97	-0.97	-1.04	-1.14	-1.28	-1.39	-1.47	-1.73	-1.88
	12	-0.68	-0.68	-0.76	-0.84	-0.97	-1.07	-1.14	-1.38	-1.51
	14	-0.38	-0.38	-0.46	-0.54	-0.66	-0.74	-0.81	-1.02	-1.14
	16	-0.09	-0.09	-0.17	-0.24	-0.35	-0.42	-0.48	-0.67	-0.78
	18	0.21	0.20	0.12	0.06	-0.03	-0.10	-0.15	-0.31	-0.41
	20	0.50	0.48	0.42	0.36	0.29	0.23	0.18	0.04	-0.04
	22	0.81	0.78	0.73	0.68	0.62	0.57	0.53	0.41	0.35
	24	1.11	1.08	1.04	1.00	0.95	0.91	0.88	0.78	0.73

Clothing clo	Ambient Temp. °C	Relative Velocity (m/s)								
		0.10	0.10	0.15	0.20	0.30	0.40	0.50	1.00	1.50
1.25	8	-0.84	-0.84	-0.91	-0.99	-1.10	-1.19	-1.25	-1.46	-1.57
	10	-0.59	-0.59	-0.66	-0.73	-0.84	-0.91	-0.97	-1.16	-1.26
	12	-0.33	-0.33	-0.40	-0.47	-0.56	-0.63	-0.69	-0.86	-0.95
	14	-0.07	-0.07	-0.14	-0.20	-0.29	-0.35	-0.40	-0.55	-0.63
	16	0.19	0.18	0.12	0.06	-0.01	-0.07	-0.11	-0.24	-0.32
	18	0.45	0.44	0.38	0.33	0.26	0.22	0.18	0.06	0.00
	20	0.71	0.69	0.64	0.60	0.54	0.50	0.47	0.37	0.31
	22	0.98	0.96	0.91	0.88	0.83	0.80	0.77	0.69	0.64
1.50	-2	-1.63	-1.63	-1.68	-1.77	-1.90	-2.00	-2.07	-2.29	-2.41
	2	-1.19	-1.19	-1.25	-1.33	-1.44	-1.52	-1.58	-1.78	-1.88
	6	-0.74	-0.74	-0.80	-0.87	-0.97	-1.04	-1.09	-1.26	-1.35
	10	-0.29	-0.29	-0.36	-0.42	-0.50	-0.56	-0.60	-0.74	-0.82
	14	0.17	0.17	0.11	0.06	-0.01	-0.05	-0.09	-0.20	-0.26
	18	0.64	0.62	0.57	0.54	0.49	0.45	0.42	0.34	0.29
	22	1.12	1.09	1.06	1.03	1.00	0.97	0.95	0.89	0.85
	26	1.61	1.58	1.56	1.55	1.52	1.51	1.50	1.46	1.44

Activity Level 120 W/m^2

Clothing clo	Ambient Temp. °C	0.10	0.10	0.15	0.20	0.30	0.40	0.50	1.00	1.50
0	18		-2.00	-2.02	-2.35					
	20		-1.35	-1.43	-1.72					
	22		-0.69	-0.82	-1.06					
	24		-0.04	-0.21	-0.41					
	26		0.59	0.41	0.26					
	28		1.16	1.03	0.93					
	30		1.73	1.66	1.60					
	32		2.33	2.32	2.31					
0.25	16		-1.41	-1.48	-1.69	-2.02	-2.29	-2.51		
	18		-0.93	-1.03	-1.21	-1.50	-1.74	-1.93	-2.61	
	20		-0.45	-0.57	-0.73	-0.98	-1.18	-1.35	-1.93	-2.32
	22		0.04	-0.09	-0.23	-0.44	-0.61	-0.75	-1.24	-1.56
	24		0.52	0.38	0.28	0.10	-0.03	-0.14	-0.54	-0.80
	26		0.97	0.86	0.78	0.65	0.55	0.46	0.16	-0.04
	28		1.42	1.35	1.29	1.20	1.13	1.07	0.86	0.72
	30		1.88	1.84	1.81	1.76	1.72	1.68	1.57	1.49
0.50	14		-1.08	-1.16	-1.31	-1.53	-1.71	-1.85	-2.32	
	16		-0.69	-0.79	-0.92	-1.12	-1.27	-1.40	-1.82	-2.07
	18		-0.31	-0.41	-0.53	-0.70	-0.84	-0.95	-1.31	-1.54
	20		0.07	-0.04	-0.14	-0.29	-0.40	-0.50	-0.81	-1.00
	22		0.46	0.35	0.27	0.15	0.05	-0.03	-0.29	-0.45
	24		0.83	0.75	0.68	0.58	0.50	0.44	0.23	0.10
	26		1.21	1.15	1.10	1.02	0.96	0.91	0.75	0.65
	28		1.59	1.55	1.51	1.46	1.42	1.38	1.27	1.21
0.75	10		-1.16	-1.23	-1.35	-1.54	-1.67	-1.78	-2.14	-2.34
	12		-0.84	-0.92	-1.03	-1.20	-1.32	-1.42	-1.74	-1.93
	14		-0.52	-0.60	-0.70	-0.85	-0.97	-1.06	-1.34	-1.51
	16		-0.20	-0.29	-0.38	-0.51	-0.61	-0.69	-0.95	-1.10
	18		0.12	0.03	-0.05	-0.17	-0.26	-0.32	-0.55	-0.68
	20		0.43	0.34	0.28	0.18	0.10	0.04	-0.15	-0.26
	22		0.75	0.68	0.62	0.54	0.48	0.43	0.27	0.17
	24		1.07	1.01	0.97	0.90	0.85	0.81	0.68	0.61

Clothing clo	Ambient Temp. °C	Relative Velocity (m/s)								
		0.10	0.10	0.15	0.20	0.30	0.40	0.50	1.00	1.50
1.00	8		-0.95	-1.02	-1.11	-1.26	-1.36	-1.45	-1.71	-1.86
	10		-0.68	-0.75	-0.84	-0.97	-1.07	-1.15	-1.38	-1.52
	12		-0.41	-0.48	-0.56	-0.68	-0.77	-0.84	-1.05	-1.18
	14		-0.13	-0.21	-0.28	-0.39	-0.47	-0.53	-0.72	-0.83
	16		0.14	0.06	0.00	-0.10	-0.16	-0.22	-0.39	-0.49
	18		0.41	0.34	0.28	0.20	0.14	0.09	-0.06	-0.14
	20		0.68	0.61	0.57	0.50	0.44	0.40	0.28	0.20
	22		0.96	0.91	0.87	0.81	0.76	0.73	0.62	0.56
1.25	-2		-1.74	-1.77	-1.88	-2.04	-2.15	-2.24	-2.51	-2.66
	2		-1.27	-1.32	-1.42	-1.55	-1.65	-1.73	-1.97	-2.10
	6		-0.80	-0.86	-0.94	-1.06	-1.14	-1.21	-1.41	-1.53
	10		-0.33	-0.40	-0.47	-0.56	-0.64	-0.69	-0.86	-0.96
	14		0.15	0.08	0.03	-0.05	-0.11	-0.15	-0.29	-0.37
	18		0.63	0.57	0.53	0.47	0.42	0.39	0.28	0.22
	22		1.11	1.08	1.05	1.00	0.97	0.95	0.87	0.83
	26		1.62	1.60	1.58	1.55	1.53	1.52	1.47	1.45
1.50	-4		-1.52	-1.56	-1.65	-1.78	-1.87	-1.95	-2.16	-2.28
	0		-1.11	-1.16	-1.24	-1.35	-1.44	-1.50	-1.69	-1.79
	4		-0.69	-0.75	-0.82	-0.92	-0.99	-1.04	-1.20	-1.29
	8		-0.27	-0.33	-0.39	-0.47	-0.53	-0.58	-0.72	-0.79
	12		0.15	0.09	0.05	-0.02	-0.07	-0.11	-0.22	-0.29
	16		0.58	0.53	0.49	0.44	0.40	0.37	0.28	0.23
	20		1.01	0.97	0.94	0.91	0.88	0.85	0.79	0.75
	24		1.47	1.44	1.43	1.40	1.38	1.36	1.32	1.29
Activity Level 140 W/m^2										
0	16			-1.88	-2.22					
	18			-1.34	-1.63					
	20			-0.79	-1.05					
	22			-0.23	-0.44					
	24			0.34	0.17					
	26			0.91	0.78					
	28			1.49	1.40					
	30			2.07	2.03					
0.25	14			-1.31	-1.52	-1.85	-2.12	-2.34		
	16			-0.89	-1.08	-1.37	-1.61	-1.81	-2.49	
	18			-0.47	-0.63	-0.89	-1.10	-1.27	-1.87	-2.26
	20			-0.05	-0.19	-0.41	-0.58	-0.73	-1.24	-1.58
	22			0.59	0.28	0.09	-0.05	-0.17	-0.60	-0.88
	24			0.84	0.74	0.60	0.48	0.39	0.05	-0.17
	26			1.28	1.22	1.11	1.02	0.95	0.70	0.53
	28			1.73	1.69	1.62	1.56	1.51	1.35	1.24
0.50	12			-0.97	-1.11	-1.34	-1.51	-1.65	-2.12	-2.40
	14			-0.62	-0.76	-0.96	-1.11	-1.24	-1.65	-1.91
	16			-0.28	-0.40	-0.58	-0.71	-0.82	-1.19	-1.42
	18			0.07	-0.03	-0.19	-0.31	-0.41	-0.73	-0.92
	20			0.42	0.33	0.20	0.10	0.01	-0.26	-0.43
	22			0.78	0.71	0.60	0.52	0.45	0.22	0.08
	24			1.15	1.09	1.00	0.94	0.88	0.70	0.59
	26			1.52	1.47	1.41	1.36	1.32	1.19	1.11

Clothing clo	Ambient Temp. °C	Relative Velocity (m/s)								
		0.10	0.10	0.15	0.20	0.30	0.40	0.50	1.00	1.50
0.75	10			-0.71	-0.82	-0.99	-1.11	-1.21	-1.53	-1.71
	12			-0.42	-0.52	-0.67	-0.79	-0.88	-1.16	-1.33
	14			-0.13	-0.22	-0.36	-0.46	-0.54	-0.79	-0.94
	16			0.16	0.08	-0.04	-0.13	-0.20	-0.42	-0.56
	18			0.45	0.38	0.28	0.20	0.14	-0.05	-0.17
	20			0.75	0.69	0.60	0.54	0.49	0.32	0.22
	22			1.06	1.01	0.94	0.88	0.84	0.70	0.62
	24			1.37	1.33	1.27	1.23	1.20	1.09	1.02
1.00	6			-0.78	-0.87	-1.01	-1.12	-1.20	-1.45	-1.60
	8			-0.54	-0.62	-0.75	-0.85	-0.92	-1.15	-1.29
	10			-0.29	-0.37	-0.49	-0.57	-0.64	-0.86	-0.98
	12			-0.04	-0.11	-0.22	-0.29	-0.36	-0.55	-0.66
	14			0.21	0.15	0.06	-0.01	-0.07	-0.24	-0.34
	16			0.47	0.41	0.33	0.27	0.22	0.07	-0.02
	18			0.73	0.68	0.60	0.55	0.51	0.38	0.30
	20			0.98	0.94	0.88	0.84	0.80	0.69	0.62
1.25	-4			-1.46	-1.56	-1.72	-1.83	-1.91	-2.17	-2.32
	0			-1.05	-1.14	-1.27	-1.37	-1.44	-1.67	-1.80
	4			-0.62	-0.70	-0.81	-0.90	-0.96	-1.16	-1.27
	8			-0.19	-0.26	-0.35	-0.42	-0.48	-0.64	-0'74
	12			0.25	0.20	0.12	0.06	0.02	-0.12	-0.20
	16			0.70	0.66	0.60	0.55	0.52	0.41	0.35
	20			1.16	1.13	1.08	1.05	1.02	0.94	0.90
	24			1.65	1.63	1.60	1.57	1.56	1.51	1.48
1.50	-8			-1.44	-1.53	-1.67	-1.76	-1.83	-2.05	-2.17
	-4			-1.07	-1.15	-1.27	-1.35	-1.42	-1.61	-1.72
	0			-0.70	-0.77	-0.87	-0.94	-1.00	-1.17	-1.27
	4			-0.31	-0.37	-0.46	-0.53	-0.57	-0.72	-0.80
	8			0.07	o.02	-0.05	-0.10	-0.14	-0.27	-0.34
	12			0.47	0.43	0.37	0.33	0.29	0.19	0.14
	16			0.88	0.85	0.80	0.77	0.74	0.66	0.62
	20			1.29	1.27	1.24	1.21	1.19	1.13	1.10

Activity Level 175 W/m^2

Clothing clo	Ambient Temp. °C	0.10	0.10	0.15	0.20	0.30	0.40	0.50	1.00	1.50
0	14				-1.92	-2.49				
	16				-1.36	-1.87				
	18				-0.80	-1.24				
	20				-0.24	-0.61				
	22				0.34	0.04				
	24				0.93	0.70				
	26				1.52	1.36				
	28				2.12	2.02				
0.25	12				-1.19	-1.53	-1.90	-2.02		
	14				-0.77	-1.07	-1.31	-1.51	-2.21	
	16				-0.35	-0.61	-0.82	-1.00	-1.61	-2.02
	18				0.08	-0.15	-0.33	-0.48	-1.01	-1.36
	20				0.51	0.32	0.17	0.04	-0.41	-0.71
	24				1.41	1.29	1.19	1.11	0.83	0.64
	26				1.87	1.78	1.71	1.69	1.45	1.32

Clothing clo	Ambient Temp. °C	Relative Velocity (m/s)								
		0.10	0.10	0.15	0.20	0.30	0.40	0.50	1.00	1.50
0.50	10				-0.78	-1.00	-1.18	-1.32	-1.79	-2.07
	12				-0.43	-0.64	-0.79	-0.92	-1.34	-1.60
	14				-0.09	-0.27	-0.41	-0.52	-0.90	-1.13
	16				0.26	0.10	-0.02	-0.12	-0.45	-0.65
	18				0.61	0.47	0.37	0.28	0.00	-0.18
	20				0.96	0.85	0.76	0.68	0.45	0.30
	22				1.33	1.24	1.16	1.10	0.91	0.79
	24				1.70	1.63	1.57	1.53	1.38	1.28
0.75	6				-0.73	-0.93	-1.07	-1.18	-1.52	-1.72
	8				-0.47	-0.64	-0.76	-0.86	-1.18	-1.36
	10				-0.19	-0.34	-0.45	-0.54	-0.83	-1.00
	12				0.10	-0.03	-0.14	-0.22	-0.48	-0.63
	14				0.39	0.27	0.18	0.11	-0.12	-0.26
	16				0.69	0.58	0.50	0.44	0.24	0.12
	18				0.98	0.89	0.82	0.77	0.59	0.49
	20				1.28	1.20	1.14	1.10	0.95	0.87
1.00	-6				-1.68	-1.88	-2.03	-2.14	-2.50	-2.70
	-2				-1.22	-1.39	-1.52	-1.62	-1.94	-2.12
	2				-0.74	-0.90	-1.01	-1.10	-1.37	-1.53
	6				-0.26	-0.39	-0.49	-0.56	-0.80	-0.93
	10				0.22	0.12	0.04	-0.02	-0.22	-0.33
	14				0.73	0.64	0.58	0.53	0.38	0.29
	18				1.24	1.18	1.13	1.09	0.97	0.91
	22				1.77	1.73	1.69	1.67	1.59	1.54
1.25	-8				-1.36	-1.52	-1.64	-1.73	-2.00	-2.15
	-4				-0.95	-1.10	-1.20	-1.28	-1.52	-1.65
	0				-0.54	-0.66	-0.75	-0.82	-1.03	-1.15
	4				-0.12	-0.22	-0.30	-0.36	-0.54	-0.64
	8				0.31	0.22	0.16	0.11	-0.04	-0.13
	12				0.75	0.65	0.63	0.59	0.47	0.40
	16				1.20	1.15	1.11	1.08	0.98	0.93
	20				1.66	1.62	1.59	1.57	1.50	1.46
1.50	-10				-1.13	-1.26	-1.35	-1.42	-1.64	-1.76
	-6				-0.76	-0.87	-0.96	-1.02	-1.21	-1.32
	-2				-0.39	-0.49	-0.56	-0.62	-0.79	-0.88
	2				-0.01	-0.10	-0.16	-0.21	-0.36	-0.44
	6				0.38	0.30	0.25	0.21	0.08	0.01
	10				0.76	0.70	0.66	0.62	0.52	0.46
	14				1.17	1.12	1.09	1.06	0.98	0.93
	18				1.58	1.54	1.52	1.50	1.44	1.40

APPENDIX 2

SUN-PATH DIAGRAMS FOR LATITUDE 40°, 50° and 60° N

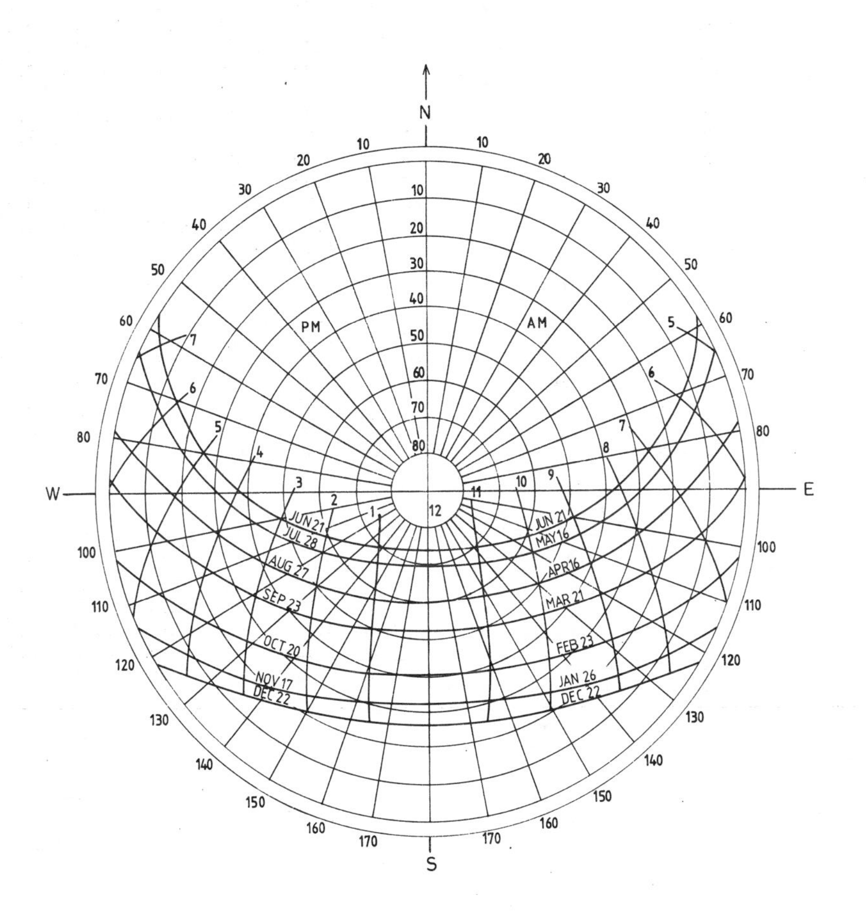

Fig. A2.1. $\phi = 40^{o}$ Latitude

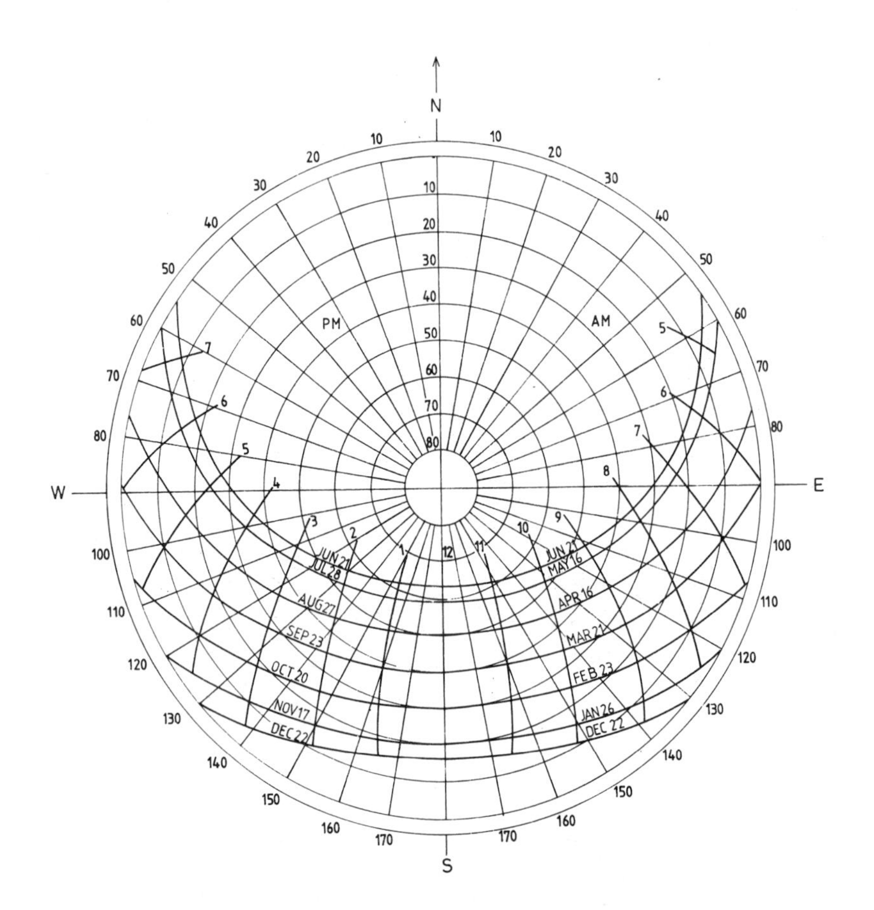

Fig. A2.2. $\phi = 50^{o}$ Latitude

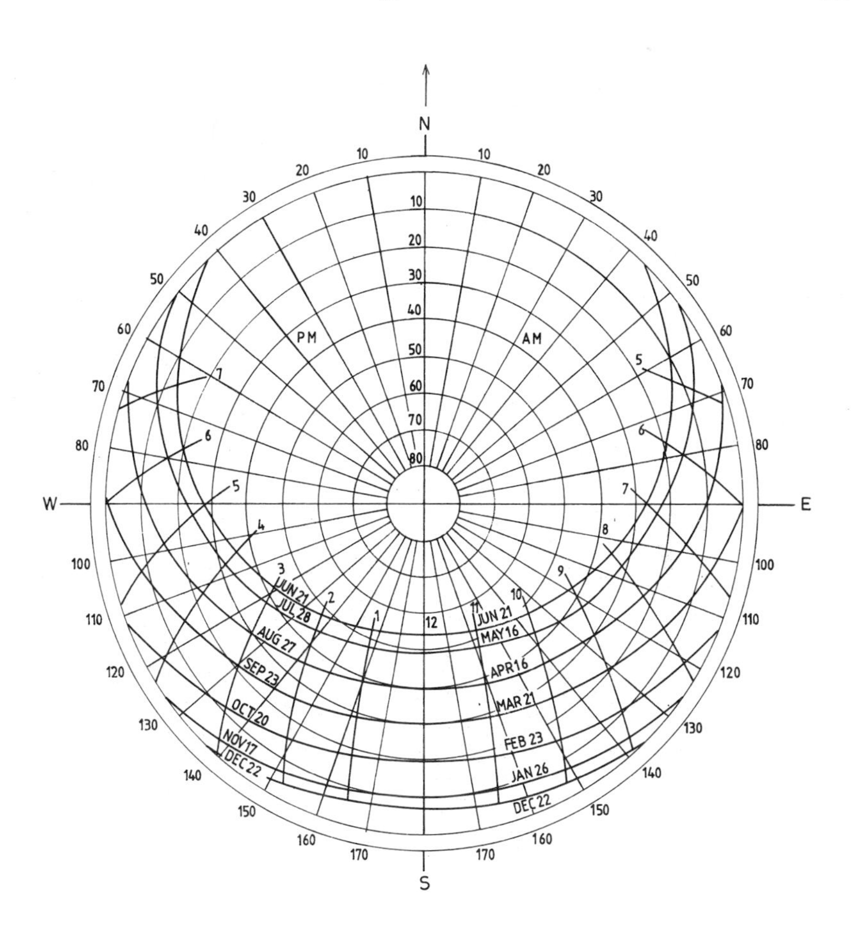

Fig. A2.3. $\phi = 60^{0}$ Latitude

APPENDIX 3

TABLES OF THERMOPHYSICAL PROPERTIES AND HEAT-TRANSFER COEFFICIENTS

TABLE A1 Thermal Properties of Building and Insulating Materials at a Mean Temperature of 24°C (After ASHRAE, 1972)

Material	Description	Density ρ	Thermal Conductivity, K	Unit Conductance, C_t	Resistivity $\frac{1}{K} = r$	Resistivity $\frac{1}{C_r} = R_t$	Specific Heat
		$\frac{kg}{m^3}$	$\frac{W}{°C.m}$	$\frac{W}{°C.m^2}$	$\frac{°C.m}{W}$	$\frac{°C.m^2}{W}$	$\frac{kJ}{kg.\ °C}$
Building board Boards, panels, sub-flooring, sheathing, woodbased panel products	Asbestos-cement board 6mm	1922	-	93.7	-	0.011	-
	Gypsum or plasterboard						
	10 mm	800	-	17.6	-	0.057	-
	13 mm	800	-	12.8	-	0.078	-
	Plywood	545	0.12	-	8.70	-	1.21
	6 mm	545	-	18.2	-	0.055	1.21
	10 mm	545	-	12.1	-	0.083	1.21
	13 mm	545	-	9.09	-	0.110	1.21
	20 mm	545	-	6.08	-	0.165	1.21
	Insulating board and sheathing						
	13 mm	288	-	4.32	-	0.232	1.30
	20 mm	288	-	2.78	-	0.359	1.30
	Hardboard, high density, standard tempered	1010	0.14	-	6.94	-	1.38
	Particle board Medium density	800	0.14	-	7.35	-	1.30
	Underlayment 16 mm	640	-	6.93	-	0.144	1.21
	Wood subfloor 20 mm	-	-	6.02	-	0.166	1.42

Building paper	Vapour-permeable felt	-	-	94.8	-	0.011	-
	Vapour-seal, two layers of mopped, 15-lb felt	-	-	47.4	-	0.021	-
Finish flooring materials	Carpet and fibrous pad	-	-	2.73	-	0.367	-
	Carpet and rubber pad	-	-	4.60	-	0.217	1.42
	Tile-asphalt, linoleum, vinyl, or rubber	-	-	113.0	-	0.009	1.26
Insulating materials Blanket and batt	Mineral fibre-fibrous form processed from rock, slag, or glass						
	Approximately 50 to 70 mm	-	-	0.812	-	1.23	0.754
	Approximately 75 to 90 mm	-	-	0.517	-	1.94	0.754
	Approximately 135 to 165 mm	-	-	0.301	-	3.32	0.754
Board and slabs	Cellular glass	144	0.058	-	17.2	-	1.0
	Glass fibre, organic bonded	64-144	0.036	-	27.8	-	8.0
	Expanded polystyrene-molded beads	16	0.040	-	25.0	-	1.2
	Expanded polyurethane-R-11 expanded	24	0.023	-	43.5	-	1.6
	Mineral fibre with resin binder	240	0.042	-	-	-	0.71
Loose fill	Mineral fibre-rock, slag, or glass						
	Approximately 75 mm	-	-	-	0.63	1.58	0.75
	Approximately 115 mm	-	-	0.44	-	2.29	0.75
	Approximately 160 mm	-	-	0.30	-	3.35	0.75
	Approximately 185 mm	-	-	0.24	-	4.23	0.75
	Silica acrogel	122	0.025	-	40.8	-	-
	Vermiculite (expanded)	122	0.068	-	14.8	-	-

TABLE A1 (cont'd)

Material	Description	Density ρ	Thermal Conductivity, K	Unit Conductance, C_t	Resistivity $\frac{1}{K} = r$	Resistivity $\frac{1}{C_r} = R_t$	Specific Heat
		$\frac{kg}{m^3}$	$\frac{W}{C.m}$	$\frac{W}{^oC.m^2}$	$\frac{^oC.m}{W}$	$\frac{^oC.m^2}{W}$	$\frac{kJ}{kg.\ ^oC}$
Roof insulation	Preformed, for use above deck						
	Approximately 13 mm	-	-	4.1	-	0.24	1.0
	Approximately 25 mm	-	-	2.0	-	0.49	2.1
	Approximately 50 mm	-	-	1.1	-	0.93	3.9
	Cellular glass	144	0.058	-	17.3	-	1.0
Masonry materials	Lightweight aggregates including expanded shale, clay, or slate; expanded slags; cinders; pumice, vermiculite; also cellular concretes	3200	0.75	-	1.32	-	-
		1600	0.52	-	1.94	-	-
		1280	0.36	-	2.77	-	-
		640	0.17	-	6.03	-	-
		320	0.10	-	10.0	-	-
	Sand and gravel or stone	2242	1.73	-	0.58	-	-
Masonry units	Brick, common	1922	0.72	-	1.39	-	-
	Brick, face	2082	1.30	-	0.77	-	-
	Concrete blocks, three-oval core-sand and gravel aggregate						
	100 mm	-	-	8.0	-	0.13	-
	200 mm	-	-	5.1	-	0.20	-
	300 mm	-	-	4.4	-	0.23	-
	lightweight aggregate (expanded shale, clay slate or slag; pumice)						

	75 mm	-	-	4.5	-	0.22	-
	100 mm	-	-	3.8	-	0.26	-
	200 mm	-	-	2.8	-	0.35	-
	300 mm	-	-	2.5	-	0.40	-
Plastering materials	Cement, plaster, sand, aggregate	1858	0.72	-	1.39	-	-
	Gypsum plaster:						
	Lightweight aggregate						
	13 mm	721	-	17.7	-	0.056	-
	16 mm	721	-	15.2	-	0.066	-
	Lightweight aggregate on metal lath						
	20 mm	-	-	12.1	-	0.083	
Roofing	Asbestos-cement shingles	1922	-	27.0	-	0.037	-
	Asphalt roll roofing	1121	-	36.9	-	0.027	-
	Asphalt shingles	1121	-	12.9	-	0.078	-
	Built-up roofing 10 mm	1121	-	17.0	-	0.059	-
	Slate 13 mm	-	-	113.0	-	0.009	-
	Wood shingles-plain	-	-	6.02	-	0.166	-
Siding materials (on flat surface)	Shingles						
	Asbestos-cement	1922	-	27.0	-	3.70	-
	Siding						
	Wood, drop 25 mm	-	-	7.21	-	0.139	1.30
	Wood, plywood, 10 mm lapped	-	-	9.03	-	0.111	1.21
	Aluminium or steel, over sheathing, hollow backed	-	-	9.14	-	0.109	-
	Insulating board-backed nominal 10 mm foil-backed	-	-	3.12	-	0.320	-
	Architectural glass	-	-	56.8	-	0.018	-

TABLE A1 (cont'd)

Material	Description	Density	Thermal Conductivity, K	Unit Conductance, C_t	Resistivity $\frac{1}{K} = r$	Resistivity $\frac{1}{C_r} = R_t$	Specific Heat
		$\frac{kg}{m^3}$	$\frac{W}{^oC.m}$	$\frac{W}{^oC.m^2}$	$\frac{^oC.m}{W}$	$\frac{^oC.m^2}{W}$	$\frac{kJ}{kg. ^oC}$
Woods	Maple, oak, and similar hard woods	721	0.159	-	6.3	-	1.26
	Fir, pine, and similar soft woods	513	0.115	-	8.67	-	1.38
	Aluminium (1100)	2739	221.5	-	0.0045	-	0.896
	Steel, mild	7833	45.3	-	0.022	-	0.502
	Steel, stainless	7913	15.6	-	0.064	-	0.456

TABLE A2 Configuration Factors for Various Geometries (After Wong, 1977)

No.	Description of system	Schematic presentation	Configuration factor formula
	RADIATION FROM ELEMENTARY AREA dA_1 IMPINGING ON FINITE AREA A_2		
1	dA_1 parallel to a retangular plane	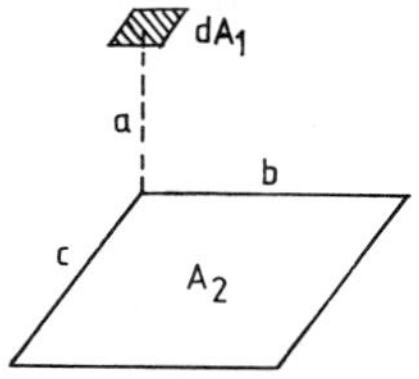	$F_{dA_1 - A_2} = \frac{1}{2\pi}\left[\frac{B}{\sqrt{1+B^2}} \tan^{-1} \frac{C}{\sqrt{1+B^2}} + \frac{C}{\sqrt{1+C^2}} \times \tan^{-1} \frac{B}{\sqrt{1+C^2}}\right]$ $B = \frac{b}{a}$, $C = \frac{c}{d}$
2	dA_1 and a rectangular plane intersection at an angle ϕ	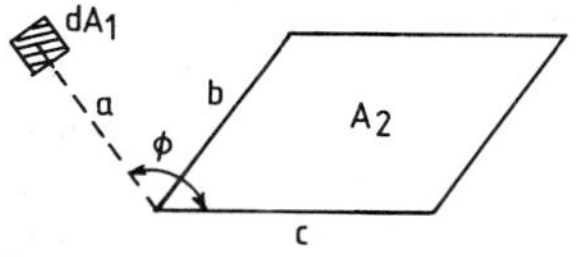	$F_{dA_1 - A_2} = \frac{1}{2\pi}\left[\tan^{-1} B + \frac{1}{X}(C\cos\phi - 1)\tan^{-1}\frac{B}{X} + \frac{B\cos\phi}{Y}\left\{\tan^{-1}\left(\frac{C-\cos\phi}{Y}\right) + \tan^{-1}\left(\frac{\cos\phi}{Y}\right)\right\}\right]$ $B = \frac{b}{a}$, $C = \frac{c}{a}$, $X = \sqrt{1 + C^2 - 2C\cos\phi}$, $Y = \sqrt{B^2 + \sin^2\phi}$
3	A spherical point source dA_1 to a rectangular plane intersecting at an angle ϕ	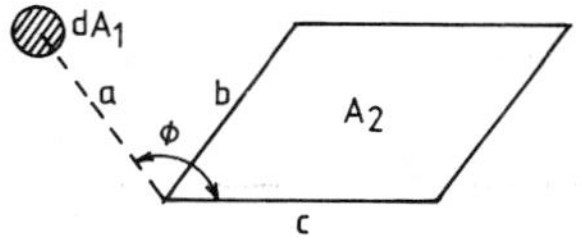	$F_{dA_1 - A_2} = \frac{1}{4\pi}\left\{\tan^{-1}\left[\frac{B(C-\cos\phi)}{\sqrt{B^2 + C^2 + 1 - 2C\cos\phi}}\right] + \tan^{-1}\left(\frac{B\cos\phi}{\sqrt{1+B^2}}\right)\right\}$ $B = \frac{b}{a}$, $C = \frac{c}{a}$

TABLE A2 (cont'd)

No.	Description of system	Schematic presentation	Configuration factor formula
4	d 1 parallel to a circular disc	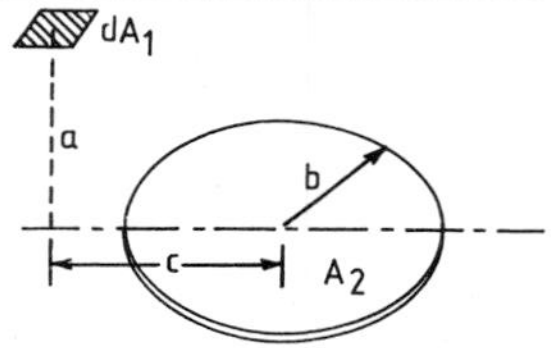 dA_1, a, b, c, A_2	$F_{dA_1 - A_2} = \frac{1}{2}\left[1 - \frac{1 + C^2 - B^2}{\sqrt{(1 + B^2 + C^2)^2 - 4B^2 C^2}}\right]$ $B = \frac{b}{a}, \quad C = \frac{c}{a}$
5	dA_1 perpendicular to circular disc	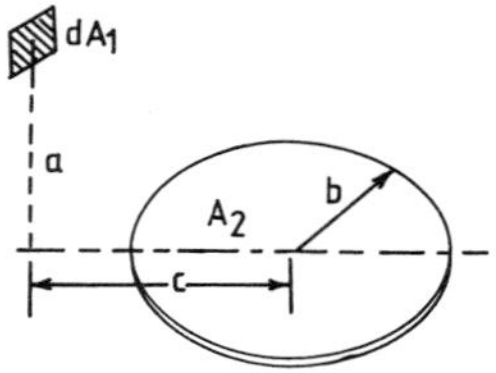 dA_1, a, A_2, b, c	$F_{dA_1 - A_2} = \frac{1}{2C}\left[\frac{1 + B^2 + C^2}{\sqrt{(1 + B^2 + C^2)^2 - 4C^2 B^2}} - 1\right]$ $B = \frac{b}{a}, \quad C = \frac{c}{a}$
6	Strip dA_1 to infinitely long cylinder	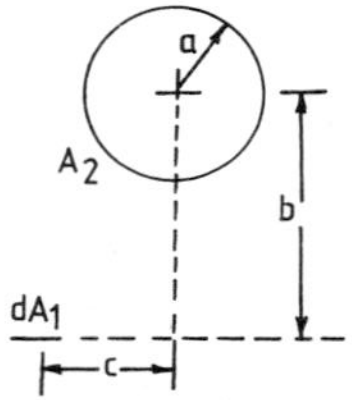 a, A_2, b, dA_1, c	$F_{dA_1 - A_2} = \frac{ab}{b^2 + c^2}$

7	dA_1 on cylinder to an infinite plane	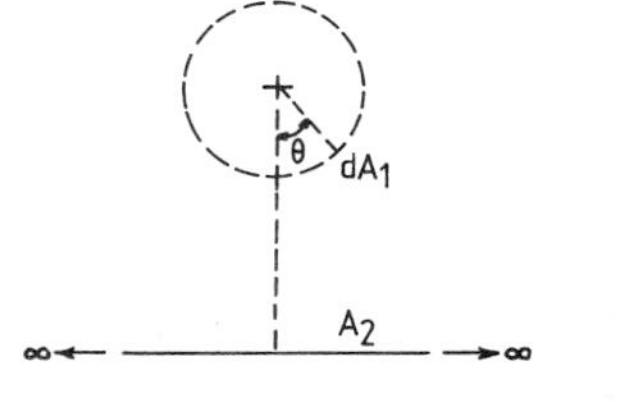	$F_{dA_1 - A_2} = \frac{1}{2}(1 + \cos\phi)$
8	Interior ring dA_1 to end of hollow cylinder	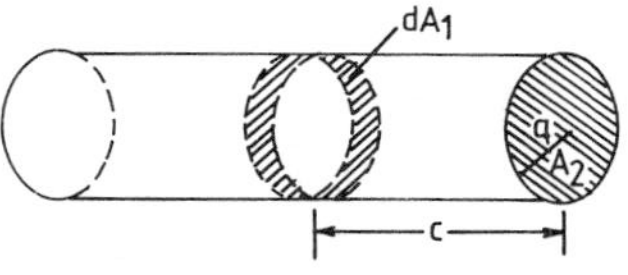	$F_{dA_1 - A_2} = \frac{1}{2}\left(\frac{C^2 + 2}{\sqrt{C^2 + 4}} - C\right)$ $C = \frac{c}{a}$
9	dA_1 at one end of cylinder parallel to cylinder axis	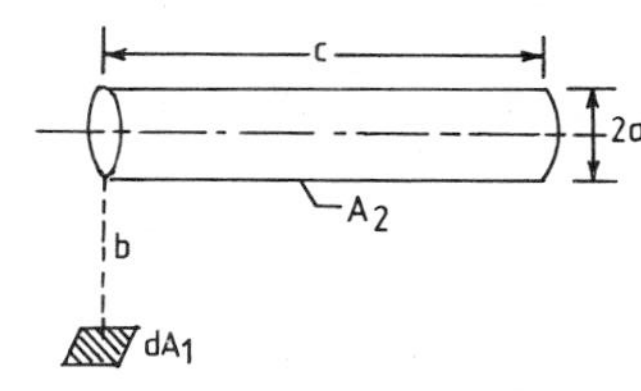	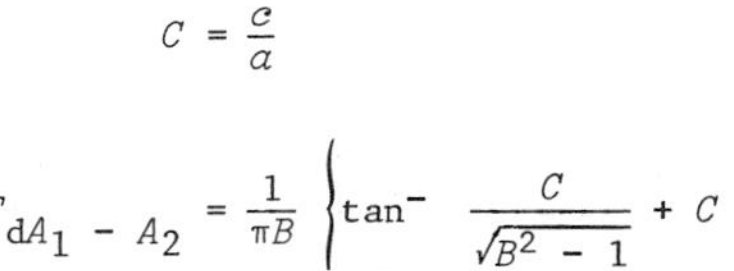 $F_{dA_1 - A_2} = \frac{1}{\pi B}\left\{\tan^{-} \frac{C}{\sqrt{B^2 - 1}} + C\left[\frac{X^2 - 2B}{XY}\right.\right.$ $\left.\left.\tan^{-1}\frac{X}{Y}\sqrt{\frac{B - 1}{B + 1}} - \tan^{-1}\sqrt{\frac{B - 1}{B + 1}}\right]\right\}$ $B = \frac{b}{a}\ ,\ C = \frac{c}{a} \quad X = \sqrt{(1 + B^2)^2 + C^2},$ $Y = \sqrt{(1 - B^2)^2 + C^2}$
10	dA_1 parallel to elliptical disc	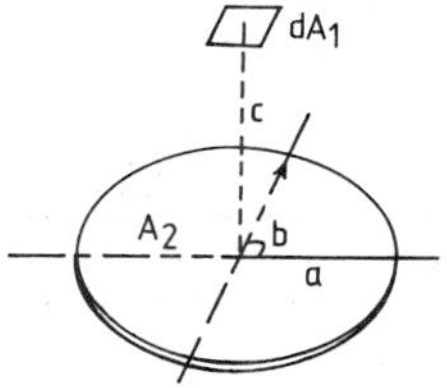	$F_{dA_1 - A_2} = \frac{ab}{\sqrt{(a^2 + c^2)(b^2 + c^2)}}$

TABLE A2 (cont'd)

No.	Description of system	Schematic presentation	Configuration factor formula
11	A line source dA_1 parallel to a rectangular plane	line source dA_1; a; c; b; A_2	$F_{dA_1 - A_2} = \frac{1}{\pi B}\left[\sqrt{1+B^2}\tan^{-1}\left(\frac{C}{\sqrt{1+B^2}}\right) - \tan^{-1} C + \frac{BC}{\sqrt{1+C^2}}\tan^{-1}\left(\frac{B}{\sqrt{1+C^2}}\right)\right]$ $B = \frac{b}{a}$, $C = \frac{c}{a}$
12	A line source dA_1 and a rectangular plane intersecting at an angle ϕ	line source dA_1; a; c; b; A_2; ϕ	$F_{dA_1 - A_2} = \frac{1}{\pi}\left\{\tan^{-1} B + \frac{\sin^2\phi}{2B}\ln\left[\frac{B + X^2}{(1+B^2)X^2}\right] - \frac{\sin 2\phi}{2B}\left[\frac{\pi}{2} - \phi + \tan^{-1}\left(\frac{C - \cos\phi}{\sin\phi}\right)\right] + \frac{Y}{B}\left[\tan^{-1}\left(\frac{C - \cos\phi}{Y}\right) + \tan^{-1}\left(\frac{\cos\phi}{Y}\right)\right] \mathrm{x} \cos\phi + \frac{C\cos\phi - 1}{X}\tan^{-1}\left(\frac{B}{X}\right)\right\}$ $B = \frac{b}{a}$, $C = \frac{c}{a}$. $X = \sqrt{C^2 - 2C\cos\phi + 1}$, $Y = \sqrt{B^2 + \sin^2\phi}$

13 dA_1 intersecting an angle ϕ with a rectangular plate with a triangular plate added to the top

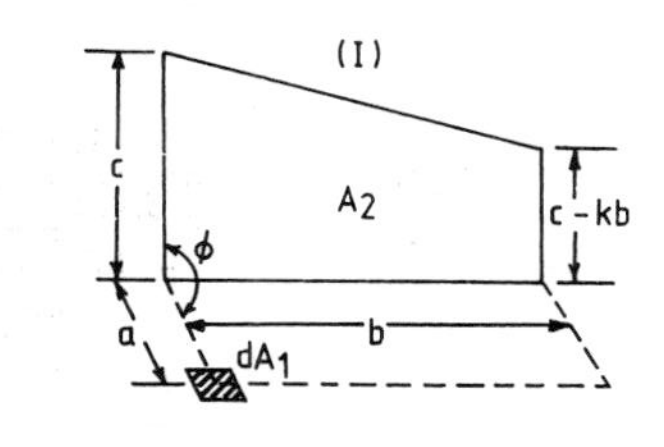

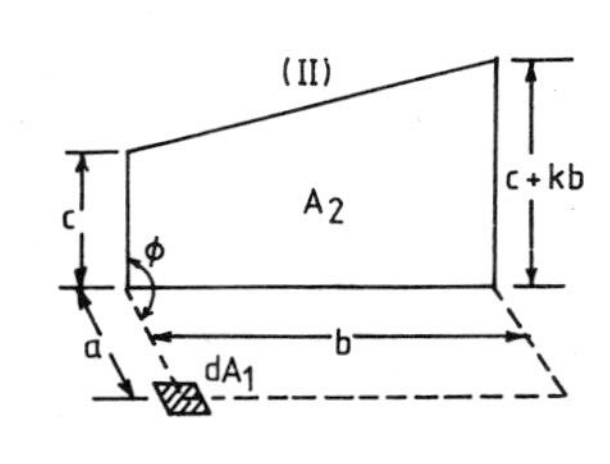

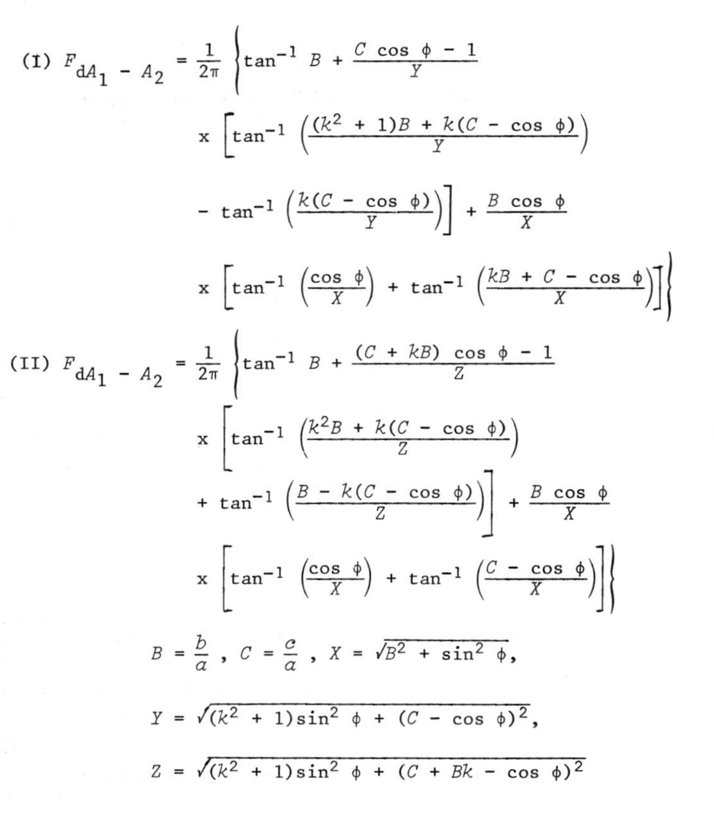

$$(\mathrm{I})\ F_{dA_1 - A_2} = \frac{1}{2\pi}\left\{\tan^{-1} B + \frac{C\cos\phi - 1}{Y}\right.$$

$$\times\left[\tan^{-1}\left(\frac{(k^2+1)B + k(C-\cos\phi)}{Y}\right)\right.$$

$$\left.- \tan^{-1}\left(\frac{k(C-\cos\phi)}{Y}\right)\right] + \frac{B\cos\phi}{X}$$

$$\left.\times\left[\tan^{-1}\left(\frac{\cos\phi}{X}\right) + \tan^{-1}\left(\frac{kB + C - \cos\phi}{X}\right)\right]\right\}$$

$$(\mathrm{II})\ F_{dA_1 - A_2} = \frac{1}{2\pi}\left\{\tan^{-1} B + \frac{(C+kB)\cos\phi - 1}{Z}\right.$$

$$\times\left[\tan^{-1}\left(\frac{k^2B + k(C-\cos\phi)}{Z}\right)\right.$$

$$\left.+ \tan^{-1}\left(\frac{B - k(C-\cos\phi)}{Z}\right)\right] + \frac{B\cos\phi}{X}$$

$$\left.\times\left[\tan^{-1}\left(\frac{\cos\phi}{X}\right) + \tan^{-1}\left(\frac{C - \cos\phi}{X}\right)\right]\right\}$$

$$B = \frac{b}{a},\ C = \frac{c}{a},\ X = \sqrt{B^2 + \sin^2\phi},$$

$$Y = \sqrt{(k^2+1)\sin^2\phi + (C-\cos\phi)^2},$$

$$Z = \sqrt{(k^2+1)\sin^2\phi + (C + Bk - \cos\phi)^2}$$

TABLE A2 (cont'd)

No.	Description of system	Schematic presentation	Configuration factor formula
	RADIATION FROM FINITE AREA A_1 IMPINGING ON FINITE AREA A_2		
14	Two parallel rectangular surfaces of equal size	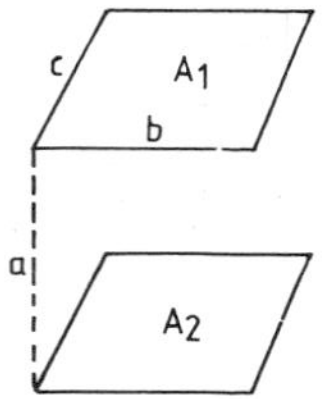	$F_{A_1 - A_2} = \frac{1}{\pi}\left[\frac{1}{BC}\ln\left(\frac{XY}{X+Y-1}\right) + \frac{2\sqrt{X}}{B}\tan^{-1}\frac{C}{\sqrt{X}} + \frac{2\sqrt{Y}}{C}\tan^{-1}\frac{B}{\sqrt{Y}} - \frac{2}{C}\tan^{-1}B - \frac{2}{B}\tan^{-1}C\right]$ $B = \frac{b}{a}$, $C = \frac{c}{a}$, $X = 1 + B^2$, $Y = 1 + C^2$
15	Two rectangular planes perpendicular to each other	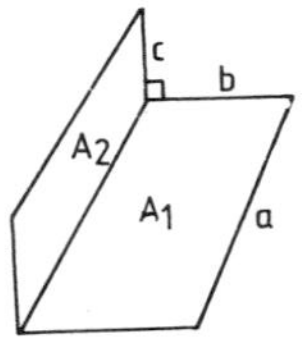	$F_{A_1 - A_2} = \frac{1}{\pi B}\left[\frac{1}{4}\ln\left\{\left[\frac{(1+B^2)(1+C^2)}{1+B^2+C^2}\right]\left[\frac{B^2(1+B^2+C^2)}{(1+B^2)(B^2+C^2)}\right]^{B^2} \times \left[\frac{C^2(1+B^2+C^2)}{(1+C^2)(B^2+C^2)}\right]^{C^2}\right\} + B\tan^{-1}\frac{1}{B} + C\tan^{-1}\frac{1}{C} - \sqrt{B^2+C^2}\tan^{-1}\left(\frac{1}{B^2+C^2}\right)\right.$ $B = \frac{b}{a}$, $C = \frac{c}{a}$ $F_{A_1 - A_2} = \frac{1}{2}\left[1 + \frac{c}{b} - \sqrt{1 + (c/b)^2}\right]$ for $a \to \infty$

16 Two parallel circular discs

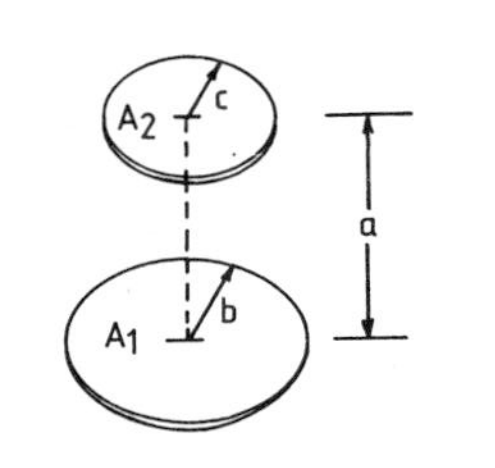

$$F_{A_1 - A_2} = \frac{1}{2B^2}\left(X - \sqrt{X^2 - 4B^2C^2}\right)$$

$$B = \frac{b}{a}, \quad C = \frac{c}{a}, \quad X = (1 + B^2 + C^2)$$

17 Two infinitely long circular cylinders

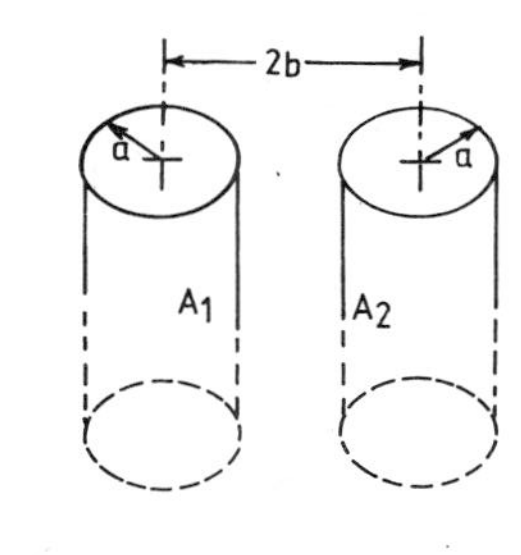

$$F_{A_1 - A_2} = \frac{2}{\pi}\left[\sqrt{B^2 - 1} + \sin^{-1}\left(\frac{1}{B}\right) - B\right]$$

$$B = \frac{b}{a}$$

18 Two concentric circular cylinders

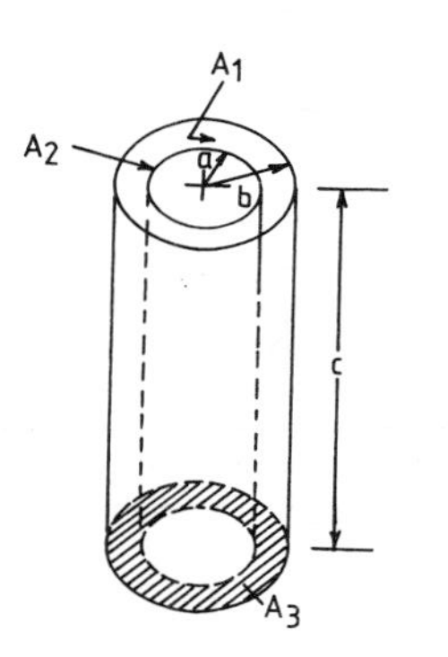

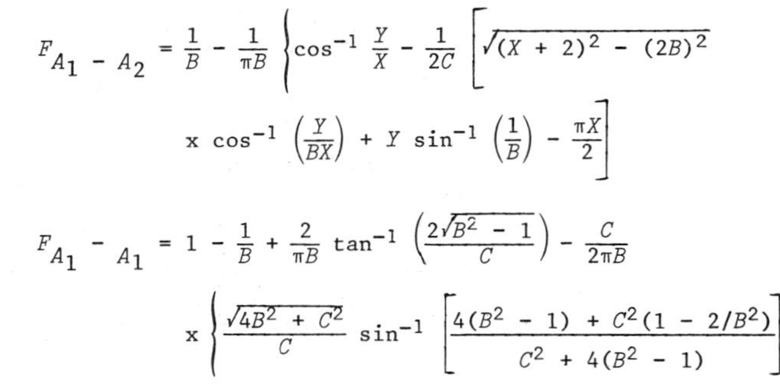

$$F_{A_1 - A_2} = \frac{1}{B} - \frac{1}{\pi B}\left\{\cos^{-1}\frac{Y}{X} - \frac{1}{2C}\left[\sqrt{(X + 2)^2 - (2B)^2}\right.\right.$$

$$\left.\left.\times \cos^{-1}\left(\frac{Y}{BX}\right) + Y\sin^{-1}\left(\frac{1}{B}\right) - \frac{\pi X}{2}\right]\right\}$$

$$F_{A_1 - A_1} = 1 - \frac{1}{B} + \frac{2}{\pi B}\tan^{-1}\left(\frac{2\sqrt{B^2 - 1}}{C}\right) - \frac{C}{2\pi B}$$

$$\times \left\{\frac{\sqrt{4B^2 + C^2}}{C}\sin^{-1}\left[\frac{4(B^2 - 1) + C^2(1 - 2/B^2)}{C^2 + 4(B^2 - 1)}\right]\right.$$

TABLE A2 (cont'd)

No.	Description of system	Schematic presentation	Configuration factor formula
18	(continued)		$- \sin^{-1}\left(1 - \frac{2}{B^2}\right) + \frac{\pi}{2}\left(\frac{\sqrt{4B^2 + C^2}}{C} - 1\right)\Bigg\}$ $B = \frac{b}{a}$, $C = \frac{c}{a}$, $X = B^2 + C^2 - 1$, $Y = C^2 - B^2 + 1$ $F_{A_1 - A_2} = \frac{1}{B}$ and $F_{A_2 - A_1} = 1$ for $c \to \infty$ $F_{A_1 - A_2} = \frac{1}{2}(1 - F_{A_1 - A_2} - F_{A_1 - A_1})$
19	Two concentric spheres	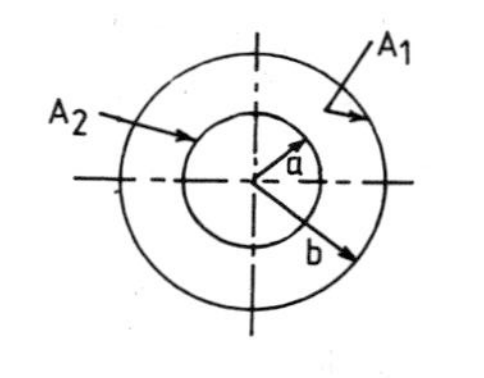	$F_{A_1 - A_2} = \left(\frac{a}{b}\right)^2$ $F_{A_2 - A_1} = 1$ $F_{A_1 - A_2} = 1 - \left(\frac{a}{b}\right)^2$

20 Two infinitely long parallel strips of different widths

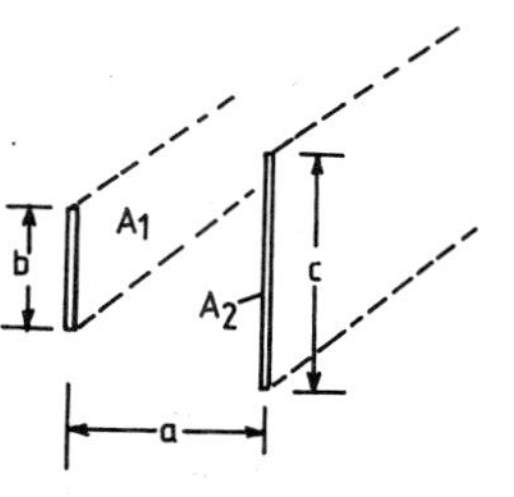

$$F_{A_1 - A_2} = \frac{1}{2B}\left[\sqrt{(B + C)^2 + 4} - \sqrt{(C - B)^2 + 4}\right]$$

$$F_{A_2 - A_1} = \frac{1}{2C}\left[\sqrt{(B + C)^2 + 4} - \sqrt{(B - C)^2 + 4}\right]$$

$$B = \frac{b}{a}\ ,\ C = \frac{c}{a}$$

$$F_{A_1 - A_2} = F_{A_2 - A_1} = \frac{1}{B}\left[\sqrt{B^2 + 1} - 1\right] \quad \text{for} \quad b = c$$

21 A strip parallel in a circular cylinder, both of infinite length

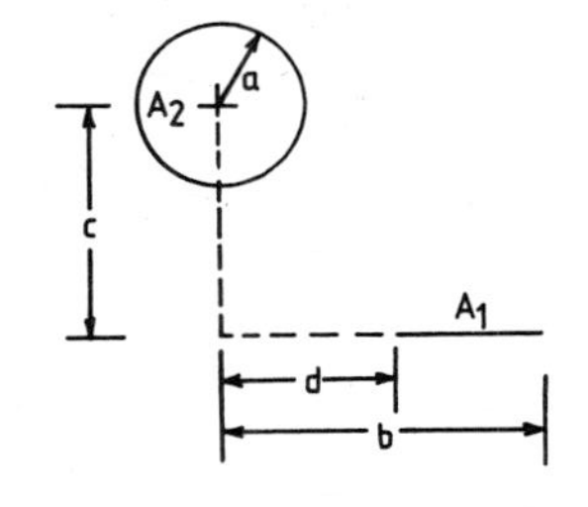

$$F_{A_1 - A_2} = \frac{a}{b - d}\left[\tan^{-1}\frac{b}{c} - \tan^{-1}\frac{d}{c}\right]$$

22 Two-dimensional wedge cavity

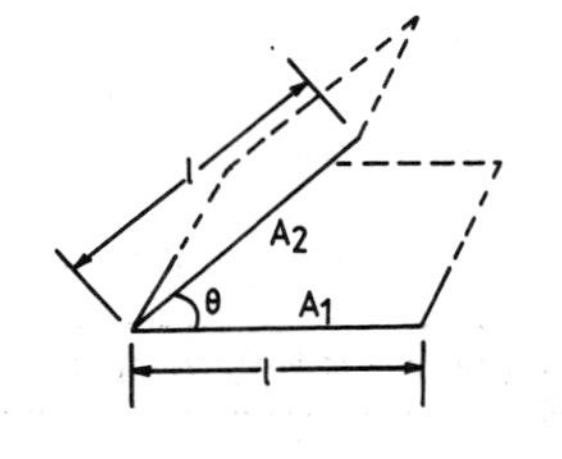

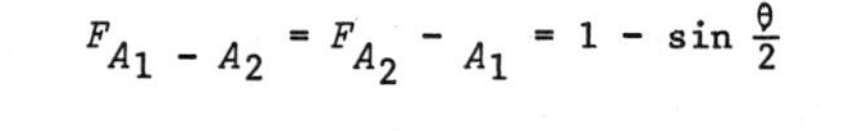

$$F_{A_1 - A_2} = F_{A_2 - A_1} = 1 - \sin\frac{\theta}{2}$$

TABLE A2 (cont'd)

No.	Description of system	Schematic presentation	Configuration factor formula
23	Between sides of an infinitely long triangular enclosure	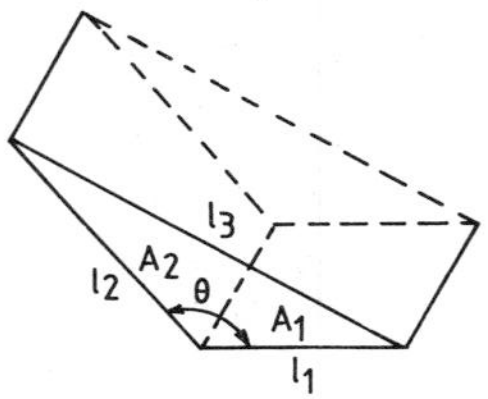	$F_{A_1 - A_2} = \frac{l_1 + l_2 - l_3}{2l_1}$ $F_{A_1 - A_2} = 1 - \sin\frac{\theta}{2}$ for $A_1 = A_2$
24	Centrally placed sphere to a plane disc	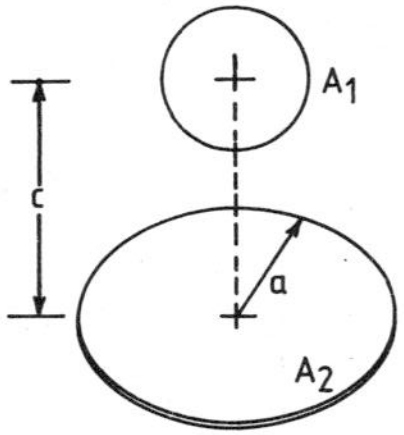	$F_{A_1 - A_2} = \frac{1}{2}\left(1 - \frac{1}{\sqrt{1 + (a/c)^2}}\right)$
25	Sphere to sector of disc	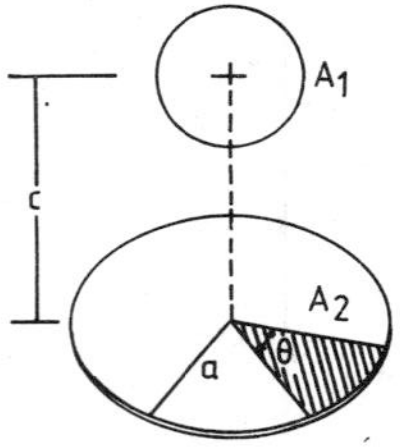	$F_{A_1 - A_2} = \frac{\theta}{4\pi}\left(1 - \frac{1}{\sqrt{1 + (a/c)^2}}\right)$

26 Sphere to segment of disc

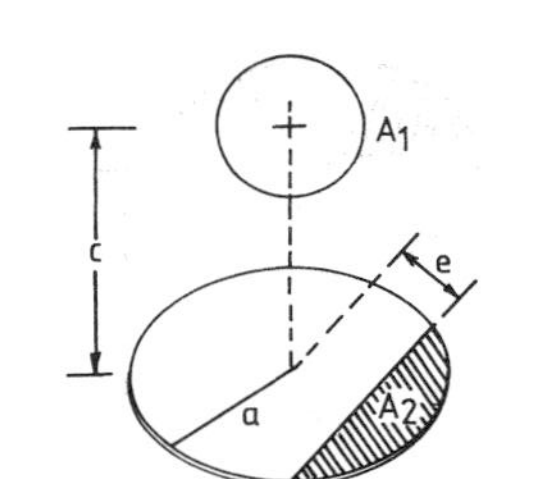

$$F_{A_1 - A_2} = \frac{1}{8} - \frac{\cos^{-1} E}{2\pi\sqrt{1 + \frac{1}{C^2}}} + \frac{1}{4}\sin^{-1}\left[\frac{C^2 - E^2 - 2E^2C^2}{C^2 + E^2}\right]$$

$$C = \frac{c}{a}, \quad E = \frac{e}{a}$$

27 Inner surface of circular cylinder to base

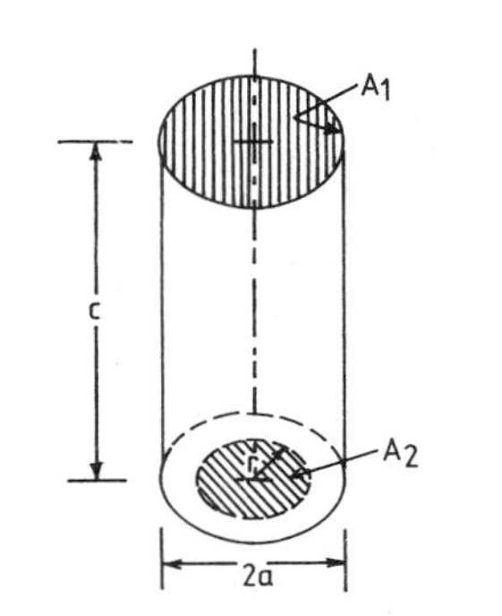

$$F_{A_1 - A_2} = \frac{1}{4C}\left[\sqrt{C^4 + 2C^2(1 + R^2) + (1 - R^2)^2} - (1 - R^2 + C^2)\right]$$

$$C = \frac{c}{a}, \quad R = \frac{r}{a}$$

28 Inner surface of cylinder between C_1 and C_2 to base ring between r_2 and r_1

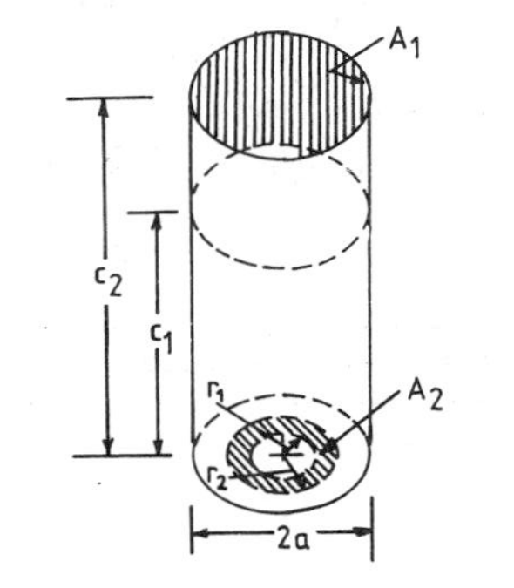

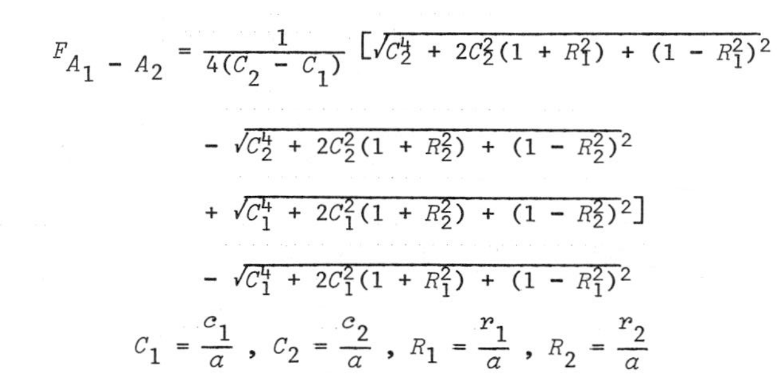

$$F_{A_1 - A_2} = \frac{1}{4(C_2 - C_1)}\left[\sqrt{C_2^4 + 2C_2^2(1 + R_1^2) + (1 - R_1^2)^2} - \sqrt{C_2^4 + 2C_2^2(1 + R_2^2) + (1 - R_2^2)^2} + \sqrt{C_1^4 + 2C_1^2(1 + R_2^2) + (1 - R_2^2)^2}\right] - \sqrt{C_1^4 + 2C_1^2(1 + R_1^2) + (1 - R_1^2)^2}$$

$$C_1 = \frac{c_1}{a}, \quad C_2 = \frac{c_2}{a}, \quad R_1 = \frac{r_1}{a}, \quad R_2 = \frac{r_2}{a}$$

TABLE A2 (cont'd)

No.	Description of system	Schematic presentation	Configuration factor formula
29	Spherical surface to base in equatorial plane	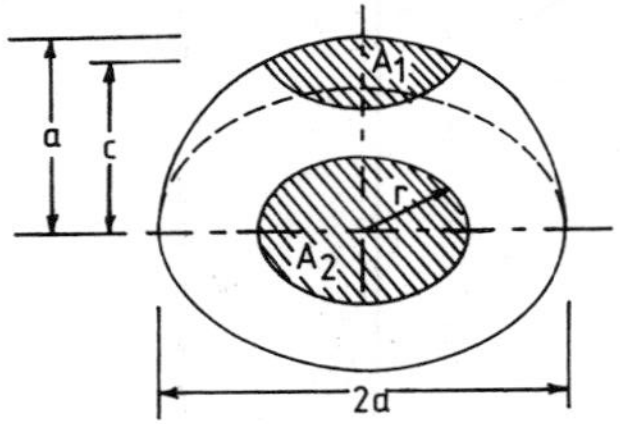	$F_{A_1 - A_2} = \frac{1}{4(1 - C)} \left[\sqrt{4R^2 + (1 - R^2)} - \sqrt{4R^2C^2 + (1 - R^2)^2} \right]$ $C = \frac{c}{a}, \ R = \frac{r}{a}$
30	Ring on hemispherical surface to ring on the base	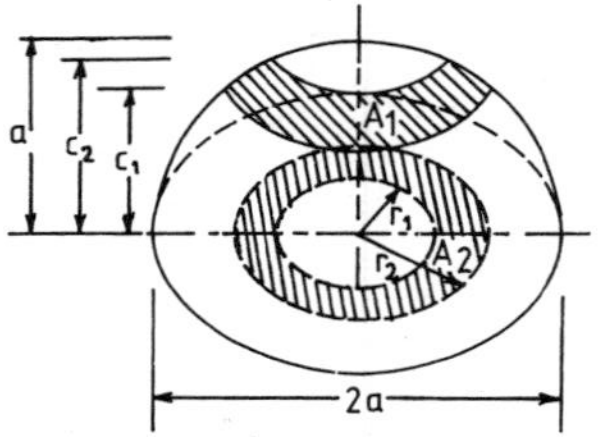	$F_{A_1 - A_2} = \frac{1}{4(C_2 - C_1)} \left[\sqrt{4R_2^2C_2^2 + (1 - R_2^2)^2} - \sqrt{4R_1^2C_2^2 + (1 - R_1^2)^2} + \sqrt{4R_1^2C_1^2 + (1 - R_1^2)^2} - \sqrt{4R_2^2C_1^2 + (1 - R_2^2)^2} \right]$ $C_1 = \frac{c_1}{a}, \ C_2 = \frac{c_2}{a}, \ R_1 = \frac{r_1}{a}, \ R_2 = \frac{r_2}{a}$

31 Cylinder to rectangular plate in plane parallel to cylinder axis

$$F_{A_1 - A_2} = \frac{2}{Y} \int_0^{Y/2} \left[\frac{X}{X^2 + \beta^2} - \frac{X}{\pi(X^2 + \beta^2)} \right.$$

$$\times \left(\cos^{-1} \frac{W}{V} - \frac{1}{2Z} \left[\sqrt{V^2 + 4Z^2} \right. \right.$$

$$\times \cos^{-1} \left(\frac{W}{V\sqrt{X^2 + \beta^2}} \right) + W \sin^{-1}$$

$$\left. \left. \left. \times \left(\frac{1}{\sqrt{X + \beta^2}} \right) - \frac{\pi V}{2} \right] \right) \right] d\beta$$

$$X = \frac{a}{r}, \quad Y = \frac{b}{r}, \quad Z = \frac{c}{r}, \quad V = X^2 + Z^2 + \beta^2 - 1,$$

$$W = Z^2 - X^2 - \beta^2 + 1$$

32 Infinite plane to a row of pipes

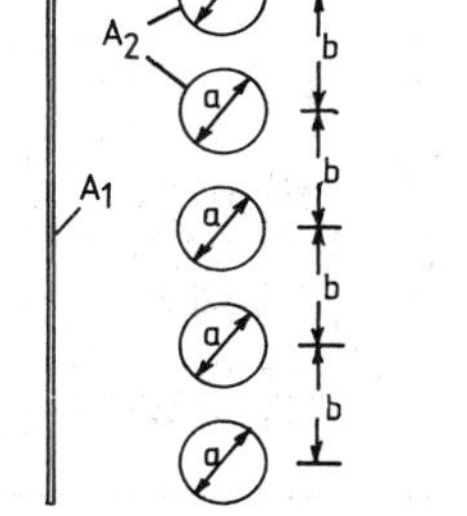

$$F_{A_1 - A_2} = 1 - \sqrt{1 - (a/b)^2} + \frac{a}{b} \tan^{-1} \sqrt{(b/a)^2 - 1}$$

$$F_{A_2 - A_1} = \frac{1}{\pi} \left[\frac{b}{a} - \sqrt{\left(\frac{b}{a}\right)^2 - 1} + \tan^{-1} \sqrt{\left(\frac{b}{a}\right)^2 - 1} \right]$$

$F_{A_1 - (A_2)_n} = 1 - (1 - F_{A_1 - (A_2)_1})^n$ for n rows of in-line pipes.

The radiation areas per unit length may be taken as

$$A_1 = b, \quad A_2 = \pi a$$

$F_{A_1 - (A_2)_1}$ is the configuration factor for one row of pipes.

TABLE A3 Standard Thermal Resistance of Air Spaces or Cavities (After Evans, 1980 and Anon., 1975)

Cavity	Conductance W/m^2 deg C	Resistance m^2 deg C/W
Unventilated 5 mm cavity; high emissivity (normal) surface on both sides		
Heat flow downwards	9.09	0.11
Heat flow horizontal or upwards	9.09	0.11
Unventilated 20 mm cavity; high emissivity (normal) surface on both sides:		
Heat flow downwards	4.76	0.21
Heat flow horizontal or upwards	5.56	0.18
Unventilated 80 mm cavity; high emissivity (normal) surface on both sides		
Heat flow downwards	5.56	0.18
Heat flow horizontal or upwards	7.69	0.13
Unventilated 5 mm cavity; low emissivity (reflective) surface on one side:		
Heat flow downwards	0.94	1.06
Heat flow horizontal or upwards	2.86	0.35
Ventilated vertical cavity:		
In brick wall	5.68	0.176
In brick wall with foil lining	2.86	0.35
Behind asbestos cement sheets	6.25	0.16
Behind asbestos cement sheets (with a foil lining)	3.33	0.30
Between hung tiles and wall	8.33	0.12
Horizontal ventilated roof cavities between a flat ceiling and a pitched roof:		
Heat flow up, roof of tiles (unsealed)	9.09	0.11
Heat flow up, tiles (with felt lining)	5.55	0.18
Heat flow up, unsealed roofing sheets	7.14	0.14
Heat flow up, aluminium sheet roof (or aluminium foil lining)	4.00	0.25
Heat flow down, normal roof in summer	4.76	0.21
Heat flow down, aluminium sheet roof (or aluminium foil lining)	2.00	0.50
High emissivitive planes and corrugated sheets in contact:		
Heat flow upwards	11.11	0.09
Heat flow downwards	9.09	0.11
Low emissivitive multiple foil insulation with air space on one side:		
Heat flow upwards	1.61	0.62
Heat flow downwards	0.57	1.76

TABLE A4 Internal Surface Coefficient (After *IHVE Guide A*, 1970)

Parameter	Values	
	0.9	for ordinary building materials
	0.2	for dull aluminium
	0.05	for polished aluminium
h_c	1.5	for floors
	3.0	for walls
	4.5	for ceiling
h_r	5.7	at T_{si} = 20°C
	5.14	at T_{si} = 10°C
	4.61	at T_{si} = 0°C

TABLE A5 Inner Surface Coefficients Recommended in Various Countries for Calculating Overall Transmittance Values (After van Straaten, 1967)

Country	Position of surface	Direction of heat flow	Surface coefficient W/m² °C
America	Vertical	Horizontal	8.3
	Horizontal	Upward	9.40
	Horizontal	Downward	6.1
	Sloping 45°	Upward	9.1
	Sloping 45°	Downward	7.5
England	Vertical	Horizontal	8.2
	Horizontal	Upward	9.6
	Horizontal	Downward	6.8
Holland	Vertical	Horizontal	7.0
	Horizontal	Upward	8.2
	Horizontal	Downward	5.6
South Africa	Vertical	Horizontal	9.5
	Horizontal	Upward	11.1
	Horizontal	Downward	6.9

TABLE A6 U-values for External Walls: Masonry Construction

Construction	Thickness (mm)	U value ($W\ m^{-2}\ {}^oC^{-1}$) Sheltered	Normal (Standard)	Severe		Thermal properties Density ($kg\ m^{-3}$)	Conductivity K ($W\ m^{-1}\ {}^oC^{-1}$)
Brickwork							
1. Solid walls, unplastered	105	3.0	3.3	3.6			
	220	2.2	2.3	2.4			
	335	1.6	1.7	1.8	Brick	1700	0.84
2. Solid wall, with 16 mm plaster on inside face							
(a) With dense plaster	105	2.8	3.0	3.2			
	220	2.0	2.1	2.2			
	335	1.6	1.7	1.8	Plaster	1300	0.50
(b) With lightweight plaster	105	2.3	2.5	2.7			
	220	1.8	1.9	2.0			
	335	1.4	1.5	1.6	Plaster	600	0.16
3. Solid wall, with 10 mm plasterboard lining fixed to brickwork plaster dabs	105	2.6	2.8	3.0			
	220	1.9	2.0	2.1			
	335	1.5	1.6	1.7	Plasterboard	950	0.16
4. Cavity wall (unventilated) with 105 mm outer and inner leaves with 16 mm plaster on inside face							
(a) With dense plaster	260	1.4	1.5	1.6	Brick (outer leaf)	1700	0.84
(b) With lightweight plaster	260	1.3	1.3	1.3	Brick (inner leaf)	1700	0.62
5. As 4, but with 230 mm outer leaf and 105 mm inner leaf							
(a) With dense plaster	375	1.2	1.2	1.2	Plaster	1300	0.50
(b) With lightweight plaster	375	1.1	1.1	1.1	Plaster	600	0.16

Brickwork/lightweight concrete b-ock							
6. Cavity wall (unventilated) with 105 mm bricker outer leaf 100 mm lightweight concrete block inner leaf and with 16 mm dense plaster on inside face	260	0.93	0.96	0.98	Concrete block	600	0.19
7. As 6, but with 13 mm expanded polystyrene board in cavity		0.69	0.70	0.71	Expanded polystyrene	25	0.033
Lightweight concrete block							
8. Solid wall, 150 mm aerated concrete block, with tile hanging externally and with 16 mm plaster on inside face		0.95	0.97	1.0	Aerated concrete block	750	0.22
9. Cavity wall (unventilated) with 75 mm aerated concrete block outer leaf, rendered externally, 100 mm aerated concrete block inner leaf and with 16 mm plaster on inside face 50 mm cavity		0.82	0.84	0.86	Aerated concrete block (outer leaf) (inner leaf)	750 750	0.24 0.22
Concrete							
10. Cast	150 200	3.2 2.9	3.5 3.1	3.9 3.4	Concrete	2100	1.40
11. Cast, 150 mm thick. with 50 mm woodwool slab permanent shuttering on inside face finished with 16 mm dense plaster		1.1	1.1	1.1	Woodwool slab	450	0.09
12. As 11, but 200 mm thick		1.1	1.1	1.1			

TABLE A6 (cont'd)

Construction	Thickness	U value ($W\ m^{-2}\ {}^{o}C^{-1}$) Sheltered	Normal (Standard)	Severe	Material	Thermal properties: Density ($kg\ m^{-3}$)	Conductivity K ($W\ m^{-1}\ {}^{o}C^{-1}$)
13. Pre-cast panels, 75 mm thick		3.9	4.3	4.8			
14. As 13, but with 50 mm cavity and sandwich lining panels, composed of 5 mm asbestos-cement sheet, 25 mm expanded polystyrene and 10 mm plasterboard		0.79	0.80	0.82	Asbestos-cement sheet	1500	0.36
15. Pre-cast sandwich panels comprising 75 mm dense concrete, 25 mm expanded polystyrene and 150 mm lightweight concrete		0.71	0.72	0.73	Concrete Lightweight concrete	2100 1200	1.4 0.38
16. Pre-cast panels 38 mm on timber battens and framing with 10 mm plasterboard lining and 50 mm glass-fibre insulation in cavity (Assumed 10% area of glass-fibre bridged by timber)		0.61	0.62	0.63	Glass fibre Timber	25 650	0.035 0.14

U-values for external walls: Curtain Wall Construction

	U value ($W\ m^{-2}\ {}^oC^{-1}$)			Material	Thermal properties	
	Sheltered	Normal (Standard)	Severe		Density ($kg\ m^{-3}$)	Conductivity K ($W\ m^{-1}\ {}^oC^{-1}$)
Composite cladding panels						
1. Comprising 25 mm expanded polystyrene between 5 mm asbestos-cement sheets set in metal framing, 50 mm cavity, 10 mm lightweight concrete block inner wall, finished with 16 mm plaster rendering on inside face (Assumed 5% area of expanded polystyrene bridged by metal framing)	0.79	0.81	0.83	Expanded polystyrene	25	0.033
2. Obscured glass, 38 mm expanded polystyrene cavity 10 mm lightweight concrete back-up wall, dense plaster	0.51	0.51	0.52	Lightweight concrete	800	0.23
3. Stove-enamelled steel, 10 mm asbestos board, cavity 100 mm lightweight concrete back-up wall dense plaster	1.1	1.1	1.1	Asbestos board	700	0.11
Curtain walling panelling with 5% bridging by metal mullions, 150 mm x 50 mm wide						
4. With mullion projecting outside, flush inside:						
Panel construction 2	0.8	0.9	0.9			
Panel construction 3	1.4	1.4	1.5			
5. With mullion projecting inside and outside:						
Panel construction 2	1.3	1.5	1.8			
Panel construction 3	1.9	2.1	2.4			

U-values for external walls: Curtain Wall Construction (cont'd)

Construction	U value ($W\ m^{-2}\ {}^{o}C^{-1}$)			Material	Thermal properties	
	Sheltered	Normal (Standard)	Severe		Density ($kg\ m^{-3}$)	Conductivity K ($W\ m^{-1}\ {}^{o}C^{-1}$)
Curtain walling panelling with 10% bridging by metal mullions 150 mm x 50 mm wide						
6. With mullion projecting outside, flush inside:						
Panel construction 2	1.2	1.2	1.3			
7. With mullion projecting inside and outside:						
Panel construction 2	2.2	2.5	3.0			
Panel construction 3	2.7	3.1	3.6			

U-values for external walls: Framed Construction

Construction	Sheltered	Normal (Standard)	Severe	Material	Density ($kg\ m^{-3}$)	Conductivity K ($W\ m^{-1}\ {}^{o}C^{-1}$)
Tile hanging						
1. On timber battens and framing with 10 mm plasterboard lining, 50 mm glass-fibre insulation in the cavity and building paper behind the battens (assumed 10% area of glass-fibre bridged by timber)	0.64	0.65	0.66	Clay tiles	1900	0.84
Weather boarding						
2. On timber framing with 10 mm plasterboard lining, 50 mm glass-fibre insulation in the cavity and building paper behind the boarding (Assumed 10% area of glass-fibre bridged by timber)	0.61	0.62	0.63	Weather boarding	650	0.14

Corrugated sheeting						
3. 5 mm thick asbestos-cement (No allowance has been made for effect of corrugations on heat loss)	4.7	5.3	6.1	Asbestos-cement sheeting	1500	0.36
4. As 3, but with cavity and aluminium foil-backed plasterboard lining	1.7	1.8	1.9			
5. Double-skin asbestos-cement with 25 mm glass-fibre insulation in between	1.1	1.1	1.1			
6. As 5, but with cavity and aluminium foil-backed plasterboard lining	0.76	0.78	0.79			
7. Aluminium						
(a) Bright surface outside and inside	2.4	2.6	2.9			
(b) Dull surface outside, bright surface inside	2.6	2.8	3.0			
8. As 7, but with cavity and aluminium foil-backed plasterboard lining:						
(a) Bright surface outside	1.7	1.8	1.9			
(b) Dull surface outside	1.8	1.9	2.0			
9. Plastic-covered steel	5.0	5.7	6.6			
10. As 9, but with cavity and aluminium foil-backed plasterboard lining	1.8	1.9	2.0			

U-value for Flat Roofs

Construction	U value ($W\ m^{-2}\ {}^oC^{-1}$) Sheltered	Normal (Standard)	Severe	Material	Thermal properties Density ($kg\ m^{-3}$)	Conductivity K ($W\ m^{-1}\ {}^oC^{-1}$)
1. Asphalt 19 mm thick or felt/bitumen layers* on solid concrete 150 mm thick (treated as exposed)	3.1	3.4	3.7	Asphalt	1700	0.50
				Concrete	2100	1.4
2. As 1, but with 50 mm lightweight concrete screed and 16 mm plaster ceiling	2.1	2.2	2.3	Lightweight concrete	1200	0.42
				Dense plaster	1300	0.50
3. As 2, but with screen laid to falls, average 100 mm thick	1.7	1.8	1.9			
4. Asphalt 19 mm thick or felt/bitumen layers* on 150 mm thick autoclaved aerated concrete roof-slabs	0.87	0.88	0.89	Aerated concrete	500	0.16
5. Asphalt 19 mm thick or felt/bitumen layers* on hollow tiles 150 mm thick	2.1	2.2	2.3	Hollow tile	R = 0.27	
6. As 5, but with 50 mm lightweight concrete screed and 16 mm plaster ceiling	1.5	1.6	1.7			
7. As 6, but with screed laid to falls, average 100 mm thick	1.4	1.4	1.5			
8. Asphalt 19 mm thick or felt/bitumen layers* on 13 mm cement and sand screed, 50 mm woodwool slabs on timber joists and aluminium foil-backed 10 mm plasterboard ceiling, sealed to prevent moisture penetration				Cement/sand	2100	1.28
				Woodwool slab	560	0.10
	0.88	0.90	0.92	Plasterboard	950	0.16

9. As 8, but with 25 mm glass-fibre insulation laid between joists	0.59	0.60	0.61	Glass fibre	25	0.035
10. Asphalt 19 mm thick or felt/bitumen layers* on 13 mm cement and sand screed on 50 mm metal edge reinforced woodwool slabs on steel framing, with vapour barrier at inside	1.4	1.4	1.5			

*Other flat roof values can be found in the IHVE Guide.

U-values for pitched roofs (35° slope)

Construction	U value ($W\ m^{-2}\ {}^{o}C^{-1}$)			Material	Thermal properties	
	Sheltered	Normal (Standard)	Severe		Density ($kg\ m^{-3}$)	Conductivity K ($W\ m^{-1}\ {}^{o}C^{-1}$)
1. Tiles on battens, roofing felt and rafters, with roof space and aluminium foil-backed 10 mm plasterboard ceiling on joists	1.4	1.5	1.6	Tiles Roofing felt	1900 960	0.84 0.19
2. As 1, but with boarding on rafters	1.3	1.3	1.3	Timber	650	0.14
3. As 2, but with 50 mm glass-fibre insulation between joists	0.49	0.50	0.51			
4. Corrugated asbestos-cement sheeting	5.3	6.1	7.2	Asbestos-cement sheeting	1500	0.36
5. As 4, but with cavity and aluminium foil-backed 10 mm plasterboard lining	1.8	1.9	2.0			
6. Corrugated double-skin asbestos-cement sheeting with 25 mm glass-fibre insulation between (No allowance has been made for effect of corrugations on heat loss)	1.1	1.1	1.1	Glass fibre	25	0.035
7. As 6, but with cavity and aluminium foil-backed 10 mm plasterboard lining; ventilated air space	0.79	0.80	0.82			
8. Corrugated aluminium sheeting	3.3	3.8	4.3			
9. As 8, but with cavity and aluminium foil-backed 10 mm plasterboard lining	1.8	1.9	2.0			
10. Corrugated plastic-covered steel sheeting	5.7	6	8.1			
11. As 10, but with cavity and aluminium foil-backed 10 mm plasterboard lining; ventilated air space	1.9	2.0	2.1			

TABLE A7 Thermal Factors for Standard Construction

EXTERNAL WALLS External Surface Resistance: R_{so} = 0.055 $m^2 °C/W$,
Internal, R_{si} = 0.123 $m^2 °C/W$

	Density	Conductivity	Specific heat	U value	Admittance		Decrement	
	(kg/m^3)	(W/m °C)	(J/kg °C)	(W/m^2 °C)	Y(W/m^2°C)	ψY(hours)	d_f	ψ_f(hour)
BRICKWORK								
1. Solid brickwork, unplastered Brick 105 mm	1700	0.84	800	3.3	4.2	1.2	0.88	2.5
2. Solid brickwork, unplastered Brick 220 mm	1700	0.84	800	2.3	4.6	1.5	0.54	6.0
3. Solid brickwork, unplastered Brick 335 mm	1700	0.84	800	1.7	4.7	1.4	0.29	9.4
4. Solid brickwork with dense plaster Brick 105 mm Dense plaster 16 mm	1700 1300	0.84 0.50	800) 1000)	3.0	6.1	2.3	0.83	3.0
5. Solid brickwork with dense plaster Brick 220 mm Dense plaster 16 mm	1700 1300	0.84 0.50	800) 1000)	2.1	6.6	2.1	0.50	6.5
6. Solid brickwork with dense plaster Brick 335 mm Dense plaster 16 mm	1700 1300	0.84 0.50	800) 1000)	1.7	6.5	2.0	0.25	9.9
7. Solid brickwork with lightweight plaster Brick 105 mm Lightweight plaster 16 mm	1700 600	0.84 0.16	800) 1000)	2.5	6.2	2.5	0.81	3.1

TABLE A7 (cont'd)

Description		Density (kg/m^3)	Conductivity (W/m^2 °C)	Specific (J/kg °C)	U value (W/m^2 °C)	Admittance Y (W/m °C)	Admittance ψ_Y (hours)	Decrement d_f	Decrement ψ_f (hour)
8. Solid brickwork, with lightweight plaster					1.9	6.6	2.0	0.44	6.7
Brick	220 mm	1700	0.84	800)					
Lightweight plaster	16 mm	600	0.16	1000)					
9. Solid brickwork with lightweight plaster					1.5	6.5	2.0	0.23	10.0
Brick	335 mm	1700	0.84	800)					
Lightweight plaster	16 mm	600	0.16	1000)					
10. Solid brickwork with plasterboard					1.9	6.6	2.0	0.45	6.7
Brick	220 mm	1700	0.84	800)					
Plasterboard	10 mm	950	0.16	840)					
CAVITY WALLS (UNVENTILATED)									
11. Cavity wall with 105 mm inner and outer brick leaves with dense plaster on inner					1.5	6.5	2.6	0.43	7.8
Brick	105 mm	1700	0.84	800)					
Cavity	>20 mm	resistance =	0.18 m^2 °C/W	)					
Brick	105 mm	1700	0.62	800)					
Dense plaster	16 mm	1300	0.50	1000)					
12. Cavity wall as 11 but with lightweight plaster					1.3	6.5	2.6	0.40	7.9
Brick	105 mm	1700	0.84	800)					
Cavity	>20 mm	resistance =	0.18 m^2 °C/W	)					
Brick	105 mm	1700	0.62	800)					
Lightweight plaster	16 mm	600	0.16	1000)					

Construction	Thickness	Density	Conductivity	Specific heat					
13. Cavity wall as 11 but with 230 mm outer leaf									
Brick	230 mm	1700	0.84	800)	1.2	6.6	2.0	0.20	11.7
Cavity	>20 mm	resistance =	0.18 m^2 °C/W	)					
Brick	105 mm	1700	0.62	800)					
Dense plaster	16 mm	1300	0.50	1000)					
14. Cavity wall as 13 but with lightweight plaster									
Brick	230 mm	1700	0.84	800)	1.1	6.6	2.0	0.18	11.8
Cavity	>20 mm	resistance =	0.18 m^2 °C/W	)					
Brick	105 mm	1700	0.62	800)					
Lightweight plaster	16 mm	600	0.16	1000)					
15. Cavity wall with brick outer and lightweight concrete block inner with dense plaster on inner									
Brick	105 mm	1700	0.84	800)	0.96	6.8	2.7	0.56	7.2
Cavity	>20 mm	resistance =	0.18 m^2 °C/W	)					
Lightweight concrete block	100 mm	600	0.19	1000)					
Dense plaster	16 mm	1300	0.50	1000)					
16. Cast concrete (150 mm) with 60 mm woodwool slabs on inner surface finished with 16 mm dense plaster									
Concrete	150 mm	2100	1.40	840)	1.2	9.1	2.1	0.51	6.4
Woodwool	50 mm	500	0.10	1000)					
Dense plaster	16 mm	1300	0.50	1000)					
17. As 21 but with 200 mm concrete									
Concrete	200 mm	2100	1.40	840)	1.2	8.9	1.8	0.36	7.8
Woodwool	50 mm	500	0.10	1000)					
Dense plaster	16 mm	1300	0.50	1000)					

TABLE A7 (cont'd)

Description	Density	Conductivity	Specific	U value	Admittance		Decrement	
	(kg/m^3)	(W/m^2 ^{o}C)	(J/kg ^{o}C)	(W/m^2 ^{o}C)	Y(W/m ^{o}C)	ψ_Y(hours)	d_f	ψ_f(hour)
18. Pre-cast sandwich consisting of 75 mm dense concrete, 25 mm expanded polystyrene and 150 mm lightweight concrete								
Concrete 75 mm	2100	1.40	840)					
Expanded polystyrene 25 mm	25	0.033	1380)	0.72	7.5	3.4	0.30	9.4
Lightweight concrete 150 mm	1200	0.38	1000)					
ROOF								
19. Asphalt 15 mm on lightweight concrete screed 75 mm on dense concrete 150 mm								
Asphalt 15 mm	1700	0.50	1000)					
Screed 75 mm	1200	0.41	840)	1.9	5.0	2.2	0.36	7.4
Dense concrete 150 mm	2100	1.40	840)					
Dense plaster 15 mm	1300	0.50	1000)					
20. Asphalt 19 mm on 150 mm autoclaved aerated concrete roof-slabs with dense plaster internally								
Asphalt 19 mm	1700	0.50	1000)					
Aerated concrete 150 mm	500	0.16	840)	0.86	3.6	3.8	0.78	4.7
Dense plaster 15 mm	1300	0.50	1000)					

APPENDIX 4

A. SUMMARY OF SOLAR HEATING AND COOLING SIMULATIONS

The following summary table and text describe the most frequently used and currently available solar analysis computer methods.

Program Name	Latest Version	Purchase($)	Availability: Time Share	Availability: Special Arrangements	Comments	User Manual	Application: Service Hot Water	Application: Space Heating	Application: Space Cooling	Application: Process Heat	Application: Active System	Application: Passive System	Intended Users: Research Engineers	Intended Users: Architect/ Engineers	Intended Users: Builders	Computation Interval: Hour	Computation Interval: Month	Computer Versions Available	Economic Analysis	Sponsor
ACCESS*	1978	10,000		•	No cost to EEI members	•	•	•	•	•	•		•	•		•		IBM	•	Edison Electric Institute (EEl)
BLAST*	1978	300	•		Training available	•	•	•	•		•		•	•		•		CDC	•	USAF, USA, GSA
DEROB	1979	Nom.				•		•	•			•	•	•		•		CDC		NSF, ERDA, DOE
DOE-2*	1979	400	•			•	•	•	•	•	•		•	•		•		CDC	•	LASL, DOE
EMPSS	1978	500		•	Consulting with ADL	•	•	•	•		•		•	•		•		IBM	•	EPRI
F-CHART	1978	100	•		Training available	•	•	•			•			•	•		•	CDC, IBM UNIVAC	•	DOE
FREHEAT	1979	150			Limited documentation			•				•	•			•		CDC	•	DOE
HISPER	1978	Avail. on request			Limited documentation		•	•	•		•		•			•		UNIVAC		NASA, MSFC
HUD-RSVP/2	1979	175	•		Based on F-CHART	•	•	•			•	Δ		•	•		•	CDC UNIVAC	•	HUD
SHASP	1978	Avail. on request				•	•	•	•		•		•			•		UNIVAC	•	DOE
SIMSHAC	1973	300					•	•			•		•			•		CDC	•	NSF
SOLCOST	1978	300	•			•	•	•			•			•	•		•	CDC, IBM UNIVAC	•	DOE
SOLOPT	1978	20				•	•	•			•		•				•	AMDAHL	•	Texas A & M Univ.
SOLTES SOLTES	1978 1978				Available Argonne Fall 1979	•				•	•		•			•		CDC		Sandia
SUNCAT SUNCAT	1979	Nom.			Limited documentation	•		•				•	•	•		•		Data General Eclipse	•	NCAT
SUNSYM*	1979		•	•		•	•	•	•		•			•		•		IBM	•	Sunworks Comp. Systems
SYRSOL SYRSOL	1978	Nom.			Avail. but not actively marketed		•	•	•			•	•	•			•	IBM		ERDA, NSF, DOE
TRACE*	1979			•		•	•	•	•	•	•	•	•	•				IBM		The Trane Co.
TRNSYS	1978	200	•		Training available	•	•	•	•		•	Δ	•			•		CDC, IBM UNIVAC		DOE
TWO ZONE	1977			•		•		•	•			•	•			•		CDC	•	LBL
UWENSOL	1978	200				•		•				•	•			•		CDC		State of Wash
WATSUN	1978	175				•	•	•			•		•			•		IBM	•	Nat'l Research Center of Canada

*Programs are primarily developed for large-scale, multi-zone applications

ΔBeing added

ACCESS

The ACCESS (*A*lternate *C*hoice *C*omparison for *E*nergy *S*ystem *S*election) program was designed to provide economic comparisons among the different energy systems which may be used in buildings. ACCESS was developed in 1972 by Seelye, Stevenson, Value, and Knetch consulting engineers under funding by the Edison Electric Institute; the current Version 6 was released in 1978. Its primary purpose is to enable decisions on energy sources rather than to design an HVAC system. The program series consists of the energy analysis program and an independent short financial analysis program. The energy analysis program can analyse up to six separate mechanical/electrical systems including consideration of various lighting schemes, terminal systems, and primary systems in a single computer run. Onsite systems are simulated on a modular basis to facilitate specifications of various combinations of HVAC terminals and primary systems including cogeneration systems. Building loads may be input on an hourly basis, or total building design loads may be input for summer and winter design conditions and steady state hourly loads, then scaled based on the hourly weather conditions. The model simulates active liquid and air type solar heating and cooling systems with stationary as well as tracking collectors. Input consists of building construction data, occupancy schedules, base energy load profiles, heating and cooling load data, description of terminal systems, hourly meteorological data, and selection of the type of output. Output consists of monthly energy usage and demand levels for each input specified meter in the building, with the option of sample calculations for selected days, hours, and zones.

Reference

Douglas, Edwin and George Reeves, "ACCESS: Alternate Choice Comparison for Energy Systems Selection," *ASHRAE Journal*, November 1971, pp. 41-43.

Contact

Edison Electric Institute
Attn: Edwin S. Douglas
90 Park Avenue
New York, NY 10016
(212)573-8773

BLAST

The BLAST (*B*uilding *L*oads *A*nalysis and *S*ystem *T*hermodynamics) program, developed by the US Army Construction Engineering Research Laboratory (CERL) with joint sponsorship by the US Air Force and the US Army, was released for public use in 1978. The program was written to permit analysis and design of energy conservation in new and existing buildings including application of liquid type active solar energy and total energy systems. Many of the methods used are based on the ASHRAE algorithms; however, new algorithms have been included relating to building shading, cooling coil modelling, and room heat balances. The program employs its own English-like input language and can perform two types of analyses: (1) Hourly energy analysis in which actual hourly weather data are used to calculate hourly heating and cooling loads for each zone of the building. Output from the hourly energy analysis is used as input to the system simulation subprogram. (2) Design data analysis in which user-supplied input weather data are used to calculate hourly heating and cooling loads for each zone of the building for the specified design days. Life-cycle costs are computed for each central plant option selected on the basis of user-supplied or default capital costs, maintenance costs, operating costs, and utility rate schedules. Output data provide monthly and daily loads and energy consumption with separate meters for different end uses. Both average and peak day load and energy profiles may be output.

References

The Building Loads Analysis and System Thermodynamics (BLAST) Program, Vol. *I*, *User's Manual*, AD-A048 982/3ST, 1977.

The Building Loads Analysis and System Thermodynamics (BLAST) Program, Vol. *II*, *Reference Manual*, AD-A048 734/8ST, 1977.

Contact

US Army Construction Engineering Research Lab.
Attn: Douglas C. Hittle
P.O. Box 4005
Champaign, IL 61820
(217)352-6511

DEROB

The DEROB (*D*ynamic *E*nergy *R*esponse *O*f *B*uildings) programs have been under development since 1972 at the University of Texas as a research and design tool relating to heat flow in complex structures. In its current release, 1979, it is capable of simulation of a variety of passive solar system designs. The series consists of three main programs: (1) development of building geometric factors relating to radiation interchanges, (2) development of material properties for elements of the structures, and (3) calculations of hourly loads and interior distributions. Interior heat flow takes place by air circulation, conduction through opaque walls, and direct radiation via transparent walls. Interior temperature control is by passive means only. Input data consist of building geometry data, thermophysical property data, use factors, and hourly SOLMET weather data. Output data for daily, monthly, and peak day load profiles are available.

Reference

Arumi, F. and D. Northrup, A Field Validation of the Thermal Performance of a Passively Heated Building as Simulated by the DEROB System. *Energy and Buildings*, Vol. 2, No. 1, January 1979, p. 65.

Contact

University of Texas
Attn: Francisco Arumi
2604 Parkview Drive
Austin, TX 78757

DOE-2 SOLAR SIMULATOR

An active solar system simulator package, designed as a stand-alone program or as a module of the DOE-2 building energy analysis program, was developed at Los Alamos Scientific Laboratories with release in early 1979. A component base approach is used whereby the user defines which components are present and the manner in which they are connected. The component library contains liquid and air collector systems (stationary or tracking), multiple storage units, and an absorption chiller. Several preconnected liquid or air solar heating systems are also included in the program library. As a stand-alone program it is applicable to residential building analysis; commercial building systems can be evaluated when the program is used as part of DOE-2. Features of the input processor include free-formatting of input data, defaults, and scaling of certain input variables by other inputs. Hourly weather data are user supplied in DOE-2 weather format. Several insolation models are available for both measured and calculated radiation. Octput options allow the user to select predefined reports or to define reports or plots according to special requirements.

References

Hunn, B. D. *et al.*, *CAL/ERDA Program Manual*, Lawrence Berkeley Lab ANG/ENG-77-04, October 1977.

Roschke, M. A., B. D. Hunn, and S. C. Diamond. Component-Based Simulator for Solar Systems. LA-UR-78-1494, Proceedings of Systems Simulation and Economic Analysis for Solar Heating and Cooling Conference, San Diego, Calif., June 1978.

Contact

Los Alamos Scientific Laboratory
Attn: March A. Roschke
Mail Stop 985
Los Alamos, NM 87545
(505)667-3348

EMPSS

EMPSS (*E*PRI *M*ethodology for *P*referred *S*olar *S*ystems) is a solar simulation program, developed by Arthur D. Little, Inc. in 1977 under funding by the Electric Power Research Institute, to provide a basis on which to select a residential heating and cooling system which minimizes total life-cycle costs associated with owning and operating the equipment and with supplying electrical energy to meet the building's energy requirements. The program is capable of simulating on an hourly basis conventional, active solar, load management, or combinations of these systems. Inputs to the model consist of a weather tape; parameters describing the thermal characteristics of the building and its occupants; data related to the utility's cost of electrical supply, rate schedules, and system-wide loads; operating characteristics of the HVAC system and its control logic; and economic indices such as equipment life, mortgage rates, inflation rates, etc. The program user is permitted latitude in describing the house and choosing the HVAC system. Building loads are based on finite difference coupling equations. Both the electrical energy use and energy costs are output in monthly and annual summaries. Energy use data for the peak residence day of the month and the utility's peak generation day of the month are also output. The output summaries give a description of the technical performance of the house and its HVAC system and break down the cost by the various utility cost components.

Reference

EPRI Methodology for Preferred Solar Systems (EMPSS) Computer Program Documentation, *User's Guide*, Arthur D. Little, Inc., EPRI ER-771, May 1978.

Contacts

Arthur D. Little, Inc.
Attn: Dan Nathanson
20 Acorn Park
Cambridge, MA 021 40
(617)854-5770

EPRI
Attn: James E. Beck
3412 Hillview Ave.
Palo Alto, CA 94304
(415)855-2000

F-CHART

A computer tool for estimating the long-term thermal performance and life-cycle cost of active solar heating systems, F-CHART was developed at the University of Wisconsin Solar Energy Laboratory in 1976 under funding by the Department of Energy. Its most current version (3.0) was released in June 1978. The F-CHART method, which models residential liquid or air solar water heating or combined solar water and space heating systems (solar cooling or heat pump systems cannot be modelled), is based on correlations of detailed computer simulations (using the TRNSYS model) in the form of "f-charts" showing the fraction of the total monthly loads carried by solar energy as a function of two dimensionless parameters (the ratio of available solar energy to the load and the ratio of a reference collector heat loss to the load). Weather data for 266 cities are built into the program with the ability offered to the user to input one additional set of weather data at time of use. The program is interactive and requires the following general input data; Collector design parameters, storage capacity, building UA and internal heat generation, domestic water temperature and use, and various economic parameters. A list of default values is provided. Output data describe the monthly thermal loads, weather parameters and solar fraction and annual cash flow, corresponding to the life-cycle cost. Calculations may be made for a specifid collector area or for an economic optimum collector area determined by the program.

References

F-CHART User's Manual, An Interactive Program for Designing Solar Heating Systems, EES Report 49-3, Solar Energy Laboratory, University of Wisconsin, June 1978.

Beckman, W. A., S. A. Klein, and J. A. Duffie, *Solar Heating Design by the F-Chart Method*, Wiley-Interscience, New York, 1977.

Contacts

Design Tool Manager
Market Development Branch
Solar Energy Research Institute
1536 Cole Boulevard
Golden, CO 80401
(303)231-1261

A version for programmable calculators is available for $150 from:
Sandy Kelin
F-CHART
P.O. Box 5562
Madison, WI 53705

FREHEAT

FREHEAT was developed at Colorado State University (CSU) under funding from the Department of Energy, with its original release in 1978 and its current release in 1979. The program is intended as a research tool for the design of a passive solar heating system at CSU. A one-room system may be modelled with a simple mass wall or with an isothermal mass wall. Heat capacity may be added to the room and floor, or to the floor only. Direct gain systems may also be simulated, with the addition of a mass floor (either single node or multi-node), or an interior mass wall. Inputs to the system include hourly weather data, type of system, and thermodynamic properties. The outputs provided by the program are hourly room temperature, node temperatures, solar fraction, energy incident on wall, energy stored in wall, auxiliary energy required and standard deviations in temperature, both daily and for the study period.

Contact

Colorado State University
Attn: Dr. C. Byron Winn
Mechanical Engineering Department
Fort Collins, CO 80523
(303)491-6783

HISPER

HISPER (*Hi*gh *S*peed *Per*formance) was developed by NASA's Marshall Space Flight Center under funding from DOE and became public in 1978. The program simulates the operation of an active solar heating and cooling system in response to loads which are based upon measured climatic data. Loads calculations are limited to single family or small commercial dwellings with attic and floor losses considered. The dwelling heat build-up from incident wall and roof radiation is accounted for in two ways: (1) the ASHRAE Sol-Air temperature method and (2) limiting ambient temperatures above or below which no heating and cooling are required (by assumptions relating to the thermostat deadband). Dwelling orientation is limited to north-south and east-west. Input data for the loads calculations are heat transfer loss coefficients for the building walls, ceiling and roof; building internally generated heat; and infiltration rates. Domestic hot water loads are taken from an interval 24-hour use table totaling 60 gallons per day. A flat-plate collector is modelled using either liquid or air as the heat transfer medium. In the liquid, the storage tank is assumed non-stratified. In the air system, rock bed storage is modelled after the NTU model of Hughes, Klein, and Close with axial conduction added. The program is available but has limited documentation.

Reference

Gray, J. C. Solar Heating and Cooling High-Speed Performance Program-HISPER, NASA/TN-240-1661, July 1976.

Contact

National Aeronautics and Space Administration
Attn: Robert Elkin
George C. Marshall Space Flight Center
Huntsville, Al 35812
(205)453-1757

RSVP/2

RSVP (*R*esidential *S*olar *V*iability *P*rogram) is an interactive computer program developed in 1977 and updated in 1979 by Booz, Allen and Hamilton, Inc. under contract with the US Department of Housing and Urban Development (HUD). The program combines a simplified solar system performance model, which is taken from the latest version of F-CHART, with a comprehensive economic analysis model for residential solar applications. System performance is calculated on a monthly average basis and includes flat-plate air and water systems for hot water and space heating. The data base contains weather data for 266 cities, including cooling degree days for use in estimating conventional air-conditioning loads. Auxiliary systems modelled include electric furnace, electric baseboard, heat pump, and gas and oil furnaces. Outputs include thermal system performance, building loads, and a variety of economic reports. The conversational interface is designed for a variety of users by allowing access to the program inputs at various levels of detail and providing default values and data checks for most input variables. It contains features such as report selection and an iteration routine that facilitates sensitivity analysis. The economic routines include an income property analysis.

References

HUD Residential Solar Economic Performance Model: User's Manual, US Department of Commerce, PB-273-199, June 1977.

HUD Residential Solar Economic Performance Model: Programmer's Manual, US Department of Commerce, PB-273-200, June 1977.

Contact

RSVP
National Solar Heating and Cooling Information Center
P.O. Box 1607
Rockville, MD 20850
(202)223-8105

SHASP

SHASP (*S*olar *H*eating and *A*ir-Conditioning *S*imulation *P*rograms) was developed within the Mechanical Engineering Department at the University of Maryland with funds provided by the Department of Energy. Originally released in 1977, SHASP consists of a set of eight programs designed to simulate the performance of solar heating, cooling, and hot water systems, subject to a variety of operational control strategies using real or stochastic weather data. Although the system configurations are built into the program, the performance characteristics of the components can be specified by the user via performance curves. Liquid, flat-plate and evacuated tube type collectors can be simulated as can solar cooling derived from an absorption machine or a Rankine cycle heat engine vapour compression system. Although the program has to date been primarily used by and for researchers, the user's manual and source code are available at no cost by contacting the authors.

Reference

SHASP Solar Heating and A/C Simulation Program, Solar Energy Project Report. Department of Mechanical Engineering, University of Maryland, December 1978.

Contact

Department of Mechanical Engineering
Attn: Dr. D. K. Anand
University of Maryland
College Park, MD 20742
(301)454-2411

SIMSHAC

SIMSHAC (*Si*mulation *M*odel for *S*olar *H*eated *A*nd *C*ooled Buildings) was developed and released in 1973 by Colorado State University under a National Science Foundation grant. The program was written primarily for control system studies associated with the development of the CSU-I solar demonstration house. The model contains a solar irradiation program, an implicit finite difference thermal analyser program to calculate house heating loads, and a system simulation program based on the components (modular) approach. Inputs to the model include specification of the components that comprise the system and how they are connected, initial conditions, data relating to the building configuration, and weather data. Four different methods of modelling incident solar radiation are included in the program. Output data consist of daily, monthly, and annual building loads, wnergy and demand rate schedules.

Reference

Winn, C. B., G. R. Johnson, and T. E. Corder, SIMSHAC - A Simulation Program for Solar Heating and Cooling of Buildings, *Simulation*, Vol. *23*, No. 6, 1974.

Contact

Mechanical Engineering Department
Attn: Dr. C. Byron Winn
Colorado State University
Fort Collins, CO 80523
(303)491-6783

SOLCOST

This program provides life-cycle cost estimates of various residential solar systems. It was developed by Martin Marietta Aerospace Corp. in 1978 under contract to ERDA. The current (1978) version is maintained by International Business Services, Inc. No detailed, hour-by-hour performance simulations are possible with SOLCOST. Statistical averages of weather data in conjunction with gross solar system performance characteristics (both from the program's data base) are used first to calculate heating and cooling loads for the specific residence and then to choose an optimum collector area and storage volume. The optimum system size is determined by the lowest life-cycle cost (compared with a conventional system) as derived from the user's choice of financial indices and type of solar system. Program output describes the optimum system in terms of collector area and storage volume and further details its economic characteristics in terms of life-cycle cost, payback period, etc. Sizing of the solar system is based on calculations for a single day in each of the twelve months of the year. Weather data consist of historical averages of maximum and minimum temperatures, degree-days, and percentage of sunshine. Heating and hot water loads for the residence may be user-specified or internally calculated based upon several simplified methods. Collector types which may be considered are stationary, flat-plate, air or liquid models in addition to the evacuated tube and single-axis tracking variations. Financial analysis incorporates the owner's tax structure and initial outlay for the solar system when computing a life-cycle cost based on constant fuel or electric (energy/demand) rates during each year with provision made for cost escalation over the equipment life.

Reference

SOLCOST: Space Heating Handbook with Service Hot Water and Heat Loads Calculations. DOE/CS-0042/3, July 1978.

Contacts

International Business Services, Inc.
1010 Vermont Avenue, NW
Suite 1010
Washington DC 20005
(202)628-1470

Martin Marietta Aerospace
Attn: Rojer Giellis
Mail Stop 50484
P.O. Box 179
Denver, CO 80201
(303)973-3853

SOLOPT

This program was developed at and sponsored by Texas A & M University. After its original release in 1978, a later version of SOLOPT appeared in the same year. Although the program is available to the public, its primary use is as a life-cycle cost optimization tool for Texas A & M students taking a solar systems design course. A variety of liquid circulant solar collectors with liquid storage medium may be simulated with this program. Flat-plate, evacuated tube, and focusing collector models are provided internally in the program. They may be of the stationary or hourly tracking type, and the user need only specify gross collector performance parameters in order to obtain annual performance predictions. Hourly calculations for the collector are performed over the entire year using weather data that need only be resolved on a monthly basis. No building heating or hot water loads are evaluated by the program and must be provided by the user as input data. Simulation of conventional HVAC systems as auxiliaries to the solar collector is limited to forced air furnaces, either gas, oil, or electric, and to an electric heat pump. Output from the program includes daily, monthly, and annual energy use provided by the solar system and by the HVAC auxiliaries. Life-cycle cost of the solar system is based upon uniform electric rates (energy only) and single metering for the residence.

Reference

Degelman, L. User's Guide to SOLOPT, Mod 2: A Computer Program for Optimization of Flat-Plate Solar Collector Rreas, Department of Agriculture, Texas A & M University, 1978.

Contact

Department of Agriculture
Attn: Larry O. Degelman
Texas A & M University
College Station, TX 77843
(713)845-1015

SOLTES

The original release of SOLTES in 1978 signalled the debut of the first general purpose computer program for simulation of industrial process heat operations and thermal electric systems. It was developed by Sandia Laboratories under contract to the Department of Energy and is available to the public through Argonne National Laboratory. Since the program was not intended for solar heating and cooling analyses, it cannot calculate building space conditioning loads. The hourly weather data required as part of program input are used soley for simulation of the solar collectors and storage and not for determining comfort requirements. Active solar systems which the program can handle include stationary flat-plate or evacuated tube collectors, with reflectors if desired. Air or liquid coolants are permitted as are seasonal adjustments or tracking mechanisms for the collectors. Several storage models are available including liquid and rock media as well as phase change and chemical storage materials. The process heat systems can be elaborate if desired since the components and configuration as well as their performance characteristics may be specified by the user. Program output for the process heat simulation includes daily, monthly, and annual load summaries, broken down by component if desired. Energy use by the nonsolar auxiliary systems is also included. Economic issues are not addressed by the program.

Reference

Edenburn, M. W. and N. R. Grandjean, Energy System Simulation Computer Program - SOLSYS, SAND 75-0048, June 1975.

Contact

Sandia Laboratory
Attn: Norman Grandjean
Kirtland East
Div. 4722
Albuquerque, NM 87185
(505264-8819

SUNCAT

The SUNCAT program was developed by the National Center for Appropriate Technology (NCAT) to simulate the thermal performance of a passive solar structure. The program consists of two mainline programs which simulate the transmissivity and shading of a user-specified glazing system and the thermal performance of a user-specified passive solar structure. Input to the program consists of hourly radiation and temperature data files and a window and building description. The present version of the program allows for interactive creation of the window and building descriptions. The output of the program is a tabulation by zone and total building of the thermal characteristics of the passive system, including solar fraction and the useful heat delivered.

Contact

National Center for Appropriate Technology
Attn: Larry Palmiter or Terry Wheeling
P.O. Box 3838
Butte, MT 59701
(406)723-5475

SUNSYM

The SUNSYM program was developed privately by Sunworks Computer Services for primary use in designing solar collectors. Two versions of the program exist: SUNSYM1, based on the general theory of flat-plate collector performance, and SUNSYM2, based on the performance equation and incident angle modifier as determined by ASHRAE 93-77 testing SUNSYM2 is particularly useful in making performance comparisons between collector manufacturers. SUNSYM1 is useful when it becomes necessary to vary a collector's design parameters. There is also a version of the program which determines the extent of shading for adjacent rows of collectors.

Contact

Sunworks Computer Services
Attn: Philip Fine
P.O. Box 1004
New Haven, CT 06508
(203)934-6301

SYRSOL

The SYRSOL (*Syr*acuse University *Sol*ar Building Energy Analysis) program for simulating the hourly thermal performance of a multizone building with a solar assisted series of commercially available water-to-air heat pumps in a closed loop was developed at Syracuse University in 1976 under funding by ERDA and NSF. The solar system consists of commercially available flat-plate collectors, a water storage tank, and a cooling tower integrated in a fixed manner with the heat

heat pump system. Building thermal loads include heat transmission through walls based on the Sol-Air method, ventilation, and internal heat inputs from lighting and building occupants, with flexible hourly building use schedules permitted. Sensible heat recovery from ventilation air and heat pump domestic water heating may be simulated. Generalized weather functions are built into the program for 13 cities, consisting of sinusoidal representations of daily dry-bulb and wet-bulb temperatures and monthly averaged "cloud cover modifiers" to estimate the incident solar flux. The output data provide daily, monthly, and annual building loads and energy uses by zone with energy costs by energy rates.

Reference

Ucar, M. *et al.*, Solar Building Energy Use Analysis. *Proceedings*, Section 6, Session D-4, American Section, ISES Conference, Winnipeg, 1977.

Contact

Department of Mechanical Engineering
Attn: Dr. Manas Ucar
Syracuse University
Syracuse, NY 13210
(315)423-3038

TRACE

The first version of TRACE, (*Tr*ane *A*ir *C*onditioning *E*conomics), developed by the Trane Company, was introduced in 1973 and intended as an aid to architects and engineers in comparing life-cycle costs of various HVAC systems for multizone structures. A third version of the program, released in 1978, incorporates a solar capability based on TRNSYS. Using the ASHRAE (1967) algorithms for calculating heat gains and losses by zone, the load phase prepares input for the design segment of the program. The system simulation phase translates design conditions into equipment loads taking into account part load operation of the HVAC systems. These equipment loads are converted to fuel and electrical usage rates in the equipment simulation phase where as many as four distinct (Trane manufacture) equipment alternatives may be considered simultaneously. Input for the program consists of a building description and operating schedules, design conditions for various zones, equipment selections, and various economic data. Weather data for a number of cities are included in TRACE's data base in the form of a "typical" 24-hour profile for each of the twelve months. The program output contains design loads for the HVAC equipment, monthly fuel and electrical usage by component, peak hourly loads, and economic comparisons of the selected equipment options. TRACE SOLAR can accommodate a variety of energy/demand electric rates. Its solar capability extends to 20 user-selected solar system configurations involving air or liquid, stationary, flat-plate collectors.

Contact

The Trane Company
Attn: Neil R. Patterson
3600 Pammel Creek Road
LaCrosse, WI 54601
(608)782-8000

TRNSYS

A program for simulating the dynamic thermal behaviour of active solar heating and cooling systems. TRNSYS was developed at the University of Wisconsin Solar Energy Laboratory in 1974 under funding from ERDA (DOE) and has been continually upgraded and maintained with its most current version (9.2) released in 1978. It is based on a modular approach whereby the user specifies which components comprise the system and the manner in which they are connected: the only system arrangement built into the model is a solar domestic water heater. The components models, which are formulated as separate FORTRAN subroutines, include collectors, controls, storage tanks, heat exchangers, furnaces, building loads, an integration procedure, etc., and are linked by an executive routine. The subroutine library allows simulation of both air and liquid type systems including stratified thermal storage units. The program also facilitates the introduction of additional user-supplied component subroutines. Inputs and outputs from the component subroutines correspond to inputs and outputs of the modelled hardware. Building heating and cooling loads may be calculated by the degree-day method, by using input response factors, or may be input from another model. Card image records of hourly weather data are required. Output data include heating and cooling loads, energy use by various system components, auxiliary energies, and overall system thermal energy balances for error checking. Plotting routines are included in the subroutine library.

References

TRNSYS: A Transient Simulation Program - Report No. 38, Solar Energy Laboratory, University of Wisconsin, February 1978.

Klein, S. A., W. A. Beckman, and J. A. Duffie, TRNSYS - A Transient Simulation Program, *ASHRAE Transactions*, Vol. *82*, Part 1, 1976.

Contact

Solar Energy Laboratory
Attn: John C. Mitchell
University of Wisconsin
1500 Johnson Drive
Madison, WI 53706
(608)263-1589

TWO-ZONE

Motivation for the development of this program in 1975 stemmed from a desire to analyse the collection efficiencies of windows in a residence. The program computes the hourly heating and cooling loads for the north and south zones ("two-zones") of a single-family dwelling and provides for thermal coupling between them through convective air exchange. Hourly weather data are used in the calculation in addition to various active controls such as thermostat setback, scheduled shade or curtain closings, and window opening for ventilation. In addition to these passive solar capabilities, the current program version (1977) provides for simulation of evaporative coolers and for some economic analysis. Program input consists of the hourly weather data, a description of the internal loads (appliances, occupants, etc.), schedules for active controls, and calculated transfer functions for the dynamic thermal response of the walls and roof. Output from the simulation includes monthly summaries of the total heating and cooling loads for the residence as well as a break down by contribution (windows, walls, infiltration, etc.). HVAC equipment simulation is limited to the evaporative cooler and a warm air furnace with a fixed efficiency. No active solar systems are simulated by TWO-ZONE. Energy use by the auxiliary systems is reported in the output, and their life-cycle costs, based upon constant annual fuel or electric (energy/demand) rates, are included if desired.

Reference

A. G. Gadgil, G. Gibson, and A. H. Rosenfeld. *User's Manual*. Lawrence Berkeley Laboratory. Report LBL-6840, March 1978.

Contact

University of California
Attn: Arthur H. Rosenfeld
Lawrence Berkeley Laboratory, Building 90
Berkeley, CA 94720
(415)843-2740, Ext. 5711

UWENSOL

The State of Washington's plans for construction of a series of buildings prompted funding of a team of architects and engineers at the University of Washington (Seattle) to develop energy conservation guidelines for those buildings. These guidelines were embodied in a pair of computer programs released in 1978 which were designed to be used in tandem to minimize the lighting and comfort maintenance requirements of a multizone structure. Reductions in these requirements are achieved by changing the building's architecture and fenestration to achieve passive control and augmentation of the two functions. Changes are directed at reducing the heating and cooling loads which would need to be met by an HVAC system selected for the building. Architectural changes which reduce these loads are feasible so long as the lighting levels and thermal insulation values stipulated in building codes are met, and there is an acceptable thermal comfort level maintained under all weather conditions. The lighting program UWLIGHT first determines the lighting distribution in the building and the steady-state heat transmission coefficients of its envelope to ensure code compliance. UWENSOL then performs a dynamic hour-by-hour simulation of the building using a finite-difference approximation of the zonal interactions and calculated the transient heating and cooling load requirements for the HVAC system. Architectural changes are then made to reduce these loads, and the attractiveness of the modifications is later checked by evaluating a thermal comfort level for each zone. Electrical load profiles or fuel requirements for the comfort control system are not determined by this program, and no active solar system (collectors) can be simulated. Input requirements for the program emphasize the passive aspects of the structure since its principal concern is with the building's reaction to solar input. Hourly data for ambient temperature and humidity, direct and diffuse solar radiation, etc., must be supplied by the user. Program output includes daily heating and cooling loads for the various thermal zones as well as average and peak day hourly load profiles.

Reference

Emery, A. F. *et al.* UWENSOL: A Computer Program for the Dynamic Simulation of the Thermal Response of Buildings with Emphasis on Human Comfort. University of Washington, 1978.

Contact

Department of Mechanical Engineering, FU-10
Attn: A. F. Emery
University of Washington
Seattle, WA 98195
(206)543-5338

WATSUN

A second version of this program (WATSUN-II) was made available in 1978 following release of the original in 1977. Both were developed by the University of Waterloo (Ontario) under contract to the National Research Council of Canada as part of NRCC's continuing research in solar water heating and space heating.

WATSUN is intended to simulate the operation of an active solar system and storage in response to specified weather conditions and the heating load profile of a residence. As such, the program does not have the capability to calculate the heating or hot water load for the structure, so these hourly load profiles must be supplied to the program as input data. WATSUN will then simulate the hourly performance of the solar collector and storage and the auxiliary (conventional) systems in meeting the designated load profiles under user-specified weather conditions. Either flat-plate or evacuated tube stationary collectors may be simulated with suitable choice of gross performance indices for either liquid or air circulants, or a detailed collector model may be specified where the standard performance indices are inadequate. The program will also handle a variety of storage systems including rock, liquid, phase change, stratified, etc. Auxiliary systems include an instantaneous-recovery water heater, electric or fossil-fueled furnaces, and an electric heat pump. Life-cycle costs of operation of the solar equipment may be calculated and are based upon a uniform electric rate (energy only).

Reference

Orgill, J. F. and K. G. T. Hollands. A Solar Heating Simulation and Economic Evaluation Program: User's Manual, University of Waterloo, Canada, April 1976. (Available from US DOE Technical Information Center, P.O. Box 62, Oak Ridge, TN 37830).

Contact

Waterloo Research Institute
Attn: W. Blair Bruce
University of Waterloo
Waterloo, Ontario, Canada N2L-3G1
885-1211, Ext. 3189

B. SHAC MANUAL DESIGN METHODS SUMMARY

The following table describes solar heating and cooling manual design methods. This table does not give all of the design methods applicable to SHAC analysis, but it does contain the most currently used and best known methods. These methods do not require access to a computer although some (e.g. F-CHART) have been implemented on computers. They vary in degree of sophistication from the simple, almost rule-of-thumb type to methods requiring programmable calculators. Some of the latter type methods are available from the source indicated as prerecorded programs on magnetic cards.

Description	Author	Availability			Application				Coil. Type		System Type		Tools Required				Basis of Method			Output		
		Cost ($)	Date	Reference/Source	Space Heat	Domestic Hot Water	Space Cooling	Economic Analysis	Liquid	Air	Active	Passive	Graphs/Tables	4 Function Calculator	Scientific Calculator	Programmable Calculator	Detailed Simulation	F-Chart Correlation	Other Correlation	Solar Fraction	Optimum Area	Other
A simplified Method for Calculating Solar Collector Array Size for Space Heating	J.D. Balcomb and J.C. Hedstrom		1976	Sharing the Sun: Solar Technology in the Seventies, Vol. 4, American Section, International Solar Energy Society, 1976, pp. 281-284	•	•			•	•	•		•	•					•		•	
A Simple Empirical Method for Estimating the Performance of a Passive Solar Heated Building of the Thermal Storage Wall Type	J.D. Balcomb and R.D. Farland		1978	Rept. No. LA-UR-78-1159, Available from NTIS, 5285 Port Royal Road, Springfield, VA 22161	•							•	•	•					•	•		
Optimal Sizing of Solar Collectors by the Method of Relative Areas	C.D. Barley and C.B. Winn		1978	Solar Energy, Vol. 21, No. 4, 1978, pp. 279-289	•	•		•	•	•	•		•		•			•			•	
MESH	Dr. John Clark		1978	Dr. John Clark Central Solar Energy Research Corp., 1200 6th Street, Room 328, Detroit, MI 48226	•			•	•	•	•				•				•	•	•	
Predicting the Performance of Solar Energy Systems	U.S. Army Construction Engineering Research Lab.		1977	Rept. No. AD-A035 608/9 ST (NTIS)	•	•	•	•	•	•	•		•	•					•	•	•	
Copper Brass Bronze Design Handbook - Solar Energy Systems	Copper Development Association	3	1978	Copper Development Association, Inc. 1011 High Ridge Road Stamford, CN 06905	•	•			•	•	•		•	•					•	•		
PEGFIX and PEGFLOAT	W. Glennie	75 both	1978	Princeton Energy Group 729 Alexander Road Princeton, NJ 08540	•							•				•	•					•
Solarcon Programs ST355 and ST365	R.W. Graeii	239 both 142 each 15 weather data	1977	Solarcon, Inc. 607 Church Street Ann Arbor, MI 48104	•	•		•	•	•	•					•		•		•		

Solarcon Program ST33	R.W. Graeii	138	1979	Solaron, Inc.	•							•				•	•			•		
Solar Heating Systems Design Manual	ITT Corp. Fluid Handling Division	2.50	1977	Bulletin TESE-576, Rev. 1 ITT Training & Education Dept. Fluid Handling Division Morton Grove, IL 60053	•	•			•	•	•		•	•			•			•		
A General Design Method for Closed-Loop Solar Energy Systems	S.A. Klein and W.A. Beckman		1977	Proceedings of the 1977 Annual Meeting, Vol. 1, American Section, International Solar Energy Society, 1977, pp. 8.1-8.5			•		•		•		•		•			•		•		
Solar Heating Design by the F-Chart Method	S.A. Klein, W.A. Beckman and J.A. Duffie	10	1977	John Wiley and Sons, New York, NY, 1977 (Publisher)	•	•		•	•	•	•		•	•				•		•		
A Design Procedure for Solar Heating Systems	S.A. Klein, W.A. Beckman and J.A. Duffie		1976	Solar Energy, Vol. 18, No. 2, 1976 pp. 113-127.	•	•			•				•		•				•	•		
TEANET	J.T. Kohier and P.W. Sullivan	95	1978	Total Environmental Action Inc. Church Hill Harrisville, NH 04350	•							•				•	•					•
The GFL Method for Sizing Solar Energy Space and Water Heating Systems	G.F. Lameiro and P. Bendt		1978	Rept. No. SERI-30 Solar Energy Research Institute 1536 Cole Boulevard Golden, CO 80401	•				•	•	•		•		•			•		•		
A Design Handbook for Direct Heat Transfer Passive Solar Systems	R.M. Lebens	10	1978	Northeast Solar Energy Association P.O. Box 541, 22 High Street Brattleboro, VT 05301	•			•				•	•			•	•			•		
A National Procedure for Predicting the Long-Term Average Performance of Flat-Plate Solar Energy Collectors	B.Y.H. Liu and R.C. Jordan		1963	Solar Energy, Vol. 7, No. 2, 1963, pp. 53-70	•	•			•	•	•		•		•		•			•		
Pacific Regional Solar Heating Handbook	Los Alamos Scientific Lab.		1976	Rept. No. TID-27630 (NTIS)	•	•			•	•	•		•		•				•	•		
Prediction of the Monthly and Annual Performance of Solar Heating Systems	P.J. Lunde		1977	Solar Energy, Vol. 20, No. 3, 1977, pp. 283-287	•	•			•		•		•	•					•	•		
SCOTCH Program	R. McClintock	195 Thermal alone, 175; econ. anal. alone, 75	1977	SCOTCH Programs P.O. Box 430734 Miami, FL 33143	•	•		•	•	•	•					•		•		•	•	
Solar Heating of Buildings and Domestic Hot Water	Naval Facilities, Engineering Com. E.J. Beck, Jr. and R.L. Field		1976	Rept. No. AD-A021 862/8ST (NTIS)	•	•		•	•	•	•		•		•			•		•	•	
PCTS	J. Schoenfelder	F-CHART Therm 35 F-CHART Econ. 35	1978	Central States Research Corp. P.O. Box 2623 Iowa City, IA 52240	•	•		•	•	•	•					•		•		•	•	

Domestic Hot Water Manual Using Sunearth Solar Collector Systems	Sunearth Corp.	3	1976	Sunearth Solar Products Corp. Technical Services R.D. 1 P.O. Box 337 Green Lane, PA 18054		•			•		•		•					•		•		
An Averaging Technique for Predicting the Performance of a Solar Energy Collector System	G.H. Stickford		1976	Sharing the Sun, Vol. 4, 1976, pp. 295-315	•	•			•	•	•	•	•		•		•			•		
Calculation of Long-Term Solar Collector Heating System Performance	S.R. Swanson and R.F. Boem		1977	Solar Energy, Vol. 19, No. 2, 1977, pp. 129-136	•				•	•	•		•		•			•		•		
Minimum Cost Sizing of Solar Heating Systems	J.C. Ward		1976	Sharing the Sun, Vol. 4, 1976, pp. 336-348	•			•	•	•			•		•			•		•	•	
Designing and Building a Solar House; Your Place in the Sun	D. Watson	9	1977	Garden Way Publishing Charlotte, VT 05445	•	•		•	•	•	•		•	•			•			•		
SEECI - Heat Load, Monthly Solar Fraction Economics	C.B. Winn	125	1976	Solar Environmental Engineering Co. Inc. 2524 East Vine Drive Fort Collins, CO 80522	•	•		•	•	•	•					•		•		•		
SEECII - Collector Optimization, Annual Solar Fraction, Economics	C.B. Winn, D. Barley, G. Johnson, J. Leflar	95	1978	Solar Environmental Engineering Co. Inc.	•	•	•	•	•	•	•					•		•			•	
SEECIII - SEECII Plus Insulation Optimization	C.B. Winn, D. Barley, G. Johnson J. Leflar	125	1978	Solar Environmental Engineering Co. Inc.	•	•		•	•	•	•					•		•			•	
SEECVI - Passive Solar Heating	C.B. Winn, D. Barley, G. Johnson, J. Leflar	125	1978	Solar Environmental Engineering Co. Inc.	•			•				•				•			•	•		
Sunshine Power Programs for Modeling Solar Energy Components and Systems	G. Shramek	30-60	1977	Sunshine Power Co 1018 Lancer Drive San Jose, CA 95129	•	•		•	•	•	•	•				•	•	•	•	•		•

C. HVAC COMPUTER PROGRAMS SUMMARY

The following table lists programs intended primarily for building heating and cooling load analysis. Some provide for solar analysis, but in such cases it is secondary to the conventional energy analysis. The programs have been generally developed and maintained by heating, ventilating, and air-conditioning (HVAC) consulting engineers for their own analysis use; however, most are available through special arrangements with the contract.

Program Name	Sponsor	Author	Contact	Original Release	Current Version	Application: Building Load	Application: HVAC System	Application: Active Solar	Application: Passive Solar	Availability: Source Cost ($)	Availability: Time Share	Availability: Special Arrangements
ECUBE III	American Gas Association (AGA)	Subcontractors	American Gas Association 1515 Wilson Blvd. Arlington, VA 22209 (203)841-8400 David S. Wood		1979	•	•		•	No info	•	•
ENERGY 1	Gibson-Yackey-Trindade Assoc.	Robert Gibson	Gibson-Yackey-Trindade Associates 311 Fulton Ave. Sacramento, CA 95811 (916)483-4369 Robert Gibson	1974	1974	•	•	•		Not Avail.	N/A	•
EP	Energy Management Services (EMS)	EMS	Energy Management Services 0435 SW Iowa Portland, OR 97201 (503)244-3613 Robert M. Helm	1976	1978	•	•	•	•	Not Avail.	•	•
ESAS	Ross F. Merlwether & Associates, Inc.	Ross F. Merlwether	Ross F. Merlwether & Associates 1600 NE Loop 410 San Antonio, TX 78209 (512)824-5302	1969	1978	•	•	•	•	Not avail.	N/A	•

ESP-1	Automatic Procedures for Engineering Consultants (APEC)	Stone and Webster	APEC, Inc. Executive Off. Grant-Doneau Tower Suite M-15 40th & Ludlow Streets Dayton, OH 45402 (513)228-2602 Doris Wallace	1977	1978	●	●	●		6,000 no cost to APEC members	N/A	●
HACE	William Tao & Associates (WTA)	WTA/Computer Services, Inc.	WTA/Computer Services Inc. 2357 59th Street St. Louis, MO 63110 (314)644-1400 Richard Lampe	1970	1978	●	●	●	●	Not avail.	●	●
NECAP	NASA	NASA	NASA, Langley Research Center Mail Stop 453 Hampton, VA 23665 (804)827-4641 Ron Jensen	1974	1975	●	●	●			N/A	N/A
SCOUT	Guard, Inc.	Guard, Inc.	Guard, Inc. 7440 N. Natchez Ave. Niles, IL 60648 (312)647-9000 Robert Henninger	1976	1978	●	●		●	See contact	N/A	●
SEE	The Singer Co. (NSF Grant)	W.S. Fleming & Assoc. Inc. The Singer Co.	The Singer Co. Climate Control Division 62 Columbus St. Auburn, NY 13201 (315)253-2771, X391 Philip Parkman	1975	1977	●	●	●	●	Not avail.	N/A	●
Westinghouse Programs	Westinghouse Electric Corp.	Westinghouse Electric Corp.	Westinghouse Electric Corp. Energy Systems Analysis 2040 Ardmore Blvd. Pittsburgh, PA 15221 (412)256-3168	1964	1978	●	●	●	●	Not avail.	N/A	●

N/A = Not applicable

APPENDIX 5

SYSTEM IDENTIFICATIONS

Water Wall Systems

Designation	Thermal Storage Capacity* kJ/m^{o}C	Wall Thickness mm	No. of glazings	Wall Surface	Night Insulation
A1	331	76.2	2	Normal	no
A2	661	152.0	2	Normal	no
A3	992	229.0	2	Normal	no
A4	1322	305.0	2	Normal	no
A5	1984	457.0	2	Normal	no
A6	2646	610.0	2	Normal	no
B1	992	229.0	1	Normal	no
B2	992	229.0	3	Normal	no
B3	992	229.0	1	Normal	yes
B4	992	229.0	2	Normal	yes
B5	992	229.0	3	Normal	yes
C1	992	229.0	1	Selective	no
C2	992	229.0	2	Selective	no
C3	992	229.0	1	Selective	yes
C4	992	229.0	2	Selective	yes

*per unit of projected area

Vented Trombe Wall Systems

Designation	Thermal Storage Capacity* (kJ/m^2C)	Nominal wall Thickness** (mm)	(ρck) $(kJ)^2/s\ m^{4\,o}C$	No. of Glazings	Wall surface	Night Insulation
A1	318	152.0	36.9×10^3	2	normal	no
A2	477	229.0	36.9×10^3	2	normal	no
A3	636	305.0	36.9×10^3	2	normal	no
A4	954	457.0	36.9×10^3	2	normal	no
B1	318	152.0	18.45×10^3	2	normal	no
B2	477	229.0	18.45×10^3	2	normal	no
B3	636	305.0	18.45×10^3	2	normal	no
B4	954	457.0	18.45×10^3	2	normal	no
C1	318	152.0	9.23×10^3	2	normal	no
C2	477	229.0	9.23×10^3	2	normal	no
C3	636	305.0	9.23×10^3	2	normal	no
C4	954	457.0	9.23×10^3	2	normal	no
D1	636	305.0	36.9×10^3	1	normal	no
D2	636	305.0	36.9×10^3	3	normal	no
D3	636	305.0	36.9×10^3	1	normal	yes
D4	636	305.0	36.9×10^3	2	normal	yes
D5	636	305.0	36.9×10^3	3	normal	yes
E1	636	305.0	36.9×10^3	1	selective	no
E2	636	305.0	36.9×10^3	2	selective	no
E3	636	305.0	36.9×10^3	1	selective	yes
E4	636	305.0	36.9×10^3	2	selective	yes
F1	318	152.0	36.9×10^3	2	normal	no
F2	477	229.0	36.9×10^3	2	normal	no
F3	636	305.0	36.9×10^3	2	normal	no
F4	954	457.0	36.9×10^3	2	normal	no
C1	318	152.0	18.45×10^3	2	normal	no
C2	477	229.0	18.45×10^3	2	normal	no
C3	636	305.0	18.45×10^3	2	normal	no
C4	954	457.0	18.45×10^3	2	normal	no

Vented Trombe Wall Systems (cont'd)

Designation	Thermal Storage Capacity* (kJ/m^2C)	Nominal wall Thickness** (mm)	(ρck) $(kJ)^2/s\ m^{4\,o}C$	No. of Glazings	Wall surface	Night Insulation
H1	318	152.0	9.23×10^3	2	normal	no
H2	477	229.0	9.23×10^3	2	normal	no
H3	636	305.0	9.23×10^3	2	normal	no
H4	954	457.0	9.23×10^3	2	normal	no
I1	636	305.0	36.9×10^3	1	normal	no
I2	636	305.0	36.9×10^3	3	normal	no
I3	636	305.0	36.9×10^3	1	normal	yes
I4	636	305.0	36.9×10^3	2	normal	yes
I5	636	305.0	36.9×10^3	3	normal	yes
J1	636	305.0	36.9×10^3	1	selective	no
J2	636	305.0	36.9×10^3	2	selective	no
J3	636	305.0	36.9×10^3	1	selective	yes
J4	636	305.0	36.9×10^3	2	selective	yes

* per unit of projected area

**for the particular case of $\rho c = 2120\ M\ J/m^3\ {}^oC$

Direct Gain Systems

Designation	Thermal Storage Capacity* kJ/m ^{o}C	Nominal mass Thickness** (mm)	Mass-to-Glazing Area Ratio	No. of Glazings	Night Insulation
A1	636	51	6	2	no
A2	636	51	6	3	no
A3	636	51	6	2	yes
B1	954	152	3	2	no
B2	954	152	3	3	no
B3	954	152	3	2	yes
C1	1272	102	6	2	no
C2	1272	102	6	3	no
C3	1272	102	6	2	yes

* per unit of projected area

**for the particular case of ρc = MJ/m^{3o}C

Sunspace Systems

(all are double-glazed)

Designation	Type	Tilt (degrees)	Common wall	End Walls	Night Insulation
A1	attached	50	masonry	opaque	no
A2	attached	50	masonry	opaque	yes
A3	attached	50	masonry	glazed	no
A4	attached	50	masonry	glazed	yes
A5	attached	50	insulated	opaque	no
A6	attached	50	insulated	opaque	yes
A7	attached	50	insulated	glazed	no
A8	attached	50	insulated	glazed	yes
B1	attached	90/30	masonry	opaque	no
B2	attached	90/30	masonry	opaque	yes
B3	attached	90/30	masonry	glazed	no
B4	attached	90/30	masonry	glazed	yes
B5	attached	90/30	insulated	opaque	no
B6	attached	90/30	insulated	opaque	yes
B7	attached	90/30	insulated	glazed	no
B8	attached	90/30	insulated	glazed	yes
C1	semi-enclosed	90	masonry	common	no
C2	semi-enclosed	90	masonry	common	yes
C3	semi-enclosed	90	insulated	common	no
C4	semi-enclosed	90	insulated	common	yes
D1	semi-enclosed	50	masonry	common	no
D2	semi-enclosed	50	masonry	common	yes
D3	semi-enclosed	50	insulated	common	no
D4	semi-enclosed	50	insulated	common	yes
E1	semi-enclosed	90/30	masonry	common	no
E2	semi-enclosed	90/30	masonry	common	yes
E3	semi-enclosed	90/30	insulated	common	no
E4	semi-enclosed	90/30	insulated	common	yes

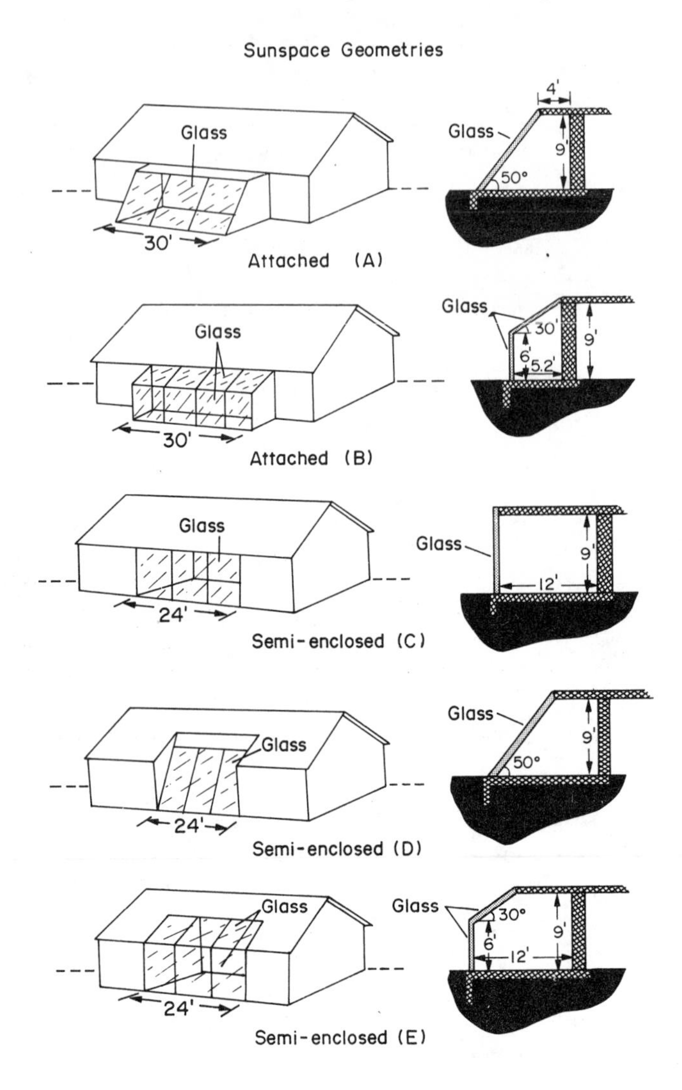

Sunspace Geometries
Glass
30'
Attached (A)
4'
Glass
9'
50°
Glass
30'
Attached (B)
Glass
30'
6'
5.2'
9'
Glass
24'
Semi-enclosed (C)
Glass
9'
12'
Glass
24'
Semi-enclosed (D)
Glass
9'
50°
Glass
24'
Semi-enclosed (E)
Glass
30°
6'
12'
9'

Fig. A5.1

APPENDIX 6

MODIFICATION TO THE ADMITTANCE PROCEDURE

The admittance procedure as described in section 6.3.2 of Chapter 6 yields results which can have deviations up to 10% from the corresponding results obtained from a more rigorous periodic solution method (see 6.3.5) (Sodha *et al.*, 1985). By incorporating the effects of higher harmonics of the frequency ω of varying parameters, admittance procedure can be suitably modified; the modified procedure yields results which are in exact agreement with the corresponding periodic solution results.

Including the effect of higher harmonics, eq. (6.55) for the nth harmonic can be written as

$$\begin{bmatrix} \theta_{on} \\ q_{on} \end{bmatrix} = \begin{bmatrix} 1 & R_{so} \\ 0 & 1 \end{bmatrix} \begin{bmatrix} E_n & F_n \\ G_n & H_n \end{bmatrix} \begin{bmatrix} 1 & R_{si} \\ 0 & 1 \end{bmatrix} \begin{bmatrix} \theta_{jn} \\ q_{jn} \end{bmatrix} \quad \text{(A6.1)}$$

where

$$\begin{bmatrix} E_n & F_n \\ G_n & H_n \end{bmatrix} = \begin{bmatrix} A_{1n} & B_{1n} \\ D_{1n} & A_{1n} \end{bmatrix} - - - \begin{bmatrix} A_{kn} & B_{kn} \\ D_{kn} & A_{kn} \end{bmatrix} \quad \text{(A6.2)}$$

k corresponding to the kth layer and

$$A_{kn} = \cos h\ (1 + i)\ \psi_{kn} \quad \text{(A6.3a)}$$

$$B_{kn} = \frac{R_k}{(1+i)\psi_{kn}} \sin h\ (1+i)\ \psi_{kn} \quad \text{(A6.3b)}$$

$$D_{kn} = \frac{(1+i)\psi_{kn}}{R_k} \sin h\ (1+i)\ \psi_{kn} \quad \text{(A6.3c)}$$

$$\psi_{kn} = \frac{n\omega L_k^{2}}{2D_k}^{\frac{1}{2}} \qquad \text{(A6.3d)}$$

$$R_k = \frac{L_k}{K_k} \qquad \text{(A6.3e)}$$

θ_o, θ_j and q_o, q_j, the temperatures and heat fluxes at the outermost and jth surface are also given by periodic series as

$$\theta_o = \theta_{oo} + \mathrm{Re} \sum_n \theta_{on} \exp(in\omega t) \qquad \text{(A6.4a)}$$

$$q_o = q_{oo} + \mathrm{Re} \sum_n q_{on} \exp(in\omega t) \qquad \text{(A6.4b)}$$

$$\theta_j = \theta_{jo} + \mathrm{Re} \sum_n {}_{jn} \exp(in\omega t) \qquad \text{(A6.4c)}$$

$$q_j = q_{jo} + \mathrm{Re} \sum_n q_{jn} \exp(in\omega t) \qquad \text{(A6.4d)}$$

θ_{oo}, q_{oo}, θ_{jo} and q_{jo} are the average values of the corresponding parameters for $n = 0$.

Writing eq. (A6.1) as

$$\begin{bmatrix} \theta_{on} \\ q_{on} \end{bmatrix} = \begin{bmatrix} E_n' & F_n' \\ G_n' & H_n' \end{bmatrix} \begin{bmatrix} \theta_{jn} \\ q_{jn} \end{bmatrix} \qquad \text{(A6.5)}$$

various quantities of interest may be evaluated as follows.

(i) Heat-flux into the Room

Assuming O and j to be the outside and inside surfaces respectively, q_j will represent the amount of heat flux entering the room, and may be given as

$$q_j = \left[\frac{\theta_o - E_n' \theta_j}{F_n'} \right] \qquad \text{A6.6)}$$

where θ_o and θ_j denote the solair temperature of the surface and room air temperature respectively.

(ii) Room Air Temperature

Room air temperature is essentially determined by the net amount of heat loss/gain by the room through all the components and its thermal mass. If M_R be the total thermal mass of the room, one can write

$$M_R \frac{d\theta_j}{dt} = \sum_{\substack{\text{all} \\ \text{components} \\ \text{of the room}}} q_j \pm \text{other gains/losses} \quad (A6.7)$$

which is a linear equation of room air temperature and can easily be solved by substituting appropriate expressions for various q's and θ's.

(iii) Heat Loss from the Room

Assuming the surfaces 0 and j to be in contact with the air of room 1 and 2 respectively, the q_o will represent the amount of heat going away from room 1 to room 2 which may be given as

$$q_o = \frac{H_n'}{F_n'} \theta_o + G_n' - \frac{H_n' E_n'}{F_n'} \theta_j \quad (A6.8)$$

(iv) Heat Exchange with the Ground

In order to find out the amount of heat exchanged between the room air and ground, the ground is assumed to be a semi-infinite medium which means that

$$\theta_G = \text{finite as } x_G \to \infty \quad (A6.9)$$

where G refers to the ground.

The other boundary condition for the floor surface which is in contact with the room air may be written as follows

$$- K_G \left. \frac{\partial \theta_G}{\partial x_G} \right|_{x_G = o} = h_{if} \left[\theta_R - \left. \theta_G \right|_{x_G = o} \right] \quad (A6.10)$$

where h_{if} is the heat transfer coefficient between the room air and floor and θ_R denotes the room air temperature.

In view of eqs. (A6.9) and (A.VI.10), it can conveniently be shown that the amount of heat flux exchanged with a multi-layered floor may be evaluated by the following expression:

$$q_G = \sum_n \frac{H_n''}{F_n''} \theta_{Rn} \exp(in\omega t) \quad (A6.11)$$

where θ_{Rn} is the Fourier part of θ_R and H_n'' and F_n'' are defined by the following equation:

$$\begin{bmatrix} E_n'' & F_n'' \\ G_n'' & H_n'' \end{bmatrix} = \begin{bmatrix} 1 & h_{if}^{-1} \\ 0 & 1 \end{bmatrix} \begin{bmatrix} A_{1n} & B_{1n} \\ D_{1n} & A_{1n} \end{bmatrix} - - -$$

$$
- - - \begin{bmatrix} A_{kn} & B_{kn} \\ D_{kn} & A_{kn} \end{bmatrix} \begin{bmatrix} 1 & (K_G \beta_{Gn})^{-1} \\ 0 & 1 \end{bmatrix}
$$

It may be noted here that there will be no heat exchange between the room air and the floor in the steady-state conditions which are represented by $n = 0$ in eq. (A.11).

REFERENCE

Sodha, M. S., Kaur, B., Kumar, B. and Bansal, N. K. (1985) A Comparison of the Admittance and Fourier Method for Predicting Heat/Cooling Loads, *Solar Energy*,

APPENDIX 7

CONVERSION FACTORS

Length

1 m = 3.281 ft = 39.37 in = 1.094 yd

1 ft = 0.3048 m = 12 in

1 yd = 0.91 m

1 mil = 0.00254 cm = 0.001 in

Area

1 acre = 0.4047 hectare = 4046.9 m^2

1 m^2 = 10.76 ft^2 = 1550 in^2

1 ft^2 = 0.0929 m^2

1 yd^2 = 0.8361 m^2

Volume

1 m^3 = 35.31 ft^3

1 ft^3 = 0.02832 m^3

Mass

1 kg = 2.2 lb

1 lb = 0.454 kg

Density

1 kg/m^3 = 0.06243 lb/ft^3 = 0.001 g/cm^3

1 lb/ft^3 = 16.018 kg/m^3

Speed/Flow rate

1 m/s = 3.6 km/h = 196.85 ft/min = 2.237 miles/h

1 miles/h = 1.6094 km/h = 0.4470 m/s

1 m^3/h = 0.589 ft^3/min

1 m^3/s = 35.3198 ft^3/s

Pressure

1 atm = 760 mm Hg = 33.9 ft H_2O = 14.696 lb/in^2
= 1.013 x 10^5 N/m^2 = 1.0332 kg/cm^2

1 bar = 750 mm Hg = 10^5 N/m^2

1 Pa = 1 N/m^2 = 1 pascal

1 psi = 6897.76 N/m^2

1 mm Hg = 1 Torr

Energy/Heat

1 kJ = 0.239 k cal = 0.948 BTU = 0.000278 kwh

1 BTU = 1.0551 kJ = 0.252 k cal

1 Cal = 4.184 J

1 Tons oil equivalent = 4.19 x 10^{10} J

1 Tons coal equivalent = 2.93 x 10^{10} J

1 Kilowatt hour (kwh) = 3.6 x 10^6 J

1 m^3 of Natural Gas at NTP = 37.1 x 10^6 J

1 hp = 745.7 W

1 W = 1 J/s

1 Langley = 1 gcal/cm^2 = 2520 kcal/m^2

1 W/m^2 = 0.3170 BTU/ft^2/hr

1 Ton refrigeration = 3516.85 W

Specific Heat/Heat Content

1 kJ/kg^{o}C = 0.2388 BTU/lb^{o}F

1 BTU/lb^{o}F = 1 k cal/kg ^{o}C = 1.163 wh/kg^{o}C

1 BTU/lb = 0.566 k cal/kg = 2.324 kJ/kg

Heat-Transfer Coefficient

1 W/m^2 ^{o}C = 0.1761 BTU/ft^2 h ^{o}F
= 0.8598 k cal/m^2 h ^{o}C

1 BTU/h ft^2 ^{o}F = 5.678 W/m^2 K = 4.882 k cal/m^2h ^{o}C

1 k cal/m^2 h ^{o}C = 1.1631 W/m^2^{o}C

Thermal Conductivity

1 W/m^{o}C = 0.861 k cal/h m ^{o}C
= 0.5778 BTU/ft h ^{o}F

1 BTU/ft h^{o}F = 1.731 W/m^{o}C

INDEX